Methods in Enzymology

Volume 314
ANTISENSE TECHNOLOGY
Part B
Applications

METHODS IN ENZYMOLOGY

EDITORS-IN-CHIEF

John N. Abelson   Melvin I. Simon

DIVISION OF BIOLOGY
CALIFORNIA INSTITUTE OF TECHNOLOGY
PASADENA, CALIFORNIA

FOUNDING EDITORS

Sidney P. Colowick and Nathan O. Kaplan

Methods in Enzymology

Volume 314

Antisense Technology

Part B
Applications

EDITED BY

M. Ian Phillips

UNIVERSITY OF FLORIDA COLLEGE OF MEDICINE
GAINESVILLE, FLORIDA

ACADEMIC PRESS

San Diego London Boston New York Sydney Tokyo Toronto

Academic Press
A Harcourt Science and Technology Company
525 B Street, Suite 1900, San Diego, California 92101-4495, USA
http://www.academicpress.com

Academic Press Limited
24-28 Oval Road, London NW1 7DX, UK
http://www.hbuk.co.uk/ap/

International Standard Book Number: 0-12-182215-X

PRINTED IN THE UNITED STATES OF AMERICA
99 00 01 02 03 04 MM 9 8 7 6 5 4 3 2 1

Table of Contents

Section I. Antisense Receptor Targets

Section II. Antisense Neuroscience Targets

v

Section III. Antisense in Nonneuronal Tissues

Section IV. Antisense in Therapy

Contributors to Volume 314

Article numbers are in parentheses following the names of contributors.
Affiliations listed are current.

FE C. ABOGADIE (10), *Wellcome Laboratory for Molecular Pharmacology, Department of Pharmacology, University College London, London WC1E 6BT, United Kingdom*

YIJIA BAO (12), *Vysis, Inc., Downers Grove, Illinois 60515*

RUTH BEATTIE (21), *Lilly Research Centre, Eli Lilly and Company Limited, Windlesham, Surrey GU20 6PH, United Kingdom*

MARGERY C. BEINFELD (8), *Department of Pharmacology and Experimental Therapeutics, Tufts University School of Medicine, Boston, Massachusetts 02111*

GAETANO BERGAMASCHI (30), *Dipartimento di Medicina Interna e Terapia Medica, Medicina Interna e Oncologia Medica, I.R.C.C.S. Policlinico San Matteo, 27100 Pavia, Italy*

E. A. L. BIESSEN (23), *Division of Biopharmaceutics, Leiden/Amsterdam Center for Drug Research, Leiden University, 2300 RA Leiden, The Netherlands*

M. K. BIJSTERBOSCH (23), *Division of Biopharmaceutics, Leiden/Amsterdam Center for Drug Research, Leiden University, 2300 RA Leiden, The Netherlands*

H. E. BLUM (37), *Department of Medicine II, University of Freiburg, D-79106 Freiburg, Germany*

FRÉDÉRIC BOST (24), *Sidney Kimmel Cancer Center, San Diego, California 92121*

DUSTIN H. O. BRITTON (12), *DuPont Pharmaceuticals, Wilmington, Delaware 19880*

WILLIAM C. BROADDUS (9), *Division of Neurosurgery and Department of Anatomy, Medical College of Virginia, Virginia Commonwealth University, Richmond, Virginia 23298-0631*

DAVID A. BROWN (10), *Wellcome Laboratory for Molecular Pharmacology, Department of Pharmacology, University College London, London WC1E 6BT, United Kingdom*

NOEL J. BUCKLEY (10), *School of Biochemistry and Molecular Biology, University of Leeds, Leeds LS2 9JT, United Kingdom*

MARINA CATSICAS (11), *Department of Physiology, University College London, London, United Kingdom*

STEFAN CATSICAS (11), *Institut de Biologie Cellulaire et de Morphologie, Université de Lausanne, CH-1005 Lausanne, Switzerland*

MALCOLM P. CAULFIELD (10), *Wellcome Laboratory for Molecular Pharmacology, Department of Pharmacology, University College London, London WC1E 6BT, United Kingdom*

MARIO CAZZOLA (30), *Dipartimento di Medicina Interna e Terapia Medica, Medicina Interna e Oncologia Medica, I.R.C.C.S. Policlinico San Matteo, 27100 Pavia, Italy*

BYUNG-MIN CHOI (34), *Department of Microbiology and Immunology, Wonkwang University School of Medicine, Iksan-shi, Chonbug 570-749, Korea*

CHUAN-CHU CHOU (29), *Department of Pathology, New Jersey Medical School, University of Medicine and Dentistry of New Jersey, Newark, New Jersey 07103*

HUN-TAEG CHUNG (34), *Department of Microbiology and Immunology, Wonkwang University School of Medicine and Medicinal Resources Research Center of Wonkwang University, Iksan-shi, Chonbug 570-749, Korea*

CATHERINE L. CIOFFI (25), *Department of Metabolic and Cardiovascular Diseases, Novartis Institute for Biomedical Research, Summit, New Jersey 07901*

ix

PETER J. CRAIG (21), *Lilly Research Centre, Eli Lilly and Company Limited, Windlesham, Surrey GU20 6PH, United Kingdom*

MONICA CURTO (18), *Department of Cytomorphology, School of Medicine, Cittadella University, Cagliari, Italy*

MARIZA DAYRELL (10), *Wellcome Laboratory for Molecular Pharmacology, Department of Pharmacology, University College London, London WC1E 6BT, United Kingdom*

NICHOLAS M. DEAN (24), *ISIS Pharmaceuticals, Inc., Carlsbad, California 92008*

ISABEL DE ANTONIO (1), *Department of Neuropathology, Cajal Institute, Consejo Superior de Investigaciónes Científicas, E-28002 Madrid, Spain*

PATRICK DELMAS (10), *Wellcome Laboratory for Molecular Pharmacology, Department of Pharmacology, University College London, London WC1E 6BT, United Kingdom*

MICHEL DE WAARD (21), *Laboratoire de Neurobiologie des Canaux Ioniques, INSERM U374, 13916 Marseille Cedex 20, France*

MICHAEL G. DUBE (13), *Department of Physiology, University of Florida Brain Institute, University of Florida College of Medicine, Gainesville, Florida 32610-0274*

MARCEL EGGER (32), *Department of Physiology, University of Bern, CH-3012 Bern, Switzerland*

ANNE FELTZ (21), *Laboratoire de Neurobiologie Cellulaire, UPR 9009 Centre National de la Recherche Scientifique, 67084 Strasbourg, France*

HELEN L. FILLMORE (9), *Division of Neurosurgery, Medical College of Virginia, Virginia Commonwealth University, Richmond, Virginia 23298-0631*

K. FLUITER (23), *Division of Biopharmaceutics, Leiden/Amsterdam Center for Drug Research, Leiden University, 2360 RA Leiden, The Netherlands*

MICHAEL B. GANZ (26), *Department of Medicine, Case Western Reserve University, Cleveland, Ohio 44106*

JAVIER GARZÓN (1), *Department of Neuropathology, Cajal Institute, Consejo Superior de Investigaciónes Científicas, E-28002 Madrid, Spain*

SAMANTHA GILLARD (21), *Department of Pharmacology, University of Chicago, Chicago, Illinois 60637*

GEORGE T. GILLIES (9), *Department of Biomedical Engineering, Health Sciences Center, University of Virginia, Charlottesville, Virginia 22908*

SUSAN GOULD-FOGERITE (29), *Department of Pathology, New Jersey Medical School, University of Medicine and Dentistry of New Jersey, Newark, New Jersey 07103*

MARIA GRAZIA ENNAS (18), *Department of Cytomorphology, School of Medicine, Cittadella University, Cagliari, Italy*

FULVIA GREMO (18), *Department of Cytomorphology, School of Medicine, Cittadella University, Cagliari, Italy*

GABRIELE GRENNINGLOH (11), *Institut de Biologie Cellulaire et de Morphologie, Université de Lausanne, CH-1005 Lausanne, Switzerland*

JANE E. HALEY (10), *Wellcome Laboratory for Molecular Pharmacology, Department of Pharmacology, University College London, London WC1E 6BT, United Kingdom*

FINN HALLBÖÖK (11), *Department of Neuroscience, Uppsala University, BMC, S-751 23 Uppsala, Sweden*

MAI HE (29), *Department of Pathology, New Jersey Medical School, University of Medicine and Dentistry of New Jersey, Newark, New Jersey 07103*

MATTHEW O. HEBB (19), *Faculty of Medicine, University of Toronto, Toronto, Ontario, Canada M5S 1A8*

MARKUS HEILIG (19), *Department of Clinical Neuroscience, Occupational Therapy and Elderly Care Research, Karolinska Institute, Stockholm, Sweden*

JULIE G. HENSLER (6), *Department of Pharmacology, University of Texas Health Science Center, San Antonio, Texas 78284-7764*

SIEW PENG HO (12), *DuPont Pharmaceuticals, Wilmington, Delaware 19880*

JEFFREY T. HOLT (35), *Departments of Cell Biology and Pathology, Vanderbilt University Medical School, Nashville, Tennessee 37232*

ALAN L. HUDSON (5), *Psychopharmacology Unit, School of Medical Sciences, University of Bristol, Bristol BS8 1TD, United Kingdom*

JOHN C. HUNTER (14), *Center for Biological Research, Roche Bioscience, Palo Alto, California 94304*

ANTTI P. JEKUNEN (36), *Department of Clinical Pharmacology, Helsinki University Central Hospital, FIN-00029 HYKS, Finland*

ROY A. JENSEN (35), *Departments of Pathology and Cell Biology, Vanderbilt University Medical School, Nashville, Tennessee 37232*

MAGDALENA JUHASZOVA (22), *Department of Physiology, University of Maryland School of Medicine, Baltimore, Maryland 21201*

CHANG-DUK JUN (34), *Department of Microbiology and Immunology, Wonkwang University School of Medicine, Iksan-shi, Chonbug 570-749, Korea*

KALEVI J. A. KAIREMO (36), *Department of Clinical Chemistry, Norwegian University of Science and Technology, N-7006 Trondheim, Norway*

PUSHPA S. KALRA (13), *Department of Physiology, University of Florida Brain Institute, University of Florida College of Medicine, Gainesville, Florida 32611*

SATYA P. KALRA (13), *Department of Neuroscience, University of Florida Brain Institute, University of Florida College of Medicine, Gainesville, Florida 32611*

JESPER KARLE (2), *Department of Psychiatry, Rigshospitalet (National Hospital), DK-2100 Copenhagen, Denmark*

MICHAEL J. KATOVICH (39), *Department of Pharmacodynamics, College of Pharmacy, University of Florida, Gainesville, Florida 32610*

JOSEPHINE LAI (14), *Department of Pharmacology, University of Arizona, Tucson, Arizona 85724*

REGIS C. LAMBERT (21), *Laboratoire de Neurobiologie Cellulaire, UPR 9009 Centre National de la Recherche Scientifique, 67084 Strasbourg, France*

PETER LIPP (32), *Laboratory of Molecular Signalling, The Babraham Institute, Cambridge CB2 3EJ, United Kingdom*

DI LU (39), *Department of Physiology, University of Florida College of Medicine, Gainesville, Florida 32610*

CLAUDIA LUCOTTI (30), *Dipartimento di Medicina Interna e Terapia Medica, Medicina Interna e Oncologia Medica, I.R.C.C.S. Policlinico San Matteo, 27100 Pavia, Italy*

DAVID L. MATTSON (27), *Department of Physiology, Medical College of Wisconsin, Milwaukee, Wisconsin 53226*

YVES MAULET (21), *Laboratoire de Neurobiologie Cellulaire, UPR 9009 Centre National de la Recherche Scientifique, 67084 Strasbourg, France*

ROBERT MCKAY (24), *ISIS Pharmaceuticals, Inc., Carlsbad, California 92008*

DAN MERCOLA (24), *Sidney Kimmel Cancer Center, San Diego, California 92121, and Center for Molecular Genetics, University of California at San Diego, La Jolla, California 92093*

RADMILA MILEUSNIC (15), *Department of Biological Sciences, The Open University, Milton Keynes MK7 6AA, United Kingdom*

DAGMARA MOHUCZY (3), *Department of Physiology, University of Florida College of Medicine, Gainesville, Florida 32610*

BRETT P. MONIA (25), *Department of Molecular Pharmacology, ISIS Pharmaceuticals, Inc., Carlsbad, California 92008*

D. MORADPOUR (37), *Department of Medicine II, University of Freiburg, D-79106 Freiburg, Germany*

ISABELLA MUSSINI (18), *CNR Center of Muscle Biology and Physiopathology, University of Padova, Padova, Italy*

INGA D. NEUMANN (16), *Max Planck Institute of Psychiatry, D-80804 Munich, Germany*

MOGENS NIELSEN (2), *Research Institute of Biological Psychiatry, St. Hans Hospital, DK-4000 Roskilde, Denmark*

ERNST NIGGLI (32), *Department of Physiology, University of Bern, CH-3012 Bern, Switzerland*

THADDEUS S. NOWAK, JR. (17), *Departments of Anesthesiology and Resuscitology, Okayama University School of Medicine, Shikata-cho, Okayama City, Japan*

TAKAHIRO OCHIYA (28), *Section for Studies on Metastasis, National Cancer Center Research Institute, Chuo-ku, Tokyo 104-0045, Japan*

S. OFFENSPERGER (37), *Department of Medicine II, University of Freiburg, D-79106 Freiburg, Germany*

WOLF-BERNHARD OFFENSPERGER (37), *Department of Medicine II, University of Freiburg, D-79106 Freiburg, Germany*

MICHAEL H. OSSIPOV (14), *Department of Pharmacology, University of Arizona, Tucson, Arizona 85724*

HYUN-OCK PAE (34), *Department of Microbiology and Immunology, Wonkwang University School of Medicine, Iksan-shi, Chonbug 570-749, Korea*

YING-XIAN PAN (4), *Memorial Sloan-Kettering Cancer Center, New York, New York 10021*

GEORGE A. PARKER (29), *Department of Pathology, New Jersey Medical School, University of Medicine and Dentistry of New Jersey, Newark, New Jersey 07103*

GAVRIL W. PASTERNAK (4), *Memorial Sloan-Kettering Cancer Center, New York, New York 10021*

BIHAI PENG (29), *Department of Pathology, New Jersey Medical School, University of Medicine and Dentistry of New Jersey, Newark, New Jersey 07103*

M. IAN PHILLIPS (3), *Department of Physiology, University of Florida College of Medicine, Gainesville, Florida 32610*

FRANK PORRECA (14), *Department of Pharmacology, University of Arizona, Tucson, Arizona 85724*

O. POTAPOVA (24), *Laboratory of Biological Chemistry, Gerontology Research Center, National Institute of Aging, National Institute of Health, Baltimore, Maryland 21224*

SUJIT S. PRABHU (9), *Division of Neurosurgery, Medical College of Virginia, Virginia Commonwealth University, Richmond, Virginia 23298-0631*

MOHAN K. RAIZADA (39), *Department of Physiology, University of Florida College of Medicine, Gainesville, Florida 32610*

ELIZABETH S. RAVECHÉ (29), *Department of Pathology, New Jersey Medical School, University of Medicine and Dentistry of New Jersey, Newark, New Jersey 07103*

PHYLLIS Y. REAVES (39), *Department of Physiology, University of Florida College of Medicine, Gainesville, Florida 32610*

EMMA S. J. ROBINSON (5), *Psychopharmacology Unit, School of Medical Sciences, University of Bristol, Bristol BS8 1TD, United Kingdom*

CHERYL ROBINSON-BENION (35), *Department of Cell Biology, Vanderbilt University Medical School, Nashville, Tennessee 37232*

VITTORIO ROSTI (30), *Laboratorio di Ricerca Area Trapianti, Unità di Immunologia Clinica, I.R.C.C.S. Policlinico San Matteo, 27100 Pavia, Italy*

E. T. RUMP (23), *Division of Biopharmaceutics, Leiden/Amsterdam Center for Drug Research, Leiden University, 2300 RA Leiden, The Netherlands*

PILAR SÁNCHEZ-BLÁZQUEZ (1), *Department of Neuropathology, Cajal Institute, Consejo Superior de Investigaciónes Científicas, E-28002 Madrid, Spain*

JOANNE M. SCALZITTI (6), *Department of Pharmacology, New York University Medical School, New York, New York 10016*

CHRISTOPH SCHUMACHER (31), *MCD Research, Novartis Pharmaceuticals Corporation, Summit, New Jersey 07901-1398*

BEAT SCHWALLER (32), *Department of Histology and General Embryology, University of Fribourg, CH-1700 Fribourg, Switzerland*

MICHAEL S. SCULLY (12), *DuPont Pharmaceuticals, Wilmington, Delaware 19880*

MARIELLA SETZU (18), *Department of Cytomorphology, School of Medicine, Cittadella University, Cagliari, Italy*

A. PAULA SIMÕES-WUEST (33), *Department of Internal Medicine, University Hospital Zurich, CH-8044 Zurich, Switzerland*

MARTIN K. SLODZINSKI (22), *Department of Physiology, University of Maryland School of Medicine, and Department of Medicine, Mercy Hospital, Baltimore, Maryland 21201*

JANET B. SMITH (38), *Departments of Microbiology and Immunology, Kimmel Cancer Center, Cardeza Foundation for Hematological Research, Thomas Jefferson University, Philadelphia, Pennsylvania 19107-5083*

VALERIA SOGOS (18), *Department of Cytomorphology, School of Medicine, Cittadella University, Cagliari, Italy*

DONG-HWAN SOHN (34), *College of Pharmacy, Wonkwang University, Iksan-shi, Chonbug 570-749, Korea*

WOLFGANG SOMMER (19), *Department of Clinical Neuroscience, Occupational Therapy and Elderly Care Research, Karolinska Institute, Stockholm, Sweden*

KELLY M. STANDIFER (7), *Department of Pharmacological and Pharmaceutical Sciences, University of Houston, Houston, Texas 77204-5515*

JULIE K. STAPLE (11), *Institut de Biologie Cellulaire et de Morphologie, Université de Lausanne, CH-1005 Lausanne, Switzerland*

XIAOPING TANG (3), *Department of Physiology, University of Florida College of Medicine, Gainesville, Florida 32610*

MIKKO TENHUNEN (36), *Department of Oncology, Helsinki University Central Hospital, FIN-00029 HYKS, Finland*

MASAAKI TERADA (28), *National Cancer Center, Chuo-ku, Tokyo 104-0045, Japan*

C. THOMA (37), *Department of Medicine, University of Freiburg, D-79106 Freiburg, Germany*

WOLFGANG TISCHMEYER (20), *Leibniz Institute for Neurobiology, D-39008 Magdeburg, Germany*

NICOLA TOSCHI (16), *Max Planck Institute of Psychiatry, D-80804 Munich, Germany*

T. J. C. VAN BERKEL (23), *Division of Biopharmaceutics, Leiden/Amsterdam Center for Drug Research, Leiden University, 2300 RA Leiden, The Netherlands*

H. VIETSCH (23), *Division of Biopharmaceutics, Leiden/Amsterdam Center for Drug Research, Leiden University, 2300 RA Leiden, The Netherlands*

DAESETY VISHNUVARDHAN (8), *Department of Pharmacology and Experimental Therapeutics, Tufts University School of Medicine, Boston, Massachusetts 02111*

STEPHEN G. VOLSEN (21), *Lilly Research Centre, Eli Lilly and Company Limited, Windlesham, Surrey GU20 6PH, United Kingdom*

F. VON WEIZSÄCKER (37), *Department of Medicine II, University of Freiburg, D-79106 Freiburg, Germany*

HONGWEI WANG (39), *Department of Physiology, University of Florida College of Medicine, Gainesville, Florida 32610*

ERIC WICKSTROM (38), *Department of Microbiology, Thomas Jefferson University, Philadelphia, Pennsylvania 19107*

SUSANNA WU-PONG (9), *Department of Pharmaceutics, Medical College of Virginia, Virginia Commonwealth University, Richmond, Virginia 23298*

YUTAKA YAIDA (17), *Departments of Anesthesiology and Resuscitology, Okayama University School of Medicine, Shikata-cho, Okayama City, Japan*

JI-CHANG YOO (34), *Department of Microbiology and Immunology, Wonkwang University School of Medicine, Iksan-shi, Chonbug 570-749, Korea*

KATSUMI YUFU (17), *Departments of Anesthesiology and Resuscitology, Okayama University School of Medicine, Shikata-cho, Okayama City, Japan*

UWE ZANGEMEISTER-WITTKE (33), *Department of Internal Medicine, University Hospital Zurich, CH-8044 Zurich, Switzerland*

ANNEMARIE ZIEGLER (33), *Laboratory for Molecular and Cellular Oncology, Cancer Program, Faculty of Medicine, Catholic University of Chile, Santiago, Chile*

Preface

Antisense technology reached a watershed year in 1998 with the FDA approval of the antisense-based therapy, Vitravene, developed by ISIS. This is the first drug based on antisense technology to enter the marketplace and makes antisense technology a reality for therapeutic applications. However, antisense technology still needs further development, and new applications need to be explored.

Contained in this Volume 314 (Part B) of *Methods in Enzymology* and its companion Volume 313 (Part A) are a wide range of methods and applications of antisense technology in current use. We set out to put together a single volume, but it became obvious that the variations in methods and the numerous applications required at least two volumes, and even these do not, by any means, cover the entire field. Nevertheless, the articles included represent the work of active research groups in industry and academia who have developed their own methods and techniques. In this volume, Part B: Applications, chapters cover methods in which antisense is designed to target membrane receptors and antisense application in the neurosciences, as well as in nonneuronal tissues. The therapeutic applications of antisense technology, the latest area of new interest, complete the volume. In Part A: General Methods, Methods of Delivery, and RNA Studies several methods of antisense design and construction are included as are general methods of delivery and antisense used in RNA studies.

Although *Methods in Enzymology* is designed to emphasize methods, rather than achievements, I congratulate all the authors on their achievements that have led them to make their methods available. In compiling and editing these two volumes I could not have made much progress without the excellent secretarial services of Ms. Gayle Butters of the University of Florida, Department of Physiology.

M. IAN PHILLIPS

METHODS IN ENZYMOLOGY

Section I

Antisense Receptor Targets

[1] *In Vivo* Modulation of G Proteins and Opioid Receptor Function by Antisense Oligodeoxynucleotides

By Javier Garzón, Isabel de Antonio,
and Pilar Sánchez-Blázquez

Introduction

Opioids promote their effects by acting on cellular receptors located in the cell membrane. The diversity of these receptors, which was initially suggested by pharmacological and biochemical studies, has been confirmed by the cloning of cDNAs encoding the μ,[1,2] δ,[3,4] and κ[5] types. The cloned opioid receptors have been found to correspond to the pharmacological subtypes μ1, δ2, and κ1. The subtypes μ2, δ1, κ2, and κ3 (and possibly others) are still to be designated molecular identities. The use of antisense technology with the opioid system has been helpful in making molecular/pharmacological correlations and the approach is currently being used to investigate uncorrelated subtypes.

Opioid receptors couple to heterotrimeric (α, β, and γ subunits) GTP-binding regulatory proteins (G proteins). Our knowledge regarding the diversity and properties of G proteins has increased greatly.[6–8] Most of the known classes are present in the nervous system and are thought to regulate various signaling pathways, e.g., adenylyl cyclases, different types of K^+ and Ca^{2+} channels, phospholipases, protein kinases, and others. Gα subunits show differences in the regions involved in the interaction with membrane receptors (see, e.g., Jones and Reed[9]). These variations seem to account

[1] Y. Chen, A. Mestek, J. Liu, J. A. Hurley, and L. Yu, *Mol. Pharmacol.* **44,** 8 (1993).

[2] R. C. Thompson, A. Mansour, H. Akil, and S. J. Watson, *Neuron* **11,** 903 (1993).

[3] C. J. Evans, D. E. Keith, H. Morrison, K. Magendzo, and R. H. Edwards, *Science* **28,** 1952 (1992).

[4] B. L. Kieffer, K. Befort, C. Gaveriaux-Ruff, and C. G. Hirth, *Proc. Natl. Acad. Sci. U.S.A.* **89,** 12048 (1992).

[5] G. X. Xie, F. Meng, A. Mansour, R. C. Thompson, M. T. Hoversten, A. Goldstein, S. T. Watson, and H. Akil, *Proc. Acad. Natl. Sci. U.S.A.* **91,** 3779 (1994).

[6] H. E. Hamm and A. Gilchrist, *Curr. Opin. Cell Biol.* **8,** 189 (1996).

[7] E. J. Neer and T. F. Smith, *Cell* **84,** 175 (1996).

[8] D. G. Lambright, J. Sondek, A. Bohm, N. P. Skiba, H. E. Hamm, and P. B. Sigler, *Nature* (*London*) **379,** 311 (1996).

[9] D. T. Jones and R. R. Reed, *J. Biol. Chem.* **262,** 14241 (1987).

for the preference displayed by agonist-bound receptors to signal only via certain G proteins present in the cell membrane.[10–13]

Receptors also vary in the amino acid sequences that interact with α subunits of trimeric G proteins, i.e., the receptor loop that links the fifth and sixth transmembrane region and the C-terminal tail.[14,15] The cloned δ-opioid receptor[3,4] and the μ-opioid receptor[1,2] differ in the intracellular territories implicated in their interaction with $G\alpha$ subunits. It is conceivable, therefore, that distinct types or subtypes of opioid receptor regulate different classes of G proteins.[16–18] Given the variations manifested by receptors and G proteins in their interacting domains, an investigation was made into the classes of G proteins regulated *in vivo* by μ and δ receptors in the promotion of supraspinal antinociception. This chapter describes the efficacy and selectivity in reducing the expression of coded signaling proteins by *in vivo* administration of antisense oligodeoxynucleotides complementary to their mRNA sequences. Functional data are also provided in order to assess the physiological relevance of opioid receptors and GTP-binding protein subtypes.

Methods

Synthesis of Oligodeoxynucleotides

Synthetic end-capped phosphorothioate antisense oligodeoxynucleotides (ODNs) are prepared by solid-phase phosphoramidite chemistry[19] using a CODER 300 DNA synthesizer (Du Pont; Wilmington, DE) at the 1-μmol scale (Tables I and II). The introduction of phosphorothioate linkages is achieved by tetraethylthiuram disulfide (TETD) sulfurization.[20]

[10] H. Ueda, Y. Yoshihara, H. Misawa, N. Fukushima, M. Ui, H. Takagi, and M. Satoh, *J. Biol. Chem.* **264,** 3732 (1989).
[11] S. F. Law, S. Zaina, R. Sweet, K. Yasuda, G. I. Bell, J. Stadel, and T. Reisine, *Mol. Pharmacol.* **45,** 587 (1994).
[12] Y. F. Lui, K. H. Jacobs, M. M. Rasenick, and P. R. Albert, *J. Biol. Chem.* **269,** 13880 (1994).
[13] S. Offermanns, T. Wieland, D. Homann, J. Sandmann, E. Bombien, K. Spicher, G. Schultz, and K. H. Jakobs, *Mol. Pharmacol.* **45,** 890 (1994).
[14] C. W. Taylor, *Biochem. J.* **272,** 1 (1990).
[15] A. D. Strosberg, *Eur. J. Biochem.* **196,** 1 (1991).
[16] J. Garzón, M. A. Castro, J. L. Juarros, and P. Sánchez-Blázquez, *Life Sci.-Pharmacol. Lett.* **54,** PL191 (1994).
[17] J. Garzón, A. García-España, and P. Sánchez-Blázquez, *J. Pharmacol. Exp. Ther.* **281,** 549 (1997).
[18] J. Garzón, M. Castro, and P. Sánchez-Blázquez, *Eur. J. Neurosci.* **10,** 2557 (1998).
[19] M. D. Matteucci and M. H. Carouthers, *J. Am. Chem. Soc.* **103,** 3185 (1981).
[20] H. Vu and B. L. Hirschbein, *Tetrahedron Lett.* **32,** 3005 (1991).

TABLE I

OLIGODEOXYNUCLEOTIDES TO μ- AND δ-OPIOID RECEPTOR mRNA

Receptor[a]	5'-Sequence-3'	Code	Ref.
μ (r/m)	C*G*CCCCAGCCTCTTCCT*C*T[b]	μ_{UN}[c]	d, e
μ (m)	C*T*GATGTTCCCTGGG*C*C	μ_{16-32}	e, f
μ (r/m)	T*T*GGTGGCAGTCTTCATTTT*G*G	$\mu_{291-311}$	d, g
μ (r/m)	T*G*AGCAGGTTCTCCCAGTAC*C*A	$\mu_{677-697}$	d, g
μ (r/m)	G*G*GCAATGGAGCAGTTTC*T*G	$\mu_{1175-1194}$	d, g
δ (m)	G*C*ACGGGCAGAGGGCACC*A*G	δ_{7-26}	f, h, i
δ (m)	A*G*AGGGCACCAGCTCC*A*T	δ_{29-46}	f, j
δ (r)	A*C*TGCAGCTCCGCA*G*G	δ_{22-37}	f

[a] The oligonucleotides correspond to those described in the code of the rat (r) or mouse (m) opioid receptor gene sequence.

[b] An asterisk (*) indicates the phosphorothioate linkages.

[c] This ODN is directed to a specific 5' untranslated region of the μ-opioid receptor clone.

[d] G. Rossi, Y.-X. Pan, J. Cheng, and G. W. Pasternak, *Life Sci.-Pharmacol. Lett.* **54,** PL375 (1994).

[e] P. Sánchez-Blázquez, M. Rodríguez-Díaz, I. DeAntonio, and J. Garzón, *J. Pharmacol. Expt. Ther.,* in press.

[f] P. Sánchez-Blázquez, A. García-España, and J. Garzón, *J. Pharmacol. Exp. Ther.* **280,** 1423 (1997).

[g] G. Rossi, L. Leventhal, Y.-X. Pan, J. Cole, W. Su, R. J. Bodnar, and G. W. Pasternak, *J. Pharmacol. Exp. Ther.* **281,** 109 (1997).

[h] J. Lai, E. J. Bilsky, R. B. Rothman, and F. Porreca, *Neuroreport* **5,** 1049 (1994).

[i] P. Sánchez-Blázquez and J. Garzón, *J. Pharmacol. Exp. Ther.* **285,** 820 (1998).

[j] K. M. Standifer, C.-C. Chien, C. Wahlestedt, G. P. Brown, and G. W. Pasternak, *Neuron* **12,** 805 (1994).

Solvents and reagents are obtained from Cruachem Ltd. (Glasgow, UK) and TETD is from Applied Biosystems (Foster City, CA). The efficiency of each base addition is higher than 98%. After synthesis the protecting groups are removed by treatment with 32% (v/v) aqueous NH_3 at 55° overnight. Crude ODNs are purified by reversed-phase chromatography using high-performance liquid chromatography (HPLC) (Kontron, Zurich, Switzerland). Evaporated ODNs (Speed Vac Plus; Savant, Farmingdale, NY) are then dissolved in 200 μl of 0.1 M triethylammonium acetate (pH 7.0) and injected into a C_{18} reversed-phase column (Spherisorb ODS-2, 5 μm; 150 $\times$ 4.6 mm) using 0.1 M triethylammonium acetate (pH 7.0) and acetonitrile as the mobile phase. The column is eluted with a 15–35% acetonitrile gradient over a 30-min period. The collected products (2 ml) are evaporated in fractions of 100 μg and stored at $-20°$ until use.

TABLE II

OLIGODEOXYNUCLEOTIDES TO Gα SUBUNIT mRNA

Gα subunit	5'-Sequence-3'[a]	Code	Ref.
i1	G*C*TGTCCTTCCACAGTCTCTTTATGACGCCG*G*C	$i1_{588-621}$	b, c
i2	A*T*GGTCAGCCCAGAGCCTCCGGATGACGCCC*G*A	$i2_{523-556}$	b, c, d
i3	G*C*CATCTCGCCATAAACGTTTAATCACGCCT*G*C	$i3_{554-587}$	b, c, d
z	C*G*TGATCTCACCCTTGCTCTCTGCCGGGCCA*G*T	$z_{330-363}$	c, e
o1	A*G*GCAGCTGCATCTTCATAGGTG*T*T	$o1_{882-906}$	d, f, g
o2	G*A*GCCACAGCTTCTGTGAAGGCA*C*T	$o2_{882-906}$	d, f, g
q	C*G*GCTACACGGTCCAAGTC*A*T	$q_{484-504}$	d, g
11	C*T*GTGGCGATGCGGTCCAC*G*T	$11_{487-507}$	d, g
q/11	C*C*ATGCGGTTCTCATTGTC*T*G	$q/11_{724-744}$	d

[a] Nucleotides correspond to those of the Gα subunit gene sequence described in the code.
[b] R. B. Raffa, R. P. Martínez, and C. D. Connelly, *Eur. J. Pharmacol.* **258,** R5 (1994).
[c] P. Sánchez-Blázquez, A. García-España, and J. Garzón, *J. Pharmacol. Exp. Ther.* **275,** 1590 (1995).
[d] P. Sánchez-Blázquez and J. Garzón, *J. Pharmacol. Exp. Ther.* **275,** 1590 (1995).
[e] J. Garzón, Y. Martínez-Peña, and P. Sánchez-Blázquez, *Eur. J. Neurosci.* **9,** 1194 (1997).
[f] C. Kleuss, J. Hescheler, C. Ewel, W. Rosenthal, G. Schultz, and B. Wittig, *Nature (London)* **353,** 43 (1991).
[g] J. Garzón, M. Castro, and P. Sánchez-Blázquez, *Analgesia* **1,** 4 (1995).

In Vivo Administration of Oligodeoxynucleotides

ODN solutions are made up in the appropriate volume of saline immediately prior to use. Various control groups of mice are used to monitor the specificity of ODN treatments. Typically, these controls include noninjected animals (naive), those that receive the vehicle used for the ODNs (saline), and animals injected with a random sequence ODN (ODN-RD) or mismatched antisense sequence. Injections are made into the right lateral ventricle. Subsequent administrations are on the same side.[21,22] Briefly, animals are lightly anesthetized with ether and injections made with a 10-μl Hamilton syringe at a depth of 3 mm, 2 mm lateral and 2 mm caudal of the bregma. The 4-μl content is infused at a rate of 1 μl every 5 sec. The needle is then maintained for an additional 10 sec. To minimize the chance of neurotoxicity caused by repetitive intracerebroventricular (icv) injections, an interval of 24 hr is allowed between administrations of the

[21] P. Sánchez-Blázquez, A. García-España, and J. Garzón, *J. Pharmacol. Exp. Ther.* **275,** 1590 (1995).
[22] E. J. Bilsky, R. N. Bernstein, V. J. Hruby, R. B. Rothman, J. Lai, and F. Porreca, *J. Pharmacol. Exp. Ther.* **277,** 491 (1996).

ODNs.[23] Each ODN treatment is performed on a distinct group of mice according to the following schedule: on days 1 and 2 with 1 nmol, on days 3 and 4 with 2 nmol, on day 5 with 3 nmol. Functional studies are carried out on day 6: opioid agonists are injected icv and their antinociceptive activity evaluated by the warm water tail-flick test.[21] The effects of the treatments on animal activity are recorded with a Digiscan animal activity monitor system (activity cage) (Omnitech Electronics, Columbus, OH). Only procedures and doses of ODNs that do not alter the behavior of the mice are employed. The animals display no noticeable behavioral changes with the described schedule.

Under anesthesia, a 25-gauge stainless steel cannula is implanted stereotaxically into the lateral ventricles of albino male Wistar rats (240–270 g) and ODNs infused chronically. A vinyl tubing connects the cannula to an osmotic minipump (Alzet; Alza, Palo Alto, CA) placed under the skin in the lumbar region. The ODNs are delivered in saline at 2.5 μl/hr (0.1 nmol/hr) for 21 days.[21] The cannula is permanently fixed to the skull by dental acrylic.[24]

Periventricular Cellular Structure after Subchronic Intracerebroventricular Oligodeoxynucleotide Treatment of Mice

To monitor any possible injury to tissue structure owing to icv delivery of the ODNs, Nissl tinction is performed on brain coronal slices that include some of the periventricular areas involved in the functional studies. According to the Paxinos and Franklin atlas, periaqueductal gray matter (PAG) is taken from a position 3400 μm postbregma and the periventricular area adjacent to the injection 140 μm before the bregma (Fig. 1). The mice are perfused via the ascending aorta with 10 mM phosphate buffer made up to 0.9% saline (PBS, pH 7.4) and the fixative, consisting of 4% (w/v) paraformaldehyde, 0.2% (v/v) picrate, and 0.35% (v/v) glutaraldehyde in 100 mM phosphate buffer (pH 7.4). Brains are quickly removed from the skull and immersed overnight in the fixative. Fixed specimens are immersed in 100 mM phosphate buffer containing 15% (w/v) sucrose. They are frozen and cut into 20-μm-thick sections. Slices are hydrated, stained with thionine, and dehydrated. Brain sections are covered with DePeX mounting medium (BDH Laboratories Supplies, Poole, U.K.) for histological observation. No alteration of the normal structure is observed in either the PAG or periventricular area.

[23] B. J. Chiasson, J. N. Armstrong, M. L. Hooper, P. R. Murphy, and H. A. Robertson, *Cell. Mol. Neurobiol.* **14,** 507 (1994).
[24] C. C. H. Yang, J. Y. H. Chang, and S. H. H. Chan, *Endocrinology* **132,** 495 (1993).

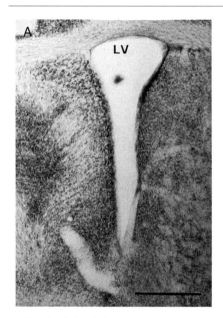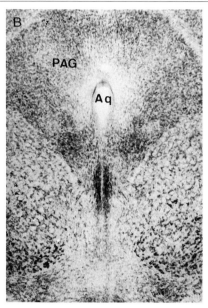

Fig. 1. Nissl staining of brain coronal slices, including the periaqueductal gray matter and periventricular area adjacent to the icv injection site. LV, Lateral ventricle; PAG, periaqueductal gray matter; Aq, cerebral aqueduct. Bar: 800 μm. See text for explanation.

Visualization of Fluorescence-Labeled Oligodeoxynucleotides in CNS

To monitor the entry of the ODNs into the CNS and their later distribution, some are labeled with fluorescein-CE phosphoramidite at the 5' end (Cruachem Ltd.). This is performed in the final synthetic cycle. Mice that have received a single icv injection of 3 nmol of a fluorescein-labeled ODN-$G_{i2}\alpha$ are sacrificed at various intervals. Brains are removed and frozen on dry ice. Coronal cryostat sections (20 μm) are cut, set on gelatin-subbed slides, and mounted in a solution of 0.1 M phosphate buffer–30% (v/v) glycerol. Sections are analyzed with a Leica TCS 4D confocal laser-scanning microscope equipped with an argon/krypton mixed-gas laser with epifluorescence illumination. Four to 5 mW of power is developed per line at 488, 568, and 647 nm. Images are collected with 16 × 0.50 PL Fluotar (625 × 625 μm^2), 40 × 1–0.50 PL Fluotar (250 × 250 μm^2), and 63 × 1.4 PL Apo (158.73 × 158.73 μm^2) 488-nm oil-immersion Plan-Neofluar objectives. The slices are scanned at a rate of ~8 μsec/pixel (0.1 μm^2), in slow scan mode. Fluorescence is observed with a standard fluorescein isothiocyanate filter. Excitation illumination is at 488 nm. Emissions are collected with a 510-

nm bandpass filter. Ten minutes after icv injection of the fluorescein-labeled ODNs the signals are detected in the PAG region and periventricular structures (Figs. 2 and 3).

Efficacy of Oligodeoxynucleotide Treatments

The depletion effect of the ODNs on the target proteins is also monitored by immunodetection studies. When possible, affinity-purified immunoglobulins (IgGs) directed to the protein of interest are labeled with ^{125}I and injected icv into mice that have received the corresponding ODN treatment. Autoradiography is then conducted on brain sections from these mice. Immunoblotting is routinely carried out in samples from brain structures of the animals undergoing either ODN treatment.

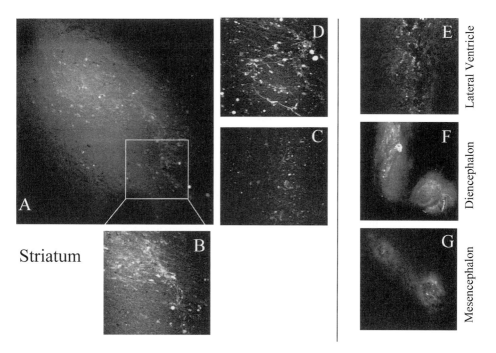

FIG. 2. Entry and progression of a fluorescein-labeled ODN to δ-opioid receptor delivered in the lateral ventricle of the mouse. Confocal images from neural areas of mouse brain. (A) Image taken with a ×16 objective (625 × 625 μm^2); striatum is shown, with a notable labeling of cells. A magnification of this image is shown in (B). (C and D) Higher resolutions of the same area, ×40 objective (250 × 250 μm^2).

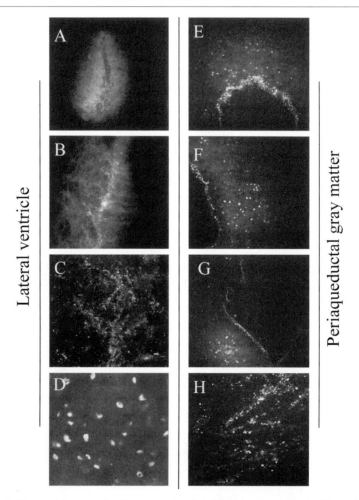

FIG. 3. Confocal images of a fluorescein-labeled ODN to $G_{i2}\alpha$ subunits: Entry into the periventricular structures and PAG. Photographs were taken with ×16 (A, B, E, F, and G), ×40 (C and H), and ×63 objectives (D). Subsequently, and in a time-dependent fashion, these signals spread through the neural tissue.

Autoradiographic Experiments

Antibodies. The antibodies used are MU/2EL[25] antiserum raised against the peptide sequence 208–216 (TKYRQGSID) of the μ receptor,

[25] J. Garzón and P. Sánchez-Blázquez, *Life Sci.-Pharmacol Lett.* **56,** PL237 (1995).

and $\Delta/1^{16}$ antiserum generated against the N-terminal peptide sequence (ELVPSARAELQSSPL) of the murine δ-opioid receptor. Anti-receptor IgGs are purified as previously described.[26] About 2 mg of the corresponding antigenic peptide is coupled to CNBr-activated Sepharose 4B (Pharmacia, Piscataway, NJ). The gel is packed in an 8-ml column and 4–6 ml of serum is then loaded. Sample recirculation in equilibration buffer [50 nM Tris-HCl, 200 mM NaCl, 0.1% (v/v) Tween 20, pH 7.7] is continued at 1 ml/min for 60 min. The column is rinsed with the same buffer, but without Tween 20, until the absorbance (280 nm) of the effluent reaches the baseline. Bound IgGs are detached by passing 0.2 M glycine hydrochloride, pH 2.5, through the column (typically 10 ml). The eluted IgGs are dialyzed and concentrated to about 300 to 500 μl in two consecutive 5-liter baths of 50 mM phosphate-buffered saline in a Micro-ProDiCon system (M_r 15,000 cutoff; Spectrum, Laguna Hills, CA). The final protein concentration of the IgGs is about 2.0 μg/μl.

Iodination of IgGs. IgGs to μ- and δ-opioid receptors are purified by affinity chromatography with antigenic peptide. Subsequent iodinization is performed according to Greenwood *et al.*[27] (using chloramine-T and Na^{125}I) with minor modifications.[28] The reaction is started by mixing 10 μl of a freshly made 0.1-mg/ml solution of chloramine-T in 50 mM sodium phosphate buffer (pH 7.4), with 65 μl of a solution containing 80 μg of purified IgGs and 500 μCi of Na^{125}I (NEZ 033A; specific activity, 17 Ci/mg) in 70 mM sodium phosphate buffer (pH 7.4). The reaction is stopped after 60 sec with 50 μl of chloramine-T stop buffer [sodium metabisulfite (2.4 mg/ml), tyrosine (saturated, 10 mg/ml), 10% (v/v) glycerol, 0.1% (v/v) xylene cyanole in 10 mM sodium phosphate (pH 7.4), 0.9% (w/v) NaCl]. Labeled IgGs are separated from free iodine in a Sephadex G-25 column (PD-10; Pharmacia) first equilibrated with 30 ml of 10 mM sodium phosphate (pH 7.4), 0.9% (w/v) NaCl, 1% (w/v) bovine serum albumin (BSA), and then with 100 ml of 10 mM sodium phosphate (pH 7.4), 0.9% (w/v) NaCl. The reaction material is eluted with 6 ml of 10 mM sodium phosphate (pH 7.4), 0.9% (w/v) NaCl and 0.5-ml fractions are collected. The IgGs are obtained in two fractions.

Autoradiography. Mice that have received either saline, ODN-RD, or ODNs to opioid receptors are injected icv with 4 μl of the ^{125}I-labeled IgGs (about 4,000,000 cpm/mouse). The radiolabeled IgGs are administered

[26] J. Garzón, J. L. Juarros, M. A. Castro, and P. Sánchez-Blázquez, *Mol. Pharmacol.* **47,** 738 (1995).

[27] F. C. Greenwood, W. M. Hunter, and J. S. Glover, *Biochem. J.* **308,** 299 (1985).

[28] P. Sánchez-Blázquez, A. García-España, and J. Garzón, *J. Pharmacol. Exp. Ther.* **280,** 1423 (1997).

bilaterally into the cerebral ventricles. After 24 hr brains are removed and frozen on dry ice. Coronal cryostat sections (20 μm) are cut at various levels of the neuraxis, mounted onto gelatin-subbed slides, and dried. Brain sections are exposed to tritium-sensitive film (Hyperfilm³H; Amersham, Arlington Heights, IL) for 20 days at $-80°$. Kodak (Rochester, NY) LX-24 developer (3 min) and Kodak AL-4 fixer (5 min) are used to develop the films. Radiolabeling of mouse brain neural structures can be observed 24 hr after injecting icv affinity-purified ¹²⁵I-labeled IgGs into μ- and δ-opioid receptors.[28] The greatest amount of labeling is localized in periventricular areas. Strong radiostaining is also found over the cortical, septal, and hippocampal regions (Fig. 4). In brain sections obtained from animals receiving denatured ¹²⁵I-labeled IgGs, or that are preabsorbed with the corresponding antigenic peptides, immunosignals are practically absent. The specificity of this labeling has been determined in previous investigations showing that these anti-opioid receptor antibodies diminish the specific binding of opioid agonists to mouse brain membranes.[16,25,26] Mice chronically treated with ODNs to μ-opioid receptors (ODN-μ_{un} and ODN-μ_{16-32}) display a substantial reduction of the immunolabeling promoted by

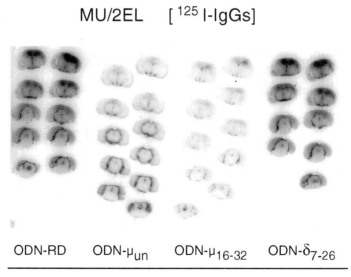

FIG. 4. *In vivo* radiolabeling of μ-opioid receptors. ODNs directed to μ- or δ-opioid receptors and a random ODN were given according to a 5-day schedule (see text). At the end of the treatment [¹²⁵I]IgGs to μ receptors were injected icv 24 hr before obtaining cryostat sections (20 μm) at various levels of the neuraxis. [Reprinted with permission from *J. Pharmacol. Exp. Ther.* **280,** 1428 (1997). Copyright © 1997 American Society for Pharmacology and Experimental Therapeutics.[28]]

MU/2EL ^{125}I-labeled IgGs. These immunosignals are preserved when using the ODN_{7-26} directed to δ-opioid receptor (Fig. 4). The $\Delta/1$ ^{125}I-labeled IgGs show a weak binding to brain sections obtained from mice injected with the ODN-δ_{29-46}.[28]

Electrophoresis and Immunoblotting

Opioid Receptors. Neural structures of the mice are obtained 6 days after commencing repeated administration of ODNs. Rats implanted with osmotic minipumps guided into the lateral ventricle are killed after 3 weeks of continuous delivery of the ODNs. Membrane fractions are then prepared and solubilized with sodium dodecyl sulfate (SDS) in a buffer containing 50 mM Tris-HCl, 3% (w/v) SDS, 10% (v/v) glycerol, 5% (v/v) 2-mercapto-ethanol, pH 6.8. About 80 μg of protein per lane is resolved by SDS–polyacrylamide gel electrophoresis (SDS–PAGE) in 8 $\times$ 11 $\times$ 0.15 cm gel slabs (7–18% acrylamide concentration/2.9% bisacrylamide cross-linker concentration) (Hoefer, San Francisco, CA) at 20-mA constant current (ISCO, Lincoln, NE). Proteins are transferred (Mini-Trans-Blot electrophoretic transfer cell; Bio-Rad, Richmond, CA) to 0.2-μm polyvinylidene difluoride Trans-Blot membranes (Bio-Rad) using Towing buffer [25 mM Tris-HCl, 192 mM glycine, 0.04% (w/v) SDS, 20% (v/v) methanol] by application of 70 V (200–300 mA) for 120 min. Unoccupied protein sites are blocked with 5% (w/v) nonfat dry milk (blocker; Bio-Rad) in Tris-buffered saline (TBS) for 1 hr at 37°. The membranes were incubated with anti-μ- and anti-δ-opioid receptor antibodies at 1:1000 dilution in TBS–0.05% (v/v) Tween 20 (TTBS) at 6° for 24 hr. After removing the antibodies the blots are washed with TTBS. Secondary antiserum [goat anti-rabbit IgG (H + L) horseradish peroxidase conjugate (Bio-Rad)] diluted 1:3000 in TTBS is left for 3 hr. The unbound secondary antiserum is then washed as before with TTBS. Antibody binding is detected with colorimetric substrate [3,3'-diaminobenzidine (1 mg/ml), 0.02% (v/v) hydrogen peroxide, 0.04% (w/v) nickel chloride in 0.1 M Trizma base buffer, pH 7.2] or by chemiluminescent detection (ECL; Amersham).

Immunoblots of SDS-solubilized membranes from mouse striatum show immunoreactive proteins at molecular masses of about 60 and 80 kDa for μ-opioid and 50 kDa for δ-opioid receptors[28] (Fig. 5). These are glycosylated proteins because the immunosignals shift to lower masses, in the range of 40 kDa, after enzymatic[26,29] or chemical removal[16] of the oligosaccharides. Glycoproteins exhibit anomalous mobility in SDS–PAGE chromatography that greatly depends on acrylamide concentration and buffer system. These

[29] L.-Y. Liu-Chen, C. Chen, and C. A. Phillips, *Mol. Pharmacol.* **44,** 749 (1993).

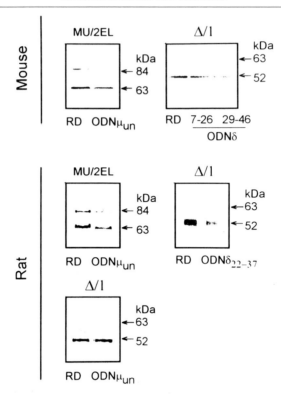

FIG. 5. Immunoblots of SDS extracts from mouse and rat striatum with anti-peptide antibodies to μ-opioid (MU/2EL) and δ-opioid (Δ/1) receptors. Mice received repeated icv injections of the ODNs to these receptors. Rats were implanted subcutaneously with osmotic minipumps guided into the lateral ventricle and the ODNs were continuously infused for 3 weeks. [Reprinted with permission from *J. Pharmacol. Exp. Ther.* **280,** 1427 (1997). Copyright © 1997 American Society for Pharmacology and Experimental Therapeutics.[28]]

considerations might apply for the diverse masses described for these glyco-sylated opioid receptors (see, i.e., Garzón *et al.*[26]).

The immunoreactivity observed in control animals receiving the random sequence ODN is comparable to that of naive mice. In mice undergoing repeated injections of the ODN-μ_{un}, a significant reduction of the μ-recep-tor-like immunoreactivity is observed[28] (Fig. 5). A greater decrease is achieved when the ODN-μ_{un} is continuously infused into the rat brain (Fig. 5). In this neural tissue the immunosignals associated with δ receptors are not altered by treatment with ODNs to μ-opioid receptors[28] (Fig. 5). The subchronic administration of ODNs to δ receptors produces small decreases in δ receptor-like immunoreactivity in mouse (Fig. 5). These immunosignals

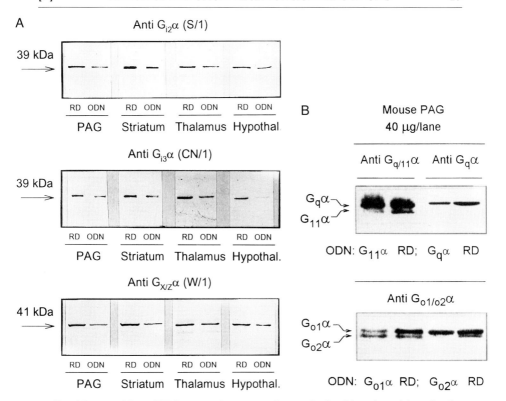

Fig. 6. Immunoblots of SDS extracts from areas of mouse brain with anti-peptide antibodies to Gα subunits. The ODNs to Gα subunits, random ODN, and mismatched ODNs were injected into the mice for five consecutive days. On day 6 the mice were killed and neural structures obtained. [Reprinted with permission from *J. Pharmacol. Exp. Ther.* **275,** 1592 (1995) and *J. Pharmacol. Exp. Ther.* **285,** 823 (1998). Copyright © 1995 and 1998 American Society for Pharmacology and Experimental Therapeutics.[21,30]]

appear notably diminished in rats receiving the ODN-δ_{22-37} chronically[28] (Fig. 5).

Gα Subunits. The antibodies used are directed to peptide sequences of Gα subunits[21,30]: anti-$G_{i1}\alpha$ internal fragment (118–124, FMTAELA), anti-$G_{i2}\alpha$ internal fragment (115–125, EEQGMLPEDLS), anti-$G_{i3}\alpha$ C-terminal fragment (345–354, KNNLKECGLY), anti-$G_z\alpha$ internal fragment (111–125, TGPAESKGEITPELL), anti-$G_q\alpha$ (371752-Q; Calbiochem, La Jolla, CA), anti-$G_o\alpha$ (NEI-804; Du Pont-New England Nuclear, Boston, MA), and anti-$G_{q/11}$ (NEI-809; Du Pont-New England Nuclear). Mice are killed

---

[30] P. Sánchez-Blázquez and J. Garzón, *J. Pharmacol. Exp. Ther.* **285,** 820 (1998).

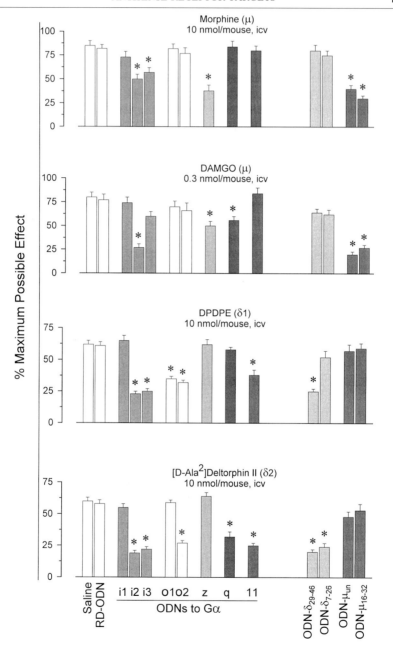

6 days after starting the subchronic administration of the ODNs and synaptosomes rich with membranes from various brain areas (P2 fraction) are solubilized and resolved by SDS–PAGE (7–18% T/2.9% C or 12.5% T/0.0625% C, with a linear gradient from 4 to 8 M urea[31]). The primary polyclonal antibodies are used typically at 1:1000 dilution. Immunoblots are analyzed by densitometry, using a Bio-Rad GS-700 imaging densitometer with reflectance capabilities. For each mouse/rat CNS structure, 30, 45, and 60 μg of protein are studied.

In the absence of urea, immunoblots show bands at molecular masses of 39 kDa for $G_{i1}\alpha$, $G_{i2}\alpha$, $G_{i3}\alpha$, and $G_o\alpha$ subunits, and 41–42 kDa for $G_z\alpha$, $G_{11}\alpha$, and $G_q\alpha$ subunits[21,30] (Fig. 6). $G_1\alpha$ and $G_{11}\alpha$ subunits can be resolved with a linear gradient of 4–8 M urea, with $G_{11}\alpha$ showing a greater electrophoretic mobility than $G_q\alpha$ subunits. An identical approach is utilized to separate $G_{o1}\alpha$ from $G_{o2}\alpha$ in immunoblots.[32] The ODNs corresponding to mRNA of $G\alpha$ subunits reduces the extent of labeling in immunoblots[21,30] (Fig. 6). Similar reductions in the expression of $G\alpha$ subunits in rodent CNS have also been reported by other groups using chronic delivery of the ODNs[33] of a single high dose.[34] The random sequence of ODN does not significantly change $G\alpha$ immunoreactivity when compared with that of naive mice. These treatments show no cross-effect on other $G\alpha$ subunits or on the immunoreactivity associated with nonrelated proteins.[21,30]

[31] B. H. Shah and G. Milligan, *Mol. Pharmacol.* **46**, 1 (1994).
[32] I. Mullaney and G. Milligan, *J. Neurochem.* **55**, 1890 (1990).
[33] K. M. Standifer, G. C. Rossi, and G. W. Pasternak, *Mol. Pharmacol.* **50**, 293 (1996).
[34] J. Shen, S. Shah, H. Hsu, and B. C. Yoburn, *Mol. Brain Res.* **59**, 247 (1988).

FIG. 7. Effect of icv administration of ODNs to different classes of α subunits of G proteins, and of ODNs directed to mRNAs encoding μ and δ receptors, on the analgesia evoked by opioids at the supraspinal level. Animals were injected for 5 days with increasing amounts of the ODNs (see text). On day 6 the antinociceptive activity of opioids was evaluated in the thermal tail-flick test. Latencies were measured 30 min after morphine, 15 min after DAMGO or DPDPE, and 10 min after [D-Ala^2]deltorphin II. Antinociception is expressed as a percentage of the maximum possible effect measurable in the warm water (52°) tail-flick test. Latencies were determined both before treatment (basal latency) and also after the administration of the substance under study. Baseline latencies ranged from 1.5 to 2.2 sec and were not affected by ODN administration. A cutoff time of 10 sec was allotted to minimize the risk of tissue damage. Values are the means ± SEM from groups of 10 to 15 mice each. *Significantly different from the control group receiving saline or the random ODN (ODN-RD) instead of the ODN to the corresponding opioid receptor type. ANOVA, Student–Newman–Keuls test, $p < 0.05$. [Reprinted with permission from *J. Pharmacol. Exp. Ther.* **275**, 1593 (1995), *J. Pharmacol. Exp. Ther.* **280**, 1425 (1997), *J. Pharmacol. Exp. Ther.* **285**, 823 (1998), and *Analgesia* **1**, 431 (1995). Copyright © 1995 and 1997 American Society for Pharmacology and Experimental Therapeutics.[21,28,30] Copyright © 1995 Cognizant Communication Corporation.[47]]

Application of in Vivo Administration of Oligodeoxynucleotide in Functional Studies

Some cases in which antisense technology has contributed considerably to *in vivo* studies of the opioid system are now presented.

Correlation of Cloned and Pharmacologically Defined Receptors. The antisense strategy has been used to impair receptor-mediated functions in *in vivo* studies.[35,36] Thus, ODNs to mRNA encoding opioid receptors are reported to selectively block antinociception evoked by agonists acting at μ receptors,[28,37–39] at δ receptors,[30,40–42] or at κ receptors.[43] In agreement with pharmacological proposals, the use of antisense technology with the cloned δ-opioid receptor also suggests the existence of different molecular forms for these receptors[22,28,30,42,44] (Fig. 7). The antisense approach has also helped determine the involvement of the cloned μ receptor in the development of morphine dependence.[28]

Assignment of G Proteins to Receptors in Production of Certain Effects: Supraspinal Analgesia. ODNs have also served to characterize the transducer system activated *in vivo* by agonist-bound receptors in the production of supraspinal analgesia. This strategy has substantiated the diversity of G proteins regulated by each type of opioid receptor in the production of this effect.[21,30,33,45–47] In addition, it has been possible to determine those G proteins regulated by only one of these opioid receptors (Figs. 7 and 8). These findings have led to new concepts such as the influence of the classes of G proteins that couple to a given receptor on the agonist–antagonist properties of its ligands.[16,18,45]

Oligodeoxynucleotides to Gα Subunits on Agonist-Evoked Stimulation of Low K_m GTPase Activity in Vitro. The role of various classes of G-

[35] C. Wahlestedt, E. M. Pich, G. F. Koob, F. Yee, and M. Helling, *Science* **259,** 528 (1993).

[36] M. Zhang and I. Creese, *Neurosci. Lett.* **161,** 223 (1993).

[37] G. Rossi, L. Leventhal, Y.-X. Pan, J. Cole, W. Su, R. J. Bodnar, and G. W. Pasternak, *J. Pharmacol. Exp. Ther.* **281,** 109 (1997).

[38] X.-H. Chen, J. U. Adams, E. B. Geller, J. K. Deriel, M. W. Adler, and L.-Y. Liu-Chen, *Eur. J. Pharmacol.* **275,** 105 (1995).

[39] P. Sánchez-Blázquez, I. DeAntonio, M. Rodríguez-Díaz, and J. Garzón, *Antisense Nucl. Acid Drug Dev.* **9,** 253 (1999).

[40] K. M. Standifer, C.-C. Chien, C. Wahlestedt, G. P. Brown, and G. W. Pasternak, *Neuron* **12,** 805 (1994).

[41] L. F. Tseng, K. A. Collins, and J. P. Kampine, *Eur. J. Pharmacol.* **258,** R1 (1994).

[42] G. C. Rossi, W. Su, H. Leventhal, and G. W. Pasternak, *Brain Res.* **753,** 176 (1997).

[43] C. C. Chien, G. Brown, Y. X. Pan, and G. W. Pasternak, *Eur. J. Pharmacol.* **253,** R7 (1994).

[44] J. Lai, E. J. Bilsky, R. B. Rothman, and F. Porreca, *Neuroreport* **5,** 1049 (1994).

[45] R. B. Raffa, R. P. Martinez, and C. D. Connelly, *Eur. J. Pharmacol.* **258,** R5 (1994).

[46] G. C. Rossi, K. M. Sandifer, and G. W. Pasternak, *Neurosci. Lett.* **198,** 99 (1995).

[47] J. Garzón, M. Castro, and P. Sánchez-Blázquez, *Analgesia* **1,** 4 (1995).

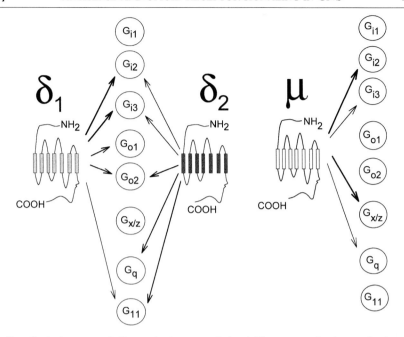

FIG. 8. Assignment of G proteins to μ- and δ-opioid receptors in the production of supraspinal analgesia. Thick lines denote significant reductions in analgesic potency in mice undergoing treatment with ODNs to these G proteins. G proteins without arrows are those regulated by neither μ- nor δ-opioid receptor. [Reprinted with permission from *Analgesia* **1,** 431 (1995) and *J. Pharmacol. Exp. Ther.* **285,** 826 (1998). Copyright © 1995 Cognizant Communication Corporation.[47] Copyright © 1998 American Society for Pharmacology and Experimental Therapeutics.[30]]

transducer proteins in the opioid-evoked activation of GTPase has been explored by using antisense ODNs to $G\alpha$ subunits. In periaqueductal gray membranes from mice administered icv an ODN to $G_z\alpha$ subunits, the agonists binding the μ-opioid receptor, [D-Ala^2,N-MePhe^4,Gly-ol^5]enkephalin (DAMGO) and morphine, show a reduced effect on the low-K_m GTPase. The agonist at $\delta 2$ receptors, [D-Ala^2]deltorphin II, displays weak activity while the agonist at $\delta 1$ receptors, [D-Pen^{2,5}]enkephalin (DPDPE), displays its full effect.[48] Thus, this approach is able to provide essential information on the classes of G proteins regulated by different opioid receptor types or subtypes.

[48] J. Garzón, Y. Martínez-Peña, and P. Sánchez-Blázquez, *Eur. J. Neurosci.* **9,** 1194 (1997).

Summary

The work in our laboratory has been designed to characterize the transducer mechanisms coupled to neurotransmitter receptors in the plasma membrane. Particular attention has been paid to the physiological/pharmacological effects mediated by the opioid system. Antisense oligodeoxynucleotides have proved useful in correlating opioid receptor clones with those defined pharmacologically. The involvement of the cloned opioid receptors μ, δ, and κ in analgesia has been determined by means of in vivo injection of ODNs directed to the receptor mRNAs. Using this strategy the classes of G-transducer proteins regulated by each type/subtype of opioid receptor in the promotion of antinociception have also been characterized. After displaying different patterns of binding to their receptors, opioids trigger a variety of intracellular signals. The physiological implications and therapeutic potential of these findings merit consideration.

Acknowledgments

This research was funded by grants from the Fondo de Investigaciones Sanitarias (FIS 97/0506) and the Comisión Interministerial de Ciencia y Tecnología (SAF98/0057).

[2] Targeting Brain GABA$_A$ Receptors with Antisense Oligonucleotides: Implications for Epilepsy

By Jesper Karle and Mogens Nielsen

Introduction

γ-Aminobutyric acid (GABA) is the principal inhibitory neurotransmitter in the central nervous system (CNS). The majority of fast neuronal inhibition is mediated via the GABA$_A$ receptor. The GABA$_A$ receptor is a member of the ligand-gated ion channel superfamily of neurotransmitter receptors.[1] In general, activation of the receptor by GABA induces a neuronal influx of chloride ions through the GABA$_A$ receptor-regulated ion channel, leading to hyperpolarization of the neuron.

GABA$_A$ receptor function is modulated by ligands that recognize different binding sites within the receptor complex. Some of these are clinically

[1] P. R. Schofield, M. G. Darlison, N. Fujita, D. R. Burt, F. A. Stephenson, H. Rodriguez, L. M. Rhee, J. Ramachandran, V. Reale, T. A. Glencorse, P. H. Seeburg, and E. A. Barnard, *Nature (London)* **328,** 221 (1987).

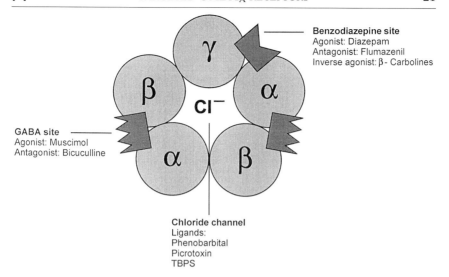

Benzodiazepine site
Agonist: Diazepam
Antagonist: Flumazenil
Inverse agonist: β- Carbolines

GABA site
Agonist: Muscimol
Antagonist: Bicuculline

Chloride channel
Ligands:
Phenobarbital
Picrotoxin
TBPS

FIG. 1. Schematic illustration of a proposed structure of the GABA$_A$ receptor complex, as seen "through" the chloride ion channel. A probable combination of receptor subunits (two α's; two β's; one γ) is shown. Also, putative recognition sites for important ligands are indicated. [Modified from J. Karle and M. Nielsen, *Rev. Contemp. Pharmacother.* **9,** 77 (1998), with permission from Marius Press.]

important compounds, e.g., the 1,4-benzodiazepines and the barbiturates, which exert their action by potentiating GABA$_A$ receptor function. Agonism of the action of GABA at the GABA$_A$ receptor leads to anxiolysis, sedation, muscle relaxation, and anticonvulsion, whereas antagonism generally results in the opposite effects.

The GABA$_A$ receptor complex has a pentamer structure of different polypeptide subunits. GABA$_A$ receptor subtypes are assembled from different combinations of receptor subunits.[2–4] Figure 1 shows a model of a GABA$_A$ receptor complex with the major putative ligand recognition sites indicated. Several subunit families and isoforms have been identified (α1–6, β1–3, γ1–3, δ, and ε).[4,5] Within the central nervous system, there are extensive regional and cellular differences in the expression of receptor subunits. The number of receptor subtypes occurring in the brain, as well

[2] G. A. R. Johnston, *Pharmacol. Ther.* **69,** 173 (1996).

[3] R. L. Macdonald and R. W. Olsen, *Annu. Rev. Neurosci.* **17,** 569 (1994).

[4] W. Sieghart, *Pharmacol. Rev.* **47,** 181 (1995).

[5] P. J. Whiting, G. McAllister, D. Vassilatis, T. P. Bonnert, R. P. Heavens, D. W. Smith, L. Hewson, R. O'Donnell, M. R. Rigby, D. J. Sirinathsinghji, G. Marshall, S. A. Thompson, and K. A. Wafford, *J. Neurosci.* **17,** 5027 (1997).

as their precise subunit composition and stoichiometry, are not known. Furthermore, the physiological implications of this receptor heterogeneity have not been elucidated. An area of great scientific importance is the investigation of the physiological roles played by individual $GABA_A$ receptor subunits/subtypes. It is possible, e.g., that individual $GABA_A$ receptor subtypes subserve different modalities of GABA-ergic neurotransmission, perhaps reflected in different behavioral modes. Also, there is widespread interest in the development of drugs that selectively affect $GABA_A$ receptor subtypes. Such compounds may have advantages compared with the present drugs used in the treatment of, e.g., seizure states, anxiety, and insomnia, and may be associated with fewer unwanted effects.

The involvement of GABA in the pathogenesis of epilepsy is a matter of long-standing interest.[6] It has been hypothesized that a dysfunction of GABA-ergic inhibitory neurotransmission via the $GABA_A$ receptor plays a central role in epileptogenesis.

Specific inhibition of the expression of neurotransmitter receptor proteins has been demonstrated after *in vivo* intracerebral administration of antisense oligodeoxynucleotides (ODNs) to rodents. Consequences of this antisense "knockdown" have been characterized by means of, e.g., behavioral experiments. Antisense knockdown may represent a rational method by which to study the physiological and possibly pathophysiological roles played by individual $GABA_A$ receptor subunit proteins. We have applied antisense technology with the purpose of selectively inhibiting the expression of $GABA_A$ receptor subunits in particular rat brain regions *in vivo*. We have focused on investigating the consequences of inhibited expression of one major $GABA_A$ receptor subunit, the $\gamma 2$ subunit. A γ subunit is essential for the presence of a high-affinity benzodiazepine binding site within a $GABA_A$ receptor complex.[7] The predominant $\gamma 2$ subunit is a constituent of the majority of $GABA_A$ receptor complexes in the brain.[8]

We have studied biochemical, morphological, electroencephalographic, and behavioral changes following continuous unilateral infusion of antisense ODN targeted to the $\gamma 2$ subunit in the rat hippocampus. The hippocampus was chosen as a target region owing to its relevance for disease states that are presumed to involve GABA-ergic mechanisms, e.g., temporal lobe epilepsy. Also, the results of intracerebroventricular (icv) and intrastriatal antisense ODN administration have been evaluated.

[6] R. W. Olsen and M. Avoli, *Epilepsia* **38,** 399 (1997).
[7] D. B. Pritchett, H. Sontheimer, B. D. Shivers, S. Ymer, H. Kettenmann, P. R. Schofield, and P. H. Seeburg, *Nature (London)* **338,** 582 (1989).
[8] R. McKernan and P. J. Whiting, *Trends Neurosci.* **19,** 139 (1996).

Intrahippocampal infusion of antisense ODN to the GABA$_A$ receptor γ2 subunit leads to a significant reduction in hippocampal γ2 subunit protein levels (50% of control)[9] and in benzodiazepine receptor radioligand binding to hippocampal membrane preparations[10,11] (Fig. 2). The results of a number of control experiments (see below) have supported the notion that the changes induced by the antisense ODN are the results of specific inhibition of the expression of the γ2 subunit. The antisense treatment appears to lead to a decrease in the number of functional GABA$_A$ receptors in the hippocampus, as reflected in decreases in the binding of radioligands to the ion channel domain and the GABA-binding site of the GABA$_A$ receptor complex.[10] Therefore, the γ2 subunit antisense ODN treatment may be viewed as a method to downregulate "benzodiazepine-site carrying" GABA$_A$ receptor complexes in particular rat brain regions.

Rats treated with a unilateral intrahippocampal infusion of γ2 subunit antisense ODN are viable and do not develop changes in spontaneous behavior, including anxiety-like behavior.[12] However, the rats experience a 10% weight loss during 6 days of antisense ODN, but not control ODN, infusion.[12] Furthermore, the hippocampal γ2 subunit antisense knockdown appears to lead to a state of diminished hippocampal GABA-ergic inhibitory neurotransmission. The rats develop spontaneous electroencephalographic seizures that evolve into profound limbic status epilepticus.[9] Antisense ODN-treated rats exhibit significant changes in induced seizure activity.[12] As an example, the elevated threshold for motor seizures induced by electrical stimulation in antisense ODN-treated rats is shown in Fig. 3. Prolonged infusion of antisense ODN results in hippocampal neurodegeneration.[10,11] In contrast, after control ODN infusion, intact hippocampal histology is found, except for a small lesion in the vicinity of the ODN infusion site.

The results of the γ2 knockdown experiments have provided direct evidence of a link between the GABA$_A$ receptor and epilepsy, supporting the hypothesis that the GABA$_A$ receptor is critically involved in the pathogenesis of seizures and status epilepticus. The described animal model is suggested as a new model of temporal lobe epilepsy and limbic status epilepticus that specifically involves the GABA$_A$ receptor and thus may be of pathophysiological relevance.

[9] J. Karle, D. P. D. Woldbye, L. Elster, N. H. Diemer, T. G. Bolwig, R. W. Olsen, and M. Nielsen, *J. Neurosci. Res.* **54,** 863 (1998).

[10] J. Karle, M.-R. Witt, and M. Nielsen, *Neurosci. Lett.* **202,** 97 (1995).

[11] J. Karle, M.-R. Witt, and M. Nielsen, *Brain Res.* **765,** 21 (1997).

[12] J. Karle, P. Laudrup, F. Sams-Dodd, J. D. Mikkelsen, and M. Nielsen, *Eur. J. Pharmacol.* **340,** 153 (1997).

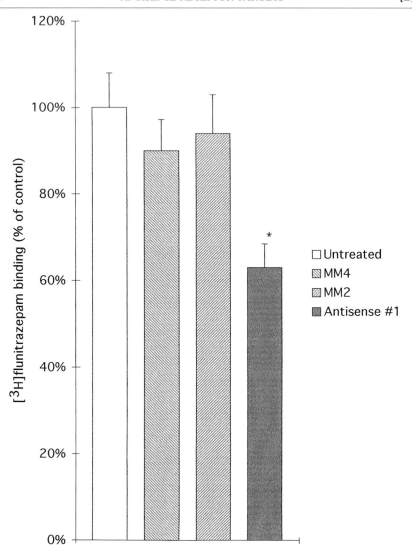

FIG. 2. Treatment of rats with an intrahippocampal infusion of antisense ODN to the GABA$_A$ receptor $\gamma2$ subunit decreases benzodiazepine receptor binding. [^{3}H]Flunitrazepam binding to hippocampal membrane preparations after continuous intrahippocampal infusion of antisense or mismatch control ODN for 5 days. MM4 is a mismatch ODN with the four central bases of the antisense ODN interchanged; in MM2 the two central bases are interchanged (see Table I). Data are expressed as a percentage of values from untreated controls (DPM/2 mg of original tissue), i.e., the contrateral hippocampi from antisense ODN-treated rats (mean $\pm$ S.D.; $n = 9$). *$p < 0.01$ versus untreated control, MM4, or MM2 (Mann–Whitney U test). [From J. Karle, M.-R. Witt, and M. Nielsen, *Brain Res.* **765,** 21 (1997), with permission from Elsevier Science.]

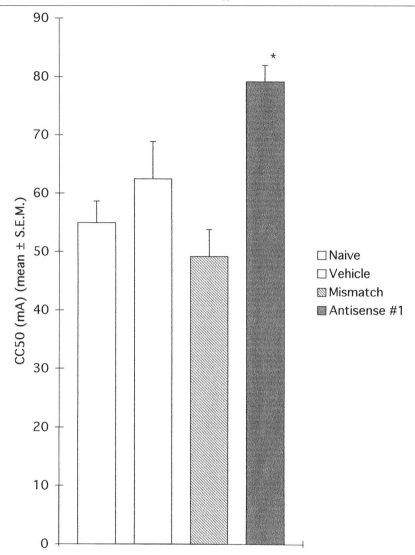

Fig. 3. Treatment of rats with an intrahippocampal infusion of antisense ODN to the GABA$_A$ receptor $\gamma2$ subunit elevates the threshold for electrically induced tonic seizures. Results of maximal electroshock seizure threshold (MEST) experiments. CC$_{50}$ is the stimulus current that induces toxic hindlimb extension in 50% of the total number of rats. *$p < 0.05$ versus naive, vehicle-treated, or mismatch ODN (MM4)-treated rats (t test). [From J. Karle, P. Laudrup, F. Sams-Dodd, J. D. Mikkelsen, and M. Nielsen, *Eur. J. Pharmacol.* **340,** 153 (1997), with permission from Elsevier Science.]

Technical Considerations

Oligodeoxynucleotide Administration

Oligodeoxynucleotides do not readily penetrate the mammalian blood–brain barrier.[13] This observation necessitates that ODNs be administered either directly into brain parenchyma or into the cerebral ventricular system. When targeting a ubiquitous brain receptor such as the $GABA_A$ receptor, it may be of high priority to ensure adequate distribution of ODN within a brain region of interest and to attempt neuroanatomical restriction of changes induced by the antisense ODN. The results of studies addressing the distribution and cellular uptake of antisense ODNs in the rodent brain[14,15] [and our own (unpublished) experiments using *in situ* hybridization with ODN complementary to the infused ODN] have suggested that the intraparenchymal route of administration offers advantages over icv administration. After icv administration, phosphorothioate ODN was detected only in structures in the near vicinity of the lateral ventricles.[15] Following intrahippocampal administration, ODN was present in the majority of the hippocampus. In terms of reaching a region of interest and, possibly, obtaining neuroanatomical restriction of effects induced by the antisense ODN, it may thus be preferred to administer ODN directly into brain parenchyma. This route of administration may, however, be associated with an increased risk of unwanted effects, e.g., tissue damage.[16]

The results of dose–response experiments have suggested that the delivered concentration of antisense ODN in experiments involving the hippocampus (1.7 μg/μl; 0.5 μl of infusion per hour) is within the window of optimal effectiveness. The duration of antisense ODN infusion should be adapted to the turnover of the targeted protein. After 2 days of continuous intrahippocampal antisense ODN infusion we have found a significant decrease in hippocampal benzodiazepine receptor radioligand binding.[11]

Antisense Oligodeoxynucleotide Sequences and Modification

The γ2 subunit antisense ODN used throughout our studies is complementary to a section spanning the putative translation initiation codon of rat $GABA_A$ receptor γ2 subunit mRNA.[17] This target region within an

[13] S. Agrawal, J. Temsamani, and J. Y. Tang, *Proc. Natl. Acad. Sci. U.S.A.* **88,** 7595 (1991).

[14] A. Szklarczyk and L. Kaczmarek, *J. Neurosci. Methods* **60,** 181 (1995).

[15] Y. Yaida and T. S. Nowak, Jr., *Regul. Pept.* **59,** 193 (1995).

[16] B. J. Chiasson, J. N. Armstrong, M. L. Hooper, P. R. Murphy, and H. A. Robertson, *Cell. Mol. Neurobiol.* **14,** 507 (1994).

[17] B. D. Shivers, I. Killisch, R. Sprengel, H. Sontheimer, M. Köhler, P. R. Schofield, and P. H. Seeburg, *Neuron* **3,** 327 (1989).

mRNA is often used in *in vivo* antisense studies involving CNS proteins and has been successful in a number of studies targeting different neuronal proteins. The nucleotide sequences of antisense ODNs used to target the γ2 subunit as well as control ODNs are shown in Table I. All experiments were carried out using fully phosphorothioate-modified ODNs. All sequences have been controlled for homologies to identified rodent gene sequences in the EMBL database.

Assessment of Changes Induced by Antisense Oligodeoxynucleotides: Control Experiments

It has not been substantially documented that antisense ODNs can be completely target gene specific.[18] Antisense ODNs may best be viewed as relatively specific, often offering a high degree of specificity. The issue of how to determine if a "true antisense" mechanism is responsible for the changes induced by an antisense ODN has raised much attention. Well-designed control experiments are warranted.

Changes induced by an antisense ODN should be estimated by a method that closely reflects the intended primary action of the ODN, i.e., measuring the levels of the targeted protein by means of immunochemical methods, e.g., Western blot[9,19] or immunocytochemistry. In receptor studies, radioligand-binding assays are often used to monitor antisense effects.

For control experiments we have applied mismatch ODNs in order to use ODN sequences that resemble that of the antisense ODN as closely as possible (Table I). Nucleotide specificity of GABA$_A$ receptor γ2 subunit antisense ODN-induced effects was supported by the observation that mismatch ODNs, in which two or four central bases of the antisense sequence were interchanged,[10–12] never induced biological effects comparable to those induced by the antisense ODN. However, a mismatch ODN with six "peripheral" base shifts, retaining a central 8-mer part of the γ2 subunit antisense ODN, induced electroencephalographic changes, although of a later onset and weaker intensity than those induced by the antisense ODN.[9] It seems plausible that this ODN is able to exert some antisense action due to the intact central nucleotide sequence.[16,20] Target gene specificity has been supported by the replication of antisense-induced changes by a second γ2 subunit antisense ODN, complementary to an adjacent part of the γ2 mRNA[9] (Table I). Sense ODN, an ODN antisense to the 5' untranslated region of the γ2 subunit mRNA as well as a dopamine D2 receptor antisense

[18] A. D. Branch, *Trends Biochem. Sci.* **23,** 45 (1998).
[19] W. J. Zhu, J. F. Wang, S. Vicini, and D. R. Grayson, *Mol. Pharmacol.* **50,** 23 (1996).
[20] A. Nicot and D. W. Pfaff, *J. Neurosci. Methods* **71,** 45 (1997).

TABLE I
OLIGODEOXYNUCLEOTIDE SEQUENCES

ODN name	Sequence	Description	Refs.[a]
γ2 subunit antisense			
1	5'-TAT-TTG-GCG-AAC-TCA-TCG-3'	Antisense to initiation codon of rat GABA$_A$ receptor γ2 subunit mRNA	1–4
2	5'-GTG-CTT-CCA-GTG-CTC-CAT-3'	Antisense to a nonoverlapping section of the γ2 mRNA in the 3' direction from the target of antisense	4
3	5'-GCC-TCT-GGT-TGC-AGA-AGA-3'	Antisense to a section of the 5' untranslated region of the γ2 mRNA	4
γ2 subunit sense	5'-CGA-TGA-GTT-CGC-CAA-ATA-3'	Complementary to antisense 1	1
Two-base γ2 subunit mismatch (MM2)	5'-TAT-TTG-GCA-GAC-TCA-TCG-3'	Two central bases of antisense 1 interchanged	2
Four-base γ2 subunit mismatch (MM4)	5'-TAT-TTG-GAA-GCC-TCA-TCG-3'	Four central bases of antisense 1 interchanged	1–4
Six-base γ2 subunit mismatch	5'-TTA-TTG-GCG-AAT-CAC-TCG-3'	Six "peripheral" changes in the sequence of antisense 1	4
Dopamine D2 receptor antisense	5'-AGG-ACA-GGT-TCA-GTG-GAT-C-3'	Antisense to dopamine D2 receptor mRNA	4, 5

[a] Key to references: (1) J. Karle, M.-R. Witt, and M. Nielsen, Neurosci. Lett. 202, 97 (1995); (2) J. Karle, M.-R. Witt, and M. Nielsen, Brain Res. 765, 21 (1997); (3) J. Karle, P. Laudrup, F. Sams-Dodd, J. D. Mikkelsen, and M. Nielsen, Eur. J. Pharmacol. 340, 153 (1997); (4) J. Karle, D. P. D. Woldbye, L. Elster, N. H. Diemer, T. G. Bolwig, R. W. Olsen, and M. Nielsen, J. Neurosci. Res. 54, 863 (1998); (5) J. M. Tepper, B.-C. Sun, L. P. Martin, and I. Creese, J. Neurosci. 17, 2519 (1997).

ODN,[21] used in different experimental set-ups, have been without effect. The observation that effects induced by the antisense ODN can be counteracted via potentiation of GABA$_A$ receptor activity with the benzodiazepine receptor agonist diazepam[9,11] has served as additional evidence of the mechanism of action of the antisense ODN.

Procedures

Overview

The procedures used for the *in vivo* administration of ODNs to rats are outlined. Also, we describe homogenate radioligand-binding assays estimating the binding of ligands to the benzodiazepine-binding site {[³H]flunitrazepam or [³H]Ro-15-1788 ([³H]flumazenil)}, the ion channel domain (*tert*-butylbicyclophosphoro[³⁵S]thionate, [³⁵S]TBPS), and the GABA-binding site ([³H]muscimol) of the GABA$_A$ receptor complex, used to measure antisense effects at the receptor level. At the level of the targeted protein we have estimated effects by means of Western blots.[9] The results of a number of experiments have shown that the effects induced by the GABA$_A$ receptor γ2 subunit antisense treatment are highly reproducible.

Oligodeoxynucleotide Manufacture

The ODNs used in our laboratory are manufactured by β-cyanoethylphosphoramidite chemistry and fully phosphorothioate modified using the Beaucage reagent (DNA Technology, Aarhus, Denmark). ODNs are purified by reversed-phase high-performance liquid chromatography (HPLC), ethanol precipitated, and finally dissolved in sterile H$_2$O. ODNs from a large number of different batches have been used with no apparent batch-to-batch variation in results.

Initial Procedures

The ODN dissolved in sterile H$_2$O is diluted to the desired concentration (see above).

Alzet osmotic minipumps for continuous infusion of ODN [e.g, Alzet 1007 (infusion, 0.5 μl/hr over 7 days) or Alzet 1003 (infusion, 1.0 μl/hr over 3 days); Alza, Palo Alto, CA] are filled with ODN solution. Correct filling with the intended volume is controlled by weighing minipumps before and after the filling. Minipumps are incubated overnight at 37°.

---

[21] J. M. Tepper, B.-C. Sun, L. P. Martin, and I. Creese, *J. Neurosci.* **17**, 2519 (1997).

Minipumps are assembled with Alzet brain infusion kits (infusion cannula and polyethylene catheter cut to desired length, e.g., 4.5 cm for intrahippocampal infusion to rats).

Surgery

Male Wistar rats (weight, 200–300 g) are anesthetized by intraperitoneal injection of sodium pentobarbital (50 mg/kg) and placed in a stereotaxic apparatus. A custom-made appliance (short metal tube plus screw) is recommended to attach the infusion cannula before the intracranial implantation.

Stereotaxic implantation of ODN infusion cannula: We have used the following coordinates (according to Paxinos and Watson[22]): right hippocampus: anterior (A), 3.0 mm; lateral (L), 4.3 mm from the aural line; 5.0 mm inferior from skull; right lateral cerebral ventricle: A, 1.5 mm; L, 1.5 mm from the bregma; 5.0 mm inferior from skull; and right corpus striatum: A, 8.8 mm; L, 2.5 mm from the aural line; 5.0 mm inferior from skull. The cannula is fixed by means of dental cement.

The osmotic minipump already connected to the infusion cannula is placed in a "poche" formed by careful dissection of subcutaneous tissue between the scapulae.

When observation of rats is desired beyond the scheduled ODN infusion time or beyond the duration of Alzet minipump function (see above), the minipumps should be removed, owing to the risk of destruction of the minipump, which may lead to tissue damage in the implantation region. The minipump and polyethylene catheter can be explanted under light ether anesthesia. The infusion cannula should be left *in situ.*

Homogenate Radioligand-Binding Assays

Following the scheduled ODN infusion time, rats are killed by decapitation. Brains are rapidly removed. The relevant brain regions (e.g., right and left hippocampus) are dissected free and weighed.

Brain tissue is homogenized with an Ultra-Turrax homogenizer in ice-cold Tris–citrate buffer (50 mM, pH 7.1). The homogenate is centrifuged at 30,000g for 10 min at 0°. The supernatant is discarded and the pellet is resuspended in Tris–citrate buffer (50 mM, pH 7.1) and centrifuged at 30,000g for 10 min. The resuspension and centrifugation procedure is repeated twice.

For specific binding of [^{3}H]flunitrazepam (0.5 nM) or [^{3}H]Ro-15-1788 (0.5 nM), the final pellet is resuspended in Tris–citrate buffer at 2 mg of

[22] G. Paxinos and C. Watson, "The Rat Brain in Stereotaxic Coordinates." Academic Press, New York, 1982.

original tissue per milliliter. Tissue samples are incubated with [^{3}H]benzodiazepine receptor ligands for 40 min at 0°. Nonspecific binding of [^{3}H]benzodiazepine receptor ligands is defined by means of, e.g., midazolam (10^{-5} M).

For specific binding of [^{35}S]TBPS (0.9 nM) the final tissue pellet is kept frozen ($-20°$) before resuspension in Tris–citrate (50 mM, pH 7.1) containing 1 mM NaCl at 5 mg of tissue per 0.5 ml. Tissue samples are incubated with [^{35}S]TBPS for 3 hr at 25°. Nonspecific binding is estimated using picrotoxin (5 × 10^{-5} M).

For specific binding of [^{3}H]muscimol (10 nM), preferably frozen tissue pellets are resuspended in Tris–citrate at 2 mg/ml. Tissue samples are incubated with [^{3}H]muscimol for 30 min at 0°. Nonspecific binding of [^{3}H]muscimol is estimated using [^{3}H]GABA (10^{-4} M).

Binding to a receptor that is not targeted by the GABA$_A$ receptor γ2 subunit antisense ODN, e.g., the muscarinic acetylcholine receptor, can be estimated using [^{3}H]quinuclidinyl benzilate ([^{3}H]QNB; 0.5 nM). Tissue samples (2 mg/ml) are incubated with [^{3}H]QNB for 60 min at room temperature; nonspecific binding is measured by means of atropine (5 × 10^{-6} M).

Following incubation, samples are added to 5 ml of ice-cold Tris–citrate buffer and rapidly filtered through Whatman (Clifton, NJ) GF/C glass fiber filters. Filters are rinsed by adding 5 ml of ice-cold Tris–citrate buffer.

The amount of ^{3}H or ^{35}S retained on the filters is quantified with conventional liquid scintillation counting equipment.

Tissue protein concentration is estimated by means of, e.g., a Bio-Rad (Hercules, CA) kit, using bovine serum albumin as standard.

All assays are carried out in duplicate.

Standard Preparation for Histological Analysis

Following the scheduled ODN infusion time rats are anesthetized with sodium pentobarbital as described above.

The left cardiac ventricle of each rat is cannulated. Rats are perfused transcardially with NaCl [0.9% (w/v) approximately 50 ml] followed by paraformaldehyde [4% (w/v) in 0.1 M phosphate buffer; pH 7.4; approximately 50 ml]. If the perfusion syringe is located correctly in the left cardiac ventricle, generalized extension of muscles and whitening of viscera should rapidly develop.

Rats are decapitated and whole brains dissected free and postfixed in paraformaldehyde (4%) for approximately 1 week. After postfixation, whole brains are quickly frozen and cut in sections (e.g., 20–30 μm), using a microtome. Sections are fixed on glass plates and stained by means of, e.g., hematoxylin–eosin according to standard procedures.

Conclusions

Increased knowledge of the contribution of individual $GABA_A$ receptor subunits to the properties of $GABA_A$ receptor complexes and to GABA-ergic neurotransmission may lead to improved understanding of the physiological roles of different $GABA_A$ receptor subtypes. Antisense technology represents a rational approach to address this important issue.

[3] Delivery of Antisense DNA by Vectors for Prolonged Effects *in Vitro* and *in Vivo*

By DAGMARA MOHUCZY, XIAOPING TANG, and M. IAN PHILLIPS

Introduction

Vectors (plasmid or viruses) can be used for delivery of the gene of interest in "sense" or "antisense" orientation. The sense orientation is to express, as a final product, active protein capable of performing physiological function. The antisense orientation is to produce antisense mRNA that interferes with normally produced "sense" mRNA and, as a result, decrease the amount of translated protein.

Sense or antisense DNA is usually at least a few hundred bases long and is subcloned into the plasmid vector containing

1. A bacterial origin of replication, so that the plasmid can be multiplied in the bacteria.
2. An antibiotic resistance gene to select bacteria containing plasmid.
3. A mammalian-type promoter to drive expression of cDNA in mammalian cells.
4. A splicing signal for efficient RNA processing.
5. A polyadenylation signal for translation of mRNA.

If the plasmid will be packaged into the virus later on, the cassette should also contain the specific virus sequences required in *cis* configuration, for instance, terminal repeats. The presence of a reporter or selection gene, which can be driven by a separate mammalian-type promoter or separated from the gene of interest by an internal ribosome entry site (IRES), is optional.

Plasmid Vectors

Plasmid vectors cause a relatively small immune response and can deliver longer DNA sequences than most viral vectors. The limitations currently are a low efficiency of gene transfer and poor long-term expression, especially when used *in vivo*. Plasmid vectors containing the gene of interest have been tested in many *in vitro* studies. Some of the examples are mentioned below.

Taniguchi *et al.*[1] used plasmid with insulin cDNA and a glucocorticoid-responsive promoter in the 3′ region of insulin cDNA in reverse orientation, so that antisense insulin mRNA is produced in response to glucocorticoids. When fibroblasts transfected with this construct were cultured in the presence of dexamethasone, they showed a reduction in proinsulin production. Smith and Prochownik[2] stably transfected murine erythroleukemia cells with the glucocorticoid-mediated c-*jun* antisense expression plasmid and achieved an 80–90% reduction in the c-Jun protein level in the presence of dexamethasone, with concomitant growth inhibition of these cells. Bradley *et al.*[3] reversed the phenotype of H-*ras*-transformed cells by transfecting them with antisense *fos* expression plasmid, inducible by dexamethasone. An insulin-like growth factor I (IGF-I) receptor antisense expression plasmid was used by Shapiro *et al.*[4] for tranfection of a human alveolar rhabdomyosarcoma cell line. Clones exhibited reduced expression of the IGF-I receptor and reduced growth rates, and failed to form tumors in immunodeficient mice.

Plasmid DNA has also been tested *in vivo,* after injection into various tissues alone or in combination with carriers. The most frequently used route of administration for plasmid DNA is probably intramuscular injection. Following are some examples of performed studies. Schlaepfer and Eckel[5] obtained significant reduction of plasma triglycerides in mice after a single intramuscular injection of the human lipoprotein lipase gene with mRNA detected for at least 21 days. Anwer *et al.*[6] observed production of significant levels of human growth hormone in muscle 2 weeks after injection into rat tibialis cranialis muscle, especially after complexing DNA with

[1] K. Taniguchi, R. Hirochika, K. Fukao, and H. Nakauchi, *Cell Transpl.* **5**(Suppl.), S55 (1996).
[2] M. J. Smith and E. V. Prochownik, *Blood* **79**(8), 2107 (1992).
[3] M. O. Bradley, S. Manam, A. R. Kraynak, W. W. Nichols, and B. J. Ledwith, *Ann. N.Y. Acad. Sci.* **660,** 124 (1992).
[4] D. N. Shapiro, B. G. Jones, L. H. Shapiro, P. Dias, and P. J. Houghton, *J. Clin. Invest.* **94**(3), 1235 (1994).
[5] I. R. Schlaepfer and R. H. Eckel, *Diabetes* **48**(1), 223 (1999).
[6] K. Anwer, M. Shi, M. F. French, S. R. Muller, W. Chen, Q. Liu, B. L. Proctor, J. Wang, R. J. Mumper, A. Singhal, A. P. Rolland, and H. W. Alila, *Hum. Gene Ther.* **9**(5), 659 (1998).

polyvinylpyrrolidone. Tokui *et al.*[7] injected plasmid with interleukin 5 (IL-5) cDNA into the soleus muscle of mice, resulting in increased serum levels of IL-5 and an increased percentage of eosinophils in white blood cells, 2 weeks after injection. Aihara and Miyazaki[8] increased IL-5 levels by 100-fold by electroporation of genes into the DNA injection site. Decrouy *et al.*[9] injected tibialis anterior muscles from mdx4cv mice (animal model for Duchenne muscular dystrophy) with plasmid DNA encoding either full-length or minidystrophin. Two weeks after injection they observed a significant amount of dystrophin-positive fibers and neuronal nitric oxide synthase (nNOS) expression. Alila and colleagues[10] detected human insulin growth factor protein in rat muscles for 4 weeks, after a single intramuscular injection of plasmid encoding human IGF-I complexed with polyvinylpyrrolidone. Vascular endothelial growth factor (VEGF) is an important angiogenic growth factor. Plasmid DNA encoding human VEGF was injected by Tsurumi *et al.*[11] into ischemic thigh muscles of rabbits; it caused significant development of collateral vessels and improved blood flow to the ischemic limb, measured 30 days after injection. This was followed by a clinical trial directed by Baumgartner *et al.,*[12] in which plasmid encoding human VEGF (hVEGF) was injected directly into the ischemic limbs of patients with nonhealing ischemic ulcers. This resulted in successful limb salvage in three of nine patients and improvement in collateral blood flow and healing of ulcers.

Another route used for introducing plasmid DNA is intravenous (iv) injection. Chao and colleagues[13] have reported success, by direct delivery of DNA plasmid carrying the gene encoding human tissue kallikrein, in reducing blood pressure of spontaneously hypertensive rats (SHRs). They used naked DNA constructs, one under the promoter control of the metallothionein metal response element, the other under the control of the Rous sarcoma virus 3' long terminal repeat (LTR). After iv injection expression of human kallikrein was identified in several tissues. Blood pressure was

[7] M. Tokui, I. Takei, F. Tashiro, A. Shimada, A. Kasuga, M. Ishii, T. Ishii, K. Takatsu, T. Saruta, and J. Miyazaki, *Biochem. Biophys. Res. Commun.* **233**(2), 527 (1997).

[8] H. Aihara and J. Miyazaki, *Nature Biotechnol.* **16**(9), 867 (1998).

[9] A. Decrouy, J. M. Renaud, J. A. Lunde, G. Dickson, and B. J. Jasmin, *Gene Ther.* **5**(1), 59 (1998).

[10] H. Alila, M. Coleman, H. Nitta, M. French, K. Anwer, Q. Liu, T. Meyer, J. Wang, R. Mumper, D. Oubari, S. Long, J. Nordstrom, and A. Rolland, *Hum. Gene Ther.* **8**(15), 1785 (1997).

[11] Y. Tsurumi, M. Kearney, D. Chen, M. Silver, S. Takeshita, J. Yang, J. F. Symes, and J. M. Isner, *Circulation* **96**(Suppl. 9), II-382 (1997).

[12] I. Baumgartner, A. Pieczek, O. Manor, R. Blair, M. Kearney, K. Walsh, and J. M. Isner, *Circulation* **97**(12), 1114 (1998).

[13] C. Wang, L. Chao, and J. Chao, *J. Clin. Invest.* **95**(4), 1710 (1995).

significantly reduced for 6 weeks and reached a maximum of −46 mmHg reduction.[13] They have also shown that iv injection of atrial natriuretic peptide (ANP) gene as naked ANP reduces hypertension in young SHRs (but not in adult SHRs).[14] To treat hypertension by decreasing activity of the renin–angiotensin system we constructed plasmid vector containing angiotensinogen (AGT) cDNA in antisense orientation under the cytomegalovirus (CMV) promoter with the green fluorescent protein (GFP) as a reporter gene. This plasmid was used to transfect H4-II-E hepatoma cells and caused a 50% reduction in the secreted angiotensinogen level. After iv injection into adult SHRs we observed a maximum decrease in the blood pressure of −22 mmHg, as compared with the sense plasmid or saline-injected animals, lasting up to 6 days.[15]

For treatment of central nervous system (CNS)-related diseases, such as Parkinson's, thalamic pain, Alzheimer's disease, and gliomas, injections into the brain could be taken under consideration. Martres *et al.*[16] injected plasmid encoding dopamine transporter (DAT) in sense or antisense orientation, complexed with polyethyleneimine, into the rat substantia nigra. Sense plasmid resulted in a significant increase in DAT expression, measured by autoradiography, and enhanced uptake of dopamine by striatal synaptosomes. Antisense plasmid caused a decrease in DAT immunolabeling and uptake of dopamine. Effects lasted up to 14 days after injection. We injected intracerebroventricularly (icv) plasmid encoding green fluorescent protein and driven by the CMV promoter. GFP mRNA was expressed 2 days after the injection, but it was undetectable on day 4. Subcutaneous injection of antisense transforming growth factor (TGF-β) expression plasmid into rats with established intracranial gliomas, by Fakhrai *et al.*,[17] increased the 12-week animal survival rate from 13 to 100%.

Another interesting possibility is to administer DNA into the respiratory system by noninvasive, intranasal delivery (inhaled or instilled). McCluskie *et al.*[18] used this method and observed maximum expression in lungs of the luciferase reporter gene by about day 4. This noninvasive method could also be used for delivery of antisense.

[14] K. F. Lin, J. Chao, and L. Chao, *Hypertension* **26**(Part 1), 847 (1995).

[15] D. Mohuczy, X. Tang, B. Kimura, and M. I. Phillips, *Miami Nature Biotechnol. Short Rep.* **10,** 139 (1999).

[16] M. P. Martres, B. Demeneix, N. Hanoun, M. Hamon, and B. Giros, *Eur. J. Neurosci.* **10**(12), 3607 (1998).

[17] H. Fakhrai, O. Dorigo, D. L. Shawler, H. Lin, D. Mercola, K. L. Black, I. Royston, and R. E. Sobel, *Proc. Natl. Acad. Sci. U.S.A.* **93**(7), 2909 (1996).

[18] M. J. McCluskie, Y. Chu, J. L. Xia, J. Jessee, G. Gebyehu, and H. L. Davis, *Antisense Nucleic Acid Drug. Dev.* **8**(5), 401 (1998).

Generally, plasmid DNA is considered safer to use than viral vectors, but expression is usually less efficient and shorter lasting. Some improvement has been shown by using different carriers and/or complexing agents, such as liposomes, polyvinylpyrrolidone, or starburst dendrimers. The starburst dendrimers, which are spherical macromolecules composed of repeating polyamidoamine subunits, are quite promising.[19] Qin *et al.*[20] complexed dendrimer with plasmid encoding IL-10 under the control of the myosin heavy chain promoter and injected it into a mouse heart transplant. Graft survival was prolonged to 39 days, using 60-fold less DNA than without the dendrimer.

Another approach to increase effectiveness of the plasmid is to introduce sequences enhancing promoter activity, facilitating correct splicing and increasing stability. Nevertheless, DNA delivered in a plasmid is not expected to last for a long time, and therefore is more useful when a transient, short-term effect is expected.

Viral Vectors

There are several viruses that have been tested for gene delivery, and each has its advantages, but none fits perfectly the description of the "ideal viral vector." To be the perfect vector, a virus should fulfill all of the following criteria.

The vector should be safe. This means that it cannot be a virus known to cause disease, or it must be reengineered to be harmless. The viral vector should not elicit an immune or inflammatory response. It should not integrate into the genome randomly; such action incurs the risk of disrupting other cellular genes and mutagenesis. The virus also must be replication deficient in order to prevent spreading to other tissues or infecting other individuals. An ideal vector would deliver a defined gene copy number into each infected cell.

In addition, the vector must be efficiently taken up in target tissue. The virus must infect the target cells with high frequency in order to achieve a biological effect.

To be practical, the vector should be easy to manipulate and produce in pure form. The virus should be able to accommodate the gene of interest,

[19] A. Bielinska, J. F. Kakowska-Latallo, J. Johnson, D. A. Tomalia, and J. R. Baker, Jr., *Nucleic Acids Res.* **24**(11), 2176 (1996).

[20] L. Qin, D. R. Pahud, Y. Ding, A. U. Bielinska, J. F. Kukowska-Latallo, J. R. Baker, Jr., and J. S. Bromberg, *Hum. Gene Ther.* **9**(4), 553 (1998).

along with its regulatory sequences, and the recombinant DNA must be packaged with high efficiency into the viral capsid proteins.

Retroviruses

Retroviruses have been used primarily because of their high efficiency in delivering genes to dividing cells.[21] Retroviruses permit insertion and stable integration of single-copy genes. Although effective in cell culture systems, they randomly integrate into the genome, which raises concerns about their safety for practical use *in vivo*. Retroviruses can act only in dividing cells, which makes them ideal for tumor therapy but less desirable where other cells are dividing that need to be protected.

Retroviral vector containing an antisense cyclin G_1 was injected by Chen *et al.*[22] into subcutaneous metastatic osteosarcoma tumors in athymic nude mice and caused a decrease in the number of cells in S and G_2/M phases of the cell cycle and accumulation of cells in the G_1 phase. In hypertension research retrovirus-mediated gene transfer is being investigated in neonatal spontaneously hypertensive rats. Our colleagues at the University of Florida tested retrovirus vector (LNSV) containing an antisense DNA to AT_{1b} receptor mRNA in astroglial and neuronal primary cell cultures and found a decrease in AT_1 receptor (AT_1-R) number.[23] Injections of this virus in the heart of 6-week-old SHRs resulted in effective long-term inhibition of AT_1 receptor mRNA and significant inhibition of the development of hypertension. Several measures indicated that the AT_1 receptors in vessels were reduced in responsiveness by the treatment.[24]

Retrovirus-mediated gene transfer has also been used in clinical trials. Some of these clinical trials include retrovirus-mediated transfer of the human multidrug resistance gene into stem cells, to protect patients receiving high-dose chemotherapy[25]; transfer of adenosine deaminase (ADA) gene to patients with ADA deficiency[26]; transfer of cDNA for glucocere-

[21] R. C. Mulligan, *Science* **260,** 926 (1993).
[22] D. S. Chen, N. L. Zhu, G. Hung, M. J. Skotzko, D. R. Hinton, V. Tolo, F. L. Hall, W. F. Anderson, and E. M. Gordon, *Hum. Gene Ther.* **8**(14), 1667 (1997).
[23] D. Lu, K. Yu, and M. K. Raizada, *Proc. Natl. Acad. Sci. U.S.A.* **92,** 1162 (1995).
[24] M. J. Katovich, D.-Lu, S. Lyer, and M. K. Raizada, *FASEB J.* **10**(3), A276 (1996).
[25] C. Hesdorffer, J. Ayello, M. Ward, A. Kaubisch, L. Vahdat, C. Balmaceda, T. Garrett, M. Fetell, R. Reiss, A. Bank, and K. Antman, *J. Clin. Oncol.* **16**(1), 165 (1998).
[26] M. Onodera, T. Ariga, N. Kawamura, I. Kobayashi, M. Ohtsu, M. Yamada, A. Tame, H. Furuta, M. Okano, S. Matsumoto, H. Kotani, G. J. McGarrity, R. M. Blaese, and Y. Sakiyama, *Blood* **91**(1), 30 (1998).

brosidase to patients with Gaucher disease;[27] IL-2[28] or p53[29] gene transfer for advanced lung cancer patients; and transfer of the herpes simplex virus thymidine kinase gene, followed by ganciclovir administration for treatment of brain tumor, metastases, and melanoma.[30–32]

Adenoviruses

Adenovirus vectors have been tested successfully in their natural host cells, the respiratory endothelia, as well as other tissues such as vascular smooth and striated muscle, and brain.[33–35] Transduction of rat pancreatic islets with adenovirus bearing an amylin antisense cDNA resulted in a 30% decrease in amylin content, with no change in insulin content.[36] Preliminary studies with adenovirus vector for delivery of AT_1-R antisense cDNA have been performed in rat neuronal and vascular smooth muscle cells, and resulted in reduction of AT_1-R protein.[37]

Adenovirus is a double-stranded DNA with 2700 distinct adenoviral gene products. The virus infects many mammalian cell types because they have a membrane receptor called the coxsackievirus and adenovirus receptor.[38] Adenovirus enters the cell by receptor-induced endocytosis and translocates to the nucleus. Most adenovirus vectors in their current form are

[27] C. Dunbar and D. Kohn, *Hum. Gene Ther.* **7**(2), 231 (1996).

[28] Y. Tan, M. Xu, W. Wang, F. Zhang, D. Li, X. Xu, J. Gu, and R. M. Hoffman, *Anticancer Res.* **16**(4A), 1993 (1996).

[29] J. A. Roth, D. Nguyen, D. D. Lawrence, B. L. Kemp, C. H. Carrasco, D. Z. Ferson, W. K. Hong, R. Komaki, J. J. Lee, J. C. Nesbitt, K. M. Pisters, J. B. Putnam, R. Schea, D. M. Shin, G. L. Walsh, M. M. Dolormente, C. I. Han, F. D. Martin, N. Yen, K. Xu, L. C. Stepehens, T. J. McDonnell, T. Mukhopadhyay, and D. Cai, *Nature Med.* **2**(9), 985 (1996).

[30] E. H. Oldfield, Z. Ram, K. W. Culver, R. M. Blaese, H. L. DeVroom, and W. F. Anderson, *Hum. Gene Ther.* **4**(1), 39 (1993).

[31] L. E. Kun, A. Gajjar, M. Muhlbauer, R. L. Heideman, R. Sanford, M. Brenner, A. Walter, J. Langston, J. Jenkins, and S. Facchini, *Hum. Gene Ther.* **6**(9), 1231 (1995).

[32] D. Klatzmann, *Hum. Gene Ther.* **7**(2), 255 (1996).

[33] S. L. Brody, H. A. Jaffe, N. T. Eissa, and C. Daniel, *Nature Genet.* **1**, 42 (1994).

[34] B. Quantin, L. D. Perricaudet, S. Tajbakhsh, and J.-L. Mandel, *Proc. Natl. Acad. Sci. U.S.A.* **89**, 2581 (1992).

[35] G. Le Gal La Salle, J. J. Robert, S. Berrard, V. Ridoux, L. D. Stratford-Perricaudet, M. Perricaudet, and J. Mallet, *Science* **259**, 988 (1993).

[36] A. Novials, J. C. Jimenez-Chillaron, C. Franco, R. Casamitjana, R. Gomis, and A. M. Gomez-Foix, *Pancreas* **17**(2), 182 (1998).

[37] D. Lu, H. Yang, and M. K. Raizada, *Am. J. Physiol.* **274**(2 Pt. 2), H719 (1998).

[38] J. M. Bergelson, J. A. Cunningham, G. Drognett, E. A. Kurt-Jones, A. Krithivas, J. S. Hong, M. S. Horwitz, R. L. Crowell, and R. W. Finberg, *Science* **275**(5304), 1320 (1997).

episomal, that is, they do not integrate into the host DNA. They provide high levels of expression but the episomal DNA will invariably become inactive. In some species, e.g., mice, this may take a long time compared with their life span, but in humans it limits the use of the virus as a vector. Repeated infections result in an inflammatory response with consequent tissue damage, because the adenovirus expresses genes that lead to immune cell attacks. This further limitation makes current recombinant adenovirus unsuitable for long-term treatment.

Several gene therapy trials using adenovirus vectors have failed to produce acceptable results, because the number of viral particles necessary to give adequate levels of gene transfer was associated with significant toxicity. Further engineering of the adenovirus may eventually avoid these limitations. Some attempts to solve this problem include redirection of adenovirus to bind to a receptor that is highly expressed on the target cells, such as fibroblast growth factor receptor,[39] or heparan-containing cellular receptors,[40] so the dose of the virus could be much smaller. Croyle et al.[41] treated Caco-2 cells with interleukin-1β to increase expression of the integrin receptor and observed a fourfold increase in transduction efficiency. An interesting study on adenovirus-mediated regulable target gene expression in vivo was done by Burcin et al.[42] They used adenovirus devoid of all viral coding sequences[43] as a carrier of the human growth hormone (hGH) gene. Expression of hGH was controlled by a mutated progesterone ligand-binding domain fused to the GAL4 DNA-binding domain and part of the activation domain of the human p65 protein; therefore expression was dependent on the presence of antiprogestin mifepristone. To achieve tissue specificity the construct contained a liver-specific promoter. Mice infected with this adenovirus and treated with mifepristone had high levels of hGH, whereas without mifepristone the hGH level was undetectable, and the animals could be reinduced up to five times over a 12-week period.[42]

[39] C. Rancourt, B. E. Rogers, B. A. Sosnowski, M. Wang, A. Piche, G. F. Pierce, R. D. Alvarez, G. P. Siegal, J. T. Douglas, and D. T. Curiel, Clin. Cancer Res. 4(10), 2455 (1998).
[40] T. J. Wickham, P. W. Roelvink, D. E. Brough, and I. Kovesdi, Nature Biotechnol. 14(11), 1570 (1996).
[41] M. A. Croyle, E. Walter, S. Janich, B. J. Roessler, and G. L. Amidon, Hum. Gene Ther. 9(4), 561 (1998).
[42] M. M. Burcin, G. Schiedner, S. Kochanek, S. Y. Tsai, and B. W. O'Malley, Proc. Natl. Acad. Sci. U.S.A. 96(2), 355 (1999).
[43] S. Kochanek, P. R. Clemens, K. Mitani, H. H. Chen, S. Chan, and C. T. Caskey, Proc. Natl. Acad. Sci. U.S.A. 93(12), 5731 (1996).

Adeno-Associated Virus

The adeno-associated virus (AAV) has been gaining attention because of its safety and efficiency.[44] It has been used successfully for delivering antisense RNA against α-globin[45] and human immunodeficiency virus (HIV) long terminal repeats,[46] and it is our vector of choice for delivering antisense targeted to the AT_1 receptor in cell culture and hypertensive rat models.

AAV is a parvovirus, discovered as a contamination of adenoviral stocks. It is a ubiquitous virus (antibodies are present in 85% of the human population in the United States), which has not been linked to any disease. It is also classified as a dependovirus, because its replication is dependent on the presence of a helper virus, such as adenovirus or herpes virus. Five serotypes have been isolated, of which AAV-2 is the best characterized. AAV has a single-stranded linear DNA that is encapsidated into capsid proteins VP1, VP2, and VP3 to form an icosahedral virion 20–24 nm in diameter.[44]

The AAV DNA is approximately 4.7 kilobases long. It contains two open reading frames and is flanked by two inverted terminal repeats (ITRs). There are two major genes in the AAV genome: *rep* and *cap*. The *rep* gene encodes proteins responsible for viral replication, whereas *cap* encodes capsid proteins VP1–3. Each ITR forms a T-shaped hairpin structure. These terminal repeats are the only essential *cis* components of the AAV for packaging the virus. Therefore, the AAV can be used as a vector with all viral coding sequences removed and replaced by the cassette of genes for delivery. Three viral promoters have been identified and named p5, p19, and p40, according to their map position. Transcription from p5 and p19 results in production of Rep proteins, and transcription from p40 produces the capsid proteins.[44] For more powerful expression we have replaced native AAV promoters by a cytomegalovirus promoter. Other promoters are being tested that are specific to certain cells; these include the arginine vasopressin (AVP) promoter for cells synthesizing AVP, neuron-specific enolase (NSE), and glial fibrillary acid protein (GFAP).

On infection of a human cell, the wild-type AAV (wtAAV) integrates to the q arm of chromosome 19.[47,48] Although chromosomal integration

[44] N. Muzyczka and S. McLaughin, *in* "Current Communications in Molecular Biology: Viral Vectors" (Y. Gluzman and S. H. Hughes, eds.), pp. 39–44. Cold Spring Harbor Laboratory Press, Cold Spring Harbor, New York, 1988.

[45] S. Ponnazhagan, M. L. Nallari, and A. Srivastava, *J. Exp. Med.* **179,** 733 (1994).

[46] S. Chatterjee, P. R. Johnson, and K. K. Wong, *Science* **258,** 1485 (1992).

[47] N. Muzyczka, *Curr. Top. Microbiol. Immunol.* **158,** 97 (1992).

[48] R. J. Samulski, X. Zhu, X. Xiao, J. D. Brook, D. E. Housman, N. Epstein, and L. A. Hunter, *EMBO J.* **10**(12), 3941 (1991).

requires the terminal repeats, the viral components responsible for site-specific integration have been targeted to the Rep proteins.[49] With no helper virus present, AAV infection remains latent indefinitely. On superinfection of the cell with helper virus the AAV genome is excised, replicated, packaged into virions, and released to the extracellular fluid. This fact is the basis of recombinant AAV (rAAV) production for research.

Several factors prompted researchers to study the possibility of using recombinant AAV as an expression vector. The first is that surprisingly little is required to package the virus. All that is necessary in *cis* is the 145-bp inverted terminal repeats, which are only 6% of the AAV genome. This leaves room in the vector to assemble up to a 4.4-kb DNA insertion. While this carrying capacity may limit the AAV to delivering large genes, it is amply suited to delivering small genes and antisense cDNA. It is sufficient to insert a gene of interest, a potent promoter, and a selective marker such as a neomycin resistance gene or reporter gene such as green fluorescent protein.

The second characteristic that makes AAV a good vector candidate is its safety. There is a relatively complicated rescue mechanism. Not only adenovirus (wild type) but also AAV genes are required to mobilize the recombinant AAV (rAAV). The spread of rAAV vectors to nontarget areas can be limited to certain tissues. AAV is not pathogenic and not associated with disease. The removal of viral coding sequences in producing an rAAV minimizes immune reactions to viral gene expression, and therefore rAAV does not evoke an inflammatory response (in contrast to the recombinant adenovirus).

AAV is also a good candidate for gene therapy because it has a broad host range. AAV infects all mammalian tissues tested, with the only exception being vascular endothelial cells.[50] The AAV remains intact for long periods of time.

The limitations of AAV are its time-consuming production and the difficulty in achieving high titers.

The advantages, particularly its safety, make AAV one of the best candidates, currently, for delivery of genes for long-term therapy. Snyder *et al.*[51] reported successful transduction of mice with AAV expressing human factor IX for at least 30 weeks after a single administration. Flotte and colleagues[52] have established gene therapy phase I trials for cystic fibrosis,

[49] R. M. Linden, E. Winocour, and K. I. Berns, *Proc. Natl. Acad. Sci. U.S.A.* **93,** 7966 (1996).
[50] J. S. Lebkowski, M. M. McNally, T. B. Okarma, and B. Lerch, *Mol. Cell. Biol.* **8,** 3988 (1988).
[51] R. O. Snyder, C. H. Miao, G. A. Patijn, S. K. Spratt, O. Danos, D. Nagy, A. M. Gown, B. Winther, L. Meuse, L. K. Cohen, A. R. Thompson, and M. A. Kay, *Nature Genet.* **16,** 270 (1997).
[52] T. R. Flotte, B. Carter, C. Conrad, W. Guggino, T. Reynolds, B. Rosenstein, G. Taylor, S. Walden, and R. Wetzel, *Hum. Gene Ther.* **7,** 1145 (1996).

using AAV gene delivery in patients. Ye and colleagues[53] used rapamycin-inducible expression of erythropoietin delivered in recombinant AAVs. The intramuscular injection into mice or rhesus monkeys caused increased hematocrit and erythropoietin levels after iv rapamycin administration for up to 4 months.

Methods for Antisense Delivery By Adeno-Associated Virus

The general concept for antisense gene delivery in the AAV vector and the steps involved are shown in Fig. 1. To illustrate these steps, a brief description is given that is applicable to hypertension. Further details are presented in the references.

Construction of Plasmids

After subcloning the target gene into the vector containing AAV terminal repeats, highly purified plasmid is needed for virus packaging. To reach this goal, we recommend the following protocol.

Large-Scale Plasmid Preparation. The step-by-step procedure for purification of the plasmid is given in the protocol for large-scale plasmid preparation.

1. Grow bacteria containing the plasmid in 1 liter of LB (Luria–Bertani) medium with ampicillin (100 μg/ml).
2. Pellet bacteria at $3000g$ at 4° for 15 min.
3. Resuspend bacteria in 20 ml of lysozyme buffer [25 mM Tris (pH 7.5), 10 mM EDTA, 15% sucrose or glucose].
4. Add 4 ml of lysozyme (12 mg/ml in lysozyme buffer); mix.
5. Incubate on ice for 5 min until the mixture becomes viscous.
6. Add 48 ml of 0.2 N NaOH–1% sodium dodecyl sulfate (SDS). Mix with a glass pipette, using it as a rod.
7. Incubate on ice for 5–10 min.
8. Add 36 ml of 3 M sodium acetate, pH 4.8–5.2; mix with the same pipette.
9. Add 0.2 ml of chloroform; mix.
10. Incubate on ice for 20 min.
11. Spin at $3000g$ at 4° for 20 min.
12. Transfer supernatant to a fresh bottle.

[53] X. Ye, V. M. Rivera, P. Zoltick, F. Cerasoli, M. A. Schnell, G. Gao, J. V. Hughes, M. Gilman, and J. M. Wilson, *Science* **283,** 88 (1999).

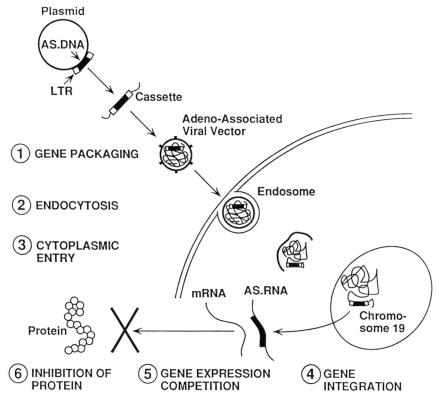

Fig. 1. Gene delivery of antisense with adeno-associated viral vector (AAV). (1) Gene packaging: The plasmid containing the terminal repeats (TRs) characteristic of AAV with cDNA subcloned in the antisense direction. The final cassette also contains promoters such as the CMV and TK promoters. The packaging cell line is transfected with AAV-based plasmid, helper plasmid with *rep* and *cap* genes, and transduced with adenovirus as a helper virus. (2) Endocytosis: The viral vector fuses with the cell membrane by binding to adhesion molecules and becomes an endosome within the bilipid layer. (3) Cytoplasmic entry: The vesicle opens in the cytoplasm, releasing the vector, which is transported to and enters the nucleus. (4) Gene integration: AAV integrates with chromosome 19. It is not known if the addition of foreign DNA interferes with this integration. (5) Gene expression competition: Genomic message in chromosome 19 produces an antisense RNA. This competes with the natural mRNA and prevents it from producing its product. (6) Inhibition of protein: A binding of antisense RNA to the mRNA prevents translation through the ribosomal assembly to produce protein. The result of this gene delivery system should be reduction of the protein, specifically targeted by the antisense DNA.

13. Add 33 ml of 40% polyethylene glycol 8000 (PEG 8000), mix, and incubate on ice for 10 min.
14. Spin at 14,000g at 4° for 10 min.
15. Discard the supernatant, dissolve the pellet in 10 ml of sterile water, then add 10 ml of 5.5 M LiCl.
16. Incubate on ice for 10 min.
17. Spin at 14,000g at 4° for 10 min.
18. Save the supernatant; transfer equal amounts (each 10 ml) into two 30-ml Corex (or plastic) tubes. Add 6 ml of 2-propanol to each tube and mix.
19. Incubate at room temperature for 10 min.
20. Spin at 10,000g at room temperature for 10 min.
21. Dissolve each pellet in 3.7 ml of TE, pH 7.4 (10 mM Tris-HCl with 1 mM EDTA).
22. Add 4.2 g of CsCl to each tube, mix, and dissolve. Add 0.24 ml of ethidium bromide (10 mg/ml) and mix.
23. Transfer the solution in each tube into a 4.9-ml Optiseal centrifuge tube (Beckman, Fullerton, CA).
24. Spin in a NVT90 rotor (Beckman) at 78,000 rpm, 15° for 4 hr. Three bands should appear: the upper one contains protein, the middle one is nicked and linear DNA, and the lower one is closed circular plasmid DNA.
25. Carefully remove the plasmid band, using a 3-ml syringe with an 18-gauge needle, and transfer it to an Eppendorf tube (two tubes, each about 0.5 ml).
26. Extract three or four times with an equal volume of isoamyl alcohol, discarding the organic phase (top layer) every time, until all pink color is removed.
27. Transfer one 30-ml Corex (plastic) tube, add 2.5 vol (2.5 ml) of H_2O, mix, and add two of the combined volumes of ethanol (7 ml).
28. Incubate on ice for 30 min.
29. Pellet the plasmid DNA at 12,000g for 15 min at 4°.
30. Discard the supernatant and dissolve the pellet in 500 μl of TE (pH 7.4).
31. Transfer to an Eppendorf tube; add 500 μl of phenol–chloroform (25:24, v/v), vortex, and spin for 2 min at top speed in a microcentrifuge.
32. Save the top aqueous layer and transfer it to a fresh Eppendorf tube.
33. Add 50 μl of 3 M sodium acetate buffer, pH 4.8–5.2, and mix; add 1 ml of ethanol and mix.
34. Pellet the plasmid DNA at top speed in a microcentrifuge for 5 min at 4°.

35. Discard the supernatant, wash the pellet with 75% ethanol, and vacuum dry.
36. Dissolve the pellet in 1 ml of TE.

To treat hypertension by decreasing the activity of the renin–angiotensin system we constructed plasmids for both AT_1 receptor antisense (pAT_1) and angiotensinogen antisense (pAGT) in the AAV-derived expression vector. We initially used a plasmid containing AAV genome and 750 bp of cDNA inserted into the AAV in the antisense direction downstream from the AAV promoter. The NG108-15 cells or hepatoma H4 cells were transfected with pAT_1-AS or pAGT-AS, respectively, using Lipofect AMINE.[54,55] In both cases, there were significant reductions in the appropriate proteins, namely AT_1 receptor and angiotensinogen. To test that the cells express AAV we used the *rep* gene product as a marker. Immunocytochemical staining with an anti-Rep protein antibody showed that the majority of cells in culture fully expressed Rep. A further development of the AAV was the insertion of a promoter more powerful than the p40 promoter: AAV with a CMV promoter and neomycin resistance gene (*neo^r*) as a selection marker was used in our next experiments. The AAV cassette contained either 750 bp of rat AT_1 cDNA in the antisense direction (Fig. 2), or markers, i.e., either GFP (Fig. 3, Ref. 56) or the *lacZ* gene.[57] The NG108-15 cells transfected with AAV plasmid containing the GFP gene under the control of the CMV promoter, and the *neo^r* gene under the control of the thymidine kinase (TK) promoter, were selected by antibiotic (G418, 600 μg/ml) and the selected clones viewed for GFP expression. Few cells died during selection. The transfection efficiency of pAAV-GFP in various cell lines including ATt20 (mouse pituitary cells), L929 (mouse fibroblasts), 239 (human embryonic kidney cells), and NG108-15 (neuroblastoma–glioma cells) was more than 50%.[58]

Preparation of pAAV-AT$_1$R-AS. The 749-bp fragment of the angiotensin receptor cDNA (positions -183 to 566) was amplified by polymerase chain reaction (PCR) and ligated to an AAV-derived vector in the antisense orientation, in place of GFP. The resulting plasmid vector (pAAV-AT$_1$R-AS) contained adeno-associated virus terminal repeats (TRs), a cytomega-

---

[54] R. Gyurko and M. I. Phillips, *FASEB J.* **9**(3), A330 (1995).

[55] R. Gyurko, P. Wu, C. Sernia, E. Meyer, and M. I. Phillips, "Antisense Expression Vector Decreases Angiotensinogen Synthesis in H-4 Hepatoma Cells" (abstract). American Heart Association 48th Annual Council for High Blood Pressure (1994).

[56] S. Zolotukhin, M. Potter, W. W. Hauswirth, J. Guy, and N. Muzyczka, *J. Virol.* **70**(7), 4646 (1996).

[57] P. Wu, B. Du, M. I. Phillips, and E. F. Terwilliger, *Soc. Neurosci.* **22,** 133.2 (1996). [Abstract]

[58] D. Mohuczy and M. I. Phillips, *FASEB J.* **10**(3), A447 (1996). [Abstract]

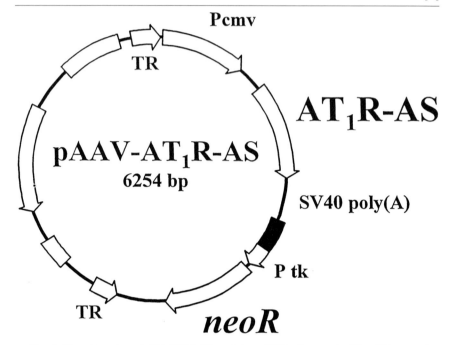

FIG. 2. Plasmid vector pAAV-AT$_1$R-AS contains a 750-bp fragment of the AT$_{1A}$ receptor cDNA in antisense orientation. TR, AAV terminal repeat; Pcmv, human cytomegalovirus early promoter; SV40 poly(A), polyadenylation signal from simian virus; Ptk, thymidine kinase promoter; *neoR*, neomycin phosphotransferase gene from Tn5. Other promoters have been substituted for CMV, including arginine vasopressin (AVP), neuron-specific enolase (NSE), and glial fibrillary acidic protein (GFAP) promoters.

lovirus promoter (Pcmv), the DNA encoding AT$_1$ receptor mRNA in the antisense direction, and a neomycin resistance gene (*neo^r*). The plasmid DNA was purified on a CsCl gradient.

pAAV-AT$_1$R-AS was tested for AT$_1$ receptor inhibition *in vitro,* using NG108-15 cells[59] and rat vascular smooth muscle cells.[60] The cells had a significant ($p < 0.01$) decrease in angiotensin II AT$_1$ receptor number and a reduced calcium response on angiotensin stimulation compared with the control cells. No effect was seen on the AT$_2$ receptor level.

AAV-based plasmids are used to prepare recombinant virus; the method is described below.

[59] T. Zelles, D. Mohuczy, and M. I. Phillips, *FASEB J.* **9**(3), A330 (1996).
[60] D. Mohuczy, C. Gelband, and M. I. Phillips, *Hypertension* **33,** 354 (1999).

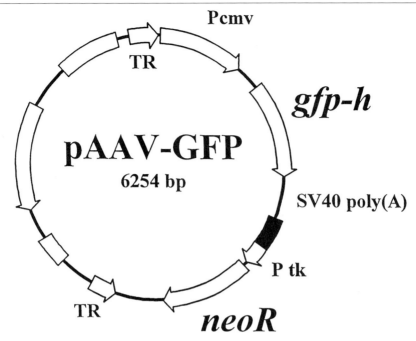

FIG. 3. Schematic diagram of a recombinant AAV vector containing *gfp* gene. In the rAAV-GFP vector almost all of the parental wtAAV genome has been deleted, except for the terminal repeats, and replaced with *gfp* (*A. victoria* green fluorescent protein gene) driven by a CMV promoter (Pcmv). Also, the neomycin resistance (*neo^r*) gene has been inserted with a thymidine kinase promoter (Ptk). *gfp* serves as a reporter gene *in vitro* or *in vivo*, and *neo^r* serves for selection *in vitro*.

Method to Prepare Recombinant Adeno-Associated Virus

To prepare antisense-expressing recombinant AAV, human embryonic kidney (HEK293) cells are transfected with plasmid vector containing the gene of interest in antisense orientation and AAV terminal repeats (pAAV-AS), together with helper plasmid delivering *rep* and *cap* genes (necessary for AAV replication) in *trans,* by the calcium phosphate method (see Protocol 1). Eight hours after transfection, adenovirus is added at a multiplicity of infection (MOI) of 5.

Protocol 1

1. Split low passage number HEK293 cells grown in Dulbecco's modified Eagle's medium (DMEM) with 10% heat-inactivated fetal bovine serum (FBS) and penicillin–streptomycin onto 15-cm cell culture plates, so they are about 70% confluent on the next day.

2. Mix (per plate): 20 μg of the plasmid, and 20 μg of helper plasmid (containing *rep* and *cap* genes) in a final volume of 876 μl of 0.1× TE.

3. Add 1 ml of 2× HBS, pH 7.05 (280 mM NaCl, 10 mM KCl, 1.5 mM Na$_2$HPO$_4$·2H$_2$O, 12 mM dextrose, 50 mM HEPES); mix.

4. Add by drops, mixing, 124 μl of 2 M CaCl$_2$.

5. Incubate for 15–20 min at room temperature to form a precipitate; mix once by pipetting.

6. Add 2 ml of the mixture to the plate with cells, swirl gently, and place the cells back into the incubator.

7. Eight hours after transfection replace the medium with fresh medium and add adenovirus at an MOI of 5.

When cells develop cytopathic effects (usually 2–3 days), they are harvested in medium and rAAV is purified (see Protocol 2).

Protocol 2

1. Harvest the cells to centrifuge-type bottles, using a rubber policeman.

2. Centrifuge the cell suspension at 900g at 4° for 15 min.

3. Resuspend the cell pellet in 1 ml multiplied by the initial amount of plates of 50 mM Tris-HCl, pH 8.4, with 150 mM NaCl and freeze–thaw three times.

4. Sonicate the cell suspension three times for 30 sec on ice.

5. Spin down the cell debris by centrifugation at 2000g at 4° for 10 min. Transfer the supernatant to a new tube.

6. Repeat freezing–thawing, sonicating, and spinning down the pellet once. Combine the supernatants.

7. Calculate 33% of the volume of supernatant, add slowly with continuous mixing this volume of ammonium sulfate, pH 7.0, saturated at 4°, and incubate the slurry at 4° for 15 min.

8. Spin the sample at 8000g, 4° for 15 min. Transfer the supernatant to a new tube.

9. Add by drops, mixing, 67% of the initial volume of supernatant of ammonium sulfate, pH 7.0, saturated at 4°.

10. Incubate on ice for 20 min. Centrifuge the solution at 17,500g at 4° for 20 min. Discard the supernatant and turn the tube upside down on a paper towel to ensure that all fluid is removed.

11. Dissolve the pellet from 10 plates in 8.5 ml of HBS, pH 7.4.

12. Mix with 5.5 g of solid CsCl and transfer to centrifuge tubes (density should be 1.39 g/ml).

13. Underlay with 1 ml of CsCl solution of 1.5-g/ml density, mark the level of the cushion, on the outside of the tube, with a marker.

14. Spin the sample in an SW41 rotor at 40,000 rpm at 18° for 40 hr.

A white diffuse band of proteins at the top and a sharp white band of adenovirus below appear.

15. Using a 6-ml syringe, pull out the fluid from a point one-third of the way from the bottom of the tube to about 2 mm below the adenovirus band.
16. Mix the sample with CsCl (1.39 g/ml) and spin in an SW41 rotor at 40,000 rpm at 18° for 40 hr.
17. Without disturbing the CsCl gradient, gently place a needle in an upper part of the tube, tape it (so you can control speed of the fluid dripping later on), puncture the bottom of the tube with a needle, and start to collect fluid. Collect five 1-ml fractions, then ten 0.5-ml fractions.

These fractions will be checked for AAV by hybridization with [^{32}P]dCTP-labeled random primed AAV probe (see Protocol 3), using a dot-blot apparatus.

Protocol 3

1. Assemble dot-blot apparatus with Whatman (Clifton, NJ) 3MM paper filter under the nylon filter presoaked in distilled H_2O, and then with 20× SSC.
2. Add 200 μl of 20× SSC per well and apply a slow vacuum to remove the fluid.
3. Add 5-μl aliquots of each fraction from the CsCl gradient and apply slow vacuum to remove the fluid.
4. Remove the nylon filter from the manifold and place it on Whatman 3MM paper filters soaked as follows: 10% SDS for 3 min, 0.5 M NaOH with 1.5 M NaCl for 5 min, and 0.5 M Tris-HCl, pH 7.2, with 1.5 M NaCl for 5 min. Air dry the filter for 10 min, briefly wash it in 4 × SSC for 10 sec, and air dry it for 10 min. Microwave the filter for 4 min with 0.5 liter of water in the beaker placed alongside.
5. Hybridize the filter with ^{32}P-labeled probe (random primed according to the kit instructions).
6. Expose the filter to X-rays at −80°. The packaged AAV virions should be present near the middle of the gradient.
7. Combine the positive fractions, and concentrate the virus by Centricon-30 (Amicon, Danvers, MA) at 4350g for 30 min at 4°.
8. Overlay the filter with concentrated virus with 0.5 ml of phosphate-buffered saline (PBS) and spin again for 2 min at 500g. Collect the virus.

Virus Titer Assay

The titer of the virus is calculated using HEK293 cells, wtAAV, and adenovirus.

1. Seed 5×10^4 cells in each well of a 24-well plate and incubate for 24 hr.
2. Infect all but one well with adenovirus at an MOI of 20.
3. Infect all but one well (different one) with wtAAV at an MOI of 4.
4. Make five serial dilutions of the rAAV and infect 8–10 wells, which already contain adenovirus and wtAAV. Infect the cells with no adenovirus or no wtAAV with undiluted rAAV. Incubate the cells for 24 hr.
5. Spin down the medium, discard the supernatant, combine the pellet with prewashed and trypsinized cells (100 μl of trypsin-EDTA per well), add 10 ml of PBS, and disperse into a single-cell suspension.
6. Transfer the cell suspension onto a nylon filter presoaked in PBS and apply low vaccum.
7. Place the filter on Whatman 2MM paper soaked in 0.5 N NaOH with 1.5 M NaCl for 5 min at room temperature.
8. Transfer the filter to the top of Whatman paper soaked in 1 M Tris-HCl, pH 7.0, with 2× SSC for 5 min at room temperature.
9. Air dry the filter.
10. Hybridize the filter with ^{32}P-labeled probe specific for the gene of interest (random primed according to the kit instructions).
11. Expose the filter to X-rays at $-80°$. Count the spots and multiply by the dilution factor for each well.

rAAV-AT$_1$R-AS was tested for AT$_1$ receptor inhibition *in vitro,* using vascular smooth muscle cells.[60] Transduced cells, without G418 selection, expressed the transgene for at least 8 weeks, and had a decreased number of AT$_1$ receptors and reduced calcium response to angiotensin II stimulation.

To test for effectiveness *in vivo,* rAAV-AT$_1$R-AS was microinfused into the lateral ventricles of adult male SHRs.[61] Control rats received AAV with GFP reporter gene but without the AS gene ("mock" vector) in vehicle, which was artificial cerebrospinal fluid. Blood pressure was measured by the tail-cuff method. There was a significant decrease in systolic blood pressure (SBP) in one group of rats that received the rAAV-AS vector. No effect was observed in the controls. SBP decreased by 23 ± 2 mmHg in the first week after administration. This drop in blood pressure was prolonged in four rats for 9 weeks, whereas controls had no reduction in blood pressure.[61] This was considerably longer than the longest effect observed with antisense oligodeoxynucleotides or plasmid vector. rAAV-GFP expression in hypothalamus of the control rat group was detectable

[61] M. I. Phillips, D. Mohuczy-Dominiak, M. Coffey, P. Wu, S. M. Galli, and T. Zelles, *Hypertension* **29**(2), 374 (1997).

by RT-nested PCR 11 months after injection. Further, intracardiac injection of rAAV-AS in SHRs significantly reduced blood pressure and slowed the development of hypertension for several weeks.[61,62] These results demonstrate that rAAV-AS, in a single application, is effective in chronically reducing hypertension. They encourage further research on gene regulation in hypertension and exploration of the most effective routes of delivery applicable to humans. To obtain tissue specificity we have tried using rAAV with promoters other than the CMV promoter. One example is the arginine vasopressin (AVP) promoter. rAAV with the *lacZ* gene under the control of the AVP promoter was constructed and expression *in vivo* was tested by direct injection into rat brain. The vector expressed β-galactosidase in neurons of the paraventricular nucleus and supraoptic nucleus. The expression was in magnocellular cells, which normally express AVP.[57] The expression was observed 1 day, 1 week, and after 1 month with no diminution of signal. These are examples of how AAV can be developed for specific tissue and/or cell gene expression and show that AAV vectors can deliver foreign genes for long periods of time.

Acknowledgment

This work has been supported by NIH (MERIT) Grant HL23774.

[62] B. Kimura, D. Mohuczy, and M. I. Phillips, *FASEB J.,* abstract 522 (1998).

[4] Antisense Mapping: Assessing Functional Significance of Genes and Splice Variants

By Gavril W. Pasternak and Ying-Xian Pan

Introduction

Establishing the functional significance of specific proteins has always been a major goal in pharmacology. The traditional approach of using inhibitors or antagonists has been helpful, but their limited availability and selectivity has hindered the overall utility of this approach. The advances in molecular biology have now provided new approaches capable of overcoming these problems. These advances have greatly simplified the manipulation of DNA and enabled the generation of mice in which specific genes have been disrupted. Commonly termed "knockout mice," these animals have been widely used. However, they do pose several problems. If the

0076-6879/99 $30.00

gene in question is crucial in development, the disruption may be lethal. Another, more subtle problem has been the appearance of compensatory changes in other systems that can take over the role of the lost protein. Although these observations can be interesting, they make simple correlations between the absence of protein and behavior difficult. Finally, knockout mice are costly in terms of resources and time.

Antisense methodologies offer an alternative approach with a number of advantages.[1–9] Short antisense probes are easily designed, relatively inexpensive, and readily available. Antisense approaches also can be used in adult animals, avoiding the developmental problems sometimes encountered in knockout models. However, the limited downregulation of the mRNA and protein does limit the approach, making it important to ensure that modest decreases in protein levels can be detected functionally. Antisense probes also have been associated with nonspecific effects, making the use of extensive controls essential. Finally, even well-designed antisense probes may lack activity for one of a number of potential reasons. Despite these difficulties, we have found antisense approaches to be extremely valuable in correlating molecular biology and behavior.

Our group has focused on the use of antisense techniques within the central nervous system (CNS). Cloning studies identified several opioid receptors,[10,11] which were readily characterized by their binding selectivity profiles. However, these studies did not address whether these same receptors were responsible for the functions associated with these receptors. We first examined the role of the cloned δ receptor gene *DOR-1* in δ analgesia in mice,[12] followed by *MOR-1*[13] and *KOR-1*,[14] which encode μ- and κ_1-opioid receptors, respectively. We also have used antisense paradigms to

[1] C. Wahlestedt, *Trends Pharmacol. Sci.* **15,** 42 (1994).
[2] S. Agrawal and R. P. Iyer, *Pharmacol. Ther.* **76,** 151 (1997).
[3] B. J. Chiasson, J. N. Armstrong, M. L. Hooper, P. R. Murphy, and H. A. Robertson, *Cell. Mol. Neurobiol.* **14,** 507 (1994).
[4] S. T. Crooke and C. F. Bennett, *Annu. Rev. Pharmacol. Toxicol.* **36,** 107 (1996).
[5] L. H. Gold, *Behav. Pharmacol.* **7,** 589 (1996).
[6] G. W. Pasternak and K. M. Standifer, *Trends Pharmacol. Sci.* **16,** 344 (1995).
[7] C. A. Stein and Y.-C. Cheng, *Science* **261,** 1004 (1993).
[8] R. W. Wagner, M. D. Matteucci, D. Grant, T. Huang, and B. C. Froehler, *BioTechnology* **14,** 840 (1996).
[9] B. Weiss, G. Davidkova, and S. P. Zhang, *Neurochem. Int.* **31,** 321 (1997).
[10] T. Reisine and G. I. Bell, *Trends Neurosci.* **16,** 506 (1993).
[11] G. R. Uhl, S. Childers, and G. W. Pasternak, *Trends Neurosci.* **17,** 89 (1994).
[12] K. M. Standifer, C.-C. Chien, C. Wahlestedt, G. P. Brown, and G. W. Pasternak, *Neuron* **12,** 805 (1994).
[13] G. C. Rossi, Y.-X. Pan, J. Cheng, and G. W. Pasternak, *Life Sci.* **54,** PL375 (1994).
[14] C.-C. Chien, G. Brown, Y.-X. Pan, and G. W. Pasternak, *Eur. J. Pharmacol.* **253,** R7 (1994).

explore the role of neuronal nitric oxide synthase in opioid tolerance[15] and the anti-opioid actions of sigma receptors.[16,17] These studies have demonstrated the great utility and selectivity of antisense paradigms. This chapter addresses our use of antisense methodologies in the central nervous system to explore the molecular biology of behavior.

Choosing Targets and Optimizing Chances for Success

Although antisense approaches can be valuable in correlating proteins and function, they have limitations and will not necessarily work in all systems. A number of factors influence the likelihood of success. The central nervous system has a number of advantages compared with systemic use of antisense techniques. Oligodeoxynucleotides administered into either brain tissue or the cerebrospinal fluid are far more stable than when given systemically. Indeed, we typically use nonmodified oligodeoxynucleotides centrally. Although phosphothioates are active centrally, we have occasionally seen more nonspecific effects with them than with simple DNA. Given into the CNS, only a small fraction of the administered antisense oligodeoxynucleotides is taken up into the cells, but once inside they are stable for at least 72 hr.[12] This is important when considering dosing paradigms. It has been suggested that the probes are taken up into vesicles intracellularly, explaining their stability. It also is important to consider the ability of the antisense to reach its target. Penetration of brain tissue by antisense oligodeoxynucleotides is limited.[1,18] If deep structures need to be targeted, it may be useful to consider direct microinjections. The opioid system is particularly amenable to intracerebroventricular antisense approaches owing to its periventricular localization, but microinjection approaches also are effective.[13,19]

In our experience, the downregulation of mRNA and protein following antisense treatments is limited, oftentimes by 30–50%.[12,20,21] It seems likely, therefore, that the effectiveness of antisense treatments would be dependent

[15] Y. A. Kolesnikov, Y.-X. Pan, A. M. Babey, S. Jain, R. Wilson, and G. W. Pasternak, *Proc. Natl. Acad. Sci. U.S.A.* **94,** 8220 (1997).

[16] M. A. King, Y.-X. Pan, J. Mei, A. Chang, J. Xu, and G. W. Pasternak, *Eur. J. Pharmacol.* **331,** R5 (1997).

[17] Y. X. Pan, J. F. Mei, J. Xu, B. L. Wan, A. Zuckerman, and G. W. Pasternak, *J. Neurochem.* **70,** 2279 (1998).

[18] F. Yee, H. Ericson, D. J. Reis, and C. Wahlestedt, *Cell. Mol. Neurobiol.* **14,** 475 (1994).

[19] G. C. Rossi, L. Leventhal, Y. X. Pan, J. Cole, W. Su, R. J. Bodnar, and G. W. Pasternak, *J. Pharmacol. Exp. Ther.* **281,** 109 (1997).

[20] K. M. Standifer, S. Jenab, W. Su, C.-C. Chien, Y.-X. Pan, C. E. Inturrisi, and G. W. Pasternak, *J. Neurochem.* **65** 1981 (1995).

[21] K. M. Standifer, G. C. Rossi, and G. W. Pasternak, *Mol. Pharmacol.* **50,** 293 (1996).

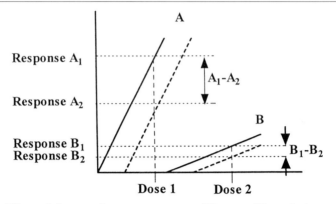

FIG. 1. Effects of slope on dose–response curve differences. Theoretical curves for two sets of dose–response curves (A and B) with different slopes. The shifts in the curves are the same (twofold). The difference in effect was different in the response by a fixed dose of drug defined by the two curves. Clearly, the difference was far more pronounced with A than with B.

on the abundance of the mRNA being targeted. The more abundant the message, the more difficult it may be to achieve greater levels of downregulation. The low abundance of opioid receptor mRNAs, therefore, may enhance their suitability in antisense studies.

Finally, it is important to ensure that the functional assays have the sensitivity to reliably detect the modest downregulation of protein levels usually seen with antisense approaches. This issue is crucial. For example, in the opioid system, antisense paradigms often shift analgesic dose–response curves only twofold.[12,19,20,22,23] Fortunately, analgesic assays can reliably detect these modest changes in sensitivity, but this may not be the case for other assay systems. The ability of an assay to detect modest changes depends on many factors. One factor that may be particularly important is the slope of the dose–response curve. This is well illustrated graphically with two sets of dose–response curves (Fig. 1). Both sets of curves show a twofold shift to the right of the dose–response curve. When looking at the changes in response to a fixed drug dose, however, the differences for the two groups are quite distinct. At dose 1, the difference in response between the two curves ($A_1 - A_2$) is relatively large. The difference in response at dose 2 between the two B dose–response curves ($B_1 - B_2$), with their more shallow slopes, is far smaller despite a similar shift in the dose–response curve. Thus, the inability to demonstrate a functional change

[22] L. Leventhal, L. B. Stevens, G. C. Rossi, G. W. Pasternak, and R. J. Bodnar, *J. Pharmacol. Exp. Ther.* **282,** 1402 (1997).
[23] G. C. Rossi, Y.-X. Pan, G. P. Brown, and G. W. Pasternak, *FEBS Lett.* **369,** 192 (1995).

following antisense treatment may simply reflect a difficulty with the sensitivity of the assay system. This can make the interpretation of negative results difficult.

Once the target has been chosen, the treatment paradigm should be optimized. As noted earlier, short antisense oligodeoxynucleotides are stable inside cells for several days.[20] Antisense can be given daily or even every other day with good results. Thus, the frequency of treatment is not a major issue, although it is probably wise to limit the total number of spinal or supraspinal injections as much as possible to minimize tissue damage unless indwelling catheters are placed. On the other hand, determining the duration of treatment can be crucial and it depends, to a large degree, on the rate of turnover of the protein being assessed. Even if new protein synthesis is immediately terminated by the antisense probe, assessing changes in function would still require the loss of the preexisting protein. Opioid affinity labels suggested a turnover time of 3–5 days for the opioid receptors.[24–26] On the basis of these results, when looking at the effects of antisense on opioid receptors, we chose to treat the animals for 5 days with antisense and to assess function on day 6. On the other hand, proteins that turn over more quickly may require far shorter treatments. An excellent example is the α subunits of the G proteins. When targeting these subunits, we saw effects 24–48 hr following a single antisense injection.[21,27]

Dosing in the central nervous system with antisense oligodeoxynucleotides remains empirical, with a wide range of doses reported in the literature. In general, try to use the lowest effective dose, particularly in behavioral assays. If there is a question about dosing, try several. We typically start with 5 μg either intracerebroventricularly or intrathecally in mice. In our experience, this dose is often effective and is generally nontoxic. If we do not see a robust functional response we will then increase the dose to 10 μg/dose. Higher doses may downregulate the mRNA and protein to a greater degree, but at the risk of an increased incidence of toxicity or side effects. The choice of dose also may depend on the goal of the study. In our studies, behavioral changes are seen at lower antisense doses than in immunohistochemical or biochemical chemicals. In many ways this is fortunate because the behavioral studies are the ones most likely to be influenced by toxic or nonspecific side effects.

[24] G. W. Pasternak, S. R. Childers, and S. H. Snyder, *Science* **208,** 514 (1980).
[25] G. W. Pasternak, S. R. Childers, and S. H. Snyder, *J. Pharmacol. Exp. Ther.* **214,** 455 (1980).
[26] E. F. Hahn, M. Carroll-Buatti, and G. W. Pasternak, *J. Neurosci.* **2,** 572 (1982).
[27] K. M. Standifer and G. W. Pasternak, *Cell. Signalling* **9,** 237 (1997).

Primer Design

Many different antisense approaches have been successfully used. We use short antisense oligodeoxynucleotides, usually 18–25 bases. Although chemically modified probes are effective in the central nervous system, we have successfully used nonmodified DNA probes, which are thought to minimize toxicity and nonspecific actions. A number of computer programs will design primers for DNA sequencing and/or polymerase chain reaction (PCR). We use the same criteria to design antisense probes. The melting temperature (T_m) is an important parameter that reflects the stability of the hybridization between the primer and its target. The nearest-neighbor thermodynamic values method is widely used to calculate the melting temperature, which is based on primer length, G/C content, and salt concentration. We typically aim for a G/C content between 50 and 70%. Selecting a G or C at either the 3′ end and/or the 5′ end of the primer may enhance the duplex stability. Primer dimers and hairpin or loop structures should be avoided. The stability of these secondary structures can be assessed by calculating their free energy, ΔG. Secondary mRNA structure also is a consideration in antisense primer design and regions with strong secondary structures, such as stem–loops or hairpins, should be avoided. Finally, it is essential that the chosen sequence be checked against a database to ensure the specificity of the chosen sequence.

Mismatch controls are essential to establish the specificity of the response, but opinions regarding the most appropriate control vary widely. Some investigators use random or unrelated primers as mismatch controls. Because even nonspecific effects can be sequence specific, we try to keep our mismatch probes closely matched to the antisense. We maintain the overall base composition of the probe and try to make as few changes in the sequence as possible. Switching the sequence of 4 bases out of the 20 is sufficient to eliminate activity in most situations, although we have occasionally changed the sequence of up to 6 bases.

Controlling for Specificity

Antisense treatments downregulate mRNA and subsequent lower protein levels. When possible, it is advisable to document these changes at both the mRNA and protein levels. However, this is often difficult, particularly when dealing with splice variants where the targeted variant mRNA accounts for only a small percentage of the entire mRNA. Mismatch controls are essential in confirming the specificity of the response. In addition, it is helpful to demonstrate that the effects of the antisense are reversible. In the opioid system, we found that animals regained their analgesic sensitivity within a few days.[12]

Nonspecific effects remain a major concern when using antisense paradigms. Some may reflect toxicity, while others are sequence specific. We have, on occasion, seen bizarre behaviors following specific antisense probes given intracerebroventricularly. The effects were sequence specific, but seem unrelated to the mRNA being targeted since other probes targeting nearby regions on the same mRNA lacked these activities. Thus, it is important to evaluate carefully all unusual or unexpected behaviors before assuming they result from specific effects on the targeted mRNA.

Knockout mice offer another approach toward assessing the functional significance of a protein. In most situations, results will be similar to those seen with antisense methods. However, there are occasionally differences. These may be due to developmental changes due to the absence of the protein during critical periods, leading to more widespread changes in other systems or to compensatory changes in which the function of the lost protein is taken over by another. Thus, different results between knockout and antisense models must be examined cautiously.

Antisense Mapping: Assessing Splice Variants and Partial cDNA Sequences

Early antisense studies assumed that targeting the region around the initiation site was important to optimize activity. However, this is not the case. When we examined the ability of antisense primers to downregulate δ-opioid receptor binding in a cell line expressing the receptor, we examined five different antisense probes targeting all three exons comprising *DOR-1*.[12] All of the probes reduced binding to a similar extent, indicating that any region along the mRNA could be targeted effectively (Fig. 2). This observation was then confirmed *in vivo* when the same antisense probes targeting the various exons all downregulated δ-opioid analgesia. Thus, it is possible to establish the importance of individual exons of a gene in specific functions, an approach we have termed *antisense mapping*.[6,23]

Antisense mapping has opened an interesting approach in general cloning approaches. When cloning new proteins, partial cDNA sequences are traditionally used to pull out the full-length clone, which is then expressed to establish its function. The ability to target anywhere along the mRNA implies that the functional significance of protein might be defined from a partial cDNA sequence. We used this approach to establish the functional significance of a partial cDNA fragment that turned out to be a novel member of the opioid receptor family.[28,29] We now routinely examine partial

[28] Y.-X. Pan, J. Cheng, J. Xu, G. C. Rossi, E. Jacobson, J. Ryan-Moro, A. I. Brooks, G. E. Dean, K. M. Standifer, and G. W. Pasternak, *Mol. Pharmacol.* **47,** 1180 (1995).

[29] Y.-X. Pan, J. Cheng, J. Xu, and G. W. Pasternak, *Regul. Peptides* **54,** 217 (1994).

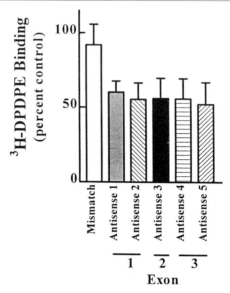

FIG. 2. Effect of antisense probes targeting various exons of *DOR-1* on δ receptor binding. NG108 cells, which naturally express δ-opioid receptors encoded by *DOR-1,* were treated with antisense probes targeting exon 1, 2, or 3. Results are modified from the literature.[12] The mismatch probe did not significantly reduce binding, but all of the other antisense probes were active. DPDPE, ■■■.

cDNA fragments by antisense approaches. Of course, this approach requires an established functional assay system and the same care in the design of the primers as noted above.

Antisense mapping has been particularly powerful in assessing the functional significance of splice variants and even in predicting their presence.[19,23] Determining whether a specific exon is present in an mRNA is relatively straightforward. Because some antisense probes may not be active even when targeting sequences present in the mRNA, a negative result may be difficult to interpret. To minimize the possibility of a false negative, we routinely examine multiple antisense probes targeting each exon. For example, when antisense mapping *MOR-1* for morphine and morphine 6β-glucuronide (M6G) analgesia, three different probes targeting exon 1 effectively blocked morphine analgesia without affecting M6G analgesia. Conversely, another three probes based on exon 2 were inactive against morphine and blocked M6G. Thus, the mechanisms of morphine and M6G analgesia are clearly separate and distinct. The power of this study was that all of the probes were active in at least one assay. Thus, the lack of activity in one assay was not simply due to the inability of the antisense

A

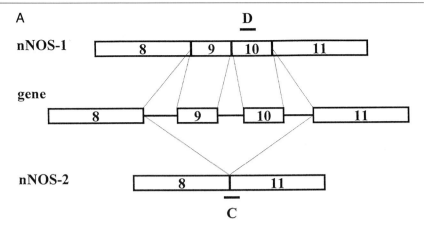

B

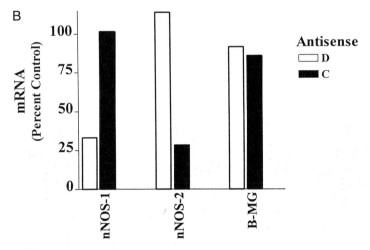

FIG. 3. Downregulation of *nNOS* isoforms using antisense. (A) A schematic of the *nNOS-1* and *nNOS-2* isoforms is presented, indicating the absence of exons 9 and 10 from *nNOS-2*. Antisense D targets only *nNOS-1,* whereas antisense C spans the splice site between exons 8 and 11, which is unique to *nNOS-2*. (B) The effect of antisense treatments on mRNA levels, as determined by RT-PCR. β-Microglobulin (B-MG) was included as an internal control. Antisense D downregulated only *nNOS-1* whereas antisense C downregulated only *nNOS-2*. Results are modified from the literature.[15]

to downregulate the mRNA and protein. These studies suggested that exon 1 was important in morphine, but not M6G, analgesia. This conclusion is supported by a study examining *MOR-1* knockout mice with selective disruptions of exon 1. These mice were insensitive to morphine, but still responded to M6G.

A more difficult situation occurs when comparing variants with more subtle differences. For example, the neuronal nitric oxide synthase gene (*nNOS*) has a number of isoforms, including one (*nNOS-2*) that differs from the major isoform (*nNOS-1*) by the lack of exons 9 and 10 (Fig. 3).[15] We functionally defined *nNOS-1* with probes targeting exons 9 and 10. However, selectively downregulating *nNOS-2* posed a problem since all of its exons also were contained within *nNOS-1*. We overcame this problem by targeting the splice site. Very short antisense fragments are not able to downregulate mRNA and protein. By limiting the length of the antisense probe on either side of the splice site, we were able to selectively downregulate *nNOS-2* without altering the expression of *nNOS-1*. This approach proved functionally valuable because *nNOS-1* and *nNOS-2* had opposite functional actions.

Conclusions

The central nervous system has proved amenable to antisense paradigms. Analgesic systems have been particularly sensitive, owing to a number of factors such as their proximity to the ventricular system and the ability to discern, easily and functionally, modest changes in protein levels. Our studies of these systems have identified a number of factors that may play a significant role in the overall success rate of antisense paradigms. Equally important, we have developed a method for selectively assessing the functional activities of splice variants of a single gene. The selectivity of these approaches far exceeds that of traditional antagonists and can provide valuable insights into the functioning of the nervous system.

Acknowledgments

The work described was supported, in part, by grants (DA02615, DA06241, and DA07242) and a Senior Scientist Award (DA00220) to G.W.P. and a Mentored Scientist Award (DA00296) to Y.-X.P. from the National Institute on Drug Abuse, and by a core grant from the National Cancer Institute (CA08748).

[5] *In Vitro* and *in Vivo* Effects of Antisense on α_2-Adrenoceptor Expression

By Emma S. J. Robinson and Alan L. Hudson

Introduction

The rationale behind the current studies was developed from the lack of highly selective agonists and antagonists available for the study of the $\alpha_{2A/D}$-adrenoceptor in the rat. On the basis of ligand-binding studies and molecular biology, four subtypes of α_2-adrenoceptors are known to exist, namely, α_{2A}, α_{2B}, α_{2C}, and α_{2D}.[1] The α_{2D} subtype is now known, from its sequence homology, to be the rat equivalent of the human α_{2A}-adrenoceptor and is pharmacologically distinct.[1] Antisense oligonucleotides provide a highly selective and reversible means for inhibiting protein expression and, thus, enabling specific receptor-mediated functions to be investigated. The experiments described in this chapter are from our investigations into the effects of a number of different antisense sequences on the expression of α_2-adrenoceptors both *in vitro* and *in vivo*. Prior to the investigation of the functional effect of antisense-mediated receptor knockdown, radioligand-binding assays were used to determine the level of receptor protein. To confirm the efficacy and specificity of the antisense sequences they were first tested in an *in vitro* assay using primary cortical neuron cultures. *In vivo* receptor expression was subsequently investigated in rat brain following an intracerebroventricular (icv) antisense infusion and receptor autoradiography was used to enable the measurement of α_2-adrenoceptors in discrete brain areas.

Effect of Antisense Treatment on α_2-Adrenoceptor Expression in Culture of Primary Cortical Neurons

The application of cell culture is a simple and relatively inexpensive assay system for characterizing the effect of several different antisense sequences. Previously this approach had been reported for a number of antisense studies where the knockdown of protein expression has been successful.[2–4] The cell types used for these investigations have included

---

[1] N. French, *Pharmacol. Ther.* **68,** 175 (1995).

[2] C. Wahlestedt, E. Golanov, S. Yamamoto, F. Yee, H. Ericson, H. Yoo, C. E. Inturrisi, and D. J. Reis, *Nature (London)* **363,** 260 (1993).

[3] C. Wahlestedt, E. Merlo-Pich, G. F. Koob, F. Yee, and M. Heilig, *Science* **259,** 528 (1993).

primary cultured cells, cell lines, and cells stably transfected with the receptor of interest.[4] One of the earliest reports of antisense oligonucleotides characterized in an *in vitro* assay prior to *in vivo* experiments, was described by Wahlestedt and co-workers.[2,3] They used a culture of primary cortical neurons to examine the effect of antisense to the neuropeptide Y-Y1 and *N*-methyl-D-aspartate R1 subunit (NMDA-R1) on receptor expression before using the most efficacious sequence for *in vivo* studies.[2,3] Given their success, we therefore also utilized cell culture as a prelude to *in vivo* work so that we could study the expression of the receptor protein known as the α_2-adrenoceptor.

The α_2-adrenoceptors are widely distributed throughout the central nervous system (CNS) and have been shown to be functionally expressed on both neuronal and glial cells in culture.[5] Furthermore, Northern analysis of RNA from both neuronal and astroglial cultures has been shown to detect the same transcript in culture and in the brain that hybridizes specifically with the RG20 cDNA probe.[6] On the basis of the evidence from previous antisense studies using primary cortical neurons and the fact that α_2-adrenoceptors are functionally expressed in these cells, this culture system was used for the *in vitro* experiments. The protocol used for the *in vitro* experiments including the time course and the oligonucleotide concentration, was adapted from the methods of Wahlestedt and co-workers[2,3] and other *in vitro* protocols.[4]

The procedure used for actually culturing the cortical neurons was adapted from that described by Sumners and co-workers,[7] which had been shown to result in cultures consisting of ~90% neurons and ~10% glial cells. Cortices were obtained from newborn Wistar rat pups of both sexes and the neurons were dissociated by trypsination. The resulting cell suspension was plated on poly(L-lysine) (Sigma, St. Louis, MO)-precoated wells at a density of 4×10^6 cells/well. The medium used was basal Eagle's medium (GIBCO, Grand Island, NY) containing 10% fetal bovine serum (heat inactivated; GIBCO), 2 mM L-glutamine (Sigma), 25 mM KCl (Sigma), and penicillin–streptomycin (50 U/50 μg; GIBCO). 1-β-D-*arabino*-furanosylcytosine (AraC, 1m M; Sigma) was added to the cells 24 hr after plating to curtail the proliferation of nonneuronal cells. The cells were

---

[4] K.-H. Schlingensiepen and M. Heilig, *in* "Antisense—from Technology to Therapy: Lab Manual and Textbook," p. 187. Blackwell Science, Berlin, 1997.

[5] J. Bockaert, V. Homburger, and F. Sladeczek, *in* "The Pharmacology of Noradrenaline in the Central Nervous System," p. 76. Humana Press, Clifton, New Jersey, 1990.

[6] M. A. Reutter, E. M. Richards, and C. Sumners, *J. Neurochem.* **68,** 47 (1997).

[7] C. Sumners, W. Tang, B. Zelezna, and M. K. Raizada, *Proc. Natl. Acad. Sci. U.S.A.* **88,** 7567 (1991).

TABLE I

SEQUENCES USED FOR *in Vitro* (PROBES 1–3) AND *in Vivo* (PROBES 1, 4, AND 5) INVESTIGATION OF EFFECT OF ANTISENSE OLIGONUCLEOTIDES TARGETING $\alpha_{2A/D}$-ADRENOCEPTORS IN RAT[a]

Probe	Oligonucleotide	Sequence						
1	Antisense	ATC	CGG	CTG	CAG	GGA	GCC	
	Sense	TAG	GCC	GAC	GTC	CCT	CGG	
	Mismatch	ATC	C**A**G	C**G**G	C**T**G	GGA	GCC	
2	Antisense	CAT	GGG	CGC	AAA	GCT	GCC	
3	Antisense	GGC	TGC	AGG	GAG	CCC	ATG	
4	Antisense	ATG	GGC	TCC	CTG	CAG	CCG	GAT
	Mismatch	ATG	**C**GC	TCC	C**A**G	C**T**G	C**G**G	GAT
5	Antisense	GCC	GTT	CCA	TCT	GCT	GAT	GC
	Mismatch	GCC	G**A**T	CCA	TCT	GCT	G**T**T	GC

[a] The sequences used are all fully modified phosphorothioate oligonucleotides and were kindly synthesized by L. Hall (Department of Biochemistry, University of Bristol). The control sequences, sense and mismatch, are also shown.

grown for 6 days without a medium change to allow the neuronal cells to develop fully before oligonucleotide treatment. On day 6 after plating, the medium was changed and replaced with medium identical to that used previously, plus the oligonucleotide. Preliminary studies used a phosphodiester oligonucleotide that, because of enzyme degradation, required the removal of serum from the cell cultures. Unfortunately, this appeared to reduce drastically the survival of the primary cortical neurons in culture (data not shown). Consequently, the oligonucleotides used for further experiments were fully modified phosphorothioate oligonucleotides that resist nuclease activity (Table I). This use of phosphorothioate sequences prevented the need for omitting fetal bovine serum (FBS) from the medium and improved neuronal survival. The oligonucleotide was added daily for 3 days at a final concentration of 1 μM and then 4 days after the start of oligonucleotide treatment the cells were prepared for ligand binding experiments to determine the level of receptor expression.

These radioligand-binding experiments were performed with intact cells using the highly selective α_2-adrenoceptor antagonist [³H]RX821002 (Amersham International, UK) to label the α_2-adrenoceptor population.[8] A single saturating concentration (10 nM) of [³H]RX821002 was used and the specific component of binding defined using 10 μM rauwolscine (Sigma). The cells were then solubilized and the bound radioactivity determined using liquid scintillation counting and the final result calculated as femto-

[8] A. L. Hudson, N. J. Mallard, R. Tyacke, and D. J. Nutt, *Mol. Neuropharmacol.* **1**, 219 (1992).

moles of bound ligand per milligram of protein. The results were adjusted using the protein concentration for each well (Coomassie blue; Pierce, Rockford, IL), to take into account any variation in the cell density or cell loss due to toxicity of the oligonucleotide. The control used in all of the *in vitro* experiments was vehicle and an oligonucleotide toxicity control was also included in the studies with the first antisense sequence. This control took the form of a sense probe and was used only in the first study with probe 1 to reduce the number of cells required. Only an antisense and a vehicle group were used for probes 2 and 3, as these experiments were originally aimed to be preliminary investigations. Had a significant reduction in binding been observed in the antisense-treated neurons with any of the sequences, the experiment would have been repeated to ensure the results were replicable and would at this stage include a toxicity control.

Treatment of neurons with the three different phosphorothioate antisense oligonucleotides resulted in an unexpected increase in specific [^{3}H]RX821002 binding of 40–66% (Fig. 1). This would indicate that the density of α_2-adrenoceptors was actually greater in these cells following incubation with the oligonucleotides (Table II). However, this increase reached a level of significance only for probe 3 compared with the vehicle control ($p < 0.05$, $n = 3$). A sense control was included in the experiment with probe 1 and a small increase in [^{3}H]RX821002 was also seen with this sequence (Fig. 1). Although this increase did not reach significance it does suggest that both sense and antisense oligonucleotides are able to increase α_2-adrenoceptor binding. That the increase in binding was seen with all three antisense sequences, and a small increase with the sense probe 1, suggests a nonspecific mechanism of action.

The lack of an antisense-mediated reduction in α_2-adrenoceptor expression in primary cortical neurons, and the surprising increase in binding observed in the antisense-treated cells, suggest this assay system is not suited to the investigation of antisense to the α_2-adrenoceptors. Studies by Wahlestedt and co-workers[2,3] using primary cortical neurons, and studies by Ishitani and Chuang[9] using cultured cerebellar granule cells, have successfully demonstrated antisense inhibition of receptor expression and function. However, our results do not show an antisense-mediated inhibition of α_2-adrenoceptor expression, which may be the result of number of factors. One explanation is a possible interaction with the mitotic inhibitor AraC, which was used in these experiments to curtail the proliferation of nonneuronal cells. In 1991, Manev et al.[10] demonstrated an interaction between antisense oligonucleotides and AraC resulting in the proliferation of non-

[9] R. Ishitani and D.-M. Chuang, *Proc. Natl. Acad. Sci. U.S.A.* **93**, 9937 (1996).
[10] H. Manev, S. Caredda, and D. R. Grayson, *Neuroreport* **10**, 589 (1991).

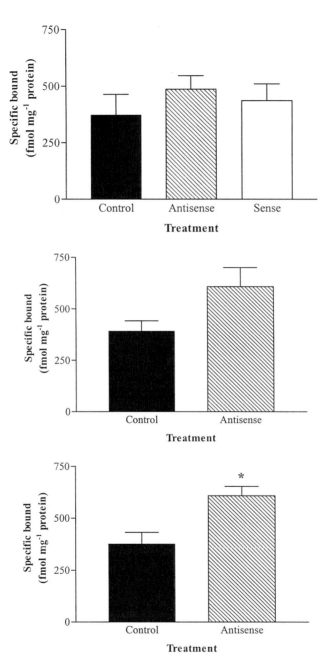

FIG. 1. Specific [³H]RX821002 binding (10 n*M*) to intact primary cortical neurons following a 3-day treatment with probe 1 (*top*), probe 2 (*middle*), or probe 3 (*bottom*). The bar charts show the mean ± SEM in femtomoles bound per milligram of protein for three or four separate cultures.

TABLE II

SPECIFIC [^{3}H]RX821002 BINDING (10 nM) TO α_2-ADRENOCEPTORS IN INTACT
PRIMARY CORTICAL NEURONS[a]

Probe	Control (fmol mg^{-1} protein)	Antisense (fmol mg^{-1} protein)	Percent increase
1	372.0 ± 92.4	486.9 ± 59.3	40.9 ± 22.4
2	391.1 ± 50.7	607.2 ± 92.3	55.1 ± 8.6
3	347.9 ± 56.8	608.6 ± 44.6[b]	65.6 ± 13.5

[a] Following a 3-day treatment with probe 1, 2, or 3. Results are shown as means ± SEM in femtomoles bound per milligram of protein for three or four separate cultures.

[b] $p < 0.05$, antisense compared with vehicle, unpaired t test.

neuronal cells. The AraC mechanism of action is not clearly understood but evidence available indicates an AraC-induced inhibition of DNA synthesis through incorporation of the metabolite of AraC, triphosphate AraC, into the DNA.[11] DNA damage may initiate the apoptotic death program of the neurons, which has been demonstrated in cultured cerebellar granule.[12] Antisense inhibition of the uptake of [^{3}H]AraC into cells has been shown with antisense sequences targeting different receptor proteins and therefore appears to be a nonselective mechanism. In our experiments the relative expression level of α_2-adrenoceptor on primary cortical neurons compared with glial cells is not known. However, if the relative expression on glial cells were higher than that on neurons, then the apparent increase in binding observed following antisense treatment may, in fact, relate to an increase in the number of glial cells in the cultures.

The growth of primary cultures of neurons requires the inclusion of a high level of KCl in the medium and there is evidence in glial cell cultures that α_2-adrenoceptor expression can be affected by the environment in which they are grown. Levels of cyclic AMP and protein kinase C (PKC) have been shown to be increased and this results in a decrease in the level of mRNA encoding the $\alpha_{2A/D}$-adrenoceptor subtype.[6] The effect on receptor mRNA is concentration dependent and, in addition to the effect in astroglial cells, an increase in mRNA for the α_{2A}-adrenoceptor subtype in response to an increase in cAMP was also seen in the HT29 cell line.[13] Overall the effects of culture conditions on receptor expression are difficult to determine owing to the nature of transformed cells. If the expression of both α_2-adrenoceptor mRNA and protein is affected by levels in cAMP

[11] D. W. Kufe and P. P. Major, *Med. Pediatr. Oncol.* **10**(Suppl.) (1982).

[12] Y. Enokido, T. Araki, S. Aizawa, and H. Hatanaka, *Neurosci. Lett.* **203**, 1 (1996).

[13] M. Sakeua and B. B. Hoffman, *Life Sci.* **54**, 1785 (1994).

and PKC, these factors may influence the response of primary cortical neurons to the culture conditions and the level of α_2-adrenoceptor expression. The mechanism by which receptor expression and mRNA production are regulated is not clear. Studies in astroglia suggest increases in cyclic AMP and PKC activation are both responsible for a decrease in $\alpha_{2A/D}$-adrenoceptor mRNA.[6]

In addition to the inclusion of high KCl, other factors in the medium, including those from the fetal bovine serum, may affect the expression of α_2-adrenoceptor. Evidence from studies using fetal bovine serum in medium used to culture the human colonic adenocarcinoma cell line HT29 suggests that the expression of α_2-adrenoceptors is decreased with increasing concentrations of fetal bovine serum.[14] Changing of the growth medium on cultured astroglia also resulted in a decrease in α_2-adrenoceptor mRNA that was not dependent on the concentration of fetal bovine serum.[6] Although these results were obtained in cell lines or astroglia there may also be an effect of culture conditions and the change in growth medium on α_2-adrenoceptor expression in neuron cultures.

The effect of phosphorothioate antisense oligonucleotide sequences on the $\alpha_{2A/D}$- and α_{2C}-adrenoceptors has previously been demonstrated in PC124D cells stably transfected with α_2-adrenoceptor subtypes. In these cells antisense to both the $\alpha_{2A/D}$ subtype and the α_{2C} subtype resulted in a significant reduction in receptor expression in their respective cell lines.[15] Antisense inhibition of α_2-adrenoceptor expression has also been shown in CHO cells stably transfected with $\alpha_{2A/D}$-adrenoceptors.[16,17] At the time of performing these experiments the most appropriate system to use seemed to be an *in vitro* assay that related to the target cell type *in vivo,* in this case rat cortex. Although primary cortical neurons have been used to demonstrate an antisense-mediated inhibition of both NPY-Y1 receptors and the NMDA-R1 receptor subunit, they do not appear to be a suitable assay system for investigating antisense to α_2-adrenoceptors.

The problems we have encountered using primary cortical neurons and antisense are numerous. The interaction of antisense oligonucleotides and the mitotic inhibitor AraC certainly warrants further investigation, as this drug is widely used in cell culture experiments. Uptake of oligonucleotides

[14] J.-C. Devedjian, M. Fargues, C. Denis-Pouxviel, D. Daviaud, H. Prats, and H. Paris, *J. Biol. Chem.* **266,** 14359 (1991).

[15] T. Mizobe, K. Maghsoudi, K. Sitwala, G. Tianzhi, J. Ou, and M. Maze, *J. Clin. Invest.* **98,** 1076 (1996).

[16] J. C. Hunter, L. R. Fontana, L. R. Hedley, J. R. Jasper, L. Kassotakis, R. Lewis, and R. M. Eglen, *Br. J. Pharmacol.* **120**(Suppl.), 229P (1997).

[17] J. C. Hunter, L. R. Fontana, L. R. Hedley, J. R. Jasper, R. Lewis, R. E. Link, R. Secchi, J. Sutton, and R. M. Eglen, *Br. J. Pharmacol.* **122,** 1339 (1997).

in vitro does not appear to be as efficient as that seen *in vivo* and may relate to the biochemical activity of the cells when they are placed under culture conditions.[18] Factors such as fetal bovine serum and high levels of KCl, which are necessary for the successful culturing of primary cortical neurons, have been reported to decrease α_2-adrenoceptor mRNA.

In summary, the treatment with three different phosphorothioate antisense sequences targeting the $\alpha_{2A/D}$-adrenoceptor failed to show an antisense-mediated knockdown in α_2-adrenoceptor expression. The potential effect of the culture conditions and the mitotic inhibitor AraC in these cultures shows that although it is preferable to screen sequences in an *in vitro* assay prior to *in vivo* experiments the choice of cell used is important. Primary cortical neurons in culture were selected for their predicted relative similarity to cortical neurons *in vivo*. However, these results have shown the complications that occur when using this system, so it is possible that stably transfected cell lines may be a more suitable *in vitro* assay system.

Effect of Antisense Treatment on α_2-Adrenoceptor Expression *in Vivo* following 7-Day Intracerebroventricular Infusion

The use of antisense oligonucleotides for *in vivo* investigation of receptor protein has been well documented.[19] At the time of commencing our *in vivo* antisense studies targeting the $\alpha_{2A/D}$-adrenoceptor, three different antisense sequences to this receptor had been published. The first sequence used (probe 1) had been reported to increase systolic blood pressure,[20] while both the second[16] and third[15] sequences (probes 4 and 5, Table I) had been shown to inhibit receptor expression *in vitro* and receptor-mediated function *in vivo*. However, none of these experiments had investigated the effect of antisense administration on the actual expression of the α_2-adrenoceptors *in vivo*. The measurement of protein expression following antisense treatment is necessary to confirm a change in receptor expression prior to functional studies and to establish the specificity of the sequence. Several factors influenced the protocol we used for our *in vivo* antisense experiments, including the time course of receptor turnover, the antisense sequence, route of administration, and controlling for potential toxicity.

The turnover rate of the α_2-adrenoceptor was considered to be important owing to the mechanism of action of antisense oligonucleotides to inhibit receptor expression. One question that still remains unanswered, is

[18] E. S. J. Robinson and A. L. Hudson, unpublished observation, 1998.

[19] E. S. J. Robinson, D. J. Nutt, H. C. Jackson, and A. L. Hudson, *J. Psychopharmacol.* **11,** 259 (1997).

[20] J. P. Nunes, *Eur. J. Pharmacol.* **278,** 183 (1995).

TABLE III
QUANTIFIED AUTORADIOGRAPHY OF SPECIFIC [^{3}H]RX821002 BINDING (1 nM) FROM RATS TREATED WITH PROBES[a]

Region	Left			Right		
	Vehicle	Antisense	Mismatch	Vehicle	Antisense	Mismatch
Probe 1						
Caudate putamen	23.1 ± 1.9	23.4 ± 1.6	26.3 ± 1.5	20.3 ± 1.9	21.2 ± 2.1	31.3 ± 2.4[b]
Lateral septal nucleus	88.2 ± 5.4	63.1 ± 4.3[c,d]	81.6 ± 5.5	88.5 ± 9.1	69.7 ± 3.4[b,e]	96.2 ± 2.9
Anterior hypothalamic area	126.7 ± 6.3	94.7 ± 1.8[c,d]	116.7 ± 5.2	127.2 ± 6.7	105.9 ± 6.3[b,d]	131.2 ± 3.3
Probe 4						
Caudate putamen	15.3 ± 1.1	13.8 ± 2.2	11.4 ± 1.1[c]	14.5 ± 1.5	16.2 ± 3.1	15.4 ± 2.0
Lateral septal nucleus	112.6 ± 9.7	70.4 ± 27.5	55.6 ± 28.0[b]	112.2 ± 7.7	61.2 ± 21.0	79.5 ± 30.0
Anterior hypothalamic area	153.5 ± 24.8	191.8 ± 45.8	122.1 ± 34.7	144.0 ± 21.6	167.9 ± 32.3	114.4 ± 45.6
Probe 5						
Caudate putamen	20.9 ± 6.5	8.3 ± 3.0	17.2 ± 4.6	20.8 ± 7.9	19.6 ± 3.9	19.5 ± 4.2
Lateral septal nucleus	78.7 ± 12.2	3.9 ± 2.9[c]	22.2 ± 21.9[b]	83.5 ± 17.2	28.8 ± 10.2[b]	29.2 ± 18.6[b]
Anterior hypothalamic area	52.6 ± 4.5	28.2 ± 10.0[b]	27.3 ± 7.7[b]	52.6 ± 2.9	39.4 ± 10.3	37.5 ± 7.4[b]

[a] Probe 1 (top), probe 4 (middle), and probe 5 (bottom). The binding values are to specific brain regions for both the left and the right side of the brain for vehicle-, antisense-, and mismatch-treated rats. The results are shown as means ± SEM in femtomoles bound per milligram of wet tissue for four rats per group.
[b] $p < 0.05$, antisense and mismatch compared with vehicle, unpaired t test.
[c] $p < 0.01$, antisense and mismatch compared with vehicle, unpaired t test.
[d] $p < 0.05$, antisense compared with mismatch, unpaired t test.
[e] $p < 0.01$, antisense compared with mismatch, unpaired t test.

why antisense oligonucleotides do not achieve a total knockdown of receptor expression. Furthermore, functional and binding data for a range of receptor proteins have reported a reduction in function of up to 90% but only a 20–30% reduction in the receptor population. This inconsistency between antisense experiments and the classic concept of receptor reserve has still not really been answered. In a review, Weiss et al.[21] proposed that antisense inhibits a functionally active pool of receptors that is rapidly turned over compared with a second pool of receptors that is not functionally coupled and has a slower turnover rate. There is some evidence to support this theory from a study by Qin and co-workers,[22] where they have examined the recovery of dopamine receptor binding in the presence of antisense, following irreversible inactivation. Although this evidence suggests that antisense can inhibit a functional response with only a relatively small reduction in receptor expression, the time course for α_2-adrenoceptor turnover was a consideration when designing the current experiment.

The turnover of α_2-adrenoceptors *in vivo* has been investigated by studying the kinetics of repopulating of α_2-adrenoceptor-binding sites following irreversible inactivation with a number of agents.[23–27] *In vivo* studies using 2-ethoxy-1(2H)-quinolinecarboxylic acid (EEDQ) and phenoxybenzamine, examining both recovery of receptor binding and function, have produced a range of values for the rate of turnover α_2-adrenoceptor. The recovery of α_2-adrenoceptor binding in the cortex, using this method, was shown to have a half-life of 12.5 hr.[26] In contrast, Adler et al.,[23] using EEDQ, showed the repopulating of α_2-adrenoceptor in the rat to be monoexponential with a half-life of 4.1 days. The recovery of α_2-mediated inhibition of the release of noradrenaline and 5-hydroxytryptamine (5-HT) differed with half-lives of 2.4 and 4.6 days, respectively.[27] A similar discrepancy between the turnover of receptor-mediated function and receptor reserve was also seen using the alkylating agent EEDQ and clonidine-induced sedation to investigate α_2-adrenoceptor turnover in the rat.[25] In this study the recovery of clonidine-induced sedation suggested a half-life of 14 hr whereas the recovery of the receptor pool was 37 hr. The turnover of α_2-adrenoceptors in the rat brain does not appear to be uniform across the whole CNS. Overall, the turnover is approximately 4 days whereas some brain regions, such as

[21] B. Weiss, G. Davidkova, and S.-P. Zhang, *Neurochem. Int.* **31**, 321 (1997).
[22] Z.-H. Qin, L.-W. Zhou, S.-P. Zhang, Y. Wang, and B. Weiss, *Mol. Pharmacol.* **48**, 730 (1995).
[23] C. H. Adler, E. Meller, and M. Goldstein, *Eur. J. Pharmacol.* **116**, 175 (1985).
[24] M. J. Durcan, P. F. Morgan, M. L. Van Etten, and M. Linnoila, *Br. J. Pharmacol.* **112**, 855 (1994).
[25] J. Pineda, J. A. Ruiz-Ortega, and L. Ugedo, *J. Pharmacol. Exp. Ther.* **281**, 690 (1997).
[26] R. M. McKernan and I. C. Campbell, *Eur. J. Pharmacol.* **80**, 279 (1982).
[27] R. M. McKernan, W. R. Strickland, and P. A. Insel, *Mol. Pharmacol.* **33**, 51 (1987).

the locus coeruleus, appear to have a higher rate of receptor turnover.[25] All told the turnover rate determined for the α_2-adrenoceptor is similar to the value obtained for dopamine receptors; therefore, the experiments using antisense to the dopamine receptors were of particular interest when deciding on the length of time to administer the oligonucleotide.

Previous antisense experiments have used a wide range of time courses for administration of the oligonucleotide. A single icv injection of antisense to the α_2-adrenoceptor has been shown to produce significant physiological changes in the antisense-treated group,[20] whereas other experiments have used a chronic infusion for 9 days to achieve an antisense-mediated change in receptor expression and function.[28] The experiments discussed in this chapter are from our investigation of three different antisense sequences (probes 1, 4 and 5; Table I), using a 7-day time course. The 7-day time course was used to take into account the proposed 4-day turnover of the α_2-adrenoceptor in the rat CNS and the three different antisense sequences chosen had all been previously reported to inhibit α_2-adrenoceptor-mediated function. For example, Nunes[20] showed that a single icv injection of antisense probe 1 in rats significantly increased systolic blood pressure. Antisense probe 4 significantly reduced the antiallodynic activity of clonidine following icv administration,[16] whereas probe 5 significantly inhibited the hypnotic response to dexmedetomidine when administered directly over the locus coeruleus.[15] However, in each of these studies no measure of *in vivo* α_2-adrenoceptor expression was reported.

The route of administration chosen was icv although antisense to α_2-adrenoceptors has been administered both icv and locally into the brain parenchyma. The icv route was preferred so that the antisense would reach areas such as the frontal cortex and hippocampus, where the presynaptic α_2-adrenoceptors play an important role in regulating neurotransmitter release. Uptake studies from a number of different sources have reported that fluorescently labeled oligonucleotides are detected in the frontal cortex, corpus callosum, hippocampus, septum, striatum, thalamus, and hypothalamus following icv administration.[29] In addition, the antisense was administered using an osmotic minipump to deliver the oligonucleotide at a constant rate. The use of the osmotic minipump and a brain infusion kit also minimizes the handling of the animals throughout the study. An icv, continuous infusion has been reported to give superior brain distribution and, furthermore, should reduce the potential toxicity associated with administering a bolus dose of the oligonucleotide. The dose of oligonucleotide administered was approximately 1 nmol μl^{-1} hr^{-1} or 8 μg μl^{-1} hr^{-1}.

[28] L.-W. Zhou, S.-P. Zhang, Z.-H. Qin, and B. Weiss, *J. Pharmacol. Exp. Ther.* **268,** 1015 (1994).
[29] Y. Yaida and T. S. Nowak, *Regul. Peptides* **59,** 193 (1995).

The specificity of the antisense and the need to control for toxicity is essential in any antisense experiment. The toxicity of antisense oligonucleotides, particularly the fully phosphorothioate-modified sequences, has been reported in a number of studies.[30-32] The control oligonucleotide chosen for all three experiments was a mismatch sequence (Table I). A mismatch sequence was chosen instead of a sense sequence as the sense sequences can, theoretically, affect the transcription of the mRNA via a triple-helix DNA formation, which can reduce the target mRNA.[33,34]

The methods used in these *in vivo* experiments were adapted from protocols in the literature. Male Wistar rats weighing 270–310 g were anesthetized with sodium pentobarbitone (60 mg kg^{-1}, intraperitoneal; Rhône Mérieux, Athens, GA) and stereotaxically implanted (0.92 mm caudal to bregma, 1.4 mm lateral, and 3.5 mm below the surface of the dura) with an icv L-shaped cannula (brain infusion kit, Alzet; Alza, Palo Alto, CA) connected to an osmotic minipump (7-day model 2001, Alzet; Alza) via a fine catheter tube. The pump was located subcutaneously in the midscapular region, and rats allowed to recover. At the end of the infusion, the rats were reanesthetized and the hearts perfused with ice-cold phosphate-buffered saline (PBS). The brains were carefully removed and immediately frozen and prepared for α_2-adrenoceptor autoradiography as previously described.[8] [^{3}H]RX821002 (Amersham International) binding (1 nM) was performed using 10 μM rauwolscine (Sigma) to define the specific component and corresponding sections were taken for histological staining with cresyl violet. As an additional control for specificity of the antisense sequence for the α_2-adrenoceptor, sections were also taken for [^{3}H]dihydroalprenolol (Amersham International) binding, to label β-adrenoceptors. β-Adrenoceptor binding was performed only when a significant reduction in α_2-adrenoceptor expression was observed and the reduction was not as a result of toxicity. β-Adrenoceptors were selected as a control receptor as they are located on noradrenergic neurons and, being another class of adrenoceptor, they control for the specificity of the antisense sequence to α_2-adrenoceptors. In some studies, neurological and behavioral

[30] B. Schöbitz, G. Pezeshli, C. P. Probst, M. H. M. Reul, T. Skutella, T. Stöhr, F. Holsboer, and R. Spanagel, *Eur. J. Pharmacol.* **331,** 97 (1997).

[31] S. M. Le Corre, P. W. J. Burnet, R. Meller, T. Sharp, and P. J. Harrison, *Neurochem. Int.* **31,** 349 (1997).

[32] A. Bourson, E. Borroni, R. H. Austin, J. Monsma, Jr., and A. J. Sleight, *J. Pharmacol. Exp. Ther.* **274,** 173 (1995).

[33] J. Georgieva, M. Heilig, I. Nylander, M. Herrero-Marschitz, and L. Terenius, *Neurosci. Lett.* **192,** 69 (1995).

[34] R. Landgraf, R. Gerstberger, A. Montkowski, J. C. Probst, C. T. Wotjak, F. Holsboer, and M. Engelmann, *J. Neurosci.* **15,** 4250 (1995).

toxicity has been observed, particularly following treatment with phosphorothioate oligonucleotides.[30–32,35]

The results from the 7-day infusion with antisense to the $\alpha_{2A/D}$-adrenoceptor revealed a significant reduction in α_2-adrenoceptor density with probe 1, whereas the other two sequences used both appeared to cause a nonspecific toxicity (Figs. 2–4; see color inserts). This is shown visually by the autoradiograms and the corresponding histology. For example, Fig. 2A shows a brain section from a vehicle-infused rat, with the distribution of α_2-adrenoceptors being dense over the septum with lower levels of binding over the striatum and frontal cortex. The adjacent section shows the corresponding histology and highlights the site of injection but otherwise shows no apparent damage to the brain (Figure 2A, right; see color insert). The lower sections (Fig. 2B; see color insert) of brain are from a rat treated with antisense probe 1. Quantitative autoradiography revealed that while α_2-adrenoceptor density was unchanged in the caudate putamen, there was a significant decrease in receptor density in the left lateral septal nucleus and left anterior hypothalamus relative to the vehicle and mismatch controls (Fig 2C, see color insert; Table III). Again the corresponding histology is shown in the adjacent tissue section and shows no damage to the brain. Reductions in binding to the hippocampus and thalamic nuclei were also observed but these results were significant only when compared with one control group (data not shown). The effects of antisense probe 1 were specific to the α_2-adrenoceptors as no change in [³H]dihydroalprenolol binding was observed in the antisense group compared with both control groups (data not shown). Furthermore, the histological staining of sections taken from all of the treatment groups was similar, with no evidence of neurological damage or glial cell proliferation in either of the oligonucleotide-treated groups (Fig. 2A–C, right; see color insert). Therefore results with probe 1, antisense and mismatch, did not reveal any indication of toxicity and a significant, specific reduction α_2-adrenoceptor expression was observed of a magnitude of 20–30%.

The significant reduction in α_2-adrenoceptor expression measured following a 7-day infusion of antisense probe 1 confirms that the $\alpha_{2A/D}$-adrenoceptor can be inhibited using antisense oligonucleotides. Functional analysis of the consequence of this receptor inhibition can now be investigated, having confirmed that the antisense sequence specifically inhibits α_2-adrenoceptor expression. In addition, [³H]dihydroalprenolol binding was not significantly different between any of the treatment groups, particularly the antisense-treated group.

In contrast to the findings with probe 1, following icv infusion, both

[35] B. J. Chiasson, J. N. Armstrong, M. L. Hooper, P. R. Murphey, and H. A. Robertson, *Cell. Mol. Neurobiol.* **14,** 507 (1994).

probes 4 and 5 resulted in nonspecific toxicity. Probe 4 did not significantly inhibit [^{3}H]RX821002 binding to α_2-adrenoceptors in the antisense-treated group compared with the vehicle and the mismatch control (Table III, Fig. 3; see color inserts). However, some of the animals in this group showed clear reductions in binding to brain areas close to the site of infusion and particularly on the left-hand side, the same side as the infusion (Fig. 3B and C; see color inserts). The variation in the response of different animals to the antisense infusion is reflected in the large standard errors associated with these sets of data (Table III). [^{3}H]RX821002 binding was significantly reduced in the mismatch group compared with the control in several brain areas, particularly those close to the site of infusion (Fig. 3C; see color inserts). The striatum, septum, thalamus, and some parts of the cortex all showed reduced α_2-adrenoceptor expression. Histological examination of the brain also revealed neurological toxicity. The area close to the site of infusion showed neuronal loss and glial cell proliferation as a result of the nonspecific toxicity of both the antisense and mismatch sequences (Fig. 3B and C, right; see color inserts). The area of damage extended from the site of infusion into the frontal cortex, the striatum, septum, and thalamus. All of the mismatch animals showed a similar pattern of toxicity whereas the antisense-treated group was less consistent. Some animals showed as severe damage as the mismatch-treated animals, whereas others were less severely damaged; hence the variation in the data (Table III). The presence of behavioral toxicity was also observed with both sequences.

Similarly, the sequences of probe 5 antisense and mismatch also produced neurological toxicity. In this case [^{3}H]RX821002 binding to the antisense and the mismatch groups was significantly reduced compared with the vehicle controls in brain areas close to the sight of the infusion (Fig. 4B and C, see color insert; Table III). The knockdown in receptor expression in the antisense-treated group was not specific to the α_2-adrenoceptor and histological examination of the brains resulted in a pattern of toxicity similar to that seen with probe 4, including neuronal loss and glial cell proliferation (Fig. 4B and C, right; see color insert). The brains from all of the oligonucleotide-treatment groups, probes 4 and 5, also showed some increase in size of the left side of the brain, suggesting edema.

The results from these studies are of interest from the point of view of observing a significant reduction in α_2-adrenoceptor expression following antisense treatment, which has not previously been reported. Furthermore, the differential toxicity observed with the three different antisense and mismatch sequences raises the question of the cause of this toxicity when the protocol used was identical for all three sequences.

The toxic side effects of the oligonucleotides have previously been reported both *in vitro* and *in vivo* and are particularly apparent when

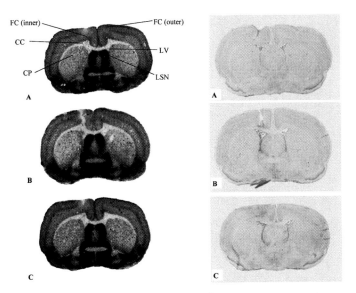

FIG. 2. Representative autoradiograms for total [³H]RX821002 binding (*left*) and cresyl violet-stained sections (*right*) from rats treated with probe 1, vehicle (A), antisense (B), or mismatch (C). FC, Frontal cortex; CC, corpus callosum; CP, caudate putamen; LV, lateral ventricle; LSN, lateral septal nucleus.

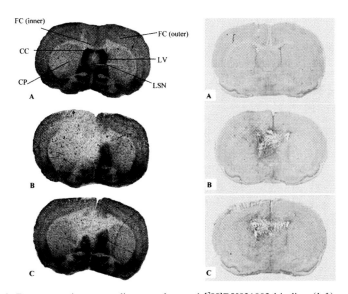

FIG. 3. Representative autoradiograms for total [³H]RX821002 binding (*left*) and cresyl violet-stained sections (*right*) from rats treated with probe 4, vehicle (A), antisense (B), or mismatch (C). FC, Frontal cortex; CC, corpus callosum; CP, caudate putamen; LV, lateral ventricle; LSN, lateral septal nucleus.

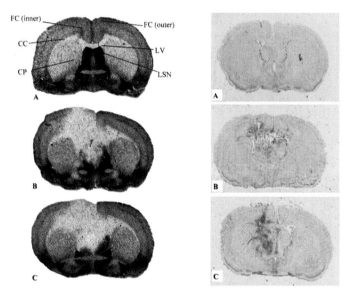

FIG. 4. Representative autoradiograms for total [³H]RX821002 binding (*left*) and cresyl violet-stained sections (*right*) from rats treated with probe 5, vehicle (A), antisense (B), or mismatch (C). FC, Frontal cortex; CC, corpus callosum; CP, caudate putamen; LV, lateral ventricle; LSN, lateral septal nucleus.

using modified oligonucleotides.[31,36-38] Nonspecific binding of the charged phosphorothioate molecules is one of the proposed mechanisms of toxicity, as is the accumulation of toxic metabolites following degradation of the oligonucleotides, which accumulate and then cause nonspecific and toxic side effects.[38] In these experiments, a differential toxic effect was seen with three different antisense and mismatch control sequences. All of the sequences were fully modified phosphorothioate oligonucleotide and, whereas probe 1 selectively inhibits α_2-adrenoceptor expression with no obvious neurological or behavioral toxicity, probes 4 and 5 both resulted in neurological damage and behavioral toxicity. The cause of the differential effects of these sequences is not clear. In some studies using antisense oligonucleotides behavioral toxicity has been reported[32] whereas others report no toxic or nonspecific effects with phosphodiester oligonucleotides[2,3] or phosphorothioate oligonucleotides.[21] In one study nonspecific effects of centrally administered oligonucleotide were examined in the rat.[30] In these experiments the administration of the oligonucleotide elevated body temperature, suppressed food intake, and inhibited nighttime activity; the authors attributed these changes to the nucleic acid structure of the oligonucleotides.

The difference between the sequences used in these experiments was the length of the oligonucleotide and the sequence itself, and one of these factors appears to be the cause of the nonspecific toxicity. The lengths of the three oligonucleotides differ by only 2 or 3 base pairs; therefore, it is the sequence that is most likely to be the cause of the toxicity. Previous studies have reported the presence of toxic motifs within the sequence, for example, a CpG motif, contiguous guanosine residues, or certain palidromes.[39,40] The presence of known and as yet unknown toxic motifs within the sequences used is the most likely explanation for the different level of toxicity observed with different sequences.

The most likely explanation for the toxicity is that the sequence itself carries toxic motifs that result in either nonspecific binding when the oligonucleotide is intact or when it has been digested into shorter fragments. The toxicity was observed to different degrees with the antisense and mismatch sequences and, in fact, with probe 3 the toxicity was greater with the mismatch sequence than the antisense. If toxicity is carried within the

[36] T. M. Woolf, C. G. Jennings, M. Rebagliati, and D. Melton, *Nucleic Acids Res.* **18,** 1763 (1990).

[37] T. M. Woolf, D. A. Melton, and C. G. B. Jennings, *Proc. Natl. Acad. Sci. U.S.A.* **89,** 7305 (1992).

[38] S. T. Cooke, *Annu. Rev. Pharmacol. Toxicol.* **32,** 329 (1992).

[39] O. Heidenreich, S.-H. Kang, X. Xu, and N. Nerenberg, *Mol. Med. Today* **1,** 128 (1995).

[40] C. A. Stein, *Nature Med.* **1,** 1119 (1995).

sequence itself this cannot be controlled for using an oligonucleotide mismatch, scrambled, or sense sequence.

Conclusion

By utilizing a primary culture of rat cortical neurons *in vitro* we were unable to demonstrate an antisense-mediated knockdown in the expression of $\alpha_{2A/D}$-adrenoceptors. In contrast, the results from our *in vivo* studies have successfully shown an antisense-mediated inhibition of α_2-adrenoceptor expression in rat brain following a continuous icv 7-day infusion. In addition, an interesting, differential toxic effect has been observed with other different oligonucleotide sequences used, for both antisense and mismatch sequences.

[6] Design and Efficacy of Serotonin-2A Receptor Antisense Oligodeoxynucleotide

By Joanne M. Scalzitti and Julie G. Hensler

Antisense oligodeoxynucleotides (ODNs) have been used extensively to decrease or prevent the synthesis of target proteins. In theory, the technique is simple yet elegant: to use the complement of a target sequence of messenger RNA (mRNA) encoding the protein of interest in order to disrupt translation or prevent translation initiation. However, in practice, the technique can be challenging and requires extensive controls. In this review, we present our unique experience with the design and use of an antisense ODN to alter the expression of serotonin-2A (5-HT_{2A}) receptors *in vitro* and *in vivo*. We discuss the mechanisms of action of antisense ODNs, potential "nonantisense" effects these ODNs may have, the design and testing of control and antisense ODNs *in vitro*, and administration of an antisense ODN intracerebroventricularly to alter expression of a targeted protein in the brain.

Mechanisms of Action

The mechanisms of action of antisense ODNs interacting with nucleic acid targets and how they may induce their biological effects are complex and potentially numerous. Several mechanisms have been demonstrated to mediate the inhibitory effects of antisense ODNs on gene expression, including "occupancy-only mechanisms" such as translational arrest or disruption of RNA secondary structure, as well as activation of RNase H, an enzyme that cleaves the RNA strand of an RNA–DNA duplex.[1] Phosphorothioated ODNs (in which a sulfur atom has been substituted for a nonbridging oxygen atom in the phosphodiester linkage of the oligonucleotide backbone), as well as phosphodiester ODNs, support RNase H cleavage of the RNA strand of an RNA–DNA duplex. It is important to note that some backbone modifications such as 2-O-methoxy or 2′-fluoro do not support or serve as substrates for RNase H activity.[1,2] Antisense ODNs complementary to the translation initiation codon and surrounding sequences are believed to inhibit the synthesis of a target protein through steric blockade of translation initiation; in many cases, the decrease in target protein appears not to be dependent on RNase H activity.[3-6] The inhibition of gene expression by antisense ODNs complementary to the coding region of the target mRNA, however, in many cases, appears to be dependent on RNase H activity[3-5,7,8] (Fig. 1).

Nonantisense Effects

A wide variety of unexpected, "nonantisense" effects of ODNs have been described, many of which are sequence dependent.[9-12] Nonantisense effects of both phosphodiester[13] and phosphorothioate ODNs have been reported.[14-18] These nonantisense effects result from numerous molecular

[1] S. T. Crooke, *Annu. Rev. Pharmacol. Toxicol.* **32,** 329 (1992).
[2] B. P. Monia, E. A. Lesnik, C. Gonzalez, W. F. Lima, D. McGee, C. J. Guinosso, A. M. Kawasaki, P. D. Cook, and S. M. Freier, *J. Biol. Chem.* **268,** 14514 (1993).
[3] C. Boiziau, R. Kurfurst, C. Cazenave, V. Roig, J. T. Thuong, and J. J. Toulmé, *Nucleic Acids Res.* **19,** 1113 (1991).
[4] M.-Y. Chiang, H. Chan, M. A. Zounes, S. M. Freier, F. L. Walt, and C. F. Bennett, *J. Biol. Chem.* **266,** 18162 (1991).
[5] C. F. Bennett, T. P. Condon, S. Grimm, H. Chan, and M.-Y. Chiang, *J. Immunol.* **152,** 3530 (1994).
[6] N. M. Dean, A. McKay, T. P. Condon, and C. F. Bennett, *J. Biol. Chem.* **269,** 16316 (1994).
[7] J. Minshull and T. Hunt, *Nucleic Acids Res.* **14,** 433 (1986).
[8] C. Cazenave, C. A. Stein, N. Loreau, N. T. Thuong, L. M. Neckers, C. Subasinghe, C. Helene, J. S. Cohen, and J. J. Toulme, *Nucleic Acids Res.* **17,** 4255 (1989).
[9] S. T. Crooke and C. F. Bennett, *Annu. Rev. Pharmacol. Toxicol.* **36,** 107 (1996).
[10] C. A. Stein, *Trends Biotechnol.* **14,** 147 (1996).
[11] A. D. Branch, *Hepatology* **24,** 1517 (1996).
[12] A. D. Branch, *Trends Biochem. Sci.* **23,** 45 (1998).
[13] J. L. Vaerman, P. Moureau, F. Deldime, P. Lewalle, C. Lammineur, F. Morschhauser, and P. Mariat, *Blood* **90,** 331 (1997).
[14] R. Bergan, Y. Connell, B. Fahmy, E. Kyle, and L. Neckers, *Nucleic Acids Res.* **22,** 2150 (1994).

A. Translational Arrest

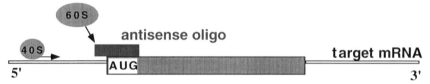

B. RNase H-mediated Degradation

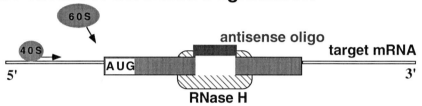

Fig. 1. Potential mechanisms of action of antisense oligodeoxynucleotides. (A) Antisense ODNs complementary to the translation initiation codon and surrounding sequences are believed to inhibit the synthesis of a target protein through steric blockade of translation initiation; in many cases the decrease in target protein appears not to be dependent on RNase H activity. (B) The activation of RNase H, an enzyme that cleaves the RNA strand of an RNA–DNA hybrid, also mediates the inhibitory effects of antisense ODNs on gene expression. The inhibition of gene expression by antisense ODNs complementary to the coding region of the target mRNA appears, in many cases, to be dependent on RNase H activity.

interactions including the aptameric properties of ODNs (interactions between ODNs and nonnucleic acid molecules such as proteins) and the formation of G quartets (four contiguous guanosine residues that form intra- or intermolecular four-stranded helices stabilized by planar Hoog-

[15] J. R. Wyatt, T. A. Vickers, J. L. Roberson, R. W. Buckheit, Jr., T. Klimkait, E. DeBaets, P. W. Davis, B. Rayner, J. L. Imbach, and D. J. Ecker, *Proc. Natl. Acad. Sci. U.S.A.* **91,** 1356 (1994).

[16] A. M. Krieg, A.-K. Yi, S. Matson, T. J. Waldschmidt, G. A. Bishop, R. Teasdale, G. A. Koretzky, and D. M. Klinman, *Nature (London)* **374,** 546 (1995).

[17] J. L. Vaerman, C. Lammineur, P. Moureau, P. Lewalle, F. Deldime, M. Blumenfeld, and P. Mariat, *Blood* **86,** 3891 (1995).

[18] L. Benimetskaya, M. Berton, S. Kolbanovsky, and C. A. Stein, *Nucleic Acids Res.* **25,** 2648 (1997).

steen-paired quartets of guanosine).[19] For instance, both Wyatt et al.[15] and Benimetskaya et al.[18] have demonstrated that four contiguous guanosine residues within an antisense ODN sequence potentiate antisense activity via the formation of a G-quartet (tetrad) structure. Bergan and colleagues[14] have demonstrated that aptameric inhibition of the protein tyrosine kinase p210[bcr–abl] by an antisense ODN results in decreased leukemic cell proliferation. Vaerman et al.[13,17] later showed that this antiproliferative effect required a 3' TAT consensus sequence. Another sequence-specific non-antisense effect is the induction of murine B cell proliferation and immunoglobulin secretion by the presence of unmethylated CpG dinucleotides in the ODN sequence.[16]

How does one prevent such nonantisense effects? Although there are no absolute rules to follow, and we certainly do not know all of the mechanisms by which these effects occur, sound design of the antisense ODN that avoids the structures or sequences known to have these effects[9] is a good starting point.

Designing Antisense Oligodeoxynucleotides

To design an antisense ODN that will be a useful tool, one must understand how they work. The most common approach is to inhibit translation of the protein of interest by using an antisense ODN to interact with the mRNA encoding that protein. The affinity of any oligonucleotide for its target nucleic acid sequence results from hybridization interactions (i.e., usually Watson–Crick base-pairing) that depend on hydrogen bonding between bases. Affinity increases with length because of hydrogen bonding and stacking between coplanar bases in the newly formed double helix. Affinity also varies as a function of sequence, since the melting temperature of an ODN (the temperature at which a population of double-stranded nucleic acid molecules is half-dissociated into single strands, T_m) is sequence dependent. The T_m depends on the proportion of G–C base pairs because G–C pairs contain three hydrogen bonds whereas A–T pairs contain only two. Therefore, the more G–C pairs in the ODN, the greater the amount of energy required to separate it from its target RNA. Specificity results from the selectivity of Watson–Crick base-pairing. The design of an antisense ODN must maximize these physical interactions.

As mentioned above, the size of the ODN is a major determinant of specificity and affinity. Although the mechanism of ODN entry into the cell has not been definitively identified, the predominant mechanism of

---

[19] D. Sen and W. Gilbert, Nature (London) **334,** 364 (1988).

ODN uptake appears to be fluid-phase endocytosis (pinocytosis).[20,21] Therefore large nucleic acids probably will not enter the cell. Cell surface, membrane-associated proteins have also been identified that specifically bind and internalize ODNs (receptor-mediated endocytosis). Size restrictions have yet to be determined for this mechanism.[22,23] The prevailing wisdom is that an ODN of at least 15 nucleotides to a maximum of 20 nucleotides in length is desirable, with 18-mers often being the most efficacious.[24,25] We designed three antisense ODNs of 15, 18, and 21 nucleotides in length and tested each for the ability to decrease 5-HT$_{2A}$ receptor-mediated hydrolysis of membrane phosphoinositides *in vitro* (see below). Only the 18-mer was effective in these preliminary studies.

We kept the G–C : A–T ratio of our ODNs below 70%, thereby avoiding long G–C repeats that can decrease specificity[24] and increase the probability of nonantisense effects, as mentioned above. We selected sequences with terminal G–C pairs that resulted in "clamps" on the ends of the ODN. These clamps result in tighter binding because, as previously mentioned, G–C pairs contain three hydorgen bonds whereas A–T pairs contain only two. We also avoided single nucleotide repeats that would greatly reduce specificity.[24] For instance, a poly(T) sequence could potentially bind to poly(A) tails of all mRNA.

It is imperative the ODNs be subjected to homology analysis with the sequences currently available in DNA databases. It is possible to restrict the homology analysis to eliminate sequences that would not be pertinent to the particular study (e.g., bacterial or viral sequences). For example, we restricted our analysis to known rodent sequences because we were targeting the rat 5-HT$_{2A}$ receptor in our study. Initially, we targeted the translation initiation site of the rat 5-HT$_{2A}$ receptor mRNA and its immediate vicinity. However, homology analysis demonstrated that these ODNs did *not* contain unique sequences and therefore could potentially interfere with the synthesis of other proteins. For example, antisense ODNs targeting the initiation codon of rat 5-HT$_{2A}$ receptor mRNA share significant sequence homology with other rat sequences, specifically rat mineralocorticoid receptor (94% identity in 16-bp overlap), rat IRS-1 mRNA for the

[20] L. A. Yakubov, E. A. Deeva, D. F. Zarytova, E. M. Ivanova, A. S. Ryte, Z. L. Yurchenko, and V. Vlassov, *Proc. Natl. Acad. Sci. U.S.A.* **86,** 6454 (1989).
[21] C. A. Stein, J. L. Tonkinson, L.-M. Zhang, L. Yakubov, J. Gervasoni, R. Taub, and S. A. Rotenberg, *Biochemistry* **32,** 4855 (1993).
[22] S. L. Loke, C. A. Stein, X. H. Zhang, K. Mori, M. Nakanishi, C. Subasinghe, J. S. Cohen, and L. M. Neckers, *Proc. Natl. Acad. Sci. U.S.A.* **86,** 3474 (1989).
[23] D. A. Geselowitz and L. M. Neckers, *Antisense Res. Dev.* **2,** 17 (1992).
[24] C. Wahlstedt, *Trends Pharmacol. Sci.* **15,** 42 (1994).
[25] A. J. Hunter, R. A. Leslie, I. S. Gloger, and M. Lawrence, *Trends Neurosci.* **18,** 329 (1995).

insulin receptor (84% identity in 18-bp overlap), and rat lens epithelial protein (94% identity in 16-bp overlap). For these reasons we did not feel that an antisense oligonucleotide targeting the translation initiation codon of 5-HT$_{2A}$ receptor mRNA would be selective for this protein.

The translation initiation codon is not the only target for antisense activity. In the screening process one may find that the most unique and potent antisense ODN corresponds to the 5' ("capping region")[26,27] or 3' untranslated region[1,4] or the coding sequence of the mRNA.[5,28,29] In our study, we targeted a sequence within the coding region of the 5-HT$_{2A}$ receptor mRNA (+57 to +74) (accession number M30705).[30] It may not be possible to design an efficacious antisense ODN that shares absolutely no sequence homology with known sequences currently found in DNA databases. It is, however, possible to test for potential effects of the antisense ODN on proteins other the target protein, thereby establishing the selectivity of the antisense ODN.

Another consideration when designing an antisense ODN is whether or not to use a modified backbone. The theory behind the replacement of the phosphodiester linkage of the DNA backbone with neutral, achiral molecules is to diminish the degradation of the antisense ODN by exogenous nucleases, and in some cases increase antisense activity. For example, unmodified phosphodiester ODNs are rapidly degraded in serum. Phosphorothioate ODNs have been shown to be resistant to nucleases and to be extremely stable in media, cells and cell extracts, urine, and various tissues (for discussion, see Refs. 1 and 31). In our *in vitro* studies described below, we chose to use phosphorothioate ODNs. As discussed above, some of the DNA analogs (e.g., phosphorothioates) serve as substrates for RNase H, whereas some backbone substitutions (e.g., 2-*O*-methoxy or 2'-fluoro) will not support RNase H activity. The use of DNA analogs can also result in increased toxicity. This is particularly the case for phosphorothioates. In our *in vitro* studies we controlled for this by determining [^{3}H]thymidine incorporation, an indication of DNA synthesis, following ODN treatment. In our *in vivo* studies described below, we addressed these issues by using a chimera in which the outer three 5' and 3' linkages were phosphorothioate bonds, and the inner 11 linkages were phosphodiester bonds.

[26] P. Westerman, B. Gross, and G. Hoinkin, *Biomed. Biochim. Acta* **48,** 85 (1989).

[27] B. J. Chiasson, M. G. Hong, and H. A. Robertson, *Brain Res. Mol. Brain Res.* **57,** 248 (1998).

[28] S. E. Barbour and E. A. Dennis, *J. Biol. Chem.* **268,** 21875 (1993).

[29] A. Osen-Sand, M. Catsicas, J. K. Staple, K. A. Jones, G. Ayala, J. Knowles, G. Grenningloh, and S. Catsicas, *Nature (London)* **364,** 445 (1993).

[30] D. Julius, K. N. Huang, T. J. Livelli, R. Axel, and T. M. Jessell, *Proc. Natl. Acad. Sci. U.S.A.* **87,** 928 (1990).

[31] R. W. Wagner, *Nature (London)* **372,** 333 (1994).

In studies utilizing an antisense strategy to regulate the expression of a specific protein, evaluation of the effects of control ODNs are necessary to exclude non-sequence-related, nonspecific effects. There is, however, some controversy as to what ODN provides the ideal control. Often the sense counterpart of the antisense sequence is used. However, sense ODNs may produce specific effects themselves (e.g., see Refs. 28 and 29), possibly due to triplex formation, primer extension, or by interference with naturally occurring antisense molecules.[32] The current consensus is to avoid using the sense ODN as a control and instead use mismatch and random sequence oligonucleotides as controls. The random ODN is a scrambled sequence, i.e., a sequence of the same length and base composition as the antisense but in a random order. A more stringent control is a mismatch ODN, in which two to four of the nucleotides in the antisense sequence are purposely mismatched and distributed along the length of the oligonucleotide.[24,32,33] Stein and Krieg[34] suggest that one or two mismatches be placed in the central section of the ODN. This guideline is a good starting point. We made four mismatches distributed along the length of the ODN and substituted G for A because this substitution can be the result of spontaneous mutation (5' to 3'): antisense, GGGCCATCACCTAATTGC; sense, GCAATTAG-GTGATGGCCC; random, CCTCGTAATGCAGACTCG; mismatch, GGACCATCGCCTAGTTAC.[35] Control ODNs must also be subjected to homology analysis with sequences currently available in DNA databases.

Testing an Antisense Oligodeoxynucleotide *in Vitro*

After designing an antisense ODN, one must decide how to test the efficacy of the ODN. We chose to test our antisense ODN *in vitro* before progressing to an *in vivo* experiment,[35] although this is certainly no guarantee of efficacy *in vivo*. We tested our antisense ODN in A_1A_1 cells, derived from 16-day rat cortical cultures immortalized by retroviral transduction of the wild-type simian virus 40 T-antigen,[36] which endogenously express the 5-HT$_{2A}$ receptor. Transfected cell lines also work for testing the efficacy of an antisense ODN,[37] although the overexpression of target mRNA in

---

[32] W. Brysch and K.-H. Schlingensiepen, *Cell. Mol. Neurobiol.* **14,** 557 (1994).

[33] M. I. Phillips and R. Gyurko, *News Physiol. Sci.* **12,** 99 (1997).

[34] C. A. Stein and A. M. Krieg, *Antisense Res. Dev.* **4,** 67 (1994).

[35] J. M. Scalzitti, K. A. Berg, S. A. Kratowicz, and J. G. Hensler, *J. Neurochem.* **71,** 1457 (1998).

[36] K. A. Berg, W. P. Clarke, Y. Chen, B. J. Ebersole, R. G. McKay, and S. Maayani, *Mol. Pharmacol.* **45,** 826 (1994).

[37] A. Ekman, H. Nissbrandt, M. Heilig, D. Dijkstra, and E. Eriksson, *Naunyn Schmiedeberg's Arch. Pharmacol.* **358,** 342 (1998).

transfected cells should be considered and may prove to be a technical challenge (see Ref. 9 for discussion). The duration of treatment of cells with antisense ODN must be considered. The best rule of thumb is to administer antisense ODNs for at least two half-lives of the protein of interest to effectively decrease the protein expression. For cells in culture, one must also consider the cell doubling rate. Thus, the dose and time course of antisense ODN administration may differ *in vitro* for different cell lines and should be determined empirically.

Because the expression of 5-HT$_{2A}$ receptors in A$_1$A$_1$ cells is low, as is often the case with endogenously expressed receptors, a functional response (i.e., the stimulation of phosphoinositide hydrolysis) to a partial agonist was used to establish the effects of antisense and control ODNs on 5-HT$_{2A}$ receptor expression.[35] For a partial agonist, response is proportional to receptor occupancy. Therefore, the maximal response to a partial agonist results from 100% occupancy of the receptors, and a decrease in 5-HT$_{2A}$ receptor number would be reflected in a decrease in the maximal response. In addition to a functional effect (e.g., the effect of antisense ODN treatment on a behavioral response or biochemical assay), the effect of antisense ODN treatment on the expression of the target protein must be established. Western blot, immunohistochemistry, or radioligand binding can be used to detect the level of a specific protein. To determine whether antisense ODN treatment results in a decrease in the levels of the targeted mRNA, RNA can be quantitated by Northern blot or ribonuclease protection assay. A decrease in target mRNA levels may be indicative of RNase H activity, but is not necessarily an indication of antisense effect as ODNs may have antisense effects without inducing RNase H activity.[4,5] Finally, the lack of effect of the antisense ODN on proteins other than the target protein should be demonstrated, thereby establishing the selectivity of the antisense ODN.

For experiments measuring the stimulation of phosphoinositide hydrolysis following antisense ODN treatment, the cells were seeded at a density of 5×10^4 cells/well in treated 24-well plates in complete medium and allowed to proliferate for 1 day. Plating in 24-well plates allowed us to assay all ODNs on one plate. Plating at a specific density is important not only for replication of results, but often affects the outcome of these experiments.[1,38] The following day, cells were placed in *serum-free* medium (to decrease the presence of exogenous nucleases) and experimental treatments with antisense, control ODNs, or vehicle begun (days 2–6, cumulative dosing). Vehicle (sterile nanopure water) or ODNs (final concentration 40 nM) were added each day for 5 days. The potential cytotoxic effect of

[38] T. Iwanaga and P. C. Ferriola, *Biochem. Biophys. Res. Commun.* **191,** 1152 (1993).

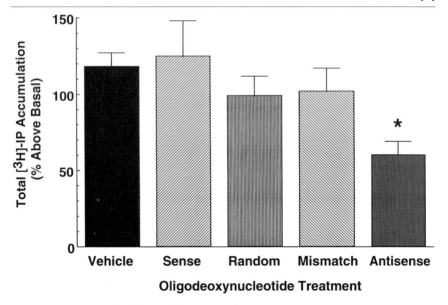

FIG. 2. Effect of oligodeoxynucleotides on 5-HT$_{2A}$ receptor-mediated phosphoinositide hydrolysis in A$_1$A$_1$ cells. Cells were treated with vehicle or antisense or control ODNs for 5 days. A single maximal concentration of the partial agonist quipazine (30 μM) was used to stimulate the accumulation of [^{3}H]inositol phosphates (IP). Plotted are the means $\pm$ SEM of $n = 5$ to 7 individual experiments. $*p < 0.01$ when compared with vehicle. [Data taken with permission from Scalzitti et al.[35]]

ODN treatment was determined by [^{3}H]thymidine incorporation as an indication of DNA synthesis.[35,39]

Treatment of A$_1$A$_1$ cells for 5 days with antisense ODN markedly decreased the maximal stimulation of inositol phosphate accumulation by the partial agonist quipazine (Fig. 2). This functional response was not altered by treatment of A$_1$A$_1$ cells with control ODNs (i.e., sense, mismatch, or random ODNs) (Fig. 2), indicating that the effect of antisense ODN in A$_1$A$_1$ cells was sequence specific. Stimulation of inositol phosphate accumulation by purinergic receptors was unaltered by treatment of A$_1$A$_1$ cells with 5-HT$_{2A}$ receptor antisense ODN, indicating that the effect of antisense ODN on 5-HT$_{2A}$ receptor-mediated stimulation of phosphoinositide hydrolysis in A$_1$A$_1$ cells was not due to nonspecific effects of the ODN on this second-messenger pathway.[35] In agreement with the functional data, a decrease in the density of 5-HT$_{2A}$ receptor sites, as measured by the

[39] R. I. Freshney, in "A Manual of Basic Technique: Culture of Animal Cells," 3rd Ed., pp. 277–289. Wiley-Liss, New York, 1994.

binding of [^{125}I]iodo-LSD, was observed in antisense-treated cells. Treatment of cells with sense ODN did not significantly alter the density of 5-HT$_{2A}$ receptor sites from vehicle control. Taken together, these data suggest that the antisense ODN decreased 5-HT$_{2A}$ receptor expression in a sequence-specific manner.[35]

Antisense ODNs directed against the coding regions of target mRNAs have been used successfully to decrease the expression of specific proteins in cultured cells. For example, antisense ODN treatment inhibits in cultured cortical neurons the expression of SNAP-25, a neuron-specific protein implicated in axonal growth and differentiation, and blocks nerve growth factor-stimulated neurite extension in PC12 cells.[29] Antisense ODNs targeting the coding region of mRNAs of endothelial cell adhesion molecules inhibit the expression of these proteins by endothelial cell lines in culture.[5] An antisense ODN, complementary to nucleotides coding for the calcium-binding loop of the murine group II phospholipase A$_2$, decreases phospholipase A$_2$ activity and reduces the release of [^{3}H]arachidonic acid and prostaglandin E$_2$ from activated cells.[28]

Administration of Antisense
  Oligodeoxynucleotides Intracerebroventricularly

Both intracerebroventricular (icv) and intraparenchymal modes of administration of antisense ODNs have been used to modify the expression of target proteins in the brain. Administration of antisense ODNs into a particular brain region allow the investigator to focus the effects of antisense treatment in a specific area of brain.[27,40,41] ODNs can be delivered via stereotaxically implanted cannulas by microinjection or Alzet osmotic pumps (Alza, Palo Alto, CA). As described below, we used Alzet osmotic pumps attached to Alzet brain infusion kits (which include cannulas, spacers, and catheter tubing) to deliver our vehicle or ODNs into the lateral cerebral ventricle. Alzet osmotic pumps have gained wide acceptance for infusions because of their continuous and consistent delivery.[25] ODNs injected or infused into the lateral cerebral ventricle of rats appear to follow the bulk flow of cerebrospinal fluid, through the ventricular system to the subarachnoid spaces and the perivascular spaces.[42] Intracerebroventricular administration of antisense ODNs has been demonstrated to alter the ex-

[40] C. P. Silvia, G. R. King, T. H. Lee, S.-Y. Xue, M. G. Caron, and E. H. Ellinwood, *Mol. Pharmacol.* **46,** 51 (1994).
[41] K. L. Widnell, D. W. Self, S. B. Lane, D. S. Russell, V. A. Vaidya, M. J. D. Miserendino, C. S. Rubin, R. S. Duman, and E. J. Nestler, *J. Pharmacol. Exp. Ther.* **276,** 306 (1996).
[42] F. Yee, H. Ericson, D. J. Reis, and C. Wahlestedt, *Cell. Mol. Neurobiol.* **14,** 475 (1994).

pression of target proteins in the cerebral cortex, striatum, and amygdala. A reduction in cortical neuropeptide Y-Y1 receptors has been observed following injection of an antisense ODN into the lateral ventricle.[43] Intracerebroventricular administration of an antisense ODN results in downregulation of cortical N-methyl-D-aspartate (NMDA)-binding sites and has been shown to protect cortical neurons from excitotoxicity, reducing the volume of ischemic infarcts.[44] Zhang and Creese[45] observed a decrease in dopamine D_2 receptors in striatum and nucleus accumbens after icv infusion of an antisense ODN.

Because the turnover of the 5-HT$_{2A}$ receptor in brain is 2.5 to 3 days[46,47] we administered our chimeric antisense ODN icv for 8 days. Adult (200–250 g) male Sprague-Dawley rats were anesthetized and a stainless steel cannula (Alzet brain infusion kits; Alza) was stereotaxically implanted into the lateral ventricle: AP, -0.8 mm; ML, 1.4 mm; DV, -3.5 mm from bregma, according to the atlas of Paxinos and Watson.[48] Antisense or control ODNs (1.9 nmol/μl; 250 μg/day), or vehicle, was administered at a rate of 1 μl/hr for 8 days by osmotic pump (model 2001; Alza) implanted subcutaneously.

Infusion of antisense ODN by osmotic minipump into the lateral cerebral ventricle of rats for 8 days (250 μg/day) resulted in a surprising *increase* in 5-HT$_{2A}$ receptor density in cortex and in striatum as measured by the binding of [^{3}H]ketanserin.[35,49] The effect of the antisense ODN was sequence specific, as infusion of mismatch ODN did not significantly alter 5-HT$_{2A}$ receptor density in these brain regions (Fig. 3). Neither 1-, 2-, nor 4-day infusions of antisense or control ODNs had an effect on cortical 5-HT$_{2A}$ receptor density.[35] Furthermore, administration of a lower concentration of antisense ODN (100 μg/day) for 8 days had no effect on cortical 5-HT$_{2A}$ receptor density. Eight-day infusion of antisense ODN, but not mismatch ODN (250 μg/day), resulted in an increase in 5-HT$_{2A}$ receptor function as indicated by a significant increase in headshake behavior induced by the 5-HT$_2$ receptor agonist DOI.[35] The increase in 5-HT$_{2A}$ receptor sites in cortex and striatum, and the increase in 5-HT$_{2A}$ receptor-mediated

[43] C. Wahlestedt, E. M. Pich, G. F. Koob, F. Yee, and M. Heilig, *Science* **259,** 528 (1993).
[44] C. Wahlestedt, E. Golanov, S. Yamamoto, F. Yee, H. Ericson, H. Yoo, C. E. Inturrisi, and D. J. Reis, *Nature (London)* **363,** 260 (1993).
[45] M. Zhang and I. Creese, *Neurosci. Lett.* **161,** 223 (1993).
[46] H. Gozlan, A. M. Laporte, S. Thibault, L. E. Schechter, R. Bolanos, and M. Hamon, *Neuropharmacology* **33,** 423 (1994).
[47] W. Pinto and G. Battaglia, *Mol. Pharmacol.* **46,** 1111 (1994).
[48] G. Paxinos and C. Watson, "The Rat Brain in Stereotaxic Coordinates," 2nd Ed. Academic Press, San Diego, California, 1986.
[49] J. M. Scalzitti, L. S. Cervera, C. Smith, and J. G. Hensler, *Pharmacol. Biochem. Behav.* **63,** 279 (1999).

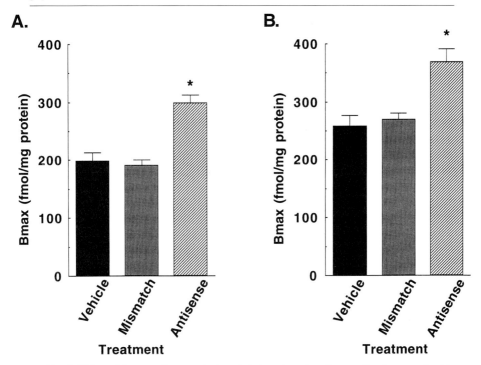

FIG. 3. Effect of intracerebroventricular administration of antisense oligodeoxynucleotide on 5-HT$_{2A}$ receptor sites in (A) cerebral cortex and (B) striatum. Vehicle, or antisense or mismatch ODNs, were administered icv for 8 days. The binding of [^{3}H]ketanserin to 5-HT$_{2A}$ receptor sites was measured in homogenates of whole cerebral cortex or striatum. B_{max} values were obtained by nonlinear regression analysis of saturation binding data. Plotted are the means ± SEM of n = 5 to 8 animals per experimental group. *p < 0.01 when compared with vehicle. [Data taken with permission from Scalzitti et al.[35,49]]

headshake behavior, indicate that icv infusion of our antisense ODN for 8 days increased the expression of this receptor in brain.[35]

As discussed above, the inhibition of gene expression by antisense ODNs directed against the coding region of a target mRNA is thought to be due to RNase H-mediated cleavage of that mRNA. RNase H-mediated hydrolysis of target mRNA is most likely the mechanism by which our antisense ODN decreased the expression of the 5-HT$_{2A}$ receptor in A$_1$A$_1$ cells (Fig. 4). We may not have observed a decrease in cortical 5-HT$_{2A}$ receptors following icv infusion of our antisense ODN because of low levels of RNase H activity in brain.[50] The stability of unmodified oligonucleotides

[50] R. J. Crouch and M.-L. Dirksen, in "Nucleases" (S. M. Linn and R. J. Roberts, eds.), pp. 222–241. Cold Spring Harbor Laboratory Press, Cold Spring Harbor, New York, 1982.

A. In vitro: RNase H-mediated Degradation

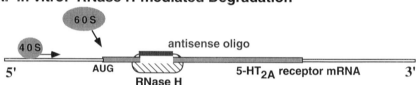

B. In vivo: Alleviation of Translational Suppression

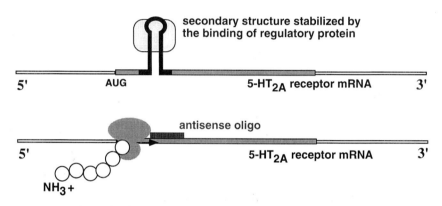

FIG. 4. Potential mechanisms of action of our serotonin-2A receptor antisense oligodeoxy-nucleotide *in vitro* and *in vivo*. (A) The inhibition of gene expression by antisense ODNs directed against the coding region of a target mRNA is thought to be due to RNase H-mediated cleavage of target mRNA. RNase H-mediated hydrolysis of target mRNA is most likely the mechanism by which our antisense ODN decreased the expression of 5-HT$_{2A}$ receptors in A$_1$A$_1$ cells. (B) The increase in cortical 5-HT$_{2A}$ receptor expression observed after 8 days of continuous icv infusion of antisense ODN may be the result of the alleviation of translational suppression. Binding of antisense ODN to the target sequence of 5-HT$_{2A}$ receptor mRNA might interfere with the formation of a secondary structure or inhibit the interaction of the RNA with regulatory proteins. The activity of RNase H in adult rat brain may not be sufficient to support the inhibition of protein expression by an antisense ODN targeting the coding region of mRNA.

in brain[42,51] and the effectiveness of unmodified antisense ODNs injected icv[43,44] are consistent with low activity of ribonucleases in brain. The activity of RNase H in adult rat brain may not be sufficient to support the inhibition of protein expression by an antisense ODN targeting the coding region of

[51] G. M. Jirikowski, P. P. Sanna, D. Maciejewski-Lenoir, and F. E. Bloom, *Science* **255**, 996 (1992).

mRNA. It is important to note that reports of downregulation of target protein expression in brain by an antisense strategy have utilized antisense ODNs designed to disrupt translation initiation.[40,41,43–45]

The increase in cortical 5-HT$_{2A}$ receptor expression observed after 8 days of continuous icv infusion of antisense ODN may be the result of the alleviation of translational suppression. RNA can assume a variety of secondary structures deriving from intramolecular base-pairing, such as stem–loop structures, which are profoundly important in determining RNA function and stability.[1,52] Because stem–loop structures may not be stable, the binding of proteins is required to stabilize these structures.[52] Computer analysis (GCG Sequence Analysis Software package) of the sequence of 5-HT$_{2A}$ receptor mRNA targeted by our antisense ODN indicates the potential existence of a stem–loop structure.[35] The authors speculate that the binding of our antisense ODN to the target sequence of 5-HT$_{2A}$ receptor mRNA may interfere with the formation of this secondary structure or inhibit the interaction of the RNA with regulatory proteins. The disruption of such a regulatory mechanism may relieve suppression of translation resulting in an increase in 5-HT$_{2A}$ receptor expression (Fig. 4).

Conclusion

In conclusion, antisense strategies have progressed rapidly. The established efficacy and specificity of this approach in animal models have resulted in the evaluation of antisense oligonucleotides in clinical trials for the treatment of cancer, Crohn's disease, human immunodeficiency virus infection, and rheumatoid arthritis.[53] Although the method can be technically challenging, the resulting data may yield information about the function of a protein that may otherwise be available only from gene knockout animals. Furthermore, antisense strategies allow researchers to knock down, transiently and inducibly, the expression of a protein, in many cases in a particular organ system or in a specific brain region, without the complications of interpreting data limited by the developmental compensation or lethality common in knockout animals.

Acknowledgments

This work was supported by United States Public Health Service Grant MH 52369 (J.G.H.), and by research funds from the National Alliance for Research in Schizophrenia and Depression (J.G.H.).

---

[52] M. Kozak, *Annu. Rev. Cell Biol.* **8**, 197 (1992).
[53] Anonymous, *Nature Biotechnol.* **15**, 519 (1997).

[7] Reduction of Neurotransmitter Receptor and G-Protein Expression *in Vivo* and *in Vitro* by Antisense Oligodeoxynucleotide Treatment

By KELLY M. STANDIFER

Introduction

The use of antisense oligodeoxynucleotides (ODNs) to reduce or elimi-nate the expression of specific proteins has become an increasingly impor-tant research tool. Although there is some disagreement about the precise mechanism of its action, few dispute the exquisite specificity of the ODN for its target.[1] Indeed, it is this specificity that makes antisense ODN technol-ogy such an attractive tool for studying the roles and regulation of select proteins. The increased popularity and utilization of this approach have also spurred the establishment of many new businesses dedicated to the custom synthesis of antisense ODNs, in any number of shapes, forms, and styles; all at increasingly more affordable rates.[2] This combination of specificity, availability, and affordability has allowed more and more investigators to utilize this molecular approach without the added expense of establishing a costly traditional molecular biology dimension within their laboratories. The use of ODNs has been especially fruitful when applied to neurotransmitter receptors and components of second-messenger sys-tems, and it has allowed us to address questions about receptor/second-messenger interactions and second messenger-mediated downstream events that have been difficult to answer.[3,4] This chapter describes how antisense ODNs may be used in cultured cell lines and in intact animals to reduce the expression of neurotransmitter receptors and G proteins.

General Considerations

Choice of Antisense Oligodeoxynucleotide

The only major limitation to the use of antisense ODNs is the availability of the sequence encoding the protein of interest, and even this may be circumvented to some extent. If possible, the antisense ODN should target

[1] C. A. Stein and Y.-C. Cheng, *Science* **261**, 1004 (1993).
[2] J. Kling, *Scientist* **12**, 18 (1998).
[3] P. R. Albert and S. J. Morris, *Trends Pharmacol. Sci.* **15**, 250 (1994).
[4] G. W. Pasternak and K. M. Standifer, *Trends Pharmacol. Sci.* **16**, 344 (1995).

a nonconserved region of the mRNA of interest to reduce the chance of cross-reacting with closely related sequences. However, if the specific sequence is not known, antisense ODNs directed against highly conserved sequences encoding functional domains such as G protein-binding sites or membrane-spanning regions of receptors may be used instead. For example, before the mRNA sequence encoding the mammalian γ-aminobutyric acid type B (GABA$_B$) receptor was known, antisense ODNs based on a highly conserved region in the amino acid sequence of other pertussis-toxin sensitive, seven-transmembrane-spanning mammalian receptors (7TMs) were generated. These ODNs specifically reduced GABA$_B$ and muscarinic receptor functionality in rat primary cerebellar cultures.[5] This approach to designing an ODN sequence also is applicable to G proteins because the sequence homology between G-protein α subunits from different species is high, as illustrated by the ability of mammalian G-protein α subunits to substitute for nonfunctional G proteins in yeast mutants.[6] If this approach is taken, however, the burden of proof required to demonstrate that the effect of the ODN is a result of the knockout of only the intended target is increased. Ultimately, inclusion of the proper controls will validate (or invalidate) the appropriateness of the sequence chosen. It is important to note that the sequence need not target only the region around the initiation codon. Although this sequence is frequently targeted, an ODN may target any region of mRNA and still be effective.[4,7]

   The decision about what type of antisense ODN to use (phosphodiester versus phosphorothioate, etc.) can be a difficult one, particularly with so many options from which to choose. Chapter 22 in volume 4[4a] discusses the advantages and disadvantages of the different modifications available. Ultimately, the type of antisense ODN chosen should reflect the properties most important for achieving the answer to the question being asked after factoring in additional considerations such as cost and method of delivery. One does not always need the more expensive modified ODN; we have had success utilizing unmodified phosphodiester ODNs to knock down specific opioid receptor subtypes and G-protein α-subunit levels in the brain, spinal cord, and cultured cells.[8–11]

[4a] P. L. Privalov and G. Protekhin, *Methods Enzymol.* **131,** 5 (1986).

[5] I. Holopainen and W. J. Wojcik, *J. Pharmacol. Exp. Ther.* **264,** 423 (1993).

[6] M. I. Simon, M. P. Strathmann, and N. Gautam, *Science* **252,** 802 (1991).

[7] R. W. Wagner, *Nature (London)* **372,** 333 (1994).

[8] K. M. Standifer, C.-C. Chien, C. Wahlestedt, G. P. Brown, and G. W. Pasternak, *Neuron* **12,** 805 (1995).

[9] K. M. Standifer, S. Jenab, W. Su, C.-C. Chien, Y.-X. Pan, C. E. Inturrisi, and G. W. Pasternak, *J. Neurochem.* **65,** 1981 (1995).

[10] G. C. Rossi, K. M. Standifer, and G. W. Pasternak, *Neurosci. Lett.* **198,** 99 (1995).

[11] K. M. Standifer, G. C. Rossi, and G. W. Pasternak, *Mol. Pharmacol.* **50,** 293 (1996).

Choice of Model System

The choice between an intact animal or cultured cell system depends, of course, on the question being asked. However, cultured cells (established lines) have the advantage of being homogeneous and having fewer delivery issues, and are an excellent system in which to test the choice of antisense ODN. For example, from a single dish of antisense-treated cells several parameters may be assayed at once (including levels of receptor and G-protein binding and function) to be certain that the ODN is performing properly. It is not always possible to assess changes to so many different parameters from the same source of tissue. Of course, delivering antisense ODNs to a particular location within the brain can be used to answer questions about circuitry that could never be answered in an *in vitro* system.

If a cultured cell system is chosen, the next question to address is whether to target a protein that is natively expressed or to use a cell line that has been transfected to express the protein of interest. There are advantages and disadvantages to both choices. If a cell line natively expresses the receptor or G protein of interest, the levels of expression are likely to be lower than in transfected cells and protein levels are often easier to downregulate.[4] A further advantage is that receptor–G protein coupling is often tighter in natively expressed systems, such that a moderate reduction in protein level can be sufficient to reduce function appreciably,[5] although this is not always the case.[8] The principal advantage of transfected cells is that one can create any combination of receptor(s) and G protein(s) of interest. However, reducing receptor or G-protein levels is more difficult if one or more of the proteins of interest are overexpressed. One also has the option of selecting colonies of transfected cells with a lower density of the protein of interest to avoid significant overexpression.

In Vitro Approaches

Experimental Design

In a review in 1994, Albert and Morris described antisense ODNs as "molecular scalpels" that could be used to tease apart interactions between various components of receptor-mediated signaling cascades.[3] This apt description emphasizes the incredible potential for insight that antisense technology can provide to the cellular signaling arena. After reducing levels of specific neurotransmitter receptors, G proteins or other GTP-binding/

regulatory proteins,[12,13] and confirming that their functional response has been reduced [by assessing changes in such areas as receptor binding,[5,8] cAMP accumulation,[5,14] or phospholipase C (PLC) activity[15]], one can examine the effect that this protein reduction has on the rest of the system. Does it affect the action of downstream mediators such as protein kinase C (PKC), intracellular calcium,[16] mitogen-associated protein (MAP) kinase, or c-Fos? Does it affect ion channel activity?[17–19] What particular effectors are modulated by interaction with this particular G-protein subunit? In some of the studies asking these questions, G-protein α-, β-, and γ-subunit levels also were significantly reduced *in vitro* by delivering antisense via microinjection[17–19] or by stable transfection of a cell line with a full-length cDNA encoding the antisense sequence.[20,21] These methods are quite effective, but require specialized equipment and molecular biological expertise.

Antisense ODNs can also be used to deplete a specific subunit within a multimeric receptor, allowing one to assess the contribution of that subunit to the binding or function mediated by different agonists.[22] It is also possible to target potential splice sites within a coding sequence, allowing one to pursue evidence of receptor subtypes and alternative splicing.[23,24] This concept, called "antisense mapping,"[4] is discussed more thoroughly in [4] in this volume.[24a]

Optimizing Conditions

Each cell line has different characteristics that will affect the uptake, distribution, and effectiveness of antisense ODNs, so several steps should

[12] N. O. Dulin, A. Sorokin, E. Reed, S. Elliott, J. H. Kehrl, and M. J. Dunn, *Mol. Cell Biol.* **19,** 714 (1999).

[13] L. Johannes, P. M. Lledo, P. Chameau, J. D. Vincent, J. P. Henry, and F. Darchen, *J. Neurochem.* **71,** 1127 (1998).

[14] K. M. Standifer, G. P. Brown, and G. W. Pasternak, *Soc. Neurosci. Abstr.* **22,** 70 (1996).

[15] M. W. Quick, M. I. Simon, N. Davisdon, H. A. Lester, and A. M. Agaray, *J. Biol. Chem.* **269,** 30164 (1994).

[16] T. Tang, J. G. Kiang, T. E. Cote, and B. M. Cox, *Mol. Pharmacol.* **48,** 189 (1995).

[17] C. Kleuss, J. Hescheler, C. Ewel, W. Rosenthal, G. Schultz, and B. Wittig, *Nature (London)* **353,** 48 (1991).

[18] C. Kleuss, H. Scherubl, J. Hescheler, G. Schultz, and B. Wittig, *Nature (London)* **358,** 424 (1992).

[19] C. Kleuss, H. Scherubl, J. Hescheler, G. Schultz, and B. Wittig, *Science* **259,** 832 (1993).

[20] T. E. Crowley, W. Nellen, R. H. Gomer, and R. A. Firtel, *Cell* **43,** 633 (1985).

[21] Y. F. Liu, K. H. Jakobs, M. M. Rasenick, and P. R. Albert, *J. Biol. Chem.* **269,** 13880 (1994).

[22] Y. Bessho, H. Nawa, and S. Nakanishi, *Neuron* **12,** 87 (1994).

[23] S. J. Morris, D. M. Beatty, and B. M. Chronwall, *J. Neurochem.* **71,** 1329 (1998).

[24] O. Ashur-Fabian, E. Giladi, D. E. Brenneman, and I. Gozes, *J. Mol. Neurosci.* **9,** 211 (1997).

[24a] G. W. Pasternak and Y.-X. Pan, *Methods Enzymol.* **314** [4] 1999 (this volume).

be taken to increase the likelihood of success. If utilizing a phosphodiester antisense ODN, it is necessary to wash the cells free of serum completely before adding antisense ODNs. Three to four washes in serum-free medium is usually sufficient. Serum contains DNases that will rapidly degrade the ODN and reduce its concentration in the extracellular medium,[25] thus affecting the final intracellular concentration of ODN. Chemical modifications (e.g., phosphorothioate) have produced ODNs that are more resistant to nucleases than phosphodiester ODNs, but these modifications can also increase nonspecific binding of the ODN to cell proteins.[26] Thus, care must be taken to ensure that the effects of the antisense ODN are due solely to a reduction of the targeted mRNA and its protein product. It is interesting to note that maintaining cultured cells in serum-free medium for several days may itself bring about changes in receptor expression in the absence of added ODNs. δ-Opioid receptor binding increased in vehicle-treated cells maintained in serum-free medium for 5 days, while μ-opioid receptor-binding levels decreased under similar conditions.[4,27] In the case of the δ receptor, increased expression was advantageous because the antisense treatment brought about a bigger drop, as protein synthesis was induced somewhat. However, it is difficult to demonstrate a significant effect of antisense on receptor levels that are already decreasing in the absence of antisense. In the latter case, cells could be maintained in medium containing serum to prevent the loss of receptors under control conditions, and nuclease-resistant antisense ODNs would need to be utilized to maintain stability in the presence of serum.

A concentration–effect curve should be generated for each antisense ODN used, to confirm that the ODN is specifically reducing the target mRNA or the protein product without reducing levels of a related protein or mRNA. This allows one not only to determine which concentration of antisense ODN is most effective, but to confirm that the antisense ODN is working as desired. One cannot assume that a measurable effect appearing after antisense treatment is due to a reduction in the expression of the targeted gene product; this must be demonstrated directly. This may be accomplished by quantifying changes in protein or mRNA levels by various means, including receptor binding, densitometric analysis of Western or Northern blots, or quantitative polymerase chain reaction (PCR). It is also helpful to show that the same effect can be produced with at least two or three different antisense ODN sequences directed against different regions

[25] S. Akhtar, R. Kole, and R. L. Juliano, *Life Sci.* **49,** 1793 (1991).
[26] D. A. Weidner, B. C. Valdez, D. Henning, S. Greenberg, and H. Busch, *FEBS Lett.* **366,** 146 (1995).
[27] K. M. Standifer and G. W. Pasternak, unpublished observations (1993).

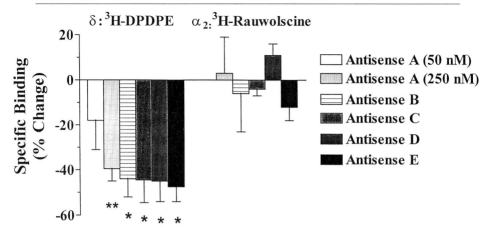

FIG. 1. Change in δ-opioid receptor binding in NG108-15 cells after a 5-day treatment with antisense to *DOR-1*. Binding in each assay was determined in triplicate as described,[8] replicated for a total of three to seven experiments per treatment, and presented as means $\pm$ SEM. Concentrations of antisense tested were 250 nM unless stated otherwise. Statistically significant changes in levels of receptor binding (*$p < 0.001$; **$p < 0.05$) were determined by analysis of variance. [Adapted from Ref. 8.]

of the same mRNA.[4,7] An example of this is illustrated in Fig. 1, where addition of five antisense ODNs (at two different concentrations) targeting different regions of the δ-opioid receptor (DOR-1)-coding sequence were used to downregulate δ-opioid receptor levels in NG108-15 cells.[8] In that study, decreases in specific radioligand ([3H]DPDPE) binding to the δ-opioid receptor as well as decreased δ-receptor immunoreactivity on a Western blot were used to demonstrate that δ-receptor levels were reduced. Consistent decreases in δ receptor-binding levels were attained at the higher concentration of antisense ODN (250 nM), but not with the 50 nM dose. Even larger reductions in δ-receptor levels were observed with a 1.25 μM concentration of antisense ODN, but it was unclear whether this reduction in receptor number was attributable directly to binding to the target mRNA or to its toxic effect on the cells, causing many to swell and lyse.[27] While all five antisense sequences were equally effective at reducing δ-receptor levels in this system,[8] this is not always the case.[28] To ensure that the effect was specific for the δ-opioid receptor, α_2-adrenoceptor binding was measured with [3H]rauwolscine (Fig. 1). No significant reductions in α_2 adrenoceptor binding levels were noted with any δ-opioid receptor-specific antisense DNA tested.[8]

[28] Y.-X. Pan, J. Cheng, J. Xu, G. Rossi, E. Jacobson, J. Ryan-Moro, A. I. Brooks, G. E. Dean, K. M. Standifer, and G. W. Pasternak, *Mol. Pharmacol.* **47**, 1180 (1995).

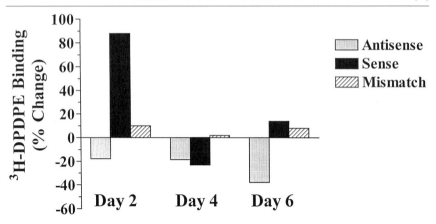

FIG. 2. Time-dependent loss of specific δ-opioid receptor binding after treatment with *DOR-1* antisense ODNs in a representative experiment. Binding to δ-opioid receptors was assessed in NG108-15 cell membranes with [^{3}H]DPDPE after a 1-, 3-, or 5-day treatment with 250 nM antisense A, sense A, or mismatch ODN, and was performed in triplicate. [Adapted from Refs. 8 and 27.]

The concentrations of antisense ODN needed will depend on the type of antisense ODN, the method of delivery, the cell type, and the number of copies of the target mRNA per cell. Levels of neurotransmitter receptors and G-protein α subunits expressed in cultured cells have been significantly reduced after treatment with phosphodiester or phosphorothioate antisense ODNs at concentrations ranging from 50 nM to 25 μM.[5,8,14,16,29] Because G proteins are normally expressed at higher levels than receptors, higher concentrations of antisense ODN may need to be used to produce the same level of reduction in G-protein levels that was attained with levels of the receptor.[3]

Knowledge of the turnover rate of the protein of interest is also important. It is not sufficient simply to stop translation; existing pools of the target protein must also be depleted before a reduction in protein levels and functional responses can be noted.[4,5,8] Figure 2 illustrates how slowly δ-opioid receptor levels decline, by comparing their change from vehicle-treated controls on days 2, 4, and 6 in a representative experiment in NG108-15 cells.[8,27] Only partial downregulation was achieved by days 2 and 4, but a considerable reduction in binding was seen by day 6, consistent with the appearance of functional analgesic deficits in the whole animal.[8]

[29] J. Lai, T. J. Crook, A. Payne, R. M. Lynch, and F. Porreca, *J. Pharmacol. Exp. Ther.* **281,** 589 (1997).

While it may require several days for preexisting neurotransmitter receptor pools to undergo degradation,[4,8] all three types of G-protein subunits appear to turn over completely within 2 days.[17–19]

If none of the antisense ODNs appear to reduce levels of the intended mRNA or protein, it may be necessary to examine the ability of the antisense to be taken up by the cells used in the study. Cellular uptake of ODN is time and concentration dependent. In the absence of additional drug or chemical treatments, it appears to follow a similar time course in two different types of neuronal cultures (primary cerebellar cultures and NG108-15 neurohybrid cells).[5,8,29] In all three studies, intracellular accumulation of antisense ODNs reached maximum levels by 24 hr. From that point, however, ODN levels differed. Levels of ^{32}P-tagged antisense ODNs remained constant in NG108-15 cells for up to 72 hr,[8] while in primary cerebellar cultures antisense ODN levels began to decrease after 24 hr.[5] Therefore, while neuronal cell lines appear to take up ODNs at approximately the same rate, their stability and disposition within the cells vary. It is not necessary to use a radioactive tag to follow the uptake and stability of the antisense ODN; the ODNs also may be conjugated to fluorescent tags, biotin, or digoxigenin to allow visualization.[29,30]

Validation of Antisense Oligodeoxynucleotide Specificity

While it is essential to demonstrate that the antisense ODN has reduced levels of the mRNA or protein of interest, it is also incumbent on the investigator to confirm that the effects are specific for the antisense sequence itself, and not its structure or charge. To this end, control experiments including the corresponding sense sequence(s), antisense sequence(s) in which three or four bases have been altered (mismatched antisense), and/ or a nonsense sequence that contains the same number and composition bases (but in random order) need to be included in the experimental design.[4,7] The use of mismatched antisense or nonsense controls may be preferable to a sense control because sense sequences have been shown to produce effects themselves.[26] For example, the representative time course of δ-receptor downregulation in Fig. 2 also demonstrates the effects of sense and mismatch treatments on receptor binding on days 2, 4, and 6.[27] Although the general trend of the antisense and mismatch ODN treatment was consistently in the same direction at all time points, changes resulting from sense ODN treatment varied wildly. This is not always the case, but one should interpret results from sense experiments with caution.

[30] F. Yee, H. Ericson, D. J. Reis, and C. Wahlestedt, *Cell. Mol. Neurobiol.* **14,** 475 (1994).

In Vivo Approaches

Experimental Design

As indicated above, the most important advantage of utilizing antisense ODNs in an animal rather than a cultured cell or isolated tissue system is that the circuitry is intact. After administering the ODN, one can study biochemical, physiological, and behavioral changes brought about by the treatment. One well-controlled approach is to administer the antisense ODN (or control ODN) into a region or ventricle on one side of the animal brain, and use the other side of the brain as its own control.[31] In this manner, Roberts and colleagues[31] demonstrated that ocular dominance plasticity, but not visual responsiveness, could be blocked by suppressing NMDA receptor function after a 5-day treatment with antisense ODN complementary to the NR1 subunit of the receptor.

Using a stereotaxic apparatus, cannulas can be implanted into many different brain regions.[32] Antisense ODNs delivered into discrete brain regions in small volumes will produce a localized loss of target protein,[4] while antisense ODNs delivered via injection into a lateral ventricle will likely flow through the brain in the cerebrospinal fluid (CSF) throughout the ventricles, achieving a greater distribution.[30] Fortunately, CSF does not contain as much nuclease activity as blood and serum,[33] and a small percentage of phosphodiester antisense ODNs directed against *DOR-1* administered directly into the CSF through the intrathecal space remained intact for up to 72 hr.[9] Although the majority of the ODNs were probably degraded, previous work by Yee *et al.*[30] indicated that ODNs are taken up from the CSF by neurons quickly, with detectable levels appearing within 15 min after administration. Therefore, despite extensive degradation, antisense ODNs were still present in a sufficient molar excess to bring about a 25–30% reduction in DOR-1 mRNA and δ-receptor binding, as well as a >50% loss of δ-opioid receptor-mediated spinal analgesia.[8,9] Similar results were obtained by others, after both intrathecal and intracerebroventricular administration.[34,35] Further, decreasing $G_i\alpha_2$ levels with phosphodiester antisense ODNs blocked morphine-mediated analgesia and reduced $G_i\alpha_2$ protein levels to the same extent as treatment with phosphorothioate anti-

[31] E. B. Roberts, M. A. Meredith, and A. S. Ramoa, *J. Neurophysiol.* **80,** 1021 (1998).

[32] G. C. Rossi, G. W. Pasternak, and R. J. Bodnar, *Brain Res.* **624,** 171 (1993).

[33] L. Whitesell, D. Geselowitz, C. Chavany, B. Fahmy, S. Walbridge, J. R. Alger, and L. M. Neckers, *Proc. Natl. Acad. Sci. U.S.A.* **90,** 4665 (1993).

[34] E. J. Bilsky, T. Wang, J. Lai, and F. Porreca, *Neurosci. Lett.* **220,** 155 (1996).

[35] E. J. Bilsky, R. N. Bernstein, V. J. Hruby, R. B. Rothman, J. Lai, and F. Porreca, *J. Pharmacol. Exp. Ther.* **277,** 491 (1996).

sense ODNs.[10,11,36,37] Therefore, although phosphorothioate-modified ODNs may be more stable in CSF than phosphodiester ODNs, the same results can be obtained by using unmodified ODNs, with less likelihood of toxicity.[33]

As described above, neurotransmitter receptors from the 7TM super-family[8,9] as well as the heteromeric ionotropic superfamily[31,38,39] have been successfully "knocked out" by antisense ODN treatment *in vivo* after administration into the brain. This disruption of receptor function has allowed investigators to study a wide range of neuroscience topics including the role of NMDA-R1 receptors in excitotoxicity and ischemia,[38] the selectivity of the γ_2 subunit of the GABA$_A$ receptor for benzodiazepine, but not GABA or muscimol binding,[39] and confirming the roles of type I metabotropic glutamate receptors in nociception[40] and the δ-opioid receptor in antinociception.[8,34,35] Utilizing the antisense ODN approach for studying the role of metabotropic glutamate receptors in nociception was important because no selective antagonists for that receptor subtype are available.[40] The antisense ODN approach allowed investigators to delineate the contribution of a single subunit of the heteromeric GABA$_A$ receptor to benzodiazepine, not GABA, binding without having to resort to labor-intensive molecular biological techniques to achieve the same goal.[39]

As noted above, administration of antisense ODNs directed against specific G-protein subunits helped to determine the specificity of coupling between different neurotransmitter receptors and G-protein subunits at the cellular level.[14–21] By administering antisense ODNs into the brain and/or spinal cord, one can determine the contribution distinct G-protein subunits make to a particular response and perhaps delineate between responses made by two closely related agonists.[10,11,41] For example, antisense ODNs directed against various G-protein α subunits selectively disrupt analgesia mediated through distinct opioid receptor subtypes.[10,11,36,37,41] Supraspinal and spinal morphine analgesia was mediated through the G$_i\alpha_2$ subunit as analgesia was blocked by administration of antisense ODN directed against G$_i\alpha_2$ mRNA. However, the same antisense treatment had no effect on analgesia mediated by the potent morphine metabolite mor-

[36] R. B. Raffa, R. P. Martinez, and C. D. Connelly, *Eur. J. Pharmacol.* **258,** R5 (1994).
[37] P. Sanchez-Blazquez, A. Garcia-Espana, and J. Garzon, *J. Pharmacol. Exp. Ther.* **275,** 1590 (1995).
[38] C. Wahlestedt, E. Glanov, S. Yamamoto, F. Yee, H. Ericson, H. Yoo, C. E. Inturrisi, and D. J. Reis, *Nature (London)* **363,** 260 (1993).
[39] T. J. Zhao, M. Li, T. H. Chiu, and H. C. Rosenberg, *J. Pharmacol. Exp. Ther.* **287,** 752 (1998).
[40] M. R. Young, G. Blackburn-Munro, T. Dickinson, M. J. Johnson, H. Anderson, I. Nakalembe, and S. M. Fleetwood-Walker, *J. Neurosci.* **18,** 10180 (1998).
[41] P. Sanchez-Blazquez and J. Garzon, *J. Pharmacol. Exp. Ther.* **285,** 820 (1998).

phine 6β-glucuronide (M6G).[10,11] This finding is interesting because it indicates that morphine and M6G may mediate analgesia through distinct pathways, or pathways that diverge before $G_i\alpha_2$ is activated. In the absence of a selective M6G antagonist, ablation of function with antisense ODNs is a useful tool with which to explore further the differences between morphine and M6G actions and pathways.

An important distinction between administering antisense into the brains of rats versus mice is that a stereotaxic apparatus is not required to prepare mice for intracerebroventricular injections,[42] as it is for rats. Utilizing mice obviates the need for surgical implantation of cannulas and the need to allow several days to a week for the animal to recover from surgery before beginning the experiment. However, one disadvantage of using mice is that one is limited to intracerebroventricular or intrathecal injections, and cannot achieve more localized delivery within distinct brain regions.

Optimizing Conditions

Optimal antisense ODN concentration and treatment time must be determined for *in vivo* studies as well because the response is both dose and time dependent. By increasing one of these parameters, it may be possible to decrease the other. For instance, δ-opioid receptor-mediated analgesia and receptor binding were decreased to approximately the same extent after two different antisense treatment paradigms. It made no difference whether the antisense was administered by single injections (5 μg/2 μl) on days 1, 3, and 5,[8] or whether it was administered twice daily (12.5 μg/4 μl) for three consecutive days.[34] In both studies, animals recovered normal analgesic responsiveness within 3 to 4 days after the phosphodiester antisense treatment was stopped, indicating that the treatment was reversible and not neurotoxic.[8,34] Significantly, the length of time required for δ-opioid receptor binding and function to be reduced was the same *in vivo* as it was *in vitro*,[8] indicating that the antisense effects are consistent between systems.

For studies in which the antisense ODN must be administered several times to achieve a desired decrease in protein levels, it may be advantageous to deliver the ODN via osmotic minipump.[40,43] The minipumps may be connected to indwelling cannulas in the brain[44] or in the spinal cord.[40] The advantage of minipump delivery is that the animal receives a constant infusion of the antisense ODN, allowing levels to be maintained throughout the course of treatment. Provided the flow rate is appropriate for the area infused, little or no tissue damage should occur. In the metabotropic

---

[42] T. J. Haley and W. G. McCormick, *Br. J. Pharmacol.* **12,** 12 (1957).
[43] S. Chun, A. Niijima, T. Shima, M. Okada, and K. Nagai, *Neurosci. Lett.* **257,** 135 (1998).
[44] L. Levesque and S. T. Crooke, *J. Pharmacol. Exp. Ther.* **287,** 425 (1998).

glutamate receptor/nociception study, the cannula was placed directly above the superficial layer of the dorsal horn. Antisense ODN infusion produced a 35% loss of receptor immunoreactivity in lamina II of the dorsal horn of the spinal cord, with no alterations of receptor levels in the motor neurons in the ventral horn.[40]

Validation of Antisense Specificity

The increased number of variables associated with animal studies, compared with cultured cell systems, makes it even more important to confirm that the effects seen are specific for the antisense ODN treatment. Of course, sense, mismatched antisense, and nonsense controls should be employed *in vivo* as they are *in vitro*. As with any study involving cannula implantation, correct placement cannot be assumed and must be confirmed. Because antisense ODNs form something of a gradient as they are taken up by neurons within the brain, structures more distal to the site of injection will likely receive less ODN than those regions located more proximally.[30] If the ODN is delivered into a discrete region such as the periaqueductal gray (PAG) or lumbar section of the spinal cord, it is possible to dissect that tissue and assay it directly for protein loss.[8,10] If injections are directed into a ventricle, diffusion of the ODN will be greater. After intracerebroventricular injection of antisense ODNs directed against several G-protein α subunits, Sanchez-Blazquez and colleagues dissected various brain regions (PAG, striatum, thalamus, and hypothalamus) to assess the extent of G-protein reduction in each of those areas by Western blot analysis.[37] As one might have predicted, the greatest loss of immunoreactivity was noted in areas most proximal to the ventricle (PAG and hypothalamus), whereas levels in the thalamus were unchanged.[37] In addition to assessing changes in the protein of interest, it is helpful to confirm that levels of other closely related proteins are not affected by the same antisense ODN treatment (Fig. 3). In Fig. 3, changes in immunoreactivity to $G_i\alpha_1$, $G_i\alpha_2$, $G_i\alpha_3$, $G_o\alpha$, and $G_s\alpha$ were assessed by Western blotting of PAG membrane homogenates from mice treated with antisense ODNs directed against the mRNA for those same G proteins.[11] Figure 3A illustrates the most thorough method of confirming antisense ODN specificity: ensuring that $G_o\alpha$ levels were not significantly altered by treatment with antisense ODNs directed against similar G-protein α subunits, but only by $G_o\alpha$ antisense ODN treatment. As indicated in Fig. 3B–D, this is not an absolute requirement.

Conclusions

Antisense ODNs have provided a new avenue by which we may explore the subtle events initiated by and through receptor–effector coupling. This

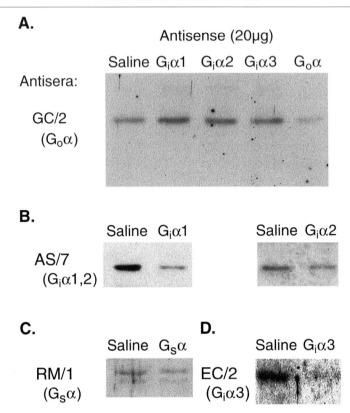

A.

Antisense (20µg)

Saline $G_i\alpha1$ $G_i\alpha2$ $G_i\alpha3$ $G_o\alpha$

Antisera:

GC/2
($G_o\alpha$)

B.

Saline $G_i\alpha1$ Saline $G_i\alpha2$

AS/7
($G_i\alpha1,2$)

C.   **D.**

Saline $G_s\alpha$ Saline $G_i\alpha3$

RM/1   EC/2
($G_s\alpha$)   ($G_i\alpha3$)

FIG. 3. Effect of antisense ODN treatment on G-protein α-subunit levels in membranes from mouse periaqueductal gray (PAG). Mice received single intracerebroventricular injections of antisense ODN (20 μg/2 μl) or saline as indicated above each gel, and membranes from the PAG region were prepared 48 hr later. Membrane proteins (10 μg) were solubilized in SDS, resolved by SDS–PAGE (10%) as described, and probed with G protein α subunit-selective antisera, as indicated to the left of each gel. Bands from the gels were digitized and quantitated by densitometric analysis, and treated groups were compared with the appropriate control by t test. (A) $G_o\alpha$ levels were reduced by 72.9 $\pm$ 13.6% after $G_o\alpha$ antisense treatment ($p <$ 0.01; $n = 5$), while antisense targeting of the other G proteins had no significant effect on $G_o\alpha$ levels. (B) $G_i\alpha_1$ and $G_i\alpha_2$ levels were reduced 74.2 $\pm$ 18% ($p < 0.05$; $n = 3$) and 65.1 $\pm$ 6.5% ($p < 0.001$; $n = 5$) after treatment with their respective antisense. (C) $G_s\alpha$ levels as indicated by the upper immunoreactive band were decreased by 74 $\pm$ 11.7% ($p < 0.01$; $n = 4$), while the lower band was not significantly affected (21.7 $\pm$ 23.4% decrease). (D) $G_i\alpha_3$ levels were depleted by 86.5 $\pm$ 7% ($p < 0.001$; $n = 5$) after treatment with $G_i\alpha_3$ antisense. [From Ref. 11.]

approach is especially useful when other tools such as selective antagonists or antisera are not available to block actions mediated through these proteins. Although beyond the scope of this chapter, it is important to note that antisense ODNs also have been generated against other components of downstream signaling pathways, and were quite successful at ablating function. Proteins as functionally diverse as protein kinase C-α^{44} and cAMP response element-binding protein[45] were depleted and their activities diminished following antisense ODN treatment *in vitro* and *in vivo,* respectively. The utility of the antisense ODN approach extends from the receptor into the nucleus.

Acknowledgment

The author gratefully acknowledges support from the National Institute on Drug Abuse (DA10738).

[45] S. B. Lane-Ladd, J. Pineda, V. A. Boundy, T. Pfeuffer, J. Krupinski, G. K. Aghjanian, and E. J. Nestler, *J. Neurosci.* **17,** 7890 (1997).

[8] Use of Expression of Antisense mRNA for Proprotein Convertases 1 and 2 in Prohormone Processing

By DAESETY VISHNUVARDHAN and MARGERY C. BEINFELD

Most of the peptide hormones are synthesized as large precursors that are proteolytically processed during their intracellular transport to produce the biologically active molecules. The processing of prohormones involves cleavages at specific basic amino acids by endoproteases such as members of the subtilisin-like serine endoprotease family in a strict temporal order. Exoproteolysis is followed further, in some cases, by the action of amino- and carboxypeptidases to remove terminal arginine and lysine residues.[1] Finally, in the case of amidated peptides the carboxyl-terminal glycine is converted into its amide by the amidating enzyme.[2] Prohormone processing is a tightly regulated process that takes place in multiple compartments of the secretory pathway.

[1] M. C. Beinfeld, *Endocrine* **8,** 1 (1998).
[2] B. A. Eipper, S. L. Milgram, E. J. Husten, H.-Y. Yun, and R. E. Mains, *Protein Sci.* **2,** 489 (1993).

A typical prohormone is cotranslationally inserted into the endoplasmic reticulum (ER) on the basis of the ability of the signal recognition particle to target the signal sequence of the growing peptide chain to the translocation complex in the ER membrane. The signal peptidase on the lumenal side of the ER cleaves the signal peptide from the prohormone, releasing it into the lumen of the ER. There, with or without the assistance of chaperones, it folds into its correct conformation. Incorrectly folded proteins are either degraded in the ER or exported from the ER to the cytosol through the translocation complex, where they are ubiquinated and degraded by proteosomes.[3] More than one endoprotease may cleave the prohormone in a strict temporal order before it is secreted from the cell. Prohormone cleavages occur mainly at specific paired basic sites, although some single and tetrabasic sites are also cleaved.

Some of the specific endoproteases that carry out these cleavages belong to the family of subtilisin-like proprotein convertases, or SPCs.[4] A major advance in the prohormone processing field was the discovery of the proprotein convertase (PC) family of enzymes, which now number seven.[5,6] These mammalian enzymes include PC1/PC3,[7,8] PC2,[9] furin/PACE,[10,11] PC4,[12] PACE4,[13] PC5/6,[14,15] and PC7/SPC7/LPC/PC8.[16] Detailed examination of the intracellular localization, catalytic activity, and tissue distribution of these PCs has yielded invaluable information.

[3] M. M. Hiller, A. Finger, M. Schweiger, and D. H. Wolf, *Science* **273,** 1725 (1996).

[4] N. Seidah and M. Chretien, *Methods Enzymol.* **244,** 175 (1994).

[5] D. F. Steiner, S. P. Smeekens, S. Ohagi, and S. J. Chan, *J. Biol. Chem.* **267,** 23435 (1992).

[6] Y. Rouille, S. J. Duguay, K. Lund, M. Furuta, Q. Gong, G. Lipkind, A. A. Oliva, Jr., S. J. Chan, and D. F. Steiner, *Front. Neuroendocrinol.* **16,** 322 (1995).

[7] S. P. Smeekens, A. S. Avruch, J. LaMendola, S. J. Chan, and D. F. Steiner, *Proc. Natl. Acad. Sci. U.S.A.* **88,** 340 (1991).

[8] N. G. Seidah, L. Gaspar, P. Mion, M. Marcinkiewicz, M. Mbikay, and M. Chretien, *DNA Cell Biol.* **9,** 415 (1990).

[9] S. P. Smeekens and D. F. Steiner, *J. Biol. Chem.* **265,** 2997 (1990).

[10] A. M. van den Ouweland, H. L. Van Duijnhoven, G. D. Keizer, L. C. Dorssers, and W. J. Van de Ven, *Nucleic Acids Res.* **18,** 664 (1990).

[11] R. J. Wise, P. J. Barr, P. A. Wong, M. C. Kiefer, A. J. Brake, and R. J. Kaufman, *Proc. Natl. Acad. Sci. U.S.A.* **87,** 9378 (1990).

[12] K. Nakayama, W.-S. Kim, S. Tori, M. Hosaka, T. Nakagawa, J. Ikemizu, T. Baba, and K. Murakami, *J. Biol. Chem.* **267,** 5897 (1992).

[13] M. C. Kiefer, J. E. Tucker, R. Joh, K. E. Landsberg, D. Saltman, and P. J. Barr, *DNA Cell Biol.* **10,** 757 (1991).

[14] J. Lusson, D. Vieau, J. Hamelin, R. Day, M. Chretien, and N. G. Seidah, *Proc. Natl. Acad. Sci. U.S.A.* **90,** 6691 (1993).

[15] T. Nakagawa, M. Hosaka, S. Torii, T. Watanabe, K. Murakami, and K. Nakayama, *J. Biochem.* **113,** 132 (1993).

[16] A. Tsuji, C. Hine, K. Mori, Y. Tamai, K. Higashine, H. Nagamune, and Y. Matsuda, *Biochem. Biophys. Res. Commun.* **202,** 1452 (1994).

Most of these enzymes cleave at dibasic pairs, such as Arg–Lys; some monobasic sites are also cleaved,[17,18] whereas furin cleaves single basic sites preferentially with an upstream arginine with a consensus sequence Arg–X–Lys–Arg or Arg–X–Arg–Arg.[19] Some of these enzymes are sorted and activated in the regulated secretory pathway (PC1, PC2, PC4, and PC5/6) and thus are candidates for processing prohormones whose secretion is regulated. Three members of this family, PC1, PC2, and PC5/6, are expressed in neural and endocrine tissues and are active in the regulated secretory pathway of these cells.[7,8,20] PC1 and PC2 have been implicated in the processing of a growing number of prohormone and proneuropeptides.

PC1 and PC2 are widely distributed in neural and endocrine tissues and cell lines, and have been shown to be involved in the processing of prohormones, such as proopiomelanocortin (POMC),[21–23] cholecystokinin (CCK),[18,24,25] insulin,[26] glucagon,[27] gastrin,[28] dynorphin,[17] enkephalin,[29] thyrotropin-releasing hormone (TRH),[30] neurotensin,[31] and somatostatin,[32,33] in the regulated pathway. PC5/6 is a good candidate for prohormone pro-

[17] A. Dupuy, I. Lindberg, Y. Zhou, H. Akil, C. Lazure, M. Chretien, N. Seidah, and R. Day, *FEBS Lett.* **337,** 60 (1994).

[18] W. Wang and M. C. Beinfeld, *Biochem. Biophys. Res. Commun.* **231,** 149 (1997).

[19] J. B. Denault and R. Leduc, *FEBS Lett.* **379,** 113 (1996).

[20] I. DeBie, M. Marcinkiewicz, D. Malide, C. Lazure, K. Nakayama, M. Bendayan, and N. G. Seidah, *J. Cell. Biol.* **135,** 1261 (1996).

[21] L. Thomas, R. Leduc, B. A. Thorne, S. P. Smeekens, D. F. Steiner, and G. Thomas, *Proc. Natl. Acad. Sci. U.S.A.* **88,** 5297 (1991).

[22] S. Benjannet, N. Rondeau, R. Day, M. Chretien, and N. G. Seidah, *Proc. Natl. Acad. Sci. U.S.A.* **88,** 3564 (1991).

[23] A. Zhou, B. T. Bloomquist, and R. E. Mains, *J. Biol. Chem.* **268,** 1763 (1993).

[24] J. Y. Yoon and M. C. Beinfeld, *J. Biol. Chem.* **272,** 9450 (1997).

[25] J. Y. Yoon and M. C. Beinfeld, *Endocrinology* **138,** 3620 (1997).

[26] S. P. Smeekens, A. G. Montag, G. Thomas, C. Albiges-Rizo, R. Carroll, M. Benig, L. A. Phillips, S. Martin, S. Ohagi, P. Gardner, H. H. Swift, and D. F. Steiner, *Proc. Natl. Acad. Sci. U.S.A.* **89,** 8822 (1992).

[27] Y. Rouille, G. Westermark, S. K. Martin, and D. F. Steiner, *Proc. Natl. Acad. Sci. U.S.A.* **91,** 3242 (1994).

[28] C. J. Dickinson, M. Sawada, Y. J. Guo, S. Finniss, and T. Yamada, *J. Clin. Invest.* **96,** 1425 (1995).

[29] M. B. Breslin, I. Lindberg, S. Benjannet, S. Mathis, C. Lazure, and N. G. Seidah, *J. Biol. Chem.* **268,** 27084 (1993).

[30] E. A. Nillni, T. C. Friedman, R. B. Todd, N. P. Birch, Y. P. Loh, and I. M. Jackson, *J. Neurochem.* **65,** 2462 (1995).

[31] C. Rovere, P. Barbero, and P. Kitabgi, *J. Biol. Chem.* **271,** 11368 (1996).

[32] A. S. Galanopoulou, G. Kent, S. N. Rabbani, N. G. Seidah, and Y. C. Patel, *J. Biol. Chem.* **268,** 6041 (1993).

[33] A. S. Galanopoulou, N. G. Seidah, and Y. C. Patel, *Biochem. J.* **309,** 33 (1995).

cessing in the brain and intestine.[14,15,34] These enzymes undergo autoactivation with the removal of the propeptide. PC1 activation is thought to occur in the ER at about pH 7–8 with no additional calcium, whereas PC2 is activated much more slowly in the trans-Golgi network at about pH 5.5–6.0 and millimolar calcium concentration.[35] This difference in the time course of activation may explain why PC1 frequently cleaves prohormones before PC2.

The question of which of these many enzymes are responsible for individual prohormone cleavages in tissues is difficult to answer. The possibility that there is considerable redundancy in their activities and that different enzymes are active on the same prohormone in different tissues cannot be excluded. In terms of central nervous system (CNS) peptides, there are at least three enzymes (PC1, PC2, and PC5/6) that have a distribution[36–38] that is consistent with a role in processing a number of proneuropeptides. Oxytocin as well as CCK is colocalized with PC1, PC2, and PC5/6 in supraoptic and paraventricular nuclei of the hypothalamus. The colocalization of CCK with these three enzymes appears to hold throughout the rat CNS.

To address which of these enzymes are responsible for prohormone cleavage in specific endocrine cells, a plasmid-based antisense strategy has been used. This is necessary because there are no specific, nontoxic inhibitors that inhibit individual members of this group specifically. Plasmid-based antisense methods use expression vectors, which generate RNA sequences complementary to the key regions of specific genes. This strategy, in which PC1 or PC2 cDNAs are expressed stably in endocrine cells in the antisense orientation, has provided evidence that PC1 and PC2 are involved in POMC,[39] pro-CCK,[24,25,40] proenkephalin,[41] and glucagon[27] processing. This is a successful strategy in general, but complete inhibition of PC expression with antisense methods has been difficult to achieve. It is a technique limited to cells that can be transfected or infected. The focus of this chapter is the

[34] T. Nakagawa, K. Murakami, and K. Nakayama, *FEBS Lett.* **327,** 165 (1993).

[35] K. I. Shennan, N. A. Taylor, J. L. Jermany, G. Matthews, and K. Docherty, *J. Biol. Chem.* **270,** 1402 (1995).

[36] M. K. H. Shafer, R. Day, W. E. Cullinan, M. Chretien, N. G. Seidah, and S. J. Watson, *J. Neurosci.* **13,** 1258 (1993).

[37] I. Lemaire, O. Piot, B. P. Roques, G. A. Bohme, and J. C. Blanchard, *Neuroreport* **3,** 929 (1992).

[38] W. Dong, B. Seidel, M. Marcinkiewicz, M. Chretien, N. G. Seidah, and R. Day, *J. Neurosci.* **17,** 563 (1997).

[39] B. T. Bloomquist, B. A. Eipper, and R. E. Mains, *Mol. Endocrinol.* **5,** 2014 (1991).

[40] J. Y. Yoon and M. C. Beinfeld, *Regul. Peptides* **59,** 221 (1995).

[41] K. Johanning, J. P. Mathis, and I. Lindberg, *J. Neurochem.* **66,** 898 (1996).

application of this methodology to the study of the role of proprotein convertases PC1 and PC2 in pro-CCK processing.

Methods

Construction of Antisense Expression Plasmids

Constitutively expressing antisense PC1 plasmid PCMV5/anti-PC1 is constructed using the first 491 base pairs of the PC1 cDNA insert from prPC1.491EX. This PC1 fragment is ligated into the pCMV5 mammalian expression vector[42] in the antisense orientation in the HindIII and XbaI sites so that the cytomegalovirus promoter drives its expression.[43] The antisense PC2 plasmid pCMV5/anti-PC2 is constructed using the first 480 base pairs of the PC2 cDNA contained within prPC2.480EK. This PC2 fragment is cloned into pCMV5 at the KpnI and XbaI sites in the antisense orientation. The orientation of the inserts is confirmed by restriction digestion. Plasmids prPC1.491.EX and prPC2.480EK are kindly provided by R. Mains (The Johns Hopkins University School of Medicine, Baltimore, MD). In our plasmid-based antisense experiments we target the proregion of the protease because it is least conserved between different members of this group of endoproteases. Besides, many studies have shown that targeting the N-terminal region produces a better antisense effect than other regions. Although there are no hard and fast rules on deciding the ideal size of the antisense fragment, it is possible that smaller fragments tend to degrade and dissociate faster, and longer fragments may be unstable and may not be efficient in hybridizing.

Maintenance and Transfection of Tissue Culture Cells

STC-1 cells are maintained in Dulbecco's modified Eagle's medium (DMEM) containing 20% (v/v) newborn calf serum, 10% (v/v) horse serum, and 1× penicillin and streptomycin or 1× gentamicin in a humidified 37° incubator at 5% (v/v) CO_2. Cells transfected with expression plasmids were maintained in the same medium with 300 μg/ml G418. Cells were split twice weekly as they reached confluency by trypsinization.

Cells are grown to 80% confluency, trypsinized, pelleted, and resuspended to 5×10^6 cells/0.8 ml of growth medium, along with 5–50 μg of plasmid DNA in a Bio-Rad (Richmond, CA) 0.4-cm electroporation cu-

---

[42] S. Andersson, D. L. Davis, H. Dahlback, H. Jornvall, and D. W. Russell, *J. Biol. Chem.* **264**, 8222 (1989).

[43] J. Sambrook, E. F. Fritsch, and T. Maniatis, "Molecular Cloning: A Laboratory Manual." Cold Spring Harbor Laboratory Press, Cold Spring Harbor, New York, 1989.

vette. The cuvette is placed on ice for 10 min prior to electroporation and again for 10 min after electroporation. Plasmids pCMV5/anti-PC1 and pCMV5/anti-PC2 are cotransfected with pMtNeo, which confers resistance to the antibiotic G418, in a molar ratio of 3:1 and 5:1, respectively, by electroporation at 200 V/500 μF (microfarads), using the Bio-Rad Gene Pulser exponential decay-type electroporator. The cells are allowed to recover for 48 hr in normal growth medium and then diluted into 96-well plates containing the appropriate antibiotic for selection. Single colonies are isolated and established as stable cell lines. Initially, antisense-transfected cells grow slower compared with nontransfected cells owing to the expression of the antisense fragment; however, after a few generations their growth rate increases.

Northern Blot Analysis

Total RNA is isolated from cells with guanidine isothiocyanate,[44] and 20–40 μg is used for analysis. Polyadenylated mRNA is isolated with a Ribosep RNA isolation kit from Collaborative Biomedical Products (Bedford, MA) and 5 μg is used for analysis. RNA is fractionated by electrophoresis through a 2.2 M formaldehyde–1% agarose gel in 1× morpholine propanesulfonic acid (MOPS) buffer, transferred overnight to a Magna nylon filter (Micron Separations, Westboro, MA) in 10× SSC transfer buffer (1× SSC is 0.15 M NaCl plus 0.015 M sodium citrate), cross-linked to the filter by ultraviolet (UV), and placed with 10 ml of hybridization solution in a glass bottle set in a hybridization oven at 42°. Prehybridization and hybridization solution consist of 50% (v/v) formamide, 5× SSC, 5× Denhardt's, 0.5% (w/v) sodium dodecyl sulfate (SDS), and 100 μg of denatured salmon sperm DNA/ml.

The radioactive hybridization probe is produced using the same fragments of PC1 or PC2 cDNA described in the antisense experiments as templates for random primed labeling (RPL), using the RPL kit from New England BioLabs (Beverly, MA). These double-stranded probes allow detection of both the sense and antisense RNA. The labeled probe is separated from unincorporated nucleotides through a Sephadex G-50 column, denatured along with 300 μg of salmon sperm DNA by heating in a boiling water bath, and placed into the hybridization bottle at a specific activity of 1×10^6 cpm/ml. Filters are washed within the hybridization bottles twice at room temperature with 2× SSC–0.1% (w/v) SDS for 10 min; twice at room temperature with 0.2× SSC–0.1% (w/v) SDS for 10 min; and twice at 42° with 0.1× SSC–0.1% (w/v) SDS for 15 min. Autoradi-

[44] P. Chomczynski and N. Sacchi, *Anal. Biochem.* **162,** 156 (1987).

ography was performed with an intensifying screen at $-80°$ for several hours to days as appropriate.

Western Blot Analysis

Tissue culture cells are grown to near confluency and harvested for protein analysis by scraping from dishes with 1 ml of phosphate-buffered saline (PBS), after which an aliquot is taken for total protein measurement by the Lowry assay.[45] After centrifugation, cells are extracted by resuspending in $1\times$ SDS protein loading buffer [25 mM Tris-HCl (pH 6.8), 2% (w/v) SDS, 10% (v/v) glycerol, 5% (v/v) 2-mercaptoethanol, 0.001% (w/v) bromphenol blue], sonication, and boiling for 5 min. Protein samples are fractionated by electrophoresis at 100 V for 2 hr through a 10% (w/v) polyacrylamide gel in protein electrophoresis buffer [0.1% (w/v) SDS, 0.025 M Tris, 192 mM glycine] using a protein minigel apparatus (Bio-Rad). The fractionated proteins are electroblotted onto nitrocellulose membrane in transfer buffer [25 mM Tris, 192 mM glycine, 20% (v/v) methanol] at $4°$ overnight at 30 V, using a minigel Trans-Blot apparatus (Bio-Rad). The nitrocellulose blot is blocked using 5% (w/v) Carnation nonfat dried milk in TBS [50 mM Tris, 200 mM NaCl, (pH 7.4)] containing 0.02% (w/v) sodium azide, and then incubated with a 1:1000 dilution of the primary antibody in 10 ml of 5% (w/v) Carnation nonfat dried milk within a sealed bag at $4°$ overnight with constant shaking. Polyclonal antibodies against PC1 and PC2 are generously provided by I. Lindberg (Louisiana State University Medical Center, New Orleans, LA). The membrane is then washed three times with 50 ml of TBS with 0.05% (v/v) Tween 20 for 10 min each and incubated with a goat anti-rabbit antibody conjugated to either alkaline phosphatase or horseradish peroxidase in 10 ml of 5% (w/v) Carnation nonfat dried milk in TBS with 0.02% (w/v) sodium azide within a sealed bag at room temperature for 2 hr. Antibodies conjugated to alkaline phosphatase (Bio-Rad) and horseradish peroxidase (Bio-Rad) are diluted 1:2000 and 1:5000, respectively. Visualization by alkaline phosphatase is performed using substrates nitroblue tetrazolium and 5-bromo-4-chloro-3-indolyl phosphate (Life Technologies, Gaithersburg, MD) at a ratio of 1:1 in visualization buffer [50 mM Tris, 3 mM MgCl$_2$ (pH 10.0)] for 30 min. Visualization by enzyme chemiluminescence, using the horseradish peroxidase enzyme, is performed according to the manufacturer protocol (ECL Western detection kit; Amersham Life Science, Arlington Heights, IL). Quantitation of protein levels in Western blots is performed by densitometry, using Image Quant software.

[45] O. H. Lowry, N. J. Rosenbrough, A. L. Farr, and R. J. Randall, *J. Biol. Chem.* **193**, 265 (1951).

Radioimmunoassay

The radioimmunoassays (RIAs) to measure CCK 8 from cell extracts are performed as described previously,[46] using the rabbit polyclonal CCK antibody (R5) that is specific for the amidated forms of CCK. The RIA utilizes ^{125}I-labeled gastrin-17 produced by iodination by the chloramine-T method.[47]

Chromatography

Cells from four to eight 10-cm plates are extracted with 0.1 N HCl, pooled, and concentrated by vacuum centrifugation. Cell extracts are neutralized and separated by Sephadex G-50 chromatography in a 35 × 1 cm column run at 4° in 50 mM Tris–100 mM NaCl, pH 7.8, containing 0.1% bovine serum albumin (BSA) and 0.05% (w/v) sodium azide. Fractions of 1.0 ml are collected and aliquots are removed for the CCK RIA.

Cholecystokinin Biosynthesis and Its Processing

Cholecystokinin (CCK) is produced by endocrine cells in the gut and by neurons in the brain. It is released from the intestine following the ingestion of food and causes the contraction of the gall bladder and stimulates the release of digestive enzymes from the pancreas.[48,49] CCK is found in higher concentrations in the brain than in the gut, where it serves as a neurotransmitter[50,51] or a neuromodulator[52–55] in a number of important neuronal systems. CCK is one of the peptides with the most dramatic tissue-specific pattern of processing. The major difference is between how pro-CCK is processed in the brain and in the gut. The brain makes mainly

[46] M. C. Beinfeld, D. K. Meyer, R. L. Eskay, R. T. Jensen, and M. J. Brownstein, *Brain Res.* **212,** 51 (1981).

[47] W. M. Hunter and F. C. Greenwood, *Nature (London)* **194,** 495 (1962).

[48] G. Adler, C. Beglinger, U. Braun, M. Reinshagen, I. Koop, A. Schafmayer, L. Rovati, and R. Arnold, *Gastroenterology* **100,** 537 (1991).

[49] R. A. Liddle, I. D. Goldfine, M. S. Rosen, R. A. Taplitz, and J. A. Williams, *J. Clin. Invest.* **75,** 1144 (1985).

[50] R. B. Innis, F. M. A. Correa, G. Uhl, B. Schneider, and S. H. Snyder, *Proc. Natl. Acad. Sci. U.S.A.* **76,** 521 (1979).

[51] M. Verhage, W. E. J. M. Ghijsen, D. G. Nicholls, and V. M. Wiegant, *J. Neurochem.* **56,** 1394 (1991).

[52] J. S. Brog and M. C. Beinfeld, *J. Pharmacol. Exp. Ther.* **260,** 343 (1992).

[53] S. P. Arneric, M. P. Meeley, and D. J. Reis, *Peptides* **7,** 97 (1986).

[54] C. Cosi, A. Altar, and P. L. Wood, *Eur. J. Pharmacol.* **165,** 209 (1989).

[55] A. S. Freeman, L. A. Chiodo, S. I. Lentz, K. Wade, and M. J. Bannon, *Brain Res.* **555,** 281 (1991).

CCK 8 amide[56] whereas the gut makes larger forms such as CCK 12, 22, 33, 39, 58, and 83 amide.[57,58] Various forms of CCK found either in the brain or the gut appear to be the result of differential processing of the same pro-CCK precursor. Modifications of pro-CCK during processing include sulfation of three carboxyl-terminal tyrosines, cleavage at dibasic or monobasic sites, the action of carboxypeptidase, and amidation of the carboxyl terminal to release bioactive forms of CCK.[59]

Endoproteolytic Processing of Procholecystokinin

Significant progress has been made in the elucidation of the mechanism and enzymology of pro-CCK cleavage, using endocrine tumor cells that process pro-CCK to amidated peptides such as CCK 8 and CCK 22.[60–62] CCK mRNA is expressed in a number of tumors including insulinomas, thyroid c-cell carcinomas, intestinal tumors, neuroblastomas, small cell carcinomas, gastric carcinomas, acoustic neuromas, and neuroepitheliomas. Most of these tumor cell lines are unable to process pro-CCK completely to amidated peptides such as CCK 8. Three tumor cells lines, STC-1[62] (mouse intestinal carcinoma) RIN5F[60] (rat insulinoma), and WE[61] (rat thyroid c-cell carcinoma), do produce and secrete CCK 8 amide. The mouse anterior pituitary tumor AtT20 cells when transfected with the CCK cDNA can also make and secrete large amounts of CCK 8 amide.[63] STC-1 and RIN5F cells also make and secrete some CCK 22 and display a processing pattern that is closer to that in the gut.[62] In terms of the active products that are made by these tumor cells, they are reasonable models of processing of pro-CCK in the brain and gut, respectively.

When the STC-1 cells were engineered to express the first 491 bases of the rat PC1 cDNA or the first 480 bases of PC2 cDNA in the antisense orientation from the cytomegalovirus promoter, the CCK levels detected by RIA were greatly reduced. The most severely affected clone (S1F2) expressing PC1 antisense cDNA was selected for further analysis in comparison with parental STC-1 cells and cells expressing anti-PC2 mRNA (S2H4) and control STC-1 cells (SCD9) transfected with the expression vector

[56] G. J. Dockray, R. A. Gregory, and J. B. Hutchinson, *Nature* (*London*) **264**, 568 (1978).
[57] J. F. Rehfeld, *J. Biol. Chem.* **253**, 4022 (1978).
[58] V. E. Eysselein, G. A. Eberlein, M. Schaeffer, D. Grandt, H. Goebell, W. Niebel, G. L. Rosenquist, H. E. Meyer, and J. R. Reeve, Jr., *Am. J. Physiol.* **258**, G253 (1990).
[59] M. C. Beinfeld, *Life Sci.* **61**, 2359 (1997).
[60] M. C. Beinfeld, *Neuropeptides* **22**, 213 (1992).
[61] R. S. Haun, M. C. Beinfeld, B. A. Roos, and J. E. Dixon, *Endocrinology* **125**, 850 (1989).
[62] J. Y. Yoon and M. C. Beinfeld, *Endocrinology* **2**, 973 (1994).
[63] W. Lapps, J. Eng, A. S. Stern, and U. Gubler, *J. Biol. Chem.* **263**, 13456 (1988).

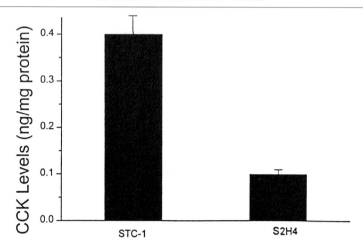

FIG. 1. CCK levels in control STC-1 cells and anti-PC2 S2H4 cells.

pCMV5 without any insert. A number of STC-1 anti-PC2 cell lines were also obtained. The most severely affected STC-1 cell line, S2H4, had about 25% of the wild-type CCK levels (Fig. 1) and expressed about 50% of the PC2 protein of parent STC-1 cells. Detection of sense and antisense PC1 messages by Northern blot revealed a 2.6-kb fragment and a 4.0-kb fragment, the two predominant endogenous PC1 transcripts in STC-1 cells (Fig. 2A). These transcripts were also observed in STC-1 cells, control SCD9 cells, and anti-PC2 S2H4 cells. In S1F2 cells, however, antisense PC1 messages was not observed and only low levels of sense PC1 messages were present, possibly owing to the degradation of the sense and antisense RNA duplex.

Northern analysis of the same cell lines using PC2 as hybridization probe revealed that all four cell lines expressed similar levels of a single endogenous PC2 transcript of 2.2 kb (Fig. 2B). In addition to the endogenous PC2 message, S2H4 cells expressed a large amount of antisense PC2 message. Although the expression of antisense PC1 message resulted in almost complete disappearance of the endogenous PC1 message in S1F2 cells, it appeared to have little effect on the expression of PC2 message. Similarly, the expression of a large amount of antisense PC2 message in S2H4 cells appeared to have little effect on the expression of PC1 message. This suggests that the effect of antisense on its targets in STC-1 cells occurs largely at the nucleic acid level.

Measurements of protein in these cell lines by Western blot analysis confirmed the inhibition of PC1 protein expression (Fig. 3A). The two predominant forms of PC1 in STC-1 cells were 87 and 66 kDa, in size;

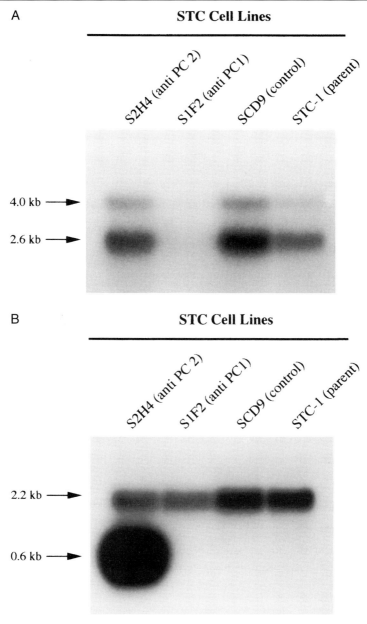

FIG. 2. PC1 and PC2 mRNA expression in wild-type, control, and anti-PC1- and anti-PC2-transfected STC-1 cells. Five micrograms of mRNA was used from each cell line for Northern blot analysis. (A) PC1 Northern blot, (B) PC2 Northern blot.

A **STC Cell Lines**

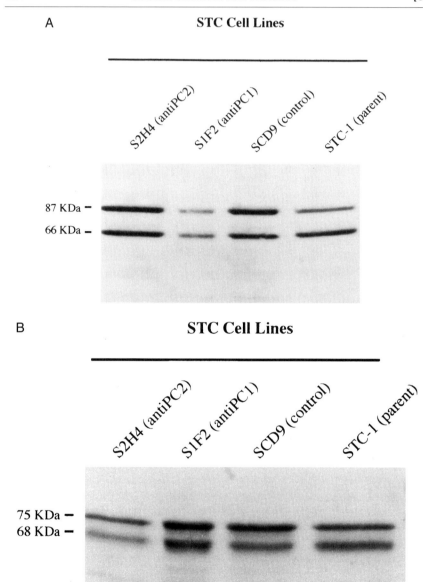

B **STC Cell Lines**

Fig. 3. PC1 and PC2 protein expression in wild-type, control, and anti-PC1- and anti-PC2-transfected STC-1 cells. Fifty micrograms of protein from each cell line was used for Western blot analysis. (A) PC1 Western blot analysis; (B) PC2 Western blot analysis.

similar forms were also seen in RIN5F cells. Both forms were found to be greatly reduced in S1F2 cells. Quantitation by densitometry showed an approximately 80% inhibition as compared with STC-1, SCD9, and S2H4. The observation that neither the 75-kDa nor the 66-kDa, PC2 protein level in S1F2 cells was decreased demonstrates that the antisense inhibition in these cells is specific (Fig. 3B). The antisense PC1 cell line S2H4, which had reduced amounts of CCK 8, also had decreased levels of PC1 protein both absolutely and relative to PC2. Interestingly, although the endogenous PC2 message was not altered after the expression of antisense PC2 in S2H4 cells, we observed a reduction in PC2 protein level of approximately 50% in S2H4 cells. The effect of expression of PC2 antisense was to produce a specific decrease in PC2 protein expression, but it did not alter PC1. Among STC-1 antisense PC2 cell lines, there was a rough correlation between CCK levels and PC2 expression. In some of the severely affected cell lines, the level of PC2 expression was between 25 and 50% of that in wild-type STC-1 cells.

Analysis of extracts of anti-PC1 S1F2 cells by Sephadex G-50 chromatography demonstrated a specific ablation of CCK 8 relative to CCK 22 (Fig. 4B). Similar analysis of extracts of control SCD9 cells did not reveal depletion of either CCK 22 or CCK8, identical to the control STC-1 cells (Fig. 4A). Chromatographic separation and RIA analysis of cellular extracts of anti-PC2 clone S2H4, when compared with the parental cell line, indicate that there was a selective depletion of CCK 22 with a comparative sparing of CCK 8 (Fig. 4C).

Concluding Remarks

Studies using antisense inhibition have been criticized because of possible nonspecific cellular effects complicating interpretation of the results. In light of these concerns, we performed several controls to ensure that the effects of antisense expression on the processing of pro-CCK were specific. The two most important controls performed were to measure the effects of antisense expression on target mRNA and target protein. Another control that was performed was to measure the effect of PC2 message and protein level in antisense PC1-transfected cells and vice versa. Although PC1 and PC2 are homologous genes within the same family of endoproteases, the effect of antisense was specific for each. These observations suggest that the effect of antisense on the RNA level is complex but specific and may occur at multiple levels to produce inhibition of the protein product.

The enzyme(s) responsible for the endoproteolytic cleavages of pro-CCK *in vivo* have not been definitively identified. It is likely that the

A

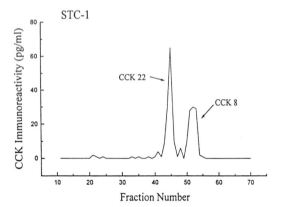

B

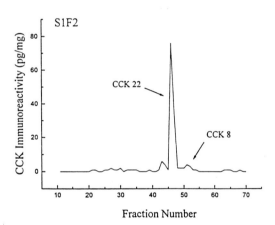

C

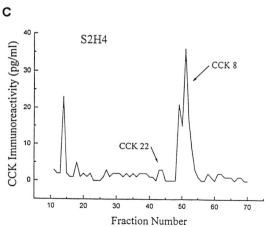

subtilisin-like proteases PC1[7,64] and PC2[9] are involved in the processing of pro-CCK in these cell lines. They are widely distributed in the brain and are present in a number of endocrine cells,[62] which express CCK mRNA and correctly process pro-CCK to amidated CCK 8.[60] The fact that the stable transfection of STC-1 cells with a PC1 antisense cDNA, with the consequent inhibition of PC1 expression, caused a selective depletion of CCK 8 with a comparative sparing of CCK 22 suggests that PC2 is not required for the production of CCK 8. In support of these observations, the AtT20 anterior pituitary cell line, which expresses PC1 but not PC2 when stably transfected with the CCK cDNA, can process pro-CCK to amidated CCK 8; this strongly suggests that PC2 is not required for production of CCK 8.[63] Inhibition of PC2 expression by a similar strategy in these cells causes a comparative depletion of CCK 22 while sparing CCK 8. The present evidence suggests that PC2 is required for the production of larger amidated peptides such as CCK 22, which are more abundant in gut. That inhibition of PC1 or PC2 expression can shift the ratio of CCK 22 to CCK 8 suggests that CCK 22 and CCK 8 arise by different pathways and do not interconvert. It also supports the hypothesis that the differences in processing of pro-CCK observed in brain and gut can be explained by differences in expression or activity of endoproteases such as PC1 and PC2.

Knowledge of the actual enzymes responsible for prohormone cleavages in specific tissues lags behind progress in other areas of the peptide hormone field owing to the technical difficulty of identification, isolation, characterization, and inhibition of these enzymes. As important as the PC enzymes appear to be, it is possible that proteases other than the PCs are involved in prohormone processing. Progress in our ability to develop specific inhibitors of these enzymes as well as to develop conditional, tissue-specific knockouts will allow us to address more directly the question of which enzymes are physiologically relevant.

[64] N. G. Seidah, M. Marcinkiewicz, S. Benjannet, I. Gaspar, G. Beaubien, M. G. Mattei, C. Lazure, M. Mbikay, and M. Chretien, *Mol. Endocrinol.* **5,** 111 (1991).

FIG. 4. Chromatographic analysis of CCK forms in STC-1, S1F2, and S2H4 cells. Cells were extracted with 0.1 *N* HCl and sonication. Concentrated extracts from several 10-cm plates were separated by Sephadex G-50 chromatography and fractions were measured for CCK immunoreactivity. (A) STC-1 cells; (B) S1F2 cells; (C) S2H4 cells.

Section II

Antisense Neuroscience Targets

[9] Strategies for the Design and Delivery of Antisense Oligonucleotides in Central Nervous System

By WILLIAM C. BROADDUS, SUJIT S. PRABHU, SUSANNA WU-PONG, GEORGE T. GILLIES, and HELEN FILLMORE

Introduction

Nucleic acid pharmaceuticals are of increasing interest to the health care community because of their almost unlimited potential in the treatment of diseases such as cancer, genetic and degenerative disorders, viral infection, and the molecular pathologic sequelae of trauma or ischemia. Modified antisense oligodeoxynucleotides (oligonucleotides; ODNs) offer the potential to block the expression of specific genes within cells. Such strategies have been shown to have antiproliferative activity in rapidly dividing cells when directed at appropriate targets, and to alter cellular interactions and behavior in other ways, depending on the role of the gene product whose expression has been reduced. The advantages of using antisense oligonucleotides include their ease of synthesis, their low molecular weight, the variety of gene targets available, and their potential for both specificity and low toxicity.

Unique features of the central nervous system (CNS) render it an area of both opportunity and challenge for the application of antisense ODN strategies. Historically, the powerful ability of the blood–brain barrier to exclude ODNs and their analogs has prevented widespread investigation of this approach. More recently, the compartmentalized nature of the CNS, coupled with innovative delivery techniques, has rekindled interest in the application of therapeutic antisense ODN strategies in the CNS. This chapter discusses technical aspects of our efforts to design and test antioncogene strategies for therapeutic application in the treatment of primary intracranial tumors. The applicability of such strategies to other CNS pathologies is also discussed briefly.

Design of Antisense Oligodeoxynucleotides

When considering the design of antisense oligonucleotides, several factors must be considered. These include the choice of target gene and nucleotide sequence within the gene, the choice of backbone modifications of the ODN, the length of the ODN, and the presence of special sequence motifs that can predispose to nonantisense effects of the ODN. There are many

METHODS IN ENZYMOLOGY, VOL. 314

excellent review articles (with references to published literature) that discuss in detail the design of antisense oligonucleotides.[1-5] The published data indicate that only a small percentage (less than 5%) of the antisense molecules tested are highly effective in the reduction of target mRNA.[3,6] Although ODNs can be designed by careful analysis of sequence and can have a certain length with specified backbone modifications, the ability of the ODN to bind target mRNA cannot be predicted owing to the structure and conformation of individual RNA species. At present, the sites of RNA accessible for ODN binding cannot be predicted. This has led to the conclusion that "rational design of antisense molecules is not possible."[3] Thus, it is important to test antisense ODNs made to different sites along the target mRNA. This highly empirical approach can be expensive but is critical until other methods of screening potential candidate antisense ODNs are developed.[6]

Below are guidelines that we have used to design potential sense, antisense, and scrambled oligonucleotides. Steps 2, 3, and 6 are done using the resources available over the Internet. Sites such the National Center for Biotechnology Information (NCBI; *http://www.ncbi.nlm.nih.gov*) are used. A good resource for the use of these and other sites on the Internet is *Guide to Human Genome Computing,* edited by M. J. Bishop.[7]

1. *Identify the target:* Some questions to ask are the following: Is there evidence that the target gene is involved in the biological process of interest? Has the target gene been sequenced and submitted to GenBank? Are antibodies and probes for the target gene product available for use in experiments to test for specific antisense effects? What is known about the half-life of the target mRNA?

2. *Obtain GenBank accession number and target sequence information:* For our experiments using rat and human glioma cell lines, we have obtained both the rat and human sequences of a variety of protooncogenes.

3. *Identify the start site and study what is known about the particular target sequence:* For example, is there an alternative start site? Where are the exon boundaries?

[1] S. T. Crooke, *Antisense Nucleic Acid Drug Dev.* **8,** 115 (1998).
[2] C. A. Stein, *Antisense Nucleic Acid Drug Dev.* **8,** 129 (1998).
[3] A. D. Branch, *Trends Biol. Sci.* **23,** 45 (1998).
[4] S. Agrawal and Q. Zhao, *Curr. Opin. Chem. Biol.* **2,** 519 (1998).
[5] S. Agrawal, *Trends Biochem. Sci.* **14,** 376 (1998).
[6] A. D. Branch, *Antisense Nucleic Acid Drug Dev.* **8,** 249 (1998).
[7] M. J. Bishop (ed.), "Guide to Human Genome Computing," 2nd ed. Academic Press, New York, 1998.

4. *Perform the first round of potential sequence analysis:* We have used a commercially available program for designing primers, called "Primer Premiere" (Premier Biosoft International, Palo Alto, CA). The results from running this program, using the imported target sequence and designating primer length, gave us a list of potential sequences to start examining. Alternatively, commercial analysis of the target sequence for candidate antisense oligonucleotides is possible. For our studies with c-*myc*[8] we employed the services of Advanced Gene Computing Technologies (AGCT, Irvine, CA).

5. *Examine potential sequences for motifs that have been shown to cause nonspecific antisense effects* (i.e., G quartets, CpGs): Until further information on the mechanisms of interaction of these motifs with cellular macromolecules is available, inclusion of such sequences should be avoided, if possible.[2] However, when a candidate sequence containing such a motif is found to be effective, additional control studies to assess the contribution of the motif to the biological effect are necessary.[8]

6. *Examine the ability of oligonucleotides to form homodimers or secondary structures:* Most ODN sequence design software assesses the potential for such autoduplex formation during the design process. If random sequences have been chosen, however, subsequent analysis of the candidate sequences using such software is advisable.

7. *Using the Blastn program at NCBI, run the candidate antisense sequences to double-check specificity for the target gene and to assess for significant homologies with other gene sequences:* It is also important that the sense and scrambled control sequences be checked for homologies with other gene sequences.

8. *Double-check the sequences:* Once the sequences are chosen, double-check for secondary structures or potential homodimerization that might reduce the efficacy of the candidate sequence.

9. *Decide on biochemical modifications of the ODN (e.g., backbone substitution or nucleoside methylation) to enhance its biological stability:* We have relied most heavily on phosphorothioate modifications, to balance biologic stability, toxicity, and cost.

10. *Order oligonucleotides from a reliable source that uses high-quality reagents:* Be cognizant of the quality control techniques utilized by the provider, and scrutinize these results for each newly synthesized batch of ODN.

---

[8] W. C. Broaddus, Z. J. Chen, S. S. Prabhu, W. G. Loudon, G. T. Gillies, L. L. Phillips, and H. Fillmore, *Neurosurgery* **41,** 908 (1997).

In Vitro Characterization of Antisense Oligodeoxynucleotides

A rapidly growing literature exists on the use of antisense ODN treatment for abrogation of the effect of a variety of gene products in neuronal and glial cells of the CNS (see Wu-Pong and Gewirtz[9] for references). Many of these have obvious potential applicability to human pathologic conditions, and thus explain the growing interest in antisense ODN treatment in the CNS. Our laboratory has focused on the potential for antitumoral effects of antisense ODNs targeted at molecular components critical for cellular proliferation in malignant glioma cells.[8] In particular, we have focused on the role of c-*myc* in a rat glioma cell line (RT2) that also forms tumors in syngeneic Fischer 344 rats after cerebral implantation.

As noted by others, several key issues must be considered in establishing the specific antisense mechanism of a candidate antisense ODN: (1) demonstration of the predicted effects with more than one antisense sequence for any given gene is important to reduce the possibility of sequence-specific nonantisense effects; (2) demonstration of stability of the oligonucleotide in the experimental system is important because ODNs may degrade rapidly under certain experimental conditions; (3) confirmation of reduced protein expression and/or reduced mRNA content for the targeted gene is critical for establishing the antisense nature of the effect; and (4) the cellular and nuclear access of the antisense ODN must also be considered, because this has been shown to vary for different cell types, growth conditions, and ODN sequences.[9]

For pragmatic reasons related to our motivation for these studies (seeking ODNs that were efficacious in preventing cell growth), we chose first to focus on the ability of candidate ODN sequences to inhibit proliferation of RT2 cells. Subsequent studies of the effects of these ODNs on c-Myc protein expression were carried out by Western blot analysis, and studies of the stability of the ODNs during cell culture were carried out by gel electrophoresis of samples from the culture medium.

Selection of c-myc Antisense Oligodeoxynucleotides

A range of 15-mer antisense phosphorothioate candidate sequences was designed as described in the preceding section, with the exception that one of the sequences had been previously described to have significant inhibitory effects in other cultured and *in vivo* cells. Sense and appropriate scrambled control sequences were also designed and studied to rule out homologies with known sequences in the appropriate genetic sequence databank. Additional studies of the possible role of a 4G sequence in the most potent of

[9] S. Wu-Pong and D. Gewirtz, *BioPharmaceuticals* **11,** 28 (1998).

these ODNs required the design of additional control sequences containing the 4G motif. The latter were designed to have the 4G portion in the same relative position as the original antisense sequence, but with sense or scrambled sequences surrounding the 4G to make up the length of the 15-mer.

Cell Culture

Cells were grown and passaged in complete medium containing 5% fetal bovine serum. For growth studies in which cells were exposed to ODNs, the cells were plated in Dulbecco's modified Eagle's medium (DMEM) containing 1% fetal bovine serum, with or without the specified concentrations of ODN. This concentration of serum represents a compromise in that ODN effects may be reduced by their binding to serum proteins, yet most cultured cells grow poorly in low serum concentrations. A concentration of 1% was found to allow acceptable growth of RT2 cells in 96-well microtiter plates.

Cell Growth Studies

For analysis of the effect on cell growth of antisense, sense, and scrambled ODNs, cells were plated at a density of 1000/well in 96-well plates in medium without added ODNs. Twenty-four hours later, the culture medium was changed to contain final concentrations of 1, 3, or 10 μM ODNs. This concentration range of ODNs was chosen after a review of the literature demonstrated this to be the typical range for antisense effects *in vitro* utilizing phosphorothioate-modified ODNs. This is as opposed to unmodified phosphodiester ODNs, which may require concentrations up to 100 μM for antisense effects *in vitro*. Control cultures received fresh culture medium without added ODNs. Growth was assessed every day thereafter for 4 to 5 days by the MTT assay. In this assay, cells growing in 96-well plates were incubated with 3-(4,5-dimethylthiazol-2 yl)-2,5-diphenyltetrazolium bromide (thiazolyl blue, MTT; Sigma, St. Louis, MO) solution [2 mg/ml phosphate-buffered saline) (PBS)] for 3 hr at 37° in a humidified 5% (v/v) CO_2 atmosphere. The dye solution was removed from the wells and replaced with 100 μl of dimethyl sulfoxide (DMSO; Sigma). The absorbance of each well was measured with a microculture plate reader at 540 nm.

c-Myc Protein Detection by Western Analysis

For the study of c-Myc protein expression, cells were plated in 60-mm petri dishes at a density of 60,000/dish. Culture medium was added and changed as described above for cell growth studies. ODNs were added to final concentrations of 10 μM. RT2 cells treated for 4 days with antisense ODN, sense ODN, scrambled ODN, or PBS were lysed in PBS containing 1% (w/v) Nonidet P-40 (Boehringer GmbH, Mannheim, Germany), 0.5% (w/v) sodium deoxycholate (Boehringer GmbH), 0.1% (w/v) sodium dodecyl sulfate (SDS; Bio-Rad, Hercules, CA), phenylmethylsulfonyl fluoride (PMSF, 0.1 mg/ml; Sigma), aprotinin (30 μl/ml; Sigma), and sodium orthovanadate (1 mM; Sigma) for 30 min at 4°. Lysates were spun in an Eppendorf microfuge at 12,000 rpm for 20 min. The supernatant was collected and an aliquot was assayed for total protein concentration by using the DC protein assay (Bio-Rad).

SDS-polyacrylamide gel electrophoresis was performed with 35 μg per lane of total protein extract and the separated proteins were transferred to nitrocellulose membranes by electroblotting. Western blot analysis was performed using a chemiluminescent detection system in which mouse monoclonal antibody (Ab-1; Oncogene Science, Cambridge, MA) or rabbit polyclonal antibody (N-262; Santa Cruz Biotechnology, Santa Cruz, CA) against c-Myc protein and rabbit polyclonal antibody to β-actin (A-5441; Sigma) were used. After transfer, the nitrocellulose membranes were blocked with 3% milk-containing PBS overnight at 4°. The blots were incubated at room temperature for 2 hr with primary antibodies to c-Myc (1:100) and β-actin (1:5000), and then with secondary antibody (peroxidase-conjugated Affini-Pure goat anti-mouse or anti-rabbit IgG (H + L); Jackson, ImmunoResearch, West Grove, PA) for 1 hr at room temperature. They were then developed by the chemiluminescent detection system (Amersham, Arlington Heights, IL) using Kodak (Rochester, NY) X-ray film. With this technique, care should be taken to avoid overexposing the bands on the radiograph, as bands that appear dark to the human eye on X-ray film may correspond to densities beyond the linear response range of standard radiographic film.

Semiquantitative Assessment of c-Myc Protein Content

One benefit of the Western blot analysis technique is that semiquantitative values for relative protein expression can be derived from the resulting radiographs, provided appropriate precautions are taken and controls are performed. As noted above, radiographs must be obtained such that band densities remain within the linear response range of the film (noted in the specifications included with the film). Confirmation of the linear response of the assay can be obtained by constructing a standard curve with varying quantities of protein extract in parallel incubations, processing these for

Western blot analysis, and plotting the densities of the resulting bands versus the quantity of extract analyzed.

For determination of c-Myc expression, relative protein concentrations after Western blot analysis were determined by computer-interfaced densitometry (MCID/M4; Imaging Research, St. Catharines, Ontario, Canada). Each film was transilluminated and captured, using a computer-interfaced video camera, and stored as a digitized image. Protein bands were measured within user-defined borders and expressed as the product of relative optical density (ROD) and area (pixels). The value of ROD × area was found to be linear with protein added ($r > 0.9$ by linear regression analysis) for both c-Myc and β-actin, using this technique. The results were normalized for sample loading by expressing the ratios of the densities of c-Myc to β-actin bands for each sample.

Results of in Vitro Studies

The results of the studies reported in this section are described and discussed in detail elsewhere.[8] They are presented here briefly to illustrate the analysis of data from such studies. Compared with cultures containing standard medium, two of six antisense ODNs (rAS-1 and rAS-3) were found to inhibit significantly the growth of glioma cells, while sense and scrambled sequence ODNs did not significantly affect cell growth at the concentrations tested (Fig. 1). The most effective of these (rAS-1) was a sequence including the translation initiation codon, as has been described by other investigators. A human c-*myc* antisense sequence (hAS-1), which differed from this rat sequence by one base substitution, also had an inhibitory effect on RT2 cells. Western blot analysis showed that expression of immunoreactive c-Myc protein was also greatly reduced in the rat antisense ODN-treated cells (and not in sense-, scrambled-, or control-treated cells). The degree of reduction of c-Myc protein expression was assessed semi-quantitatively using densitometry of the Western blots and normalization to β-actin expression. These values correlated well with the decreases in cell growth seen using the MTT assay with several antisense ODNs (Fig. 2). When assessed for increased electrophoretic mobility as an indication of oligonucleotide cleavage, the ODNs were stable in cell culture medium for at least 5 days.

Central Nervous System Delivery of Antisense Oligonucleotides

A variety of methods have been investigated for the delivery of therapeutic agents to the central nervous system.[9] *Systemic administration* of therapeutic agents intended for the central nervous system (CNS) has long

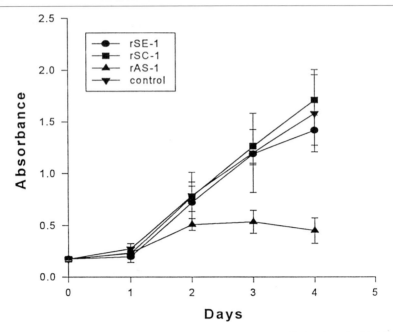

Fig. 1. Antiproliferative effects of antisense ODNs on RT2 cells incubated with 10 μM ODN added to the culture medium relative to control cultures. The series corresponds to ODNs targeting the rat c-*myc* gene exon 2 (rAS-1), the corresponding sense ODN (rSE-1), the corresponding scrambled ODN (rSC-1), and cell culture medium (control). [W. C. Broaddus, Z. J. Chen, S. S. Prabhu, W. G. Loudon, G. T. Gillies, L. L. Phillips, and H. Fillmore, *Neurosurgery* **41,** 908 (1997).]

been known to be limited by the function of the blood–brain barrier. Even low molecular weight compounds may be rendered ineffective for CNS applications by blood–brain barrier (BBB) exclusion. The intact blood–brain barrier effectively excludes antisense ODNs, with molecular weights ranging from 4500 to 8000. One approach to enhancing access of therapeutic agents to the CNS has been to transiently disrupt the blood–brain barrier. Indeed, selective intraarterial infusion of macromolecular agents in conjunction with blood–brain barrier disruption holds significant promise for delivery of a variety of agents.[10] Limitations of this approach relate to the constraint that agents be delivered to arterial territories rather than anatomic parenchymal targets, and that adverse effects related to blood–brain barrier

[10] J. A. Williams, S. Roman-Godstein, J. R. Crossen, A. D'Agostino, S. A. Dahlborg, and E. A. Neuwelt, *Adv. Exp. Med. Biol.* **331,** 273 (1993).

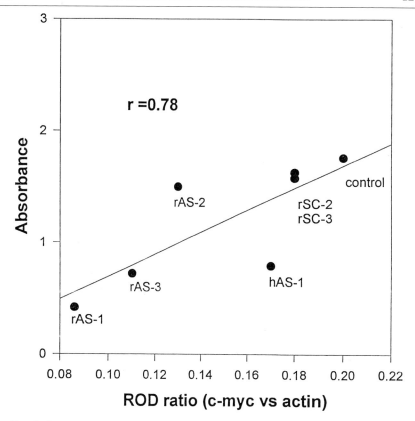

FIG. 2. Correlation of ODN effects on cellular proliferation with effects on c-Myc protein expression. There is a strong correlation between these two parameters, although two sequences (rAS-2 and hAS-1) seemed to fit this correlation less well. [W. C. Broaddus, Z. J. Chen, S. S. Prabhu, W. G. Loudon, G. T. Gillies, L. L. Phillips, and H. Fillmore, *Neurosurgery* **41,** 908 (1997).]

disruption, such as increased cerebral edema, must be contended with during each treatment session. *Modifications of antisense ODNs* to prodrug forms, which are more permeable to the BBB, are also under investigation, but have yet to demonstrate favorable combinations of permeability and systemic toxicity for consideration of general implementation. *Direct intraparenchymal administration* of small volumes of antisense ODN solutions is feasible when localized delivery of the agent is desired. For instance, slow-release biopolymers implanted at the time of surgical resection have been used for administration of chemotherapeutic agents to the margins of a tumor cavity.[11] However, distribution from the site of implantation is dependent on diffusion, which is limited by the concentration that can be safely placed in the depot site, and by the molecular weight of the compound. This approach would, therefore, offer little hope for the effective penetration of the parenchyma by larger molecules such as antisense ODNs. *Intrathecal (cerebrospinal fluid, CSF) infusion modalities* also limit delivery of the ODNs to the pial surface and 1–2 mm underneath it.

Positive pressure controlled-rate infusion involves delivery of a therapeutic agent directly into the brain parenchyma, utilizing positive pressure infusion.[12–14] The fluid and its solutes move radially outward from the cannula, and under ideal conditions can produce a "square-shaped" distribution profile for slowly degraded molecules in the perfused tissue.[15] Thus, it is possible to have significant control over the interstitial concentration delivered and the region of parenchyma to which it is delivered. The technique depends on convective flow, defined as the movement of material down a pressure gradient, distinct from the process of diffusion, by which material moves down a concentration gradient. Work reported by Oldfield, Laske and colleagues, as well as work in our own laboratory, has shown this technique to be quite versatile. Infusion rates are varied according to the size of the brain to be infused, from fractions of a microliter per hour in rats to fractions of a milliliter per hour in humans. Because infused solutes perfuse the interstitial space, a much larger volume of brain parenchyma may be perfused (V_d) relative to the volume infused (V_i). The travel of solutes through the brain parenchyma is impeded by the "tortuosity" of the interstitial space, and the magnitude of this impedance is related to the size of the solute.[16] Bobo *et al.*[12] showed a ratio of volume of distribution to volume of infusion (V_d/V_i) for [^{14}C]sucrose (molecular weight, 350) of 13:1 and for ^{111}In-labeled transferrin (molecular weight, 80,000) of 6:1. The V_d containing $\geq 1\%$ of the infusion concentration increased linearly with volume infused for both sucrose and transferrin in this paradigm. Thus, in addition to the greatly enhanced volume of brain to which drug

[11] H. Brem, M. G. Ewend, S. Piantadosi, J. Greenhoot, P. C. Burger, and M. Sisti, *J. Neurooncol.* **26**, 111 (1995).

[12] R. H. Bobo, D. W. Laske, A. Akbasak, P. F. Morrison, K. S. Bankiewicz, and E. H. Oldfield, *J. Neurosurg.* **82**, 1021 (1994).

[13] W. C. Broaddus, S. S. Prabhu, G. T. Gillies, J. Neal, W. S. Conrad, Z. J. Chen, H. Fillmore, and H. F. Young, *J. Neurosurg.* **88**, 734 (1998).

[14] S. S. Prabhu, W. C. Broaddus, G. T. Gillies, W. G. Loudon, Z. J. Chen and B. Smith, *Surg. Neurol.* **50**, 367 (1998).

[15] D. M. Lieberman, D. W. Laske, P. F. Morrison, K. S. Bankiewicz, and E. H. Oldfield, *J. Neurosurg,* **82**, 1021 (1995).

[16] P. F. Morrison, D. W. Laske, H. Bobo, E. H. Oldfield, R. L. Dedrick, *Am. J. Physiol.* **266**, R292 (1994).

is delivered, positive pressure infusion can permit pharmacological control over the volume treated. By avoiding the steep concentration gradients required for diffusion-based methods of intraparenchymal delivery, potential toxicity at the site of deposition (or in regions proximal to it) may also be reduced.

Our laboratory has begun to investigate the use of high-flow microinfusion for delivery of a variety of agents into the CNS, including phosphorothioate (PS)-ODNs. A discussion of the techniques used in a study of the distribution and stability of infused PS-ODN in rat brain is presented in the following sections.[13]

Preparation of Phosphorothioate Oligonucleotide

A hybrid 18-mer phosphorothioate oligonucleotide was generously provided by S. Agrawal (Hybridon, Cambridge, MA). Synthesis of the 18-mer was performed using deoxynucleoside phosphoramidites and 2'-O-methyl ribonucleoside phosphoramidites on an automated synthesizer (Biosearch 8800; Biosearch Technologies, Novato, CA). To prepare the ^{35}S-labeled hybrid ODN, synthesis was carried out in a similar manner except that the last three couplings were done using 2'-O-methyl ribonucleoside H-phosphonate. After the assembly, the controlled pore glass-bound oligonucleotide containing three H-phosphonate linkages was oxidized with elemental ^{35}S (Amersham, 0.5 to 2.5 Ci/milliatom) and deprotected. Further details of the synthetic techniques are beyond the intended scope of this chapter. Solutions of 18-mer oligonucleotide were prepared in mock cerebrospinal fluid (composition: 124 mM NaCl, 5 mM KCl, 1.25 mM KH$_2$PO$_4$, 2 mM MgSO$_4$, 3 mM CaCl$_2$, 26 mM NaHCO$_3$, 10 mM D-glucose, pH 7.4).

Animal Preparation

Female Fischer 344 rats weighing between 150 and 160 g were anesthetized with ketamine (50–80 mg/kg) and xylazine (5–10) mg/kg) intraperitoneally, and were positioned in a sterotactic frame (David Kopf Instruments, Tujunga, CA). Using a 2-mm diamond bit, a burr hole was made 3.5 mm lateral and 1 mm posterior to the bregma. Using a precision mechanical slide, a 25-gauge needle was inserted to a depth of 4.5 mm to permit infusion into the caudate putamen.

Positive Pressure Infusion

Previous work by Lieberman et al.[15] showed that infusion rates between 0.01 and 0.2 μl/min did not result in any retrograde flow along the needle tract; however, in larger species infusion rates with sequential increases up to 4 μl/min are well tolerated. We have found that infusion rates in increments from 0.1 to 2 μl/min are well tolerated in rats with no retrograde flow along the needle tract.[14] However, it seems clear that specific microstructural changes in brain parenchyma occur during the infusion process that allow for the distribution of infusate for substantial distances around the site of infusion. Further study will be required to determine the dependence of these changes (and the extent of the changes) on parameters such as infusion rate, volume, and pressure, and on the molecular/particulate size of the infused agent. In addition, the potential for long-term effects in the brain related to these microstructural changes exists and will require careful assessment. For small animal brains, rates less than 1 μl/min appear to be preferable. For larger brains, where higher infusion rates are essential (and where the compliant response of the bulk brain tissues is greater), there may be an advantage in starting the infusion at a relatively low rate for 5 to 10 min, followed by increasing the infusion rate sequentially thereafter. For the present study, 20 μl of a 100 μM solution containing 0.5 μCi of the 18-mer [35S]PS-ODN was infused over 1 hr at a rate of 0.3 μl/min. Rats were sacrificed 1, 6, 12, 24, and 48 hr after the start of the infusion. The brains were removed and snap-frozen in isopentane–dry ice and stored at $-80°$ for further analysis. To monitor the appearance of [35S]PS-ODN in the CSF, samples were collected from the cisterna magna of selected rats 30 min and 1, 2, 4, 6, 12, and 24 hr after the start of the infusion.

Cryosectioning and Autoradiographic Analysis

Coronal 16-μm cryotome sections were obtained serially from each frozen brain specimen. The first 4 of every 20 sections were applied to microscope slides for autoradiography, while the remaining 16 sections were reserved for additional analyses, such as extraction of oligonucleotide. The sections for autoradiography were air dried and then apposed to Kodak Biomax MR film for 18 to 36 hr. The X-ray films were developed and the resulting autoradiograms (Fig. 3) were analyzed by computer-assisted image analysis. Cryosections (16 μm) of a series of concentrations of [35S]sulfate mixed in known amounts of homogenized brain were used as brain paste standards and included in each autoradiogram.

Quantitative Analyses

The image analysis system was used to generate a standard curve based on the brain paste sections, from which tissue contents (concentrations and total amounts) of [35S]PS-ODN could be calculated for individual brain section autoradiographs. The accuracy of this method for measurement of

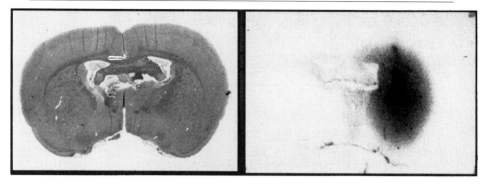

Fɪɢ. 3. Selected autoradiogram (right) and comparable brain section (left) obtained from an animal sacrificed 6 hr after starting [^{35}S]PS-ODN infusion. [W. C. Broaddus, S. S. Prabhu, G. T. Gillies, J. Neal, W. S. Conrad, Z. J. Chen, H. Fillmore, and H. F. Young, *J. Neurosurg.* **88,** 734 (1998).]

^{35}S activity was confirmed by liquid scintillation counting of selected brain sections. The image analysis system also allowed calculation of the volume of distribution of [^{35}S]PS-ODN for each brain specimen. For each brain specimen, values for average ^{35}S-labeled oligonucleotide concentration and volumes of distribution were used to calculate the total activities within the brain parenchyma. The total activities recovered in CSF were calculated by liquid scintillation counting of an aliquot of CSF and assuming a CSF volume of 100 μl in the rat. The resulting values of activity in the brain parenchyma and CSF for each time point were then summed to calculate the total activity recovered, as well as the percentages that these represented of the original infused activity (0.5 μCi).

Extraction and Analysis of Nucleic Acids from Brain and Cerebrospinal Fluid Samples

Brain sections obtained in parallel with those used for autoradiography were transferred to a prechilled mortar and pestle on dry ice and ground into powder. The powdered tissue was then suspended in digestion buffer [proteinase K (2 mg/ml), 0.5% (w/v) sodium dodecyl sulfate, 10 mM NaCl, 20 mM Tris-HCl (pH 7.6), 10 mM ethylenediaminetetraacetic acid], using 1.2 ml/100 mg of tissue. The sample was incubated at 37° for 2 hr, followed by two extractions with phenol–chloroform–isoamyl alcohol (25:24:1) and one extraction with chloroform. The [^{35}S]PS-ODN was then precipitated with ice-cold ethanol. The resulting pellet was dried, resuspended in water, and applied to a 20% (w/v) polyacrylamide–7 M urea gel for electrophoresis. Following electrophoresis, the gel was covered with plastic film and exposed to Kodak Biomax MR film for 24 to 48 hr to visualize the ^{35}S-labeled material.

Results of Infusion Studies

The results of the preceding studies have been presented and discussed in detail.[13] A brief summary of these findings is thus presented here (Table I). At 1 hr, the infused ODN was uniformly distributed in brain tissue, with a minimum tissue concentration greater than 15 μM. Because the ODN can be expected to be confined predominantly to the extracellular space, the concentration of ODN in the extracellular space may in fact be greater than 75 μM. The volume of distribution after a 20-μl infusion was approximately 100 μl at the end of the infusion, increased to 170 μl by 6 hr after the start of the infusion (Fig. 3), and increased further to 440 μl at 48 hr. By this time, the minimum tissue concentration of ODN had decreased to 3.2 μM, which may correspond to an extracellular space concentration of approximately 16 μM. The value of approximately 5 for V_d/V_i (the ratio of volume of distribution to volume of infusion) at the end of the infusion (1 hr) agrees well with the work of other investigators with similar infusates. A small amount of the ^{35}S]PS-ODN activity appeared in the CSF, peaking at the end of the 1-hr infusion, and then undergoing an exponential decay, stabilizing at a low level after 12 hr. Undegraded ODN was observed by gel electrophoresis of extracts from infused brain throughout the 48-hr

TABLE I

VOLUME OF DISTRIBUTION AND BRAIN TISSUE CONTENT OF ^{35}S]PS-ODN
AFTER DIRECT INTRAPARENCHYMAL INFUSION[a]

| | | [35]S]PS-ODN content | |
Time point (hr)[b]	Volume of distribution (μl)	Tissue content (μCi/g)	Minimum concentration (μM/liter)[c]
1	105 ± 7.9	3.8 ± 0.3	15.2
6	173 ± 26.5	2.0 ± 0.4	8.0
12	268 ± 12.2	1.3 ± 0.14	5.2
24	289 ± 36.1	1.0 ± 0.2	4.0
48	443 ± 62.3	0.8 ± 0.14	3.2

[a] Values of volume distribution and brain tissue content are expressed as means ±SD. [Reproduced from Broaddus et al.[13] by permission.]
[b] Time after start of infusion.
[c] Calculated from tissue content, using the specific activity of ^{35}S]PS-ODN and assuming the specific gravity of the brain to be approximately 1 g/ml.

period, indicating that the high levels of ^{35}S activity maintained in rat brain up to 48 hr after infusion are in the form of undegraded biologically active material. It is worth noting that the concentrations of ODN attained in the studies described are at levels expected to have substantial, if not maximal, activity against their selected targets. Even assuming that the ODN distributes homogeneously throughout the infused brain, the minimum concentration values in Table I are above the maximal effective concentration (10 μM) for c-*myc* antisense ODN at the end of the infusion, and are above the half-maximal effective concentration at 48 hr.

Conclusions

A growing knowledge of the roles of specific molecular cellular components in the physiology and pathophysiology of the CNS offers a myriad of potential targets for the application of antisense ODN strategies, for experimental as well as therapeutic purposes. We and others have demonstrated the potential for antisense ODNs directed at protooncogenes to interfere with tumor cell growth in a way that may have therapeutic applications. Besides neoplastic processes, other diseases with focal or regional pathological distributions would also be candidates for this approach. Obvious possibilities would include Parkinson's disease, with therapy directed at the substantia nigra and/or basal ganglia, and selected seizure disorders, with therapy directed at the seizure focus. In addition, growing information on the role of secondary processes in mediating the deleterious sequelae of brain injury (stroke and trauma) suggests that the infusion of antisense ODNs might be an ideal addition to the armamentarium of techniques aimed at disrupting such processes.

Nevertheless, the potential application of ODN-based therapies to the CNS has been largely ignored to date, primarily owing to the difficulty of delivering ODNs to the brain parenchyma. There is now evidence that positive pressure controlled-rate infusion techniques, such as that described in this chapter, may help solve this problem and open the way to a variety of applications of ODN-based therapeutics in the CNS.

[10] Use of Antisense Expression Plasmids to Attenuate G-Protein Expression in Primary Neurons

By NOEL J. BUCKLEY, FE C. ABOGADIE, DAVID A. BROWN, MARIZA DAYRELL, MALCOLM P. CAULFIELD, PATRICK DELMAS, and JANE E. HALEY

Introduction

Unraveling signaling pathways in primary neurons has been a central issue for neurobiologists and pharmacologists, but analysis has been hampered by the paucity of specific molecular tools with which to dissect signal transduction cascades.

The advent of gene cloning demonstrated a degree of molecular diversity unknown, and largely unsuspected, in most gene families. This was nowhere more apparent than in G protein-coupled receptor (GPCR)-mediated signal transduction. This signaling cascade starts with activation of a GPCR and consequent catalytic GDP/GTP exchange on the α subunit of a trimeric G protein, causing dissociation into a GTP-bound α subunit and a $\beta\gamma$ dimer, either of which can act on a myriad of channels or enzymes. Currently genetic diversity accounts for more than 1000 GPCRs, 17 Gα subunits, 5 Gβ subunits, and 12 Gγ subunits and hundreds of ion channel and enzyme effectors.[1,2] This diversity is further increased by the existence of splice variants. A key task has subsequently become to sort through this molecular maze and work out "who talks to whom?" The fact that recognition of this molecular diversity largely relied on molecular cloning bears testament to the degree of similarity between the gene products that was not discernible using conventional biochemical or pharmacological agents. This similarity offers a challenge to develop specific molecular tools, because the degree of amino acid identity between cognate members of a gene family can be so great as to preclude the design of agents that will readily distinguish individual relatives.

Against this background many workers have turned to antisense approaches to produce specific molecular tools that can be used to unravel mechanism, and to provide potential therapeutic agents to specifically target individual gene products. The fundamental tenet underlying all antisense approaches is that nucleic acid sequence variation is greater than amino

[1] E. J. Neer, *Cell* **80**, 249 (1995).

[2] T. Gudermann, F. Kalkbrenner, and G. Schultz, *Annu. Rev. Pharmacol. Toxicol.* **36**, 429 (1996).

acid primary sequence diversity and that this variation can be exploited to manipulate the steady state levels of mRNA in a cell, which, ideally, will be reflected by a corresponding change in protein levels.

This was largely the imperative behind our efforts to use antisense approaches to dissect specific components of GPCR signaling operative in primary neurons in culture. We focus largely on our work, and that of others, aimed at manipulating G-protein levels in postmitotic differentiated neurons using antisense RNA probes, but the principles should apply in equal measure to any antisense strategy.

General Considerations

In contrast to the more commonly used antisense oligonucleotides, antisense RNA action is not RNase H dependent. One proposed means of action is that the antisense RNA hybridizes to the mRNA transcript and the resultant duplex becomes a target for unwindase/modificase,[3,4] followed by covalent modification of the transcript that renders the mRNA refractory to translation. DEAD box helicase[5] and dsRNase activities[3,6] have also been suggested as possible mediators of mRNA modification or degradation. However, numerous alternative sites of action have been proposed for antisense RNA including transcription, nuclear RNA,[7] and transport.[8] Reviews covering the mode of action of antisense RNA in detail can be found in Refs. 9–12. In construction of an antisense RNA vector, a cDNA is cloned into a vector that allows transcription of an antisense inside the cell. Usually, transcription is driven by a strong viral promoter such as cytomegalovirus (CMV) or simian virus 40 (SV40) to guarantee high levels of antisense transcripts, but it is not clear whether such a strong promoter is necessary. If weaker promoters are sufficient, then the possibility of using cell/tissue-specific promoters may add an extra dimension to control specificity of action by directing antisense expression to specific subpopulations of cells. The ideal length of a transcript is unclear but, *in vitro* at least, transcripts as short as 20 bases have been shown to be efficacious.[13]

[3] C. Helene and J. J. Toulme, *Biochim. Biophys. Acta* **1049,** 99 (1990).
[4] M. R. Rebagliati and D. A. Melton, *Cell* **48,** 599 (1987).
[5] S. R. Schmid and P. Linder, *Mol. Microbiol.* **6,** 283 (1992).
[6] P. R. Albert, *Vitam. Horm.* **48,** 59 (1994).
[7] B. Feng and D. T. Denhardt, *Ann. N.Y. Acad. Sci.* **660,** 280 (1992).
[8] D. T. Denhardt, *Ann. N.Y. Acad. Sci.* **660,** 70 (1992).
[9] G. Sczakiel, *Antisense Nucleic Acid Drug Dev.* **7,** 439 (1997).
[10] W. Nellen and G. Sczakiel, *Mol. Biotechnol.* **6,** 7 (1996).
[11] W. Schuch, *Symp. Soc. Exp. Biol.* **45,** 117 (1991).
[12] W. Nellen and C. Lichtenstein, *Trends Biochem. Sci.* **18,** 419 (1993).
[13] J. D. Meissner *et al., Exp. Cell Res.* **225,** 112 (1996).

However, more commonly, antisense probe lengths are on the order of a few hundred bases. Ideally, design of a suitable antisense probe would take into account the secondary structure of the RNA and sites of potential RNA–protein interaction. There are public domain programs (such as Mfold and FoldRNA within the GCG package[14] that can be used for predicting RNA secondary structure, and all of them use algorithms that rely on predicting one or more low-energy conformations. Unfortunately, all such energy-minimizing algorithms have only a limited value in predicting actual secondary structures because there are usually a large number of potential structures possessing similar energies. This problem is proportionally exacerbated as the size of the input sequence increases. Indeed, it is possible to derive several overlapping constructs and frequently to find that some constructs are more effective than others in lowering protein levels (see Fig. 1b). So although rational antisense design is of fundamental importance, it is heavily colored by empiricism and pragmatism. Irrespective of constraints due to considerations of secondary structure, antisense design is often hedged by the limited regions of unique sequence available across members of a gene family. This is particularly the case when considering sequence homology across $G\alpha$ subunits. One strategy we have found useful is to target the untranslated regions of an mRNA transcript where sequence divergence among members of a gene family is usually greater than within coding regions, i.e., it is not necessary to target the coding region or, even more specifically, it is not necessary to focus exclusively on the initiation codon—although if the N terminal presents a unique domain, it may be useful to target this region as this may add the potential of contributing a translational block. Several studies have reported the use of short (<45 bases) antisense RNA derived from the 5' untranslated regions of $G\alpha_{i2}$,[15] $G\alpha_{13}$,[16,17] $G\alpha_{12}$ and $G\alpha_{13}$,[17,18] $G\alpha_o$,[17] and $G\alpha_q$ [19] to attenuate gene expression, and in the case of the $G\alpha_{12}$ and $G\alpha_{13}$ constructs the antisense did not extend to the initiation codon. In one case expression constructs containing as little as 20 bp of antisense sequence was used to specifically attenuate $G\alpha_s$ expression in HL-60 cells.[13] Use of such small constructs offers two further advantages: shorter sequences should minimize adventitious overlap with other targets and such short constructs can be cloned directly using synthetic oligonucleotides, thereby obviating cloning of any insert sequence. It is unclear what degree of mismatch can be tolerated and it is therefore

[14] M. Zuker, *Methods Enzymol.* **180,** 262 (1989).
[15] T. Voisin *et al., J. Biol. Chem.* **271,** 574 (1996).
[16] T. Voisin, A. M. Lorinet, and M. Laburthe, *Biochem. Biophys. Res. Commun.* **225,** 16 (1996).
[17] D. C. Watkins, G. L. Johnson, and C. C. Malbon, *Science* **258,** 1373 (1992).
[18] E. H. Jho and C. C. Malbon, *J. Biol. Chem.* **272,** 24461 (1997).
[19] P. A. Galvin-Parton *et al., J. Biol. Chem.* **272,** 4335 (1997).

prudent to target only molecules derived from species where the sequence is known or can be easily derived.

Means of Delivery

Intranuclear injection of antisense plasmids is our preferred mode of delivery into neurons in culture, because like most highly differentiated cells, postmitotic neurons are refractory to gene transfer techniques such as transfection by calcium phosphate or lipofection or electroporation. There are two limitations to gene delivery by intranuclear injection: (1) subsequent analysis is limited to single cells because it is not feasible to inject enough neurons to carry out biochemical assays; (2) only relatively large, robust cells will withstand intracellular injection of an antisense construct followed 1 or 2 days later by subsequent impalement or patching for electrophysiological analysis. In our hands this does not present a problem using superior cervical ganglion neurons but many CNS neurons may be refractory to this approach.

There are several means of tracking injected cells. One crude but effective method is simply to photograph the field of cells plated onto gridded coverslips and visually identify injected cells several days later. There are several problems with this approach. First, neuronal cells can change shape and migrate during the course of culture. Second, there is no indicator of the success of the microinjection. Third, it is tedious and time consuming. Coinjection of a marker overcomes all of these problems. Two types of markers can be used. Simple coinjection of a fluorescent marker such as Dextran Green allows unambiguous identification of successfully injected cells and the marker protein is stable for several days and retains its fluorescence when aldehyde or acetone fixed. This latter point is especially useful when the injected cells must be eventually processed for immunohistochemistry. A second type of marker is offered by constructs encoding fluorescent markers such as green fluorescent protein (GFP) driven by a CMV promoter. This offers an additional advantage in that not only is the success of injection monitored but the competence of the injected cell to transcribe the injected DNA is also assayed.

Assessment of Specificity of Action

The greatest burden on the antisense approach is proof of specificity of action. The most stringent requirement is to demonstrate a specific attenuation of protein levels and preferably to demonstrate concomitantly that levels of related gene products remain unchanged. This attenuation must only be apparent when using the appropriate antisense. Common

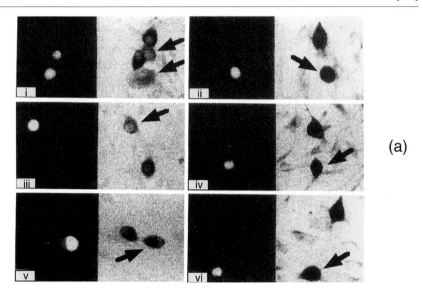

(a)

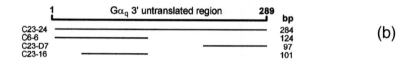

(b)

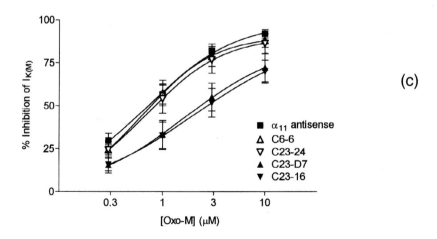

(c)

controls include the use of sense or nonsense sequences. In the case of antisense RNA, the usual control is to use sense RNA, because most vectors allow transcription of both sense and antisense RNA strands from opposite ends of the vector. If a specific antibody is available, then this represents the easiest means of demonstrating a specific change in protein levels. If no specific antibody is available, then difficulties arise and a more distal function of protein levels must be used.

Ultimately, antisense attenuation of protein levels suffers from the same pitfalls of any other metabolic blocker; efficacy demonstrates *involvement* of that molecule but not necessarily the directness of involvement of that molecule in the response. Pleiotropic effects can also give rise to questions concerning the directness of action of a given antisense. The burden of proof for demonstration of *direct* interaction requires molecular approaches such as yeast one-hybrid analysis, immunoprecipitation, or glutathione S-transferase (GST) pull-downs.

In Fig. 1 the results of a study of the involvement of $G\alpha_q$ in agonist-mediated inhibition of an M-type potassium current in dissociated sympathetic neurons are shown. Injected cells can be identified by their nuclear fluorescence in adjacent micrographs. Figure 1a(i) and (ii) shows cultures of dissociated superior cervical ganglia immunostained with an anti-$G\alpha_o$ antibody consequent to injection of $G\alpha_o$ antisense [Fig. 1a(i)] and $G\alpha_q$ (clone C23-D7) antisense [Fig. 1a(ii)] on $G\alpha_q$ levels—levels of $G\alpha_o$ are attenuated in the injected cells but not in the uninjected cell. Figure 1a(iii) to (vi) shows similar cultures immunostained with anti-$G\alpha_q$ antibody after injection of $G\alpha_q$ (clone C23-D7) antisense, $G\alpha_o$ antisense [Fig. 1a(iii)], $G\alpha_q$ (clone C23-24) antisense, and $G\alpha_{11}$ (clone 97-4). Only cells injected with $G\alpha_q$ (clone C23-D7) showed a decrease in $G\alpha_q$ immunoreactivity. The complementarity of the effects of antisense $G\alpha_q$ and $G\alpha_o$ constructs on $G\alpha_q$ and $G\alpha_o$ immunoreactivity demonstrate the specificity of this approach. The immunocytochemical data are mirrored by the dose–response curve of the agonist inhibition of $I_{K(M)}$. An example of the efficacy of different

FIG. 1. Effect of antisense treatment on endogenous levels of $G\alpha_o$ and $G\alpha_q$ in ganglion cells of the superior cervical ganglion in dissociated culture. (a) Effect of treatment on levels of $G\alpha_o$ and $G\alpha_q$ as adjudged by immunostaining. (b) Sequence relationship between four different $G\alpha_q$ antisense constructs. (c) Oxotremorine dose–response curves for neurons injected with constructs described above and with a $G\alpha_{11}$ antisense construct. Fluorescent micrographs show injected cells visualized with coinjected FITC–dextran. Light-field micrographs show immunostaining visualized with alkaline phosphatase as described in text. Arrows indicate injected cells, which can also be seen in adjacent fluorescent micrographs. Dose–response curves in (c) obtained with C23-D7 and C23-16 are significantly different from those obtained with $G\alpha_o$ and $G\alpha_{11}$ ($p < 0.004$). [Data are reproduced with permission from Ref. 22.]

antisense constructs can be seen with antisense $G\alpha_q$ (clone C23-24). This clone produced no change in $G\alpha_q$ immunoreactivity and no change in agonist inhibition of $I_{K(M)}$ yet its sequence completely overlapped with two clones that were effective both immunohistochemically and functionally. Although we have not investigated causes of such variation this nevertheless serves to underline the necessity of producing several independent antisense constructs.

Materials and Reagents

Reagents

> Tri Reagent: RNA/DNA/protein isolation reagent (Molecular Research Centre Oxford, Oxford, UK)
> Enzymes, corresponding buffers, and oligo(dT) primer (Promega, Chilworth Research Centre, Southampton, UK)
> Amersham DNA ligation system (Amersham International, Little Chalfont, Buckinghamshire, UK)
> Eukaryotic TA cloning kit (bidirectional) (InVitrogen, Leek, The Netherlands)
> Qiagen plasmid kit (Qiagen, Hilden, Germany)

Injection and Recording Equipment

> Eppendorf (Hamburg, Germany) micromanipulator: Allows positioning and movement of injection needle
> Eppendorf transjector: Provides pressure through injection needle, can use constant pressure or short pressure bursts
> Inverted microscope (e.g., a Nikon, Garden City, NY) with fluorescence attachment and lamp and a movable stage: Both lower power ($\times 4$ or $\times 5$) and high-power ($\times 40$) lenses are needed
> Incubator surrounding the microscope stage, with CO_2 supply/regulation and temperature control: To maintain cells in CO_2 and provide heat

Injection Needles

> Electrodes: Made from standard wall 1.2-mm glass and pulled using a three-stage pull on a horizontal puller to make shanks as short as possible while still retaining a sharp tip; this reduces the frequency of occurence in the tip of bubbles, which prevent adequate injection. Electrodes have a series resistance of 50–80 MΩ. Alternatively, one

can purchase prepulled needles from Eppendorf for intranuclear injections

Borosilicate electrode glass (GC120TF10; Clark Electromedical, Pangbourne, Berks, UK)

Preparation of Solutions

For intranuclear injections prepare about 50 μl of a solution containing plasmid (final concentration, 25 to 400 μg/ml) plus 0.5% (w/v) fluorescein isothiocyanate (FITC)-conjugated dextran (MW 70,000) in Ca^{2+}-free and glucose-free Krebs solution. Make a 10% (v/v) stock solution of FITC–dextran and keep at 4° (FITC–dextran is hygroscopic, so appropriate care must be taken for storage). DNA/FITC–dextran solutions should be filter centrifuged at 13,000 rpm for 5 min through a Vectaspin polysulfone microfilter (0.2 μm; Whatman, Clifton, NJ).

Fluorescein isothiocyanate (FITC)-labeled dextran: 70 kDa, lysine fixable (Molecular Probes, Eugene, OR)

Methods

Cloning Strategy

Numerous cloning strategies can be used. The following protocol is a suggestion for inexperienced cloners. Experienced molecular biological laboratories should continue to use their own (less expensive) protocols. Given the proven efficacy of short antisense RNA constructs, cloning annealed complementary oligonucleotides into a eukaryotic expression vector represents the simplest and quickest option. Because most commonly used vectors have two polymerase priming sites at opposing ends of the multiple cloning site, this is most easily achieved by designing the primers with an additional 3' A to facilitate cloning of the annealed oligonucleotides into a T-tailed vector such as pCDNA (InVitrogen). No vector preparation is necessary and background is low.

1. Design two suitable complementary 20- to 40-mers, each with an extra 3' adenosine. Purify via high-performance liquid chromatography (HPLC) or polyacrylamide gel electrophoresis (PAGE).

2. Mix and dilute oligonucleotides to 10 μM in TE (Tris–EDTA, pH 8.0) in an Eppendorf tube, place in boiling water, and allow to cool to room temperature to anneal.

3. Ligate approximately 1 pmol of annealed oligonucleotides with 10 fmol of pCDNA (corresponding to 13 ng of insert and 50 ng of vector, respectively) according to the manufacturer instructions.

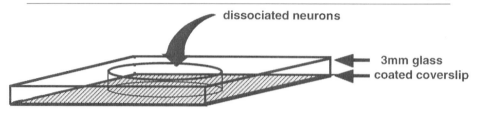

FIG. 2. Cell culture plates.

4. Select transformants at random and screen directly by polymerase chain reaction (PCR), using each of the complementary oligonucleotides and a vector primer to derive orientation. Do not screen by blue/white selection, as the small size of the insert will allow readthrough of the *lacZ* gene.

5. Confirm identity by sequencing.

Plasmid Purification

Both cesium chloride banding and some commercial ion-exchange columns deliver high-quality DNA that is nontoxic when injected into cells. We routinely use Qiagen Midi or Maxi columns according to the manufacturer instructions. DNA should be concentrated by NaCl–ethanol precipitation and resuspended in $0.1\times$ TE (pH 7.4) at about 1 mg/ml and stored at $-20°$. If toxicity appears to be a problem, then try a direct comparison of the same plasmid DNA purified by both Qiagen and cesium chloride banding.

Cell Culture

Cells are plated on custom-made microwells, using 22×22 mm coverslips coated with poly(L-lysine). Culturing on glass permits subsequent use of acetone as a fixative, which we find superior to aldehyde fixation for immunostaining using anti-$G\alpha$ antibodies (alternatively, if no subsequent fixation is required, then cells can be plated directly onto 35-mm plastic tissue culture dishes). Use of thin coverslips also enables high numerical aperture objectives to be used, thereby increasing resolution. Microwells are made by cutting a 15-mm-diameter hole in a 3-mm-thick square glass cell culture dish and affixing the cleaned coverslip to the lower surface with a thin smear of sterile vacuum grease (see Fig. 2). The lower surface of the coverslip can then be gridded (with a permanent fine marker) to aid subsequent localization of injected cells (this latter step is not necessary if a marker has been coinjected). Cells are plated at about 50 cells/mm². Cells are microinjected in culture medium while being provided with 5% CO_2

and maintained at 37°. If using neurons derived from superior cervical ganglia, 500 nM tetrodotoxin (πx) is present in the culture medium to prevent repetitive firing of action potentials and possible calcium loading of cells. Following microinjection, cells are refed by replacing all culture medium with medium not containing TTX.

> Glass coverslips (No. 1$\frac{1}{2}$, 22 × 22 mm; Chance Propper, West Midlands, UK.)

Intranuclear Injection

1. Fill the tip of the injection needle with 1–3 μl of solution, using an Eppendorf microloader attached to a P20 Gilson (Middleton, WI) pipette or, alternatively, back fill the electrode with a 5-μl Hamilton-type gas-tight syringe (for 1.5-mm glass electrodes).

2. (*Note:* Steps 2–4 refer to the use of an Eppendorf microinjector.) After filling, injection needles are secured in a holder that is connected to a transjector providing constant pressure. The holder is held tightly in the micromanipulator arm while the injection needle is lowered into the culture medium and brought close to the cells in the dynamic movement mode (needle remains in position when the micromanipulator control joystick is released). Manual injections are performed in proportional mode (needle returns to same preset vertical level above the cells when the manipulator joystick is released). Proportional mode is calibrated to attain the required fine or coarse movement control.

3. Constant pressure is applied through the injection needle tip by the transjector (we use pressure settings of about 80–100 hPa to achieve a steady flow of plasmid solution from the tip of the injection needle, but this depends very much on the properties of the needle). Cells are injected by gently lowering the injection needle into the nucleus (or cytoplasm) with the micromanipulator in proportional mode and holding the injection needle still for a short time (1–2 sec) to allow the cells to fill with the plasmid solution. Filling can be observed as a slight clearing of the nucleus/cytoplasm as well as a slight swelling (a large amount of swelling is, however, detrimental to the cell).

4. The injection needle is then removed by allowing it to spring back gently to the preset vertical level when the joystick is released. A successful injection results in bright FITC fluorescence in either the nucleus (with the cytoplasm dark) or the cytoplasm (in which case the nucleus remains dark). Occasionally the injection needle blocks, and this can be cleared by applying a short burst of high pressure from the transjector. For fast, easy injections it is best to bring the cell to the tip of the needle by moving the microscope stage, rather than bringing the needle to the cell.

5. If intranuclear injections are to be carried out on the same rig that is to be used for electrophysiological recording, then the electrode is positioned over the nucleus and lowered until it contacts the cell surface. Impalement of the neuron can be facilitated by passing hyperpolarizing current (about 0.3 nA, depending on the resistance of the electrode) into the electrode once the electrode resistance has been compensated. Contact will be registered as a negative-going voltage deflection during the hyperpolarization pulse. If the electrode is lowered further, penetration of the nucleus is apparent by a sudden reduction in the amplitude of the voltage deflection and a slowing of the kinetics of the voltage response.

Immunohistochemistry

Any of a variety of standard immunostaining protocols can be used. We prefer using alkaline phosphatase to visualize antigen as it is easy to control the intensity of the reaction product, which is important for assessing the efficacy of the antisense construct in attenuating gene expression. In addition, staining does not interfere with visualization of the fluorescently labeled nuclei of injected cells.

1. Cells are washed in TBS-IC (5.5 mM Tris-HCl, 137 mM NaCl, pH 7.4) to remove culture medium.
2. Fix the cells in either acetone (20 min at room temperature) or acetone–methanol (1 : 1 by volume for 1 min at −20°). In our hands, use of acetone instead of formaldehyde produces lower background staining.
3. Wash the cells three times (5 min each) with TBS-IC at room temperature.
4. Block nonspecific binding of secondary antibody by incubating with 10% (v/v) serum from the animal in which the secondary antibody was raised for 30 min at room temperature.
5. Incubate with primary antibody. These conditions need to be determined empirically, but usually overnight incubations at 4° give the lowest background. If background is not a problem, then incubations at room temperature for 1 hr will suffice.
6. Wash the cells three times (5 min each) with TBS-IC at room temperature.
7. Incubate with secondary antibody. We have not usually found this step to be a significant source of background, so incubation at room temperature for 1 hr is carried out.
8. Wash the cells three times (5 min each) with TBS-ABC [50 mM Tris-HCl (pH 7.4), 150 mM NaCl] at room temperature.
9. Incubate with ABC reagent for 30 min at room temperature.

TABLE I

Use of Antisense RNA in Analyzing G-Protein Actions

G-protein subunit	Effect of antisense RNA	Ref.
$G\alpha_{12}$	P19 apoptosis	18
$G\alpha_{13}$	P19 apoptosis	18
$G\alpha_o$	Stimulation of PLC and transformation in IIC9 cells	21
$g\alpha_q$	Hyperadiposity in transgenic mice	19
$G\alpha_{i2}$	Attenuation of PPY inhibition of cAMP	16
$G\alpha_s$	Induction of myeloid differentiation in HL-60 cells	13
$G\alpha_{i2}$	Potentiation of stimulated PLC in F9 and 17/2.8 cells	17
$G\alpha_{i2}$	Increase in cAMP in transgenic mice	22
$G\alpha_{i2}$	Differentiation of F9 cells	17
$G\alpha_s$	Attenuation of cAMP production in GH3 cells	23
$G\alpha_o$	Attenuation of agonist inhibition of N-type Ca^{2+} current in SCG neuron	24
$G\alpha_q$	Attenuation of agonist inhibition of $I_{K(M)}$ in SCG neuron	20

PLC, Phospholipase C.

10. Wash the cells three times (5 min each) with TBS-ABC at room temperature.

11. Incubate with BCIP/NBT 5-bromo-4-chloro-3-indolylphosphate toluidinium/nitroblue tetrazolium) at room temperature. The reaction is stopped with distilled water. The time of incubation must be determined empirically but the usual range is 2–30 min.

12. Cells are viewed under bright-field and fluorescence optics to visualize the immunostaining and identify injected cells.

Concluding Remarks

Use of antisense RNA specifically to attenuate gene expression has been successful in identifying specific G-protein subunits involved in neuronal signal transduction (Table I[13,16–24]). Particular attention must be paid to probe design and especially to use of controls. Antisense RNA has several advantages over the more widely used antisense oligonucleotide approach, chiefly in specificity and longevity of action. The principal limitations lie in gene delivery, but potential solutions to this problem involve the use of alternatives such as biolistic delivery and the use of adenoviral

[20] J. E. Haley et al., J. Neurosci. **18,** 4521 (1998).
[21] J. Cheng et al., J. Biol. Chem. **272,** 17312 (1997).
[22] C. M. Moxham, Y. Hod, and C. C. Malbon, Dev. Genet. **14,** 991 (1993).
[23] R. H. Paulssen et al., Eur. J. Biochem. **204,** 413 (1992).
[24] P. Delmas et al., Eur. J. Neurosci. **10,** 1654 (1998).

vectors. The latter approach has been used successfully to deliver antisense RNA constructs both *in vivo*[25] and to cultured cells[26] and offers great promise for the future.

Acknowledgments

The authors' work was supported by grants from the Wellcome Trust and the U.K. Medical Research Council.

[25] A. Kammesheidt *et al., Neuroreport* **8,** 635 (1997).
[26] D. Lu, H. Yang, and M. K. Raizada, *Am. J. Physiol.* **274,** H719 (1998).

[11] Gene Functional Analysis in Nervous System

By Finn Hallböök, Marina Catsicas, Julie K. Staple, Gabriele Grenningloh, and Stefan Catsicas

Introduction

Functional Genomics

Functional genomics comprises a variety of different approaches to address the precise role of selected gene products. With developments in molecular genetics, we will face more and more situations where novel gene products with unknown functions are identified. Establishing the precise function of the identified gene is an essential step for the understanding of the physiological or pathological mechanisms to which it may contribute.

Measuring the effects of inhibition of gene expression is a key step in functional gene analysis and antisense oligonucleotides have been used successfully for this, *in vitro* and *in vivo.*[1] Specific inhibition is accomplished by virtue of base pair hybridization between an mRNA and a complementary antisense oligonucleotide.[2] Phosphodiester oligodeoxynucleotides are nuclease sensitive and are rapidly digested in the cell. Phosphorothioate oligodeoxynucleotides (PONs), which have a sulfur atom substituted for

[1] R. W. Wagner, *Nature Med.* **1,** 1116 (1995).
[2] P. C. Zamecnik and M. L. Stephensson, *Proc. Natl. Acad. Sci. U.S.A.* **75,** 280 (1978).

one of the phosphate oxygen atoms at a nonbridging position, are nuclease resistant[3–5] and have been shown to inhibit gene expression.[6–10]

Neuronal Differentiation

We have addressed mainly early events occuring during neuronal differentiation. In the developing central nervous system, neurite outgrowth as well as the elimination of neurons are essential processes underlying the formation of precise patterns of synaptic connectivity (reviewed in Ref. 11). In an attempt to identify molecules and mechanisms involved in these processes, we have either selected relevant candidate genes, such as members of the neurotrophin family (see below), or employed a subtractive cloning strategy to identify cDNAs differentially expressed during specific developmental stages.[12] Two subtracted clones, encoding SNAP-25 and stathmin (see below), fulfilled a series of selection criteria (M. Catsicas and G. Grenningloh, unpublished data, 1998) and we have done extensive work to characterize both molecules. The use of antisense oligonucleotides has been central to our initial functional studies. It was possible to use antisense olgonucleotides because the two proteins, and changes in their levels of expression, are readily detectable with antibodies that we and others have raised.[6,10,13] In subsequent studies on nerve growth factor (NGF), initiated by one of us (F.H.), the lack of appropriate antibodies to evaluate the efficacy of the antisense oligonucleotides has raised important problems. We have therefore developed a strategy that may be applied when appro-

[3] S. Spitzer and F. Eckstein, *Nucleic Acids Res.* **16,** 11691 (1988).

[4] C. Cazenave, C. A. Stein, N. Loreau, N. T. Thuong, L. M. Neckers, C. Subasinghe, C. Helene, J. S. Cohen, and J. J. Toulme, *Nucleic Acids Res.* **17,** 4255 (1989).

[5] F. Eckstein and G. Gish, *Trends Biochem Sci.* **14,** 97 (1989).

[6] A. Osen-Sand, M. Catsicas, J. K. Staple, K. A. Jones, G. Ayala, J. Knowles, G. Grenningloh, and S. Catsicas, *Nature (London)* **364,** 445 (1993).

[7] R. R. Ji, Q. Zhang, K. Bedecs, J. Arvidsson, X. Zhang, X. J. Xu, H. Z. Wiesenfeld, T. Bartfai, and T. Hokfelt, *Proc. Natl. Acad. Sci. U.S.A.* **91,** 12540 (1994).

[8] C. Ramazeilles, R. K. Mishra, S. Moreau, E. Pascolo, and J. J. Toulme, *Proc. Natl. Acad. Sci. U.S.A.* **91,** 7859 (1994).

[9] K. Sainio, M. Saarma, D. Nonclercq, L. Paulin, and H. Sariola, *Cell. Mol. Neurobiol.* **14,** 439 (1994).

[10] M. Catsicas, A. Osen-Sand, J. Staple, K. A. Jones, G. Ayala, J. Knowles, G. Grenningloh, E. Merlio-Pich, and S. Catsicas, *in* "Methods in Molecular Medicine: Antisense Therapeutics" (S. Agrawal, ed.), pp. 57–85. Humana Press, Totowa, New Jersey, 1995.

[11] C. S. Goodman and C. J. Shatz, *Cell/Neuron* **72**(10), 77 (1993).

[12] M.-C. Lebeau, G. Alvarez-Bolado, W. Wahli, and S. Catsicas, *Nucleic Acids Res.* **19,** 4778 (1991).

[13] G. Di Paolo, V. Pellier, M. Catsicas, B. Antonsson, S. Catsicas, and G. Grenningloh, *J. Cell. Biol.* **133,** 1383 (1996).

priate antibodies are missing. In the following sections we describe our protocols for the studies done with SNAP-25,[6,10] stathmin,[13] and NGF.[14]

Functional Characterization of Snap-25

SNAP-25 is involved in transmitter release through its role in vesicle fusion.[15] Since vesicle fusion is also necessary for membrane expansion during axonal growth, the protein could also be involved in this latter process. We have used antisense oligonucleotides to inhibit its expression and address this possibility.

Oligonucleotide Design for SNAP-25

We considered several parameters, including proximity of the sequence chosen to the translation start site, GC content, length, melting temperature, and secondary structure. Potential oligonucleotides were checked using the program Primer (Scientific and Educational Software) and those sequences likely to form homodimers or that had internal complementary regions were eliminated. Specific sequences, known to be toxic owing to interactions with proteins, such as guanine tetramers, were avoided.[16] We chose one oligonucleotide that includes the translation start site of the mouse SNAP-25 protein (nucleotides 1–20) (PSNAP1, 5′-ATG TCT GCG TCC TCG GCC AT-3′) and one oligonucleotide complementary to a site toward the 3′ end of the protein (nucleotides 533–552) that is identical in the chick and mouse cDNA sequences (SNAP2AS, 5′-CTT CTC CAT GAT CCT GTC AA-3′).[17] Several nonsense oligonucleotides were used, including RANDOM1.SNAP (5′-ATC CCT CCG TGT AGC GCG TT) and PKC8GMN (5′-ACT GCT ACA CCT CAC GTG TT-3′).

In addition to unmodified phosphodiester oligonucleotides, it is possible to synthesize and use different types of "modified oligonucleotides," such as phosphorothioates, methylphosphonates, and 2′-*O*-alkyl- and 2′-*O*-fluoro-oligonucleotides, where either the phosphodiester backbone or the sugar moiety has been modified to confer better nuclease resistance. Unmodified oligonucleotides are degraded by nucleases and therefore

---

[14] F. Hallböök, A. Sahlén, and S. Catsicas, *Antisense Nucleic Acid Drug Dev.* **7,** 89 (1997).

[15] T. Söllner, M. K. Bennett, S. W. Whiteheart, R. H. Scheller, and J. E. Rothman, *Cell* **75,** 409 (1993).

[16] P. Yaswen, M. Stampfer, K. Ghosh, and J. Cohen, *Antisense Res. Dev.* **3,** 67 (1993).

[17] G. A. Oyler, G. A. Higgins, R. A. Hart, E. Battenberg, M. L. Billingstey, F. E. Bloom, and M. C. Wilson, *J. Cell. Biol.* **109,** 3039 (1989).

high concentrations are usually required,[18] whereas oligonucleotides composed of phosphorothioate-substituted nucleotides (PONs) are nuclease resistant[19] but sometimes have nonspecific toxic effects such as the non-sequence-specific inhibition of RNase H.[20] To determine which would be most effective and least toxic in our system, we tried unmodified oligonucleotides, PONs, and oligonucleotides that had only three phosphorothioate-modified residues at each end of the 20-mer. All oligonucleotides were synthesized on an automated DNA synthesizer (394-8; Applied Biosystems, Foster City, CA) using standard phosphoramidate chemistry. The oxidation step in the synthesis of PONs was achieved using tetraethyl-thiuram disulfide (TETD).[21] We found that the method of purification of the final product was critical in reducing nonspecific oligonucleotide toxicity. Some batches of oligonucleotides purified by *n*-butanol precipitation caused degeneration and death of neurons in culture within 24–48 hr. Further purification over Sephadex G-50 columns (Pharmacia, Piscataway, NJ) considerably reduced these problems, although they were not entirely eliminated. Also, coevaporation of oligonucleotides, after elution from Sephadex columns, with three 0.2-ml volumes of deionized water or dialysis against deionized water for 48 hr consistently resolved problems of toxicity.

Each type of oligonucleotide was tested by adding varying concentrations of the antisense PSNAP1 to cultures of rat cortical neurons immediately after plating and readding oligonucleotides every day for 1 week. The cells were then lysed for use in Western blot quantitation of SNAP-25 and neuron-specific enolase (NSE). Antisense PONs downregulated SNAP-25 by 95% within 7 days at a final concentration of 2 μM, while the other types of antisense oligonucleotides used had no effect at 2 μM or at higher concentrations (not shown).

Determination of Oligonucleotide Uptake by Cells

We used PONs that were labeled at the 5' end with fluorescein isothiocyanate (FITC) to visualize the accumulation of oligonucleotides into primary neurons from rat cortex. FITC was added to the 5' end of oligonucleotides via an amino group that was introduced using Aminolink II (Applied Biosystems). Fluorescence was visible in all cells 24 hr after addition of PONs–FITC at a final concentration of 2 μM and increased in intensity after 48 hr of incubation. When FITC alone was added to cultures no

[18] A. Caceres, J. Mautino, and K. S. Kosik, *Neuron* **9,** 607 (1992).
[19] S. T. Crooke, *Annu. Rev. Pharmacol. Toxicol.* **32,** 329 (1992).
[20] C. A. Stein and Y.-C. Cheng, *Science* **261,** 1004 (1993).
[21] H. Vu and B. Hischbein, *Tetrahedron Lett.* **32,** 3005 (1991).

cellular fluorescence was observed. Solubilized protein extracts of neuronal cultures treated with antisense PONs–FITC for 7 days contained undetectable levels of SNAP-25 compared with untreated cultures.

Evaluation of Dose–Response and Time Course

To optimize the dose of PONs to be used in our assays, various concentrations of oligonucleotides were added to cultured neurons and the effects on SNAP-25 levels were evaluated by immunoblotting. PONs were added directly to culture medium once a day at final concentrations which varied between 0.25 and 10 μM. Doses of PONs of 5 or 10 μM were toxic to both neurons and glial cells within 3–4 days. The glial cells appeared more sensitive to the toxic effects than the neurons and can be used as indicators of toxic effects (not shown). Small decreases in SNAP-25 were detectable after 3 days of treatment with 2 μM PONs. After 5 and 7 days of treatment with 2 μM PONs, SNAP-25 was decreased to 10% and undetectable levels, respectively, compared with untreated or nonsense PON-treated cultures. Cells looked healthy and expressed normal levels of NSE for at least 7 days of treatment (Fig. 1A).

Schedule of Oligonucleotide Addition

Because SNAP-25 is expressed at high levels by many cortical neurons when they are dissociated from newborn rats, it was essential to downregulate the protein as quickly as possible to evaluate changes in neuronal development. Because addition of 2 μM PONs to cultures each day efficiently downregulated SNAP, and without apparent toxicity, we chose to work with this concentration further to determine the most efficient schedule of PON addition. On the first day after dissociation and plating of neurons, PONs were added at a final concentration of either 2 or 10 μM. PONS were then added at 2 μM every day, every other day, or every 2 days. Immunoblots of proteins extracted from treated cultures after 7 days showed that initial addition of a 10 μM final concentration of PONs and subsequent 2 μM addition every other day results in optimal SNAP-25 downregulation with the least amount of total PONs added (Fig. 1B).

Functional Effects on Primary Neurons

To assess further the specificity of the inhibition we checked the level of expression of a series of control proteins, which included neuron-specific proteins as well as general cytoskeletal markers. The only protein tested that showed a reduction of expression (more that 60% in this case) was

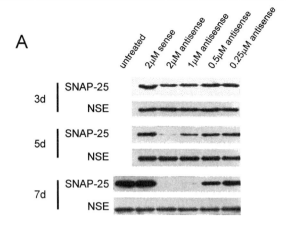

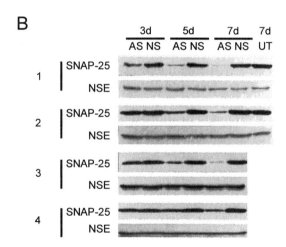

Fig. 1. Design of PON treatment in rat primary cortical neurons. (A) Time course and effective concentrations of PONs: cortical neurons were plated in 24-well dishes at 4×10^4 cells/well. PSNAP1 PON was added at a final concentration of 0.25, 0.5, 1, or 2 μM immediately after plating and the same dose of PON was readded each day for up to 7 days. After 3, 5, or 7 days of treatment cells were solubilized and protein extracts were used for immunoblots probed with SNAP-25 and NSE (Seralab) antibodies. (B) Effective schedule of PON addition: cortical neurons were plated as described above and PSNAP1 PONs were added at a final concentration of either 2 μM (#3 and #4) or 10 μM (#1 and #2) immediately after plating. Further addition of 2 μM PONs occurred every day (#1 and #3) or every other day (#2 and #4). After 7 days of treatment cells were solubilized and protein extracts were used for immunoblots probed with SNAP-25 and NSE antibodies. [Modified from Ref. 10, with permission.]

synaptophysin. Interestingly, the effects on the level of expression of synaptophysin were less pronounced if the density of neurons in culture was increased (not shown). Synaptophysin is a marker of differentiated synapses,[6,10] and so the effects on its level of expression, which cannot be attributed to a general effect on protein synthesis, suggested that inhibition of SNAP-25 expression reduced synapse formation. In addition, qualitative analysis of the cultures showed a strong reduction in the network of thin, axon-like neurites after antisense treatment (not shown).

Functional Effects on PC12 Cells

After exposure to NGF, rat pheochromocytoma (PC12) cells stop proliferating, differentiate, and assume a neuron-like morphology.[22] All PC12 cells express SNAP-25 and all of the NGF-induced neurites are SNAP-25 positive.[23] In addition, and most important, the cells can be incubated with PONs before the induction of neurite growth with NGF. Therefore, it was possible to study a situation where the levels of SNAP-25 were already low before the onset of neurite outgrowth.

The first step was to establish a dose–response curve in this new culture system. The cells were plated in medium containing NGF (25 ng/ml) and 10 μM PONs. PONs (0.2 to 2 μM final concentration each time) were then added every day or every 2 days. Higher concentrations were also tested but were found to have nonspecific effects on protein synthesis (not shown). Total proteins were extracted and antibody binding visualized and quantified on Western blots. On the basis of these data, for the neurite growth assays cells were plated on collagen-coated 24-well Costar (Cambridge, MA) plates with medium containing 10 μM PONs and 2 μM PONs was added every 2 days. The same PONs used for primary neurons were used for the PC12 cells. An additional, unrelated PON was used as control (5'-ATGCTGTGCTGTATGAGAAG-3'). NGF (25 ng/ml) was added to the cultures for the first time on day 2 and then every 2 days. Each plate contained wells with different culture conditions in duplicate (antisense, sense, and unrelated PONs); controls comprised NGF-stimulated as well as non-NGF-stimulated wells that were not treated with PONs. In all cultures, the cells started to extend neurites 3 days after NGF induction. After four additional days, the cultures were fixed and analyzed as described.[6,10] Both SNAP-25 antisense PONs, but not the control PONs, specifically reduced SNAP-25 expression by more than

[22] L. A. Greene and A. S. Tischler, *Proc. Natl. Acad. Sci. U.S.A.* **73,** 2424 (1976).
[23] P.-P. Sanna, F. E. Bloom, and M. C. Wilson, *Dev. Brain Res.* **59,** 104 (1991).

75% and this effect was dose dependent (Fig. 2A). Neuritogenesis was clearly inhibited when protein expression was decreased (Fig. 2B). We observed that the inhibition of neurite extension by antisense PONs was not only dose dependent, but also dependent on cell density. The efficiency of antisense PONs dramatically decreased with increasing number of cells per culture well. In our standard conditions, we plated 6000 cells/cm^2. Under these conditions, the antisense specifically reduced SNAP-25 expression by more than 75%. By increasing the number of cells to more than 10,000 cells/cm^2, inhibition of SNAP-25 expression was no longer achieved, nor was the concomitant block of neuritogenesis.

These results, entirely based on the use of antisense oligonucleotides, represent the first evidence that SNAP-25 is involved in axonal growth.

Functional Characterization of Stathmin

Stathmin, also known as p18, p19, prosolin, 19K, and Op18 (see references cited in Ref. 13), is a cytosolic protein whose phosphorylation is correlated with the action of multiple extracellular stimuli regulating proliferation and differentiation.[24] As a first step toward a functional analysis of stathmin in neurons, we have used antisense oligonucleotides to decrease protein levels in rat PC12 cells.

Oligonucleotide Design for Stathmin

On the basis of our experience with SNAP-25 (see above), in order to block stathmin expression we chronically added to the culture medium PONs complementary to positions 1 to 20 (AS2) and 19 to 36 (AS1) relative to the translation initiation site of the rat stathmin-coding region.[25] Controls included a nonsense (NS) and an unrelated (UR) PON (antisense PON AS1, 5'-CTCCAGCTCTTTCACCTG-3'; antisense PON AS2, 5'-TGAATATCAGAAGATGCCAT-3'; nonsense PON NS, 5'-CTTCCCACTGCTCGCTAT-3'; unrelated PON UR, 5'-GCATGTTCTT-GGTCA-3'). Western blot experiments and subsequent measurements of protein levels by optic densitometry of the autoradiograms allowed us to monitor stathmin expression after addition of PONs at different concentrations. The two antisense PONs were able to decrease stathmin protein levels significantly, to between 50 (AS2) and 75% (AS1), in a specific and dose-dependent manner (Fig. 3). Figure 3A and B shows that the strong decrease in expression of stathmin after 1 week of antisense treatment was

[24] A. Sobel, *Trends Biochem. Sci.* **16**, 301 (1991).
[25] U. K. Schubart, M. Das Banerjee, and J. Eng, *DNA* **8**, 389 (1989).

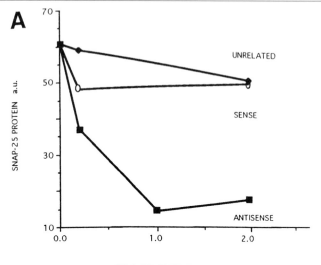

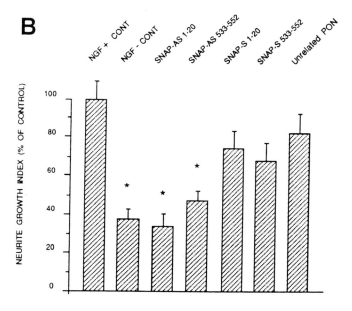

A

B

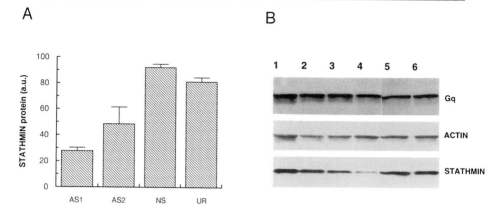

FIG. 3. Blockade of stathmin expression with antisense PONs. PC12 cells were maintained in RPMI 1640 supplemented with 1% horse serum and treated in the following way: on the first day, 10 μM antisense or control PONs was added, followed by 1, 2, or 4 μM every day in parallel samples. The second day and every other day, NGF (50 ng/ml) was added and the cells were incubated for another 5 days. (A) Effect of 4 μM antisense PONs AS1 and AS2 and 4 μM control PONs (NS, nonsense and UR, unrelated) on stathmin protein levels in PC12 cells Normalized mean gray values of the signals in Western blots (from three independent experiments) are shown in arbitrary units. Values denote means ± SEM. AS1 differs significantly from NS and UR ($p < 0.01$); AS2 differs from NS ($p < 0.01$) and from UR ($p < 0.05$). (B) Western blot of protein extracts from cultures incubated in the absence of PON (lane 1), in the presence of antisense PON AS1 at 1 μM (lane 2), 2 μM (lane 3), or 4 μM (lane 4), and nonsense PON at 1 μM (lane 5) and 4 μM (lane 6). Blots were incubated with antibodies specific for stathmin, actin, and $G_q\alpha$. Inhibition of stathmin expression after antisense treatment is specific, dose dependent, and reaches about 75% of control values. [From Ref. 13, with permission.]

FIG. 2. Effect of PON treatment on neurite extension by PC12 cells. (A) Dose–response curve of SNAP-25 expression inhibition in PC12 cells with antisense, control sense, and unrelated PONs. a.u., Normalized mean gray value of the signal in Western blots in arbitrary units. (B) Inhibition of neurite extension by PC12 cells. SNAP-25 antisense PONs prevent NGF-induced neurite elongation, whereas control PONs do not. To display the data together, the index of neurite growth (see text) is shown as a percentage of the NGF-stimulated control for all culture conditions tested. The asterisks indicate significant differences between the nonstimulated and NGF-stimulated controls ($p < 0.01$) and between the NGF-stimulated cultures in the absence (controls) and presence of PONs ($p < 0.05$), as calculated with the original data (t test). Error bars indicate the standard error of the mean. NGF + CONT, NGF-stimulated control; NGF − CONT, non-NGF-stimulated control. [From Ref. 6, with permission.]

not observed in cells treated with control PONs and that the expression of control proteins was not affected by the antisense PONs.

Uptake and Schedule of Oligonucleotide Addition

PC12 cell lines were cultured as described previously[22] and plated in 24-well collagen-coated multiwell dishes at a density of 10,000 cells per well in RPMI 1640 supplemented with 1% (v/v) horse serum. The uptake of PONs by PC12 cells was tested with three different FITC-labeled PONs (two 18-mers and one 20-mer) as described above for SNAP-25. Fluorescence was visible in more than 90% of the cells 24 hr after addition of PONs. To establish the dose–response, 10 μM high-performance liquid chromatography (HPLC)-purified antisense or control PONs, respectively, were added on the first day followed by 1, 2, or 4 μM every day. The second day and every other day, NGF (50 ng/ml) was added and cells were incubated for another 5 days. The cells were photographed using an Axiovert 135 microscope (Zeiss, Thornwood, NY). For Western blot analysis, the cells were removed from the culture dish, pelleted, resuspended in sample buffer for polyacrylamide gel electrophoresis (PAGE), and processed for immunodetection.

Functional Effects of Blockade of Stathmin Expression

Visual inspection of cultures treated with the antisense PONs revealed an apparent inability of the PC12 cells to respond to NGF. The most dramatic effects were observed with PON AS1, which correlated with its efficiency in inhibiting stathmin expression by 75%. Whereas in control PON-treated cultures, cells extended processes after 5 days of NGF exposure, no neurite outgrowth was detectable in cultures treated with antisense PON AS1. In contrast to what we observed after inhibition of SNAP-25 expression (see above), the cells maintained the characteristic round shape and phase-bright features of proliferating PC12 cells (Fig. 4).

To measure cell proliferation, we used the bromodeoxyuridine (BrdU) incorporation method, which allows the detection of cells synthesizing DNA. PC12 cells were treated for 6 days with 4 μM PON and incubated in the presence of BrdU for the last 40 hr of PON treatment. In control cultures, we found that about 50% of the cells incorporated BrdU after 5 days of NGF treatment, whereas about 90% were positive in the absence of NGF (see Ref. 13). After blockade of stathmin expression with antisense PON AS1, the rate of BrdU incorporation was highly increased ($p < 0.001$) as compared with NGF-treated controls in the presence or in the absence of nonsense PON. Also, PON AS2 had a highly significant effect on NGF-stimulated cessation of proliferation, although less prominent than

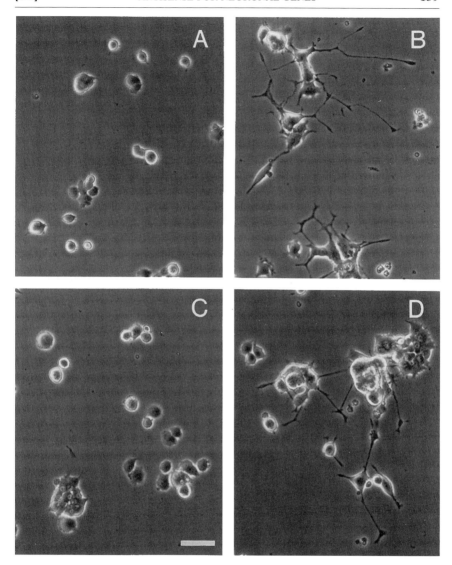

FIG. 4. Effect of stathmin antisense PON AS1 on NGF-induced neurite outgrowth in PC12 cells. Phase-contrast micrographs of PC12 cells maintained in RPMI 1640 supplemented with 1% horse serum for 6 days without any addition (A), with NGF (B), with NGF and 4 μM antisense PON AS1 (C), or with NGF and 4 μM nonsense PON (D). Neurite outgrowth in response to NGF is seen only in (B) and (D). Antisense-treated cells (C) are similar to cells not induced with NGF (A). Bar: 25 μm. [From Ref. 13, with permission.]

that of PON AS1, which is consistent with its lower efficiency in downregulating stathmin expression.[13] This result indicates that antisense-treated cells did not withdraw from the cell cycle, but rather continued to divide at a rate comparable to undifferentiated cells. In control experiments, nonsense PON (NS) did not affect BrdU incorporation in NGF-treated cultures.

These data, entirely based on the use of antisense oligonucleotides, represent the first evidence of involvement of stathmin in the growth arrest promoted by NGF in PC12 cells.[13] Despite considerable previous interest in the expression and phosphorylation of stathmin in neurons, before this work there was no direct evidence implicating this protein in specific neuronal functions. This example therefore illustrates the efficacy of using antisense oligonucleotides as initial tools to study gene function in neuronal cells.

Functional Characterization of Nerve Growth Factor

During development, NGF is synthesized in minute amounts in the target tissues of NGF-dependent peripheral neurons.[26] Methods to measure physiological levels of NGF are based on mRNA analysis, antibodies to mouse NGF, or the biological activity of NGF. However, these antibodies do not work in immunohistochemistry analyses for chicken NGF. The absence of reliable methods to measure physiological levels of chicken NGF protein (0.01–0.5 ng/mg wet tissue) makes the evaluation of antisense chicken NGF oligonucleotides difficult in systems expressing physiological levels of NGF. Therefore, we have used a system based on transient expression of high levels of NGF in COS cells.

We have previously developed a system for NGF production based on transient expression of chicken NGF in COS cells,[27] and we used this system to investigate whether antisense PONs could block NGF synthesis. COS cells were transfected by electroporation with a transient expression vector[28] construct (pECN) containing DNA fragments encoding chicken NGF.[27,29,30] To show that the inhibiting capacity of the antisense PONs was specific, we also used the COS cell expression vector with a *Xenopus* neurotrophin 4 fragment. Typically, electroporated cells were grown in Dulbecco's modified Eagle's medium (DMEM) with 10% (v/v) fetal calf serum (FCS) for

[26] H. Thoenen and Y. A. Barde, *Physiol. Rev.* **60,** 1284 (1980).
[27] F. Hallböök, T. Ebendal, and H. Persson, *Mol. Cell. Biol.* **8,** 452 (1988).
[28] G. G. Wong, J. S. Witek, P. A. Temple, A. C. Wilkens, D. P. Leary, S. S. Luxenberg, E. L. Jones, R. A. Brown, E. C. Kay, C. Orr, D. W. Shoemaker, R. J. Golde, and R. M. Kaufman, *Science* **228,** 810 (1985).
[29] F. Hallböök, C. Ibàñez, and H. Persson, *Neuron* **6,** 845 (1991).
[30] K. Kullander and T. Ebendal, *J. Neurosci. Res.* **39,** 195 (1994).

8–10 hr, washed, and grown for 4 days in DMEM with 0.5% (v/v) FCS supplemented with the PONs. Five antisense NGF PONs were designed, covering different regions of the chicken NGF mRNA including the translation initiation and stop codons (see below). The COS cells, which synthesize chicken NGF, were treated with nonsense and antisense NGF PONs and the amount of chicken NGF protein released in the medium was analyzed using a bioassay and sodium dodecyl sulfate (SDS) gel electrophoresis combined with Western blotting. The various oligonucleotides were tested individually for nonspecific cell toxicity.

Oligonucleotide Design for Nerve Growth Factor

On the basis of the experience gained with SNAP-25 and stathmin, several parameters were considered in our design of oligonucleotides: proximity of the sequence to the translation initiation site, GC content, oligonucleotide length, melting temperature, and secondary structure. Candidate oligonucleotides were examined using the software Oligo (v. 4.0; Primer Analysis Software, Plymouth, MN), and sequences that contained internal stem–loops or complementary regions, or could possibly form homodimers, were eliminated. Specific sequences that have been shown to be toxic, such as guanine tetramers, were avoided.[16,20] Five different NGF 18-mer antisense oligonucleotides were initially selected from the NGF mRNA sequence[31] (AS 1, GGACATTACGCTATGCAC; AS 2, TCTGACTTTG-GAGCTGCC; AS 3, TGCGTGCTGAACAGGACC; AS 4, CGTACTCT-GTCCTTTCCT; and AS 5, CTCAGGGTCTCCCCGATT) in addition to two nonsense oligonucleotides (NS 1, CATAGTGCATAGCGATGC; and NS 2, CACGTATCGCATTACAGG). The oligonucleotides were synthesized as described above for SNAP-25.

Administration and Toxicity of Phosphorothioate-Substituted Oligonucleotides

Phosphorothioate oligodeoxynucleotides were tested for toxicity on COS cells in the concentration range of 5–45 μM, using two methods: ocular inspection (cell shape, cell number, and floating cells) and the MTT assay for cell viability (see Ref. 14). When observed, none of the PONs were visibly toxic to the COS cells when supplemented at a concentration of 5–25 μM. When supplemented at 30 μM, the PONs caused an increase in the number of cells that were rounded after 4 days in culture. When supplemented at 40 μM the PONs were visibly toxic to the COS cells. In addition, the MTT assay showed that the viability was similar in cells

[31] T. Ebendal, D. Larhammar, and H. Persson, *EMBO J.* **5,** 1483 (1986).

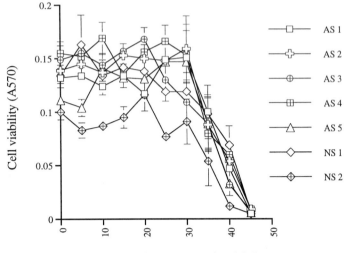

FIG. 5. Assessment of cell toxicity of PONs. COS cells were grown in a 96-well plate and were supplemented by increasing concentrations of PON (AS 1–5, NS 1 and 2). After 24 hr, cell viability was tested using the MTT assay. This assay spectrophotometrically measures the activity of dehydrogenases in living cells as a result of conversion of a substrate to a purple product. The absorbance was measured at 570 nm. Each of the PONs was assayed three times ($n = 3$) and the absorbance (viability) is plotted against the concentration of the PON. Error bars represent the SD. [From Ref. 14, with permission.]

treated with each PON at concentrations of 0–30 μM and decreased at concentrations above 30 μM (Fig. 5). A small variation in the toxicity between the different PONs could be seen.

Therefore, in all experiments, PONs were supplemented at 25 μM final concentration after the wash, and then PONs at 2 μM final concentration were added every day for 4 days. Each PON was tested twice except for AS 1, AS 5, and NS 1, which were tested three times. Media from electroporated COS cells grown in the presence of the various PONs were collected, spun for 10 min (4°, 6000 rpm) to remove cell debris, and used for the bioassays or Western blotting.

Inhibition of Nerve Growth Factor Expression

The amount of synthesized NGF and neurotrophin 4 was quantified using the ability of conditioned media from NGF- and neurotrophin 4-producing COS cells to stimulate neurite outgrowth from explanted sympathetic or sensory dorsal root ganglia of 9-day-old chicken embryos.[32,33] Dissected ganglia were placed in a gel of collagen-containing BME cell culture medium and dilutions of COS cell conditioned medium were added to the cultures. The ganglia were cultured for 2 days and neurite outgrowth was scored using low-magnification dark-field microscopy. Serial dilutions of conditioned medium were assayed and the fiber outgrowth was scored for each dilution and sample.[33] The scoring included estimation of length and density of neurite outgrowth from the explanted peripheral ganglia in comparison with neurite outgrowth produced by a serially diluted mouse NGF standard. The scoring was done on a scale from 0 to 1. The extent of neurite outgrowth is proportional to the concentration of growth factor.[33] This method of measuring the amount of biologically active NGF and neurotrophin 4 has been used extensively (see Ref. 14 for references). The scoring of neurite outgrowth was performed on a blind-test basis and the result was plotted against the corresponding dilution as shown in Fig. 6.

The inhibitory capacity of PONs AS 1 to 5 was investigated after direct addition to the medium. A final concentration of 25 μM of the PONs was used as this concentration represented the highest possible nontoxic concentration (see above). The bioassay showed that PONs AS 1 and 5 clearly resulted in less bioactivity in the conditioned media (e.g., inhibited NGF synthesis) as compared with AS 2–4 and NS 1 PONs. AS 1 and AS 5 reduced the amounts of NGF in the medium by more than 80 and 70%, respectively (Fig. 6).

To visualize the amounts of NGF synthesized by the COS cells and confirm the data from the bioassay, we analyzed samples of conditioned medium using Western blotting. The amounts of chicken NGF in the conditioned media were too low (approximately 100 ng/ml) to be analyzed directly by Western blotting, and so we concentrated the samples by ultrafiltration. The conditioned media from the two independent series for each oligonucleotide were pooled, each giving 8 ml of medium. The samples were passed through a Centriplus-100 to remove large proteins and then remaining proteins, including NGF (nucleotide weight 13,000), were concentrated to a final volume of 200 μl using a Centriplus-3 concentrator. The concentrated samples (25 μl) were processed for gel electrophoresis and Western blotting.

Consistent with the bioassay, Western blot analysis showed that AS 1 and AS 5 efficiently decreased the amount of chicken NGF protein in the conditioned media. The result also showed that the inhibition is dependent

[32] T. Ebendal, G. Norrgren, and K. O. Hedlund, *Med. Biol.* **61**, 65 (1983).
[33] T. Ebendal, *in* "Nerve Growth Factors" (R. A. Rush, ed.), pp. 81–93. John Wiley & Sons, New York, 1989.

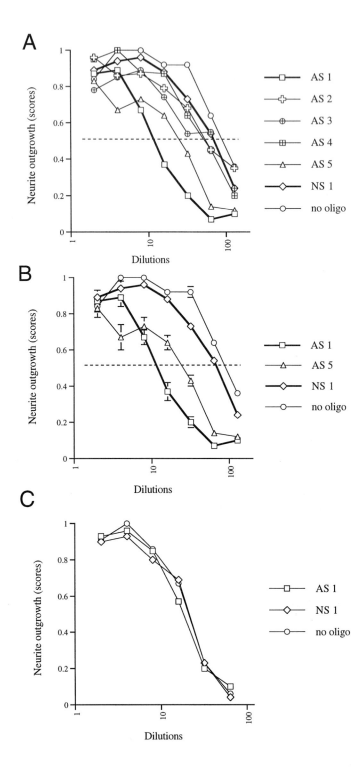

on the concentration of AS 1. Addition of 1 or 10 μM AS 1 was less efficient than addition of 25 μM. AS 2, 3, and 4 were less efficient in inhibiting expression than were AS 1 or 5.

Controls for Protein Synthesis

In addition to the nonsense oligonucleotide control (NS 1 and NS 2), unspecific inhibition of protein synthesis by the PONs was tested in COS cells electroporated with the same expression vector containing a fragment encoding *Xenopus laevis* neurotrophin 4[29] instead of chicken NGF. This control is similar to a mismatched target control,[34] demonstrating that the NGF antisense PONs used in this study are NGF sequence-specific and do not affect the synthesis of the related neurotrophic factor, neurotrophin 4. The COS cells were treated with 25 μM AS 1 and nonsense control PONs and the media were subsequently bioassayed for neurotrophin 4 activity on sensory ganglia or analyzed using SDS–PAGE. Neurotrophin 4 has an overall nucleotide identity of 57% to chicken NGF and has 67% identity in the region that is targeted by AS 1. COS cells were electroporated with the neurotrophin 4 (NT-4) expression vector and the cells were treated with the NGF AS 1 and the NS 1 PONs. The medium from the NS 1- and AS 1-treated cells contained an equal amount of neurotrophin 4 activity compared with medium from untreated COS cells (Fig. 6C). This shows that AS 1 PON does not affect neurotrophin 4 expression.

[34] C. A. Stein and A. M. Krieg, *Antisense Res. Dev.* **4**, 67 (1994).

FIG. 6. Bioassay quantitation of NGF and neurotrophin 4 after PON treatment of COS cells. (A) Serial dilutions of conditioned medium from NGF-producing COS cells were assayed in the NGF bioassay using chicken embryonic day 9 sympathetic ganglia. The amount of biologically active NGF protein in each dilution was determined by the neurite outgrowth from sympathetic ganglia. The outgrowth is proportional to the concentration of NGF and is scored after 48 hr in culture on the basis of fiber length and density on a scale between 0 and 1, using low-power dark-field microscopy. The average score in two independent assays for each dilution is plotted in the diagram ($n = 2$). The dilution for which the score is 0.5 (the half-maximum outgrowth) was used when comparing NGF concentration in the different media. Media from COS cells treated with oligonucleotides AS 1–5 and the NS as well as medium from untreated NGF-producing COS cells were assayed. Note the shift of the AS 1 and AS 5 curves to the left, showing that the NGF concentration is lower in these samples. (B) Diagram showing AS 1 and AS 5 compared with the NS 1 PON, with error bars included, indicating $\pm$SD ($n = 3$). (C) Neurotrophin 4-producing cells were treated with the NGF AS 1 and NS PONs and the neurotrophin 4 content was assayed using the bioassay with sensory ganglia. The mean from two assays is plotted against the dilutions ($n = 2$). Cells treated with AS 1 or NS 1 synthesized the same amount of neurotrophin 4 as did untreated cells (no oligonucleotide). [From Ref. 14, with permission.]

To control further for unspecific translational or cytotoxic effects on protein synthesis by the PONs, we grew NGF- and NT-4-expressing COS cells in [^{35}S]cysteine- and [^{35}S]methionine-containing medium. The COS cells were electroporated with the NGF and NT-4 vectors, plated in 35-mm dishes, and subsequently supplemented directly with AS 1, NS 1, and NS 2 PONs. On the second day after electroporation the cells were washed and the medium was exchanged for new medium free from cysteine and methionine but containing [^{35}S]cysteine and [^{35}S]methionine and supplemented with PONs. This medium (200 μl) was added and to avoid drying of the cells they were placed on a rocking platform in an incubator. After 6 hr, aliquots of medium from the different treatments, containing the radioactively labeled secreted proteins, were loaded on a 10–20% gradient polyacrylamide gel (Bio-Rad, Hercules, CA) and were electrophoresed. The gel was dried and exposed to a PhosphoImager plate (Molecular Dynamics, Sunnyvale, CA) and subsequently scanned and visualized (not shown; see Ref. 14). This way, we could visualize all secreted proteins including NGF and NT-4. This method has previously been used to quantitate the synthesis of a variety of neurotrophin mutants (see Ref. 14 for references). The concentration of the oligo nucleotides in this control experiment was 25 μM. The result showed that the cells treated either with AS 1, NS 1, or NS 2 secrete equal amounts of proteins except for a protein with a molecular mass similar to that of NGF (13–14 kDa). Treatment with either of the PONs did not affect the synthesis of neurotrophin 4 (not shown; see Ref. 14).

For many gene products that are difficult to detect, including chicken NGF, it is difficult to assess the efficiency of antisense oligonucleotides. Our assay, and related transient expression systems,[35] represent convenient methods to identify antisense oligonucleotides that efficiently downregulate a gene of interest. Indeed, our results clearly suggest that AS 1 and AS 5 should have the capacity to inhibit NGF expression in physiological systems.

Conclusions

This chapter describes three separate attempts to downregulate the expression of genes involved in neuronal development with antisense oligonucleotides. In the case of SNAP-25, these initial functional observations were confirmed later with different techniques.[36] Similarly, the system that we propose to identify oligonucleotides efficient in downregulating gene

[35] C. Marcus-Sekura, A. M. Woerner, K. Shinozuka, G. Zon, and G. J. Quinnan, *Nucleic Acids Res.* **15,** 5749 (1987).
[36] A. Osen-Sand, J. K. Staple, E. Naldi, G. Schiavo, S. Petitpierre, A. Malgaroli, C. Montecucco, and S. Catsicas, *J. Comp. Neurol.* **367,** 222 (1996).

products that are difficult to visualize has also proved its efficacy in subsequent *in vivo* studies on NGF.[36a] This clearly demonstrates that antisense oligonucleotides may be valuable tools with which to test the function of specific gene products in physiological processes. However, they cannot give direct information on the mechanisms of action of the proteins, or on their molecular environment.

Despite numerous encouraging results, the use of antisense oligonucleotides still meets strong skepticism in the scientific community. Perhaps one reason for this is the initial overuse of the technique, before identification of all the pitfalls and necessary controls. As a consequence of this, unreliable data, obtained in good faith, have reached the scientific literature. In particular, nonspecific toxic effects of oligonucleotides have been difficult to detect. More recently, important progress has been made to improve the toxicity and understand the pharmacokinetics of antisense oligonucleotides.[37] Some of the factors that underlie toxicity *in vitro* or *in vivo* have been identified.[37] While the polyanionic nature of the molecules is a likely cause of toxicity *in vitro,* CG dinucleotides have been shown to cause toxicity *in vivo* via a strong stimulation of the immune system.[37] Chemical modifications can prevent both problems, resulting in reliable research as well as safer molecules for therapeutic applications.

The goal of most experiments, or therapeutic interventions, based on the use of antisense oligonucleotides has been to reach maximal inhibition of expression of specific genes. However, only in a few cases can inhibition reach highly significant levels, and even then the remaining protein may carry on its function effectively. An alternative strategy, particularly for therapeutic interventions, would be to select regulatory proteins for which even a small decrease in expression may have important effects. This would also allow the use of lower concentrations of oligonucleotides and perhaps even the use of combinations of oligonucleotides targeting different gene products. Finally, gain of function in a given signaling pathway could also be achieved by partial downregulation of inhibitory proteins. An impressive example is the stimulation of activity of the *p53* tumor suppressor gene by inhibition of the expression of *MDM2,* the oncogene that inhibits *p53* in a negative feedback loop.[38]

Such studies could evolve toward a systematic use of combinations of nontoxic antisense oligonucleotides at low concentrations that would down- or upregulate specific signaling cascades via small alterations of the level of expression of key regulatory proteins.

[36a] F. Hallböök, in preparation (1999).

[37] S. Agrawal and Q. Zhao, *Curr. Opin. Chem. Biol.*, in press (1998).

[38] L. Chen, S. Agrawal, W. Zhou, R. Zhang, and J. Chen, *Proc. Natl. Acad. Sci. U.S.A.* **95,** 195 (1998).

[12] RNA Mapping: Selection of Potent Oligonucleotide Sequences for Antisense Experiments

By SIEW PENG HO, DUSTIN H. O. BRITTON, YIJIA BAO,
and MICHAEL S. SCULLY

Introduction

The use of short oligonucleotides complementary to a portion of an RNA molecule to downregulate gene expression was first demonstrated in 1978.[1] Since the publication of that seminal report, there has been a steady increase in the use of antisense oligonucleotides as tools for the study of protein function. In addition to laboratory applications, the "antisense approach" has been extended to encompass the concept of oligonucleotides as a new therapeutic modality. Rather than interfering with protein function by using small molecule inhibitors, the same therapeutic benefit may be achieved by inhibiting the expression of a specific protein. This therapeutic concept has been validated. The first antisense drug, an oligonucleotide for the treatment of cytomegalovirus-induced retinitis in patients with acquired immunodeficiency syndrome (AIDS), received Food and Drug Administration (FDA) approval in 1998. In addition, a dozen or more oligonucleotides are currently in various phases of clinical trials.

An enormous volume of sequence information has become available from efforts in human genome sequencing. About 80,000 genes will have been sequenced by the time the genome sequencing initiative is completed in the next 3–4 years. The function of the majority of these genes, however, will not be known. Knowledge of the function of a protein and its role in a particular disease state are of particular importance. In the pharmaceutical industry, such knowledge is critical to the process of evaluating the suitability of a protein as a target for drug development. Antisense oligonucleotides are one of several tools that can be brought to bear on such target validation experiments.

Despite the seemingly widespread application of antisense oligonucleotides, several hurdles still hinder their ease of use. The foremost of these difficulties are identification of potent antisense sequences, and delivery and transport of oligonucleotides to their cellular targets.[2,3] This chapter focuses on a molecular technique that addresses the first hurdle.

[1] P. C. Zamecnik and M. L. Stephenson, *Proc. Natl. Acad. Sci. U.S.A.* **75,** 280 (1978).
[2] A. M. Gewirtz, C. A. Stein, and P. M. Glazer, *Proc. Natl. Acad. Sci. U.S.A.* **93,** 3161 (1996).
[3] F. C. Szoka, *Nature Biotechnol.* **15,** 509 (1997).

Outline of Method

In general, only a fraction of the antisense oligonucleotides that are synthesized and tested are active (producing a greater than 50% reduction in protein or RNA expression). The percentage of active sequences varies from one mRNA target to another but is typically between 10 and 30%. One reason for this variability in activity is that mRNA molecules have a significant amount of structure. Secondary structure in the form of double-stranded regions, hairpin loops, internal bulges and loops, as well as tertiary structure resulting when secondary structural elements fold on top of one another, create "inside" and "outside" regions in an RNA molecule.[4] "Outside" regions are available for interaction with other molecules, including small oligonucleotides, but "inside" regions may not be.

RNA folding programs are limited in their ability to identify active antisense sequences. Consequently, we decided to adopt an experimental approach to address this issue. The basic idea involves the chemical synthesis of an oligonucleotide library containing every possible sequence, and the interaction of such a library with the target RNA of interest[5,6] (Fig. 1). RNA sites that are accessible for hybridization with antisense oligonucleotides should find their complementary sequences from among the library members and hybridize to these sequences. Once hybridization has occurred, a method is needed to identify such sites. Ribonuclease H (RNase H), an enzyme whose substrates are heteroduplex regions consisting of one RNA strand and one DNA strand, is used to identify these sites. This enzyme endonucleolytically cleaves the RNA strands in the heteroduplexes, producing RNA fragments that can then be sequenced to identify the location of the cleavages. Cleavage sites correspond to regions in the RNA that hybridized with members of the oligonucleotide library. Antisense sequences directed against such regions should therefore be capable of eliciting potent inhibitory effects. Indeed, oligonucleotide sequences selected using this assay exhibited potent antisense effects in cell culture and in animals.[5-7]

Library Design

The kinetics of hybridization between the RNA target and an oligonucleotide is dependent on the concentration of both reacting entities. Most

[4] J. A. Latham and T. R. Cech, *Science* **245,** 276 (1989).
[5] S. P. Ho, D. H. O. Britton, B. A. Stone, D. L. Behrens, L. M. Leffet, F. W. Hobbs, J. A. Miller, and G. L. Trainor, *Nucleic Acids Res.* **24,** 1901 (1996).
[6] S. P. Ho, Y. Bao, T. Lesher, R. Malhotra, L. Y. Ma, S. J. Fluharty, and R. R. Sakai, *Nature Biotechnol.* **16,** 59 (1998).
[7] S. P. Ho, V. Livanov, W. Zhang, J. H. Li, and T. Lesher, *Mol. Brain Res.* **62,** 1 (1998).

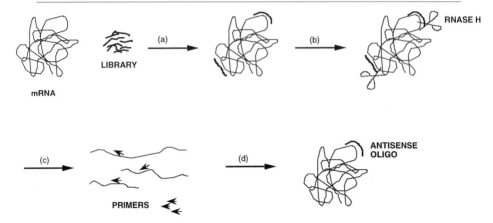

FIG. 1. RNA mapping for antisense sequence selection. (a) A library of oligonucleotides is allowed to hybridize with an RNA molecule of interest. (b) RNA regions accessible to hybridization form DNA–RNA hybrid duplexes, which are cleaved in the RNA strand using RNase H. (c) Extension of primers hybridized to the RNA fragments generated allows the location of the cleavage sites to be determined. (d) These sites correspond to regions in the RNA that are accessible to antisense oligonucleotide hybridization.

oligonucleotides used in antisense experiments are 15–21 nucleotides in length. A library whereby 10 nucleotide positions are randomized consists of 1.05×10^6 unique sequences; a library with 15 randomized nucleotide positions contains 1.07×10^9 unique sequences. There is a practical upper limit to the total concentration of the library that can be used in these experiments because high concentrations of single-stranded oligonucleotides inhibit RNase H activity. Therefore, the length of oligonucleotides in the library was chosen to be 11 nucleotides to minimize the number of unique sequences in the library.

To ensure that 11 nucleotide sequences are capable of forming stable duplexes with the RNA transcript, 2'-methoxyribonucleotide residues were incorporated because they increase the thermodynamic stability of the resulting duplexes.[8,9] Hybrid duplexes consisting of an RNA strand and a 2'-methoxyribonucleotide RNA strand are, however, not substrates of

[8] S. M. Freier, *in* "Antisense Research and Applications" (S. T. Crooke and B. Lebleu, eds.), pp. 67–82. CRC Press, Boca Raton, Florida, 1993.
[9] H. Inoue, Y. Hayase, A. Imura, S. Iwai, K. Miura, and E. Ohtsuka, *Nucleic Acids Res.* **15**, 6131 (1987).

RNase H.[10,11] There is increasing evidence supporting the involvement of RNase H in the antisense effects of phosphorothioate oligonucleotides in cells,[7,12–14] justifying the use of the enzyme in this assay. A minimum of three or four contiguous 2′-deoxyribonucleotide residues in the DNA strand of the hybrid duplex is required to elicit cleavage activity from the *Escherichia coli* RNase H that is used in these experiments. Therefore, to maintain the ability to use RNase H in this assay, only 7 of the 11 nucleotides are converted to 2′-methoxyribonucleotide residues. The remaining four residues are left as 2′-deoxyribonucleotide residues (Fig. 2B). Phosphodiester internucleotidic linkages are used in these libraries to increase the thermodynamic stability of the hybrid duplexes.

Another advantage to using chimeric libraries, as opposed to regular 2′-deoxyribonucleotide libraries, stems from the fact that chimeric oligonucleotides of the type shown in Fig. 2B exhibit simplified cleavage patterns.[10,15] While 2′-deoxyribonucleotide oligonucleotides mediate RNase H cleavage at multiple locations along the hybrid duplex, chimeric oligonucleotides mediate cleavage primarily at one location—the downstream 2′-deoxyribonucleotide–2′-methoxyribonucleotide junction (Fig. 2C). Side-by-side comparisons between chimeric libraries and 2′-deoxyribonucleotide libraries confirmed those published observations (data not shown).

The libraries used in this assay are not completely random but consist of a series of four semirandom libraries, each containing a single constrained nucleotide. Experiments performed with the completely random library produced results that were similar, but less optimal than with the semirandom libraries (data not shown). Cleavage bands were less intense and more diffuse, raising concerns about self-hybridization that could potentially occur between members of the completely random library. As experiments with various libraries indicate (Fig. 3), reducing the number of constraints increases the complexity of the libraries, resulting in the identification of larger numbers of accessible sites. Use of the four single constrained libraries should produce data that are equivalent to data obtained from a completely random library.

[10] H. Inoue, Y. Hayase, S. Iwai, and E. Ohtsuka, *FEBS Lett.* **215**, 327 (1987).

[11] B. P. Monia, E. A. Lesnik, C. Gonzalez, W. F. Lima, D. McGee, C. J. Guinosso, A. M. Kawasaki, P. D. Cook, and S. M. Freier, *J. Biol. Chem.* **268**, 14514 (1993).

[12] R. V. Giles, D. G. Spiller, and D. M. Tidd, *Antisense Res. Dev.* **5**, 23 (1995).

[13] T. P. Condon and C. F. Bennett, *J. Biol. Chem.* **271**, 30398 (1996).

[14] S. P. Ho, Y. Bao, T. Lesher, D. Conklin, and D. Sharp, *Mol. Brain Res.* **65**, 23 (1999).

[15] S. T. Crooke, K. M. Lemonidis, L. Neilson, R. Griffey, E. A. Lesnik, and B. P. Monia, *Biochem. J.* **312**, 599 (1995).

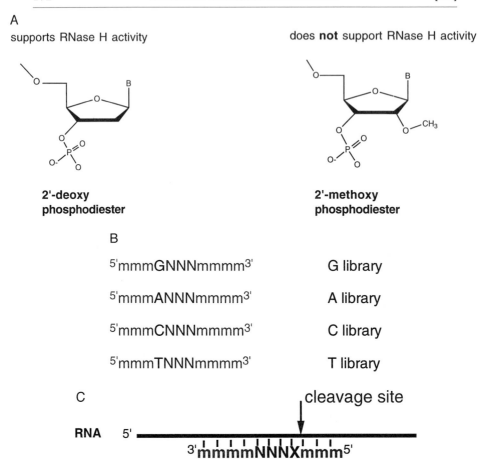

FIG. 2. (A) Chemical structure of the nucleotide residues used to construct the libraries. (B) Sequence of the chimeric libraries. m and N denote a random mixture of 2′-methoxyribonucleotide and 2′-deoxyribonucleotide residues, respectively. G, A, C, and T are 2′-deoxyribonucleotide bases. (C) Location of the primary site of RNase H cleavage when mediated by a chimeric oligonucleotide. X represents the constrained nucleotide.

Synthesis and Purification of Semirandom, Chimeric Libraries

Libraries are chemically synthesized on an automated RNA/DNA synthesizer. Controlled pore glass (CPG) supports and phosphoramidites of 2′-methoxyribonucleotide residues can be obtained from ChemGenes (Waltham, MA). Thoroughly mix equimolar amounts of CPG of the four bases (1.0 μmol of each CPG) and pack them into four 1.0-μmol scale columns for the synthesis of the A, G, C, and T libraries. Owing to the

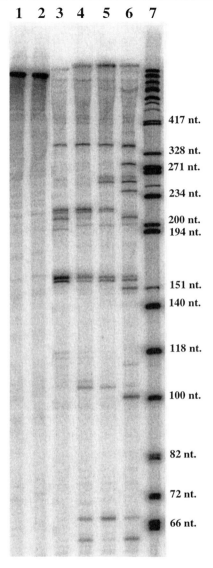

FIG. 3. Autoradiogram of RNA fragments produced by the action of oligonucleotide libraries and RNase H. A ^{32}P 5′-end-labeled transcript of the human multidrug resistance gene was allowed to interact with various libraries in the presence of RNase H. Lane 1, transcript; lane 2, transcript and library; lane 3, transcript, RNase H, and library with three constraints, $^{5′}$mmNNGACNNmm; lane 4, transcript, RNase H, and library with two constraints, $^{5′}$mmNNGANNNmm; lane 5, transcript, RNase H, and library with two constraints, $^{5′}$mmmGANNNmmm; lane 6, transcript, RNase H, and library with one constraint, $^{5′}$mmmGNNNmmmm; lane 7, molecular size markers.

different chemical reactivities of the phosphoramidites, we mix the amidites in a molar ratio of $1.5:1.25:1.15:1.0$ $(A:C:G:U/T)$ in an attempt to produce a mixture of sequences that is as random as possible. Syntheses of the libraries are performed using (1) a mixture of 2'-methoxyribonucleotide amidites, (2) a mixture of 2'-deoxyribonucleotide amidites, and (3) four separate 2'-deoxyribonucleotide amidites, using standard DNA synthesis protocols.

The libraries are not purified by high-performance liquid chromatography (HPLC) but are extracted with butanol (two or three times) to remove the organic protecting groups. Precipitation with ethanol in the presence of 0.3 M sodium acetate (three times) serves to exchange the countercation to sodium. Finally, the libraries are treated with proteinase K to remove any potential ribonuclease contamination. Quantitation of the libraries is done by UV spectroscopy.

In Vitro Transcription

The Promega (Madison, WI) Ribomax large-scale RNA production kit can be used to generate transcripts for the mapping experiment. The protocol below describes the preparation of transcripts with 5'-OH ends[16] that are ready for polynucleotide kinase treatment.

Add the following to a sterile, RNase-free microcentrifuge tube:

Transcription buffer (5×)	40.0 μl
ATP, 25 mM	15.0 μl
CTP, 25 mM	15.0 μl
GTP, 25 mM	15.0 μl
UTP, 25 mM	15.0 μl
Nuclease-free water	60.0 μl
DNA template	5–20 μg
Guanosine solution, 100 mM	10.0 μl
Enzyme mix (RNA polymerase/RNasin/yeast inorganic pyrophosphatase)	20.0 μl
Final volume:	~200 μl

Vortex the microcentrifuge tube gently before incubation at 37° for 2 hr. Digest the DNA template with RQ1 RNase-free DNase (Promega; 1 U/μg DNA template) at 37° for 15 min.

Purification of 5'-OH Transcript

Extract the transcription mixture as follows:

1. Phenol–chloroform–isoamyl alcohol [25 : 24 : 1, by volume; saturated with Tris buffer (Sigma, St. Louis, MO)]

---

[16] J. W. Harper and N. Logsdon, *Biochemistry* **30**, 8060 (1991).

2. Phenol–chloroform (25:24, v/v), 1× volume
3. Chloroform–isoamyl alcohol (24:1, v/v), 1× volume
4. Chloroform, 1× volume

Vortex for 1 min and spin at 14,000g. Filter the aqueous solution through a Sephadex G-50 or G-25 RNA spin column (Boehringer Mannheim, Indianapolis, IN) (following the manufacturer instructions). Combine the spin column-purified solutions and, if necessary, concentrate down to a 200-μl volume. Precipitate the RNA by adding a one-tenth volume of 3.0 M sodium acetate, followed by 3 volumes of 100% ethanol. Vortex the mixture, cool on dry ice for 30 min, and then spin down at 14,000g at 4° for 20 min. A large pellet should be obtained. If no pellet is visible, add more ethanol and store at −20° overnight, then in dry ice for 20 min before spinning down.

Decant the supernatant and wash the pellet with 400 μl of 80% (v/v) ethanol–water (chilled on dry ice) before spinning down for 5 min. Discard the wash and air dry the pellet for 2 min before dissolving in 100.0 μl of 1× Tris–EDTA (TE) buffer.

Quantitation of Purified Transcript

Dilute 5.0 μl of the dissolved transcript to 1000 μl with 1 × TE buffer. Record a UV spectrum and measure the absorbance of the sample at 260 nm using 1 × TE buffer as the blank. One optical density unit (ODU) at 260 nm is approximately equivalent to 40 μg of RNA per milliliter. The large-scale *in vitro* transcription reaction typically produces 100–200 pmol of RNA. Store the remaining 95 μl of RNA solution at −70°.

5'-End Labeling of Transcript

Add the following to a sterile, RNase-free microcentrifuge tube:

RNA transcript	10.0 pmol
Nuclease-free water	21.0 μl
Kinase buffer (10×): 10× One-Phor-All Plus buffer, 100 mM Tris–acetate, 100 mM magnesium acetate, 500 mM Potassium acetate (Pharmacia, Piscataway, NJ)	5.0 μl
[γ-^{32}P]ATP: 6000 Ci/mmol, 10.0 μCi/μl (NEN/ DuPont, Boston, MA)	15.0 μl
T4 polynucleotide kinase: 9.2 U/μl (Pharmacia)	3.0 μl
Final volume:	~50 μl

Vortex the solution gently before incubation at 37° for 30 min. Purify the end-labeled RNA by passage through a Sephadex G-50 spin column. Store the labeled transcript at −70°. Approximately 80–90% of the transcript is recovered after spin column filtration.

Mapping RNA Transcript

Materials

5' End-labeled RNA transcript: ~1.0 pmol/reaction	RNA
Library: ~10 pmol/reaction	A, G, C, or T
RNase H: 1.5 U/μl (Promega)	RH
RNase H buffer (10×): 400 mM Tris (pH 7.5), 40 mM MgCl$_2$, 10 mM dithiothreitol (DTT)	Buffer
Formamide loading buffer: 10.0 ml of formamide, 10 mg of bromphenol blue, 10 mg of xylene cyanol, 200 μl of 0.5 M EDTA (pH 8.0)	FLB

Denature the libraries by heating at 96° for 5 min and place on ice until use. Add RNA, library, and 1.0 μl of buffer solution to each reaction tube according to Table I and bring the volume up to 9.0 μl with nuclease-free water. Add 0.5 μl of RNase H and vortex each tube gently before incubating at 37° for 5 min. Reactions 1 to 3 are controls. After addition of 10 μl of FLB, heat the mixtures at 96° for 5 min. Place the samples on ice until they are ready to be loaded onto a 6% polyacrylamide gel (35 × 43 cm gel, 0.4 mm in thickness), which is run on a GIBCO-BRL (Gaithersburg, MD) S2 gel apparatus at 85 W (1700 V) for 1 hr.

The amounts of library, enzyme, and reaction time may need to be optimized so that no more than 50% of the starting transcript is digested. Ideally, an RNA molecule should be digested only once, so that the RNA fragments produced correspond to RNase H cleavage at primary accessible sites and not to sites that become exposed subsequent to the primary cleavage event. RNase H from several different vendors were compared; enzymes from GIBCO-BRL, Boehringer Mannheim, and Promega were found to have the least amount of nonspecific cleavage activity. Figure 3 shows the results of a mapping experiment that was conducted with libraries of increasing complexity.

TABLE I
PLAN FOR RNA MAPPING EXPERIMENT

Component	Reaction						
	1	2	3	4	5	6	7
RNA	✔	✔	✔	✔	✔	✔	✔
Library	—	A	—	A	G	C	T
RH	—	—	✔	✔	✔	✔	✔
Buffer	✔	✔	✔	✔	✔	✔	✔

TABLE II
PLAN FOR SEQUENCING EXPERIMENT[a]

Parameter	Lane								
	1	2	3	4	5	6	7	8	9
Reaction type	PE	PE	PE	PE	PE	SEQ	SEQ	SEQ	SEQ
Nucleotide mix	N	N	N	N	N	G	A	T	C
RNA	Full	G frag.	A frag.	T frag.	C frag.	Full	Full	Full	Full

[a] Primer extension (PE) or dideoxy Sanger sequencing (SEQ) reactions are performed. The RNA template for these reactions consists of either full-length transcript or RNA fragments generated from a prior mapping experiment. Therefore, "A frag." corresponds to fragments obtained from a mapping experiment with the A library. Refer to Table III for the composition of the nucleotide mixtures.

Sequencing of Accessible Sites

The mapping experiment described above provides a low-resolution map of the accessible sites. Identification of RNase H cleavages to single-nucleotide resolution is obtained through a reverse transcriptase-mediated primer extension experiment. Primers may be selected at regularly spaced intervals (100–150 nucleotides) across the entire length of the transcript. Alternatively, primers may be designed 20–30 nucleotides downstream of accessible regions whose locations are estimated from the mapping gel. Primers are labeled at the 5' end with $[\gamma\text{-}^{32}P]ATP$ prior to use in the sequencing experiment.

Two different reactions are performed in order to locate the accessible regions (Table II). In the first reaction, primer extension is carried out using RNA fragments generated from the mapping experiment as the template. In the second reaction, dideoxy Sanger sequencing of full-length RNA transcript provides an RNA sequence ladder from which the location of cleavage sites can be precisely determined.

Nucleotide Mixture Preparation

The nucleotides for dideoxy sequencing and primer extension can be obtained from Boehringer Mannheim. Prepare nucleotide mixtures from 10 mM stocks of dNTP and 1 mM stocks of ddNTP according to the volumes specified in Table III.

Primer Extension Reactions

To generate RNA fragments for the primer extension experiment (Fig. 4A, lanes 2 to 5) scale up the mapping reactions 4 to 7 (Table I) by fourfold

TABLE III
PREPARATION OF NUCLEOTIDE MIXTURES

Mixture	dATP (μl)	dTTP (μl)	dCTP (μl)	dGTP (μl)	ddNTP (μl)	Water (μl)
G	5	5	5	7	ddGTP: 13	65
A	5	5	5	5	ddATP: 13	67
T	5	7	5	5	ddTTP: 25	53
C	5	5	5	5	ddCTP: 13	67
N	5	5	5	5	—	80

and use nonradiolabeled RNA. After incubation with RNase H, quench each reaction by adding 40 μl of 10 mM EDTA. Extract the reaction mixtures once with phenol–chloroform and once with chloroform. Precipitate the RNA using sodium acetate and ethanol (four times the volume of ethanol), then dissolve the resulting RNA pellet in 20 μl of nuclease-free water before filtering through Sephadex G-50 spin columns to remove the library. Take 2.0 μl of each purified solution and add ~0.5 pmol of radiolabeled primer (total volume, ~3 μl). Denature the mixtures at 96° for 5 min and then anneal at 42° for 5 min. Add the following to each of the four RNA–primer solutions:

> Reverse transcriptase (RT) buffer: 250 mM Tris-HCl (pH
> 8.3 at 42°), 250 mM KCl, 50 mM MgCl$_2$, 50 mM DTT,
> 2.5 mM spermidine   1.2 μl
> Actinomycin D: 250 μg/ml   1.0 μl
> Avian myeloblastosis virus (AMV) reverse transcriptase:
> 9.5 U/μl (Promega)   0.5 μl

Gently vortex the microcentrifuge tubes, spin down, and add 1.0 μl of the N mixture (Table III). These tubes correspond to lanes 2 to 5 in Fig. 4A. Gently vortex the tubes before spinning them down and incubate at 42° for 10 min. Quench the reactions by adding 7 μl of FLB. Heat denature the samples at 96° for 5 min and place on wet ice until they are loaded on a 6% polyacrylamide gel.

Sequencing Reactions

Add 1.0 pmol of nonradiolabeled, full-length RNA and ~1.2 pmol of 5′ end-labeled primer to a microcentrifuge tube (total volume, 7 μl). Denature the mixture at 96° for 5 min and then anneal at 42° for 5 min. Add the following to the RNA–primer solution:

> RT buffer: 250 mM Tris-HCl (pH 8.3 at 42°), 250 mM KCl,
> 50 mM MgCl$_2$, 50 mM DTT, 2.5 mM spermidine   5.0 μl
> Actinomycin D: 250 μg/ml   5.0 μl

Nuclease-free water   4.0 μl
AMV reverse transcriptase: 9.5 U/μl (Promega)   2.0 μl

Gently vortex the microcentrifuge tube and spin down. Aliquot 4 μl of this solution into five microcentrifuge tubes labeled N, ddG, ddA, ddT, and ddC (corresponding to Fig. 4A, lanes 1 and 6–9) and add 1.0 μl of the N mixture, G mixture, A mixture, T mixture, and C mixture, respectively, to the corresponding tubes. Gently vortex all five tubes, spin down, and incubate at 42° for 10 min. Quench the reactions by adding 7 μl of FLB. Heat denature the samples at 96° for 5 min and place on wet ice until they are loaded on a 6% polyacrylamide gel (see Fig. 4A, lanes 6–9). If the bands on the gel are not sharp, degrade the RNA template with an RNase mix [RNase A (20 μg/ml) and herring sperm DNA (100 μg/ml) dissolved in 1× TE buffer containing 100 mM NaCl] prior to loading on the gel.

Design of Antisense Oligonucleotides from Sequencing Data

Bands in lanes 2 to 5 (Fig. 4A) are products resulting from primer extension off of the RNA fragments. Alignment of these bands with bands in sequencing lanes 6 to 9 (Fig. 4A) allows the location of points of cleavage to be precisely determined. These cleavages correspond directly to regions of accessibility. Lane 1 (Fig. 4A) represents an important control for the experiment. Primer extension off of the full-length RNA transcript should theoretically produce only a full-length primer extension product, which should run at the top of the gel. Other bands appearing in this lane (marked by X in Fig. 4A) represent products resulting from termination of extension due to secondary structure in the RNA template. Identical bands seen in lanes 2 to 6 of Fig. 4A should therefore be discounted, as they do not represent genuine accessible sites.

Owing to the design features of these libraries, RNase H should mediate cleavage at the nucleotide position immediately following the constrained nucleotide (Fig. 2C). Because we know the identity of the constrained nucleotide in the library, we can confirm this property. Indeed, complementarity between the constrained nucleotide and the RNA at the point of cleavage is observed 95% of the time. Figure 4B shows sequence information derived from the bottom half of the gel (starting from the position indicated by the arrowhead). Superimposed on to this sequence are the location of cleavages produced by the four libraries. In this region of the RNA, complementarity between the constrained nucleotide and the RNA at the cleavage point was found at all of the cleavage locations. For example (Fig. 4B), library T mediated the cleavage of the RNA immediately after nucleotide A. This information allows us to deduce precisely how the oligonucleotide in the library hybridized to the RNA. If the library were not

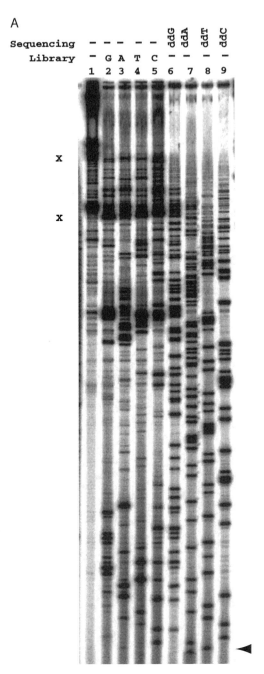

FIG. 4. (A) Autoradiogram identifying the location of RNase H-mediated cleavages. Lane 1, primer extension of the full-length transcript; lanes 2–5, primer extension of RNA fragments (obtained from reaction of RNA with libraries and RNase H); lanes 6–9, ddNTP-terminated primer extension of full-length RNA. *X* denotes bands resulting from termination of extension due to RNA secondary structure.

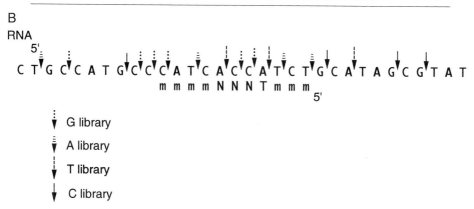

FIG. 4. (*continued*) (B) Location of cleavages mediated by the four libraries. The RNA sequence derived from the bottom of (A) (starting from the arrowhead) is depicted, together with arrows indicating points of cleavage by the libraries. Hybridization by a member of the T library is shown. [Part (A) has been adapted with permission, from *Nature Biotechnology* **16,** 59 (1998).]

chimeric in nature, but instead consisted entirely of 2′-deoxyribonucleotide residues, such oligonucleotides would elicit multiple cleavages of the RNA, not only in the hybrid duplex region but also in the single-stranded region immediately downstream of the hybrid duplex.[17] In such instances, information on the precise hybridization mode of that oligonucleotide would be lost.

Closely spaced cleavages, such as that shown in Fig. 4B, are grouped together to arbitrarily constitute an accessible site. Antisense sequences can be selected such that the 5′ ends of the antisense oligonucleotides are located three nucleotides downstream of major cleavage points (corresponding directly to the sequence information obtained above). Alternatively, antisense sequences can be selected so that the 5′ ends of the selected oligonucleotides are either in close proximity to or within the accessible site. It is important to ascertain that the antisense sequence selected has minimal propensity to self-hybridize, either through hairpin formation or dimerization. This can be determined through programs such as Oligo Primer Analysis Software (National BioSciences, Plymouth, MN). In addition, it is equally important to ensure that the RNA target site does not have significant sequence homology to other known RNA sequences. FASTA (Genetics Computer Group, Madison, WI) or BLAST (web site for National Center for Biotechnology Information, www.ncbi.nlm.nih.gov) searches can be used to make such determinations.

It is often possible to target numerous antisense sequences to any one

---

[17] W. F. Lima and S. T. Crooke, *J. Biol. Chem.* **272,** 27513 (1997).

accessible site. In our studies on the angiotensin type 1 receptor, three or four antisense oligonucleotides were selected for each accessible site examined.[6] Direct comparison of the oligonucleotides targeted to the same site revealed that the highest antisense activity was generally obtained from the oligonucleotide having the highest calculated melting temperature (T_m).

Protein Binding and Other Issues

The RNA mapping assay is performed in the absence of intracellular RNA-binding proteins, which may bind to the RNA and therefore obscure many sites that may otherwise be accessible for hybridization. Consequently, there was initial concern that antisense oligonucleotides targeted to accessible sites under cell-free conditions may turn out to be inactive. The experimental results, however, indicate otherwise. Numerous oligonucleotides selected using the RNA mapping assay have been tested in cell culture and found to exhibit high potency.[5,6] Validation of this assay has also been obtained from experiments conducted in rodents.[6,7] In all, we have successfully used this assay to identify potent antisense sequences for cell culture experiments against five different RNA targets. Others have also used related techniques with equal success.[18–20] In addition, in a panel of 10 antisense oligonucleotides systematically selected by walking a 200-nucleotide portion of the human multidrug resistance mRNA, a strikingly good correlation ($r = 0.92$) was found between the ability of the oligonucleotides to mediate RNase H cleavage of the RNA transcript in the cell-free assay and their ability to produce antisense inhibition in cell culture.[5] Therefore, there is mounting evidence suggesting that although protein binding may obscure a fraction of the accessible sites, the majority of sites may still be available for oligonucleotide targeting.

Another point of concern involves the possibility that *in vitro*-transcribed RNA may have an altogether different structure from intracellularly transcribed RNA, particularly if introns or 5' untranslated region (UTR) sequences are absent in the *in vitro* transcript. We have previously established that additional sequence at the 3' end of the transcript did not alter accessible sites in the 5' end of the RNA. This was determined by comparing the mapping gels of 800-, 1200-, and 3000-nucleotide transcripts of the multidrug resistance gene.[21] To date, studies of RNA with differing 5'

---

[18] A. Lieber and M. Strauss, *Mol. Cell. Biol.* **15,** 540 (1995).

[19] W. F. Lima, V. Brown-Driver, M. Fox, R. Hanecak, and T. W. Bruice, *J. Biol. Chem.* **272,** 626 (1997).

[20] O. Matveeva, B. Felden, S. Audlin, R. F. Gesteland, and J. F. Atkins, *Nucleic Acids Res.* **25,** 5010 (1997).

[21] Britton, D., unpublished data (1993).

sequences have not revealed any significant differences in accessibility between the related transcripts. For example, the first 318 nucleotides of the corticotropin-releasing factor receptor subtype 2α transcript are significantly different in sequence from the same region of the 2β transcript. In spite of this difference, mapping and sequencing experiments have shown that accessible sites downstream of nucleotide 319, where the two transcripts share a common sequence, are identical.[22] Similar results were observed with multidrug resistance gene transcripts with and without a 73-nucleotide polylinker sequence at the 5' end.[21] These data indicate the robustness of RNA structure, and suggest that although 5' UTR sequences or introns may be absent in the *in vitro* transcript, relevant accessibility data can still be obtained. In addition, we have applied accessibility data obtained with transcripts from one species, to cell culture or *in vivo* experiments in another species. For example, we have mapped and sequenced accessible sites using the mouse transcripts of several proteins, and successfully downregulated the human proteins in cell culture.

Summary

The importance of finding good antisense sequences cannot be underestimated. Poor inhibition of the targeted protein can compromise the final outcome of an antisense experiment, making it difficult to arrive at a definitive understanding of the function of the protein of interest. In antisense therapeutics, identification of potent sequences becomes even more important. RNA mapping greatly increases the odds of finding active sequences. When antisense sequences are selected randomly or by gene walking, a substantial number of the oligonucleotides have little to no activity. In contrast, oligonucleotides selected by RNA mapping typically produce an antisense inhibition of greater than 50%. Oligonucleotides targeted to 60% of the accessible sites in the 5' portion of the multidrug resistance transcript inhibited P-glycoprotein function with high potency.[5] In the angiotensin type 1 receptor system, oligonucleotides to the eight accessible sites examined inhibited AT_1 receptor binding by at least 50%, with oligonucleotides to four of the sites producing at least 70% inhibition.[6] The RNA mapping assay, which is based on standard molecular techniques, therefore provides an easy and reliable method for potent antisense sequence selection.

---

[22] Lesher, T., unpublished data (1996).

[13] Effects of Centrally Administered Antisense Oligodeoxynucleotides on Feeding Behavior and Hormone Secretion

By PUSHPA S. KALRA, MICHAEL G. DUBE,
and SATYA P. KALRA

Introduction

Neuropeptides in the central nervous system (CNS) have gained increasing recognition as regulators of vital life processes. Maintenance of both reproduction and energy homeostasis is now known to be profoundly affected by several families of neuropeptides that, although structurally unrelated, are produced and secreted within the hypothalamus where their receptors are also located. The 36-amino acid peptide, neuropeptide Y (NPY), belonging to the pancreatic family of polypeptides, is the best known and characterized of these pleiotropic peptides, with multiple actions in the hypothalamus including stimulation of appetite and modulation of the secretion of reproductive hormones.[1] NPY is the most abundant peptide in the hypothalamus, and is synthesized primarily in neurons located in the arcuate nucleus (ARC)[2]; in addition, catecholamine-containing neurons in the brainstem also produce and contribute as much as 40–50% to the NPY concentration in various hypothalamic nuclei.[3] Administration of NPY into the cerebroventricle invariably stimulates feeding in satiated rats in a dose-dependent manner[4] and repeated central injections or continuous infusion causes unsatiated feeding without any evidence of tolerance.[5] Intrahypothalamic microinjection studies have established the major sites of NPY action to stimulate feeding to be the paraventricular nuclei (PVN) and the perifornical hypothalamus (PFH).[6] NPY synthesis, concentration, and release are elevated under conditions of high-energy demand that induce hyperphagia such as lacta-

[1] S. P. Kalra and P. S. Kalra, *Front. Neuroendocrinol.* **17**, 371 (1996).
[2] B. M. Chronwall, *Peptides* **6**, 1 (1985).
[3] A. Sahu, S. P. Kalra, W. R. Crowley, and P. S. Kalra, *Brain Res.* **457**, 376 (1988).
[4] J. T. Clark, P. S. Kalra, W. R. Crowley, and S. P. Kalra, *Endocrinology* **115**, 427 (1984).
[5] S. P. Kalra, M. G. Dube, and P. S. Kalra, *Peptides* **9**, 723 (1988).
[6] B. G. Stanley, A. S. Chin, and S. F. Leibowitz, *Brain Res. Bull.* **14**, 521 (1985).

tion,[7,8] experimentally induced diabetes,[9,10] fasting, and food restriction.[11–13]

In addition to its powerful orexigenic effects, NPY has profound effects on the release of reproduction-related hormones. NPY stimulates the release of hypothalamic luteinizing hormone-releasing hormone (LHRH), which in turn stimulates gonadotropin release from the pituitary.[14] In the median eminence (ME)–ARC region of the hypothalamus, NPY gene transcription is upregulated and the peptide levels fluctuate in parallel with the changes in levels of LHRH that occur before the preovulatory surge release of the hypothalamic releasing hormone and pituitary LH.[15,16] Furthermore, NPY is released in pulses that correlate with the pulsatile release of LHRH.[17,18] These converging lines of evidence suggest that NPY is intimately involved in both the basal pulsatile and preovulatory modes of LHRH secretion.

Pharmacological and ligand-binding studies with NPY and related peptides have suggested the presence of at least six receptor subtypes that presumably mediate the multiple effects of NPY. Several of these G protein-coupled receptor subtypes have been cloned in mammals, including the Y_1, Y_2, Y_4, and Y_5 subtypes; however, the specificity of these receptors in terms of ligand binding and function remains to be resolved.[19] The development of specific antagonists, long hampered by the lack of knowledge of receptor function, continues to be actively pursued for their obvious therapeutic potential but remains elusive. In the absence of available antagonists, the antisense oligodeoxynucleotide technology for the study of NPY and its receptors appeared to be a viable and useful tool *in vivo* to understand and characterize the diverse actions of NPY and its receptors. As detailed

[7] P. Ciofi, J. H. Fallon, D. Croix, J. M. Polak, and G. Tramu, *Endocrinology* **128**, 823 (1991).

[8] M. S. Smith, *Endocrinology* **133**, 1258 (1993).

[9] G. Williams, J. S. Gill, Y. C. Lee, H. M. Cardoso, B. E. Okpere, and S. R. Bloom, *Diabetes* **38**, 321 (1989).

[10] A. Sahu, C. A. Sninsky, P. S. Kalra, and S. P. Kalra, *Endocrinology* **126**, 192 (1990).

[11] A. Sahu, P. S. Kalra, and S. P. Kalra, *Peptides* **9**, 83 (1988).

[12] L. S. Brady, M. A. Smith, P. W. Gold, and M. Herkenham, *Neuroendocrinology* **52**, 441 (1990).

[13] A. Sahu, J. D. White, P. S. Kalra, and S. P. Kalra, *Brain Res. Mol. Brain Res.* **15**, 15 (1992).

[14] W. R. Crowley and S. P. Kalra, *Neuroendocrinology* **46**, 97 (1987).

[15] A. Sahu, W. R. Crowley, and S. P. Kalra, *J. Neuroendocrinol* **7**, 291 (1995).

[16] A. Sahu, W. Jacobson, W. R. Crowley, and S. P. Kalra, *J. Neuroendocrinol.* **1**, 83 (1989).

[17] M. J. Woller, J. K. McDonald, D. M. Reboussin, and E. Terasawa, *Endocrinology* **130**, 2333 (1992).

[18] K. Y. Pau, O. Khorram, A. H. Kaynard, and H. G. Spies, *Neuroendocrinology* **49**, 197 (1989).

[19] A. A. Balasubramaniam, *Peptides* **18**, 445 (1997).

below, this approach has proved useful for establishing the role of hypothalamic NPY in regulating the two patterns of LH secretion; however, it has yielded mixed results with regard to the appetite-stimulating action of NPY.

Galanin (GAL), a 29-amino acid peptide in the central and peripheral nervous systems, has diverse biological activities including stimulation of appetite and regulation of reproductive hormone secretions. Like NPY, GAL stimulates food intake in satiated rats after intracerebroventricular administration or intrahypothalamic injections.[20] Whereas restricted feeding and fasting decrease GAL mRNA in the PVN, GAL levels are elevated in the PVN of obese Zucker rats.[21–23] GAL also exerts an excitatory effect on hypothalamic LHRH and pituitary LH secretions.[24] However, unlike NPY, GAL-producing neurons are widely distributed in various hypothalamic nuclei, including the ARC, dorsomedial nucleus (DMN), PVN, and the medial preoptic area (MPOA).[25,26] Two GAL receptors have been cloned from the mammalian brain and another, GAL-R3, is found in the periphery, but their specific functions have not been defined. Pharmacological blockade of GAL receptors attenuated LH pulses and blocked the surge release of gonadotropins, demonstrating an influence of endogenous GAL on the LHRH–LH axis.[27] The pharmacological agents also blocked GAL-induced feeding but were ineffective in reducing natural feeding.[28,29]

In general, as summarized above, there are striking similarities in the function and activity of hypothalamic NPY and GAL during various states of energy homeostasis. However, we have detected a departure from this parallelism in an experimental model of hyperphagia. Microinjection of the neurotoxin colchicine (COL) into the ventromedial nucleus (VMN) of the rat hypothalamus disrupts neurotransmission, resulting in rapid onset of transient hyperphagia. These rats consume 30–40% more chow than saline-injected controls for 4–5 days, resulting in a higher rate of body weight gain.[30] Quite unexpectedly, we noted that whereas GAL levels and mRNA in several hypothalamic sites are increased, NPY levels and mRNA are

[20] S. E. Kyrkouli, B. G. Stanley, R. D. Seirafi, and S. F. Leibowitz, *Peptides* **11,** 995 (1990).
[21] M. W. Schwartz, A. J. Sipols, C. E. Grubin, and D. G. Baskin, *Brain Res. Bull.* **31,** 361 (1993).
[22] B. Beck, A. Burlet, J. P. Nicolas, and C. Burlet, *Brain Res.* **623,** 124 (1993).
[23] J. N. Crawley, *Regul. Pept.* **59,** 1 (1995).
[24] F. J. Lopez and A. Negro-Vilar, *Endocrinology* **127,** 2431 (1990).
[25] T. Melander, T. Hokfelt, and A. Rokaeus, *J. Comp. Neurol.* **248,** 475 (1986).
[26] I. Merchenthaler, F. J. Lopez, and A. Negro-Vilar, *Prog. Neurobiol.* **40,** 711 (1993).
[27] A. Sahu, B. Xu, and S. P. Kalra, *Endocrinology* **134,** 529 (1994).
[28] R. L. Corwin, J. K. Robinson, and J. N. Crawley, *Eur. J. Neurosci.* **5,** 1528 (1993).
[29] R. L. Corwin, P. M. Rowe, and J. N. Crawley, *Am. J. Physiol.* **269,** R511 (1995).
[30] P. S. Kalra, M. G. Dube, B. Xu, and S. P. Kalra, *Regul. Pept.* **72,** 121 (1997).

reduced.[30–32] Nevertheless, it is likely that these neuropeptides are responsible for the hyperphagia because intracerebroventricular (icv) administration of either neuropeptide elicited a dose-related stimulation of food intake, and on a molar basis the intake in COL-injected rats after NPY and GAL were markedly higher than in control saline-injected rats.[30,31] In the absence of easy access to and availability of specific pharmacologic blockers, we utilized antisense oligodeoxynucleotides directed against the neuropeptide itself or its receptor in order to assess the contribution of each of these two neuropeptides in COL-induced hyperphagia. This chapter summarizes the results of our studies in this experimental model and various other findings resulting from the use of antisense oligodeoxynucleotides to characterize the functions of NPY and GAL in the regulation of appetite and reproduction.

Roles of Neuropeptide Y and Galanin in Modulating Luteinizing Hormone Secretion

In mammals, on stimulation by hypothalamic LHRH, pituitary LH is characteristically secreted in a pulsatile manner. Superimposed on this rhythmic pattern of secretion is the massive surge release of LH that occurs in females to stimulate ovulation from ovarian follicles. In cycling female rats, this surge occurs on the afternoon of proestrus during the estrous cycle and it can be replicated experimentally in ovariectomized (ovx) rats by sequential treatment with the ovarian steroids estradiol and progesterone. Interestingly, NPY levels in the ME and hypothalamic NPY gene expression increase abruptly before the rise in LHRH and LH in proestrous rats and in ovx rats pretreated with ovarian steroids to elicit a LH surge.[15,16,33] In view of the known stimulatory effects of NPY on the release of LHRH and LH, these temporally correlated increases in the levels of NPY mRNA and peptide suggested a crucial role of *de novo* NPY synthesis in regulation of the surge release of LH. We tested this possibility in ovx rats treated with estrogen and progesterone by administering antisense oligodeoxynucleotides to block NPY synthesis before the LH surge.[34]

[31] P. S. Kalra, S. Pu, T. G. Edwards, and M. G. Dube, *Ann. N.Y. Acad. Sci. U.S.A.* **863,** 432 (1999).

[32] S. Pu, M. G. Dube, T. G. Edwards, S. P. Kalra, and P. S. Kalra, *Brain Res. Mol. Brain Res.* **64,** 85 (1999).

[33] A. Sahu, W. R. Crowley, and S. P. Kalra, *Endocrinology* **134,** 1018 (1994).

[34] P. S. Kalra, J. J. Bonavera, and S. P. Kalra, *Regul. Pept.* **59,** 215 (1995).

TABLE I

NEUROPEPTIDE Y ANTISENSE OLIGODEOXYNUCLEOTIDE SEQUENCES[a]

Bases	Sequence	Modifi-cation	Effect	Ref.
74–93	5'-CCA TTC GTT TGT TAC CTA GC-3'	—	Steroid-induced LH surge ↓	34
		—	NPY levels ↓	34
		—	NAL-induced LH surge ↓	36
		—	LH pulses ↓	37
		S	FI ↓	This report
68–86	5'-TTT GTT ACC TAG CAT CTG-3' ⎱	—	FI ↓; ARC NPY ↓	39
207–224	5'-AGC GGA GTA GTA TCT GGC-3' ⎰			
68–86	5'-TTT GTT ACC TAG CAT CTG-3'	S_t	Fast-induced FI ↓; ARC and PVN NPY ↓	40
78–95	5'-CCC CAT TCG TTT GTT ACC-3'	S	FI ↓	38
64–83	5'-GTT ACC TAG CAT CAT GGC GG-3'	—	FI ↔	41
		S	FI ↓	41
		P	FI ↓	41
−6 to +14	5'-GCA TCA TGG CGG GCG GG-3'	S	FI ↔	45

[a] S, All bases phosphorothioated; S_t, terminal phosphorothioated; P, 3'-propyl; —, unmodified phosphodiester; FI, food intake; ↓, decrease; ↔, no change.

A 20-mer antisense oligodeoxynucleotide sequence (Table I) complementary to the rat NPY mRNA sequence (GenBank accession no. 15880) near the initiation codon encompassing bases 74–93 was prepared. The antisense and a missense oligodeoxynucleotide with the same G and C composition were prepared at the DNA Core Laboratory of the University of Florida Interdisciplinary Center for Biotechnology Research (ICBR, Gainesville, FL). The sequences were analyzed to ensure lack of homology with any known DNA sequence in GenBank and for absence of hairpin loops and self-complementarity. These unmodified phosphodiester sequences were used in three experimental paradigms to delineate the role of hypothalamic NPY in regulating LH secretion.

Ovariectomized rats primed with estradiol followed by an injection of progesterone 48 hr later to elicit an afternoon LH surge received three injections of the sense, missense, or antisense NPY oligodeoxynucleotides (4 nmol/5 μl of saline) into the lateral cerebroventricle at 2-hr intervals (10:00, 12:00, and 14:00); plasma LH levels were analyzed in serial blood samples withdrawn during the anticipated afternoon LH surge. As shown in Fig. 1, control rats injected icv with saline displayed the expected robust LH surge that was not altered by injections of the sense or missense NPY

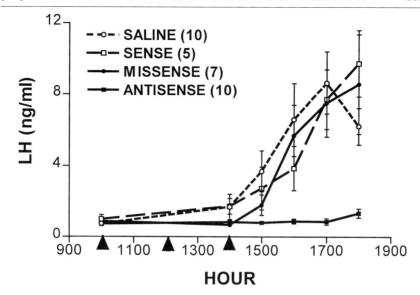

FIG. 1. Blockade of the LH surge by NPY antisense oligodeoxynucleotide treatment. Ovariectomized rats were treated with estradiol and progesterone to induce LH surge release in the afternoon. Three icv injections (denoted by arrowheads) of a NPY antisense oligonucleotide completely blocked the LH surge, whereas similar treatment with the sense or missense sequence was ineffective. Numbers in parentheses represent number of rats per group here and in Figs. 2–4.

oligodeoxynucleotide. However, in rats that received the NPY antisense oligodeoxynucleotide the LH surge was completely blocked and plasma LH levels remained basal throughout the afternoon surge period.[34]

The effect of this NPY antisense oligodeoxynucleotide treatment on NPY levels in microdissected hypothalamic nuclei was assessed in a parallel experiment. Ovx rats, similarly pretreated with estradiol and progesterone, received two icv injections of the NPY missense or antisense oligodeoxynucleotide (10:00 and 12:00) and were killed at 14:00; an additional group of rats was killed at 10:00 before the progesterone injection for comparison of NPY levels. As anticipated, NPY levels at 14:00 were significantly elevated in the ME–ARC, MPOA, and lateral POA of missense oligodeoxynucleotide-injected rats as compared with the control morning levels at 10:00 (Fig. 2). This progesterone induced rise in NPY concentrations in specific hypothalamic sites, relevant to the regulation of LHRH and LH secretion, was completely prevented by the central administration of the NPY antisense oligodeoxynucleotide.[34] NPY levels in other hypothalamic nuclei, including the PVN, were not affected by the antisense treatment. These results confirmed our earlier demonstrations of rapid activation of

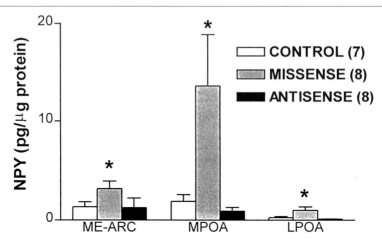

FIG. 2. The progesterone-induced increase in NPY levels in the median eminence–arcuate (ME–ARC), medial preoptic area (MPOA), and lateral POA (LPOA), as seen in rats treated with the missense sequence, is blocked by two icv injections of an NPY antisense oligodeoxynucleotide. *$p < 0.05$ versus control rats.

NPY gene expression in the ARC after progesterone treatment and increase in NPY levels in the ME during the presurge period.[21] Blockade of both NPY accumulation in the ME and the subsequent LH surge in rats injected with the antisense NPY oligodeoxynucleotide lends credence to the thesis that progesterone induces *de novo* synthesis of hypothalamic NPY and that this newly synthesized pool is critical for stimulation of the LHRH and LH surges.[34]

In a second experimental paradigm we examined the significance of upregulation of hypothalamic NPY gene expression induced by blockade of endogenous opiate receptors. Analogous to the effects of progesterone, blockade of opiate receptors by infusion of the opiate antagonist nalaxone (NAL) stimulates pituitary LH secretion and upregulates hypothalamic NPY levels and gene expression in ovx, estrogen-pretreated rats.[35,36] Interestingly, three icv injections of the unmodified NPY antisense oligodeoxynucleotide (Table I) prevented the NAL-induced rise in NPY levels in the ME–ARC as seen in the control saline- or missense oligodeoxynucleotide-injected groups of rats and it completely suppressed the rise in LH at 15:00.[36] The results confirmed our earlier finding that an increase in hypothalamic NPY activity precedes the LHRH–LH surges induced by either progesterone or NAL.[33,35] Preventing these increases in NPY levels and

[35] A. Sahu, W. R. Crowely, and S. P. Kalra, *Endocrinology* **126,** 876 (1990).
[36] B. Xu, A. Sahu, P. S. Kalra, W. R. Crowley, and S. P. Kalra, *Endocrinology* **137,** 78 (1996).

TABLE II
GALANIN ANTISENSE OLIGODEOXYNUCLEOTIDE SEQUENCES[a]

Bases	Sequence	Modification	Effect	Ref.
149–166	5'-GGA TAA CGC TGC CCC TGG-3'	—	LH pulses ↔	37
204–221	5'-ATC CCG AGC CCC AGA GTG-3'	—		
147–164	5'-ATA ACG CTG CCC CTG GCC-3'	—	FI ↓ ; PVN GAL ↓	42
277–294	5'-GTT GTC AAT GGC ATG TGG-3'	—		
		S_t	Hyperphagia ↔	This report

[a] S_t, terminal phosphorothioated; —, unmodified phosphodiester; FI, food intake; ↓, decrease; ↔, no change.

release by the antisense oligodeoxynucleotide blocked LH hypersecretion, thus revealing once again the critical requirement for the stimulatory effects of newly synthesized NPY to elicit the LH surge.

To understand how NPY and GAL participate in the episodic pattern of LH secretion, the availability of these neuropeptides for release was altered by the central administration of antisense oligodeoxynucleotides to NPY or GAL.[37] The NPY missense and antisense sequences were the same as used above in studies of the LH surge.[34,36] For GAL, two short 18-mer sequences within the GAL signal peptide codon encompassing bases 149–166 and 204–221 of the rat GAL mRNA sequence (GenBank accession no. J03624) were selected (Table II). Antisense oligodeoxynucleotides to these two sequences, designated GAL antisense oligodeoxynucleotide I and GAL antisense oligodeoxynucleotide II, and a missense oligodeoxynucleotide with similar G and C composition, were prepared at the DNA Core Laboratory (ICBR). The sequences were analyzed to ensure that no self-hybridization or hairpin loops existed.

Ovariectomized rats, preimplanted with a guide cannula into the third cerebroventricle, were injected icv with either the missense or sense oligodeoxynucleotide at 09:00, 11:00, and 13:00 and blood samples were withdrawn continously from a jugular vein cannula from 13:00 to 16:00 for LH analyses. NPY antisense or missense oligodeoxynucleotides were administered at the dose of 3.8 nmol/3 μl per injection. The GAL antisense oligodeoxynucleotide I or II, or the missense oligodeoxynucleotide, was injected icv at two dose levels (4.2 or 6.7 nmol/3 μl) in separate groups of rats. In another group of rats GAL oligodeoxynucleotides I and II were administered in combination (6.7 nmol each). The NPY antisense oligodeoxynucleotide decreased all parameters of LH pulsatility. The number

[37] B. Xu, S. Pu, P. S. Kalra, J. F. Hyde, W. R. Crowley, and S. P. Kalra, *Endocrinology* **137**, 5297 (1996).

of LH pulses, the magnitude of each pulse, and mean LH levels over the 3-hr sampling period were significantly reduced as compared with control saline-injected or missense oligodeoxynucleotide-injected rats ($p < 0.05$), demonstrating that the characteristic pattern of LHRH pulsatility is critically dependent on the excitatory action of hypothalamic NPY.[37] On the other hand, neither the high nor the low dose of GAL oligodeoxynucleotides I and II injected separately or in combination adversely affected any parameter of LH pulsatility. This was an unexpected observation since in the same study passive immunoneutralization of endogenous GAL by the icv administration of a GAL antibody markedly suppressed LH pulsatility.[37] There are several possible explanations to account for the conflicting conclusions derived from these two different experimental approaches. Unlike NPY, GAL-producing neurons are widely distributed in various hypothalamic nuclei[25,26] and antisense oligodeoxynucleotide administered into the third ventricle may not access these diverse sites in adequate amounts. Thus, it is likely that the magnitude of reduction of GAL levels may be insufficient in hypothalamic sites relevant for LH regulation to affect LH secretion adversely. Alternatively, a longer time period may be required for the oligodeoxynucleotide administered into the third ventricle to access the various subpopulations of GAL-producing neurons. Perhaps direct microinjection of GAL antisense oligodeoxynucleotides into hypothalamic nuclei may be needed to adequately reduce GAL levels in the specific sites involved in regulation of LHRH–LH (e.g., MPOA and/or ME). Finally, because GAL levels were not measured at the end of the sequential bleeding period we cannot rule out the possibility of inefficacy of the antisense sequences. Whatever may be the cause of these discrepant results, it is obvious from this series of studies that whereas positive results with the antisense technology may be accepted with confidence, negative results must be verified by alternative approaches.

Neuropeptide Y Antisense Oligodeoxynucleotides and Appetite

Our initial venture with the antisense technology was to characterize the role of endogenous hypothalamic NPY in stimulation of appetite. Unlike the situation described above for the regulation of LH surge release, stimulation of appetite has not been correlated with rapid *de novo* synthesis of NPY. We speculated that to suppress appetite it may be necessary to inhibit ongoing NPY synthesis over a longer period of time. Therefore, to protect from rapid degradation *in vivo,* all bases of the NPY antisense oligodeoxynucleotide (Table I) and the corresponding sense sequence were phosphorothioated. Groups of rats received once-daily injections of the antisense oligodeoxynucleotide either for 2 days (4 nmol/day) or 3 days (8

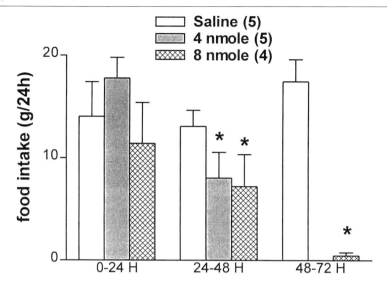

FIG. 3. Effect of a daily icv injection of NPY antisense oligodeoxynucleotide (4- or 8-nmol dose) on 24-hr food intake. The antisense oligodeoxynucleotide treatment inhibited food intake after the second injection. $*p < 0.05$ versus saline-injected rats.

nmol/day) into the third cerebroventricle. Control rats were similarly injected with either the sense oligodeoxynucleotide at an equivalent dose or vehicle saline alone. Body weights and 24-hr consumption of lab chow were recorded. As shown in Fig. 3, daily food intake in the saline-injected group remained constant for 3 days. After the first injection of the antisense oligodeoxynucleotide, rats consumed similar amounts of chow. However, during the second 24-hr period food intake was reduced by ~50% ($p < 0.05$) after both the high and low dose of the antisense oligodeoxynucleotide, but there was no evidence of a dose-related response. The third injection of NPY antisense oligodeoxynucleotide (8-nmol group only) almost completely inhibited food intake. There was no marked effect of either of these treatments on body weight over the 3-day experimental period. Consistent with the earlier demonstrations of the potent appetite-stimulating effects of centrally administered NPY and the demonstrations that endogenous NPY is a natural appetite transducer,[1,4,5] these results lead one to conclude that blockade of NPY synthesis in the ARC by icv administration of an antisense oligodeoxynucleotide causes a dramatic reduction in appetite. However, untoward behavioral responses and the general state of the experimental rats leaves the validity of these results open to question. Several of the antisense-treated rats displayed symptoms such as scruffiness, huddling, tremors, and lack of grooming, so that three of the seven rats injected with the higher dose of the antisense oligodeoxynucleotide were euthanized. The adverse effects were even more apparent in rats injected with the sense oligodeoxynucleotide, which also displayed seizures and succumbed (data not shown). Thus, we cannot rule out the possibility that the dramatic loss of appetite may be the nonspecific result of the toxic effects of the fully phosphorothioated oligodeoxynucleotide employed in this study. These results clearly interject a note of caution in the interpretation of data using antisense technology. Lower doses or nonphosphorothioated sequences may circumvent some of these difficulties.

Hulsey et al.[38] utilized a lower dose of a fully phosphorothioated sequence that was similar, but not identical, to that used by us (bases 78–95 versus 74–93; Table I). Rats received 5 μg of this antisense and a corresponding missense oligodeoxynucleotide daily in the lateral cerebroventricle for 7 days. As compared with the missense-treated rats, cumulative food intake, meal size, and meal duration were significantly suppressed within 1 day after the NPY antisense oligodeoxynucleotide administration, and this magnitude of inhibition was maintained for 6 days with no evidence of sickness or taste aversion. However, it is not known whether the missense oligodeoxynucleotide itself suppressed appetite because NPY levels in the PVN of the two groups were similar and no vehicle-treated controls were included for comparison. Akabayashi et al.[39] reported inhibition of macronutrient intake as well as total caloric intake in rats injected with a mixture of two unmodified antisense NPY oligodeoxynucleotides. One of these sequences encompassed the initiation codon (bases 68–86) and the other corresponded to bases 207–224 of the rat NPY mRNA (Table I). Daily injections of these antisense oligodeoxynucleotides for 4 days into the ARC caused a small reduction in ARC NPY levels and a 20–30% reduction in daily intake; no adverse effects were reported. Direct delivery of the oligodeoxynucleotide into the ARC, where NPY-producing neurons are concentrated, may have contributed to the efficacy of this unmodified oligodeoxynucleotide in reducing NPY levels and appetite for an extended period. Schaffhauser et al.[40] have reported that three icv injections of this NPY antisense oligodeoxynucleotide (bases 68–86), with only the terminal bases phosphorothioated, effectively prevented the fasting-induced increase in ARC and PVN NPY levels and reduced the subsequent food intake. These effects were specific to the antisense-injected rats, because in rats

[38] M. G. Hulsey, C. M. Pless, B. D. White, and R. J. Martin, Regul. Pept. 59, 207 (1995).
[39] A. Akabayashi, C. Wahlestedt, J. T. Alexander, and S. F. Leibowitz, Brain Res. Mol. Brain Res. 21, 55 (1994).
[40] A. O. Schaffhauser, A. Stricker-Krongrad, L. Brunner, F. Cumin, C. Gerald, S. Whitebread, L. Criscione, and K. G. Hofbauer, Diabetes 46, 1792 (1997).

injected similarly with the corresponding missense sequence, the fasting-induced increases in NPY levels and feeding were unaffected.[40]

A study by Dryden et al.[41] may help to explain some of these variable results. These authors evaluated the hypothalamic distribution and relative efficacies of three different structural modifications of an NPY antisense oligodeoxynucleotide sequence encompassing bases 64–83 at the initiation codon (Table I). The unmodified sequence was apparently rapidly degraded because it was undetectable in the hypothalamus despite 7 days of continuous infusion into the third cerebroventricle. The 3'-propyl-protected form, on the other hand, although widely dispersed in the hypothalamus, caused only modest reductions of 9% in food intake and 6% in body weight as compared with the saline controls but not versus the missense-treated rats. The antisense oligodeoxynucleotide with all bases phosphorothioated caused a larger, 15% suppression in food intake and body weight was reduced by 12% after 7 days of treatment. However, similar reductions in body weight and daily intake were also seen in rats treated with the phosphorothioated missense oligodeoxynucleotide. As we had noted (see above), both antisense- and missense-injected rats displayed signs of toxicity (listlessness, hunched posture, no grooming) and their condition deteriorated progressively. Thus, the nonselectivity of the 3'-propyl and phosphorothioated forms along with the toxicity caused by the latter modification led the investigators to question the validity of this approach. This series of investigations reiterates that in long-term studies when the issues of oligodeoxynucleotide degradation are countered by phosphorothioation of the bases, the likelihood of toxicity increases. Use of a similarly modified control missense sequence is critical to address this possibility as well as the issue of nonspecificity. A comparative analysis of the reports summarized above suggests that sequences in which only the terminal bases are phosphorothioated achieve the desired protection of the oligodeoxynucleotide *in vivo* without producing toxicity in the animal.

Galanin Antisense Oligodeoxynucleotides

Central administration of GAL has also been shown to stimulate ingestive behavior, albeit to a lesser extent than after NPY administration. A role for endogenous GAL was demonstrated by microinjection of GAL antisense oligodeoxynucleotides directly into the PVN of the hypothalamus.[42] Two unmodified 18-mer sense or antisense sequences corresponding

[41] S. Dryden, L. Pickavance, D. Tidd, and G. Williams, *J. Endocrinol.* **157,** 169 (1998).
[42] A. Akabayashi, J. I. Koenig, Y. Watanabe, J. T. Alexander, and S. F. Leibowitz, *Proc. Natl. Acad. Sci. U.S.A.* **91,** 10375 (1994).

to bases 147–164 and 277–294 (Table II) were injected together daily for 4 days directly into the PVN. GAL levels in the PVN were significantly reduced and there was a marked reduction in fat ingestion and body weight.[42] It is noteworthy that, in this study, the sense sequence did not adversely affect either the PVN GAL levels or feeding behavior.

Our attempts to suppress feeding in an experimental model of hyperphagia with the same oligodeoxynucleotide sequences yielded less specific results. Rats microinjected with the neurotoxin colchicine (COL) into the ventral medial nucleus (VMN) of the hypothalamus rapidly develop transient hyperphagia and body weight gain.[30] These rats display significant increases in GAL concentration and GAL mRNA levels in several hypothalamic nuclei during the period of hyperphagia.[31,32] To test the possibility that the increased endogenous GAL activity may account for this hyperphagia, we used GAL antisense oligodeoxynucleotides to prevent the increase in hypothalamic GAL. At the time of COL microinjection into the VMN, a guide cannula into the third cerebroventricle, or bilateral guide cannulas aimed at the PVN, were implanted into the rats. Starting on the next day, rats were injected daily for 3 days with the sense or antisense oligodeoxynucleotides either directly into the PVN (250 ng/0.33 μl of saline on each side) or icv (27 μg/3 μl of saline). The two antisense sequences were the same as used by Akabayashi et $al.$[42] (Table II), with the modification of terminal capping with phosphorothioate to retard degradation. As shown in Fig. 4, intra-PVN injections of the two antisense GAL oligodeoxynucleotides rapidly reduced food intake within the first 24-hr period and this low level of intake was maintained for the next 2 days. An even more marked reduction in intake occurred after injection into the third ventricle, suggesting that GAL-producing neurons in areas other than the PVN also stimulate feeding. In this group food intake was reduced progressively over the next 2-day experimental period, demonstrating a cumulative effect. Rats displayed no signs of toxicity or sickness; however, the issue of specificity of the antisense oligodeoxynucleotides was unresolved. As seen in Fig. 4, analogous declines in food intake were seen in COL rats injected with the sense oligodeoxynucleotides into the PVN or third ventricle. These results raise a disquieting issue regarding the most desirable feature of antisense technology, namely, specificity of antisense oligodeoxynucleotides. Despite the dramatic reduction in food intake after the administration of GAL antisense oligodeoxynucleotides the nonspecificity of the results prevents us from unequivocally concluding that the increase in endogenous hypothalamic GAL contributes to the hyperphagia displayed by these VMN COL-injected rats.

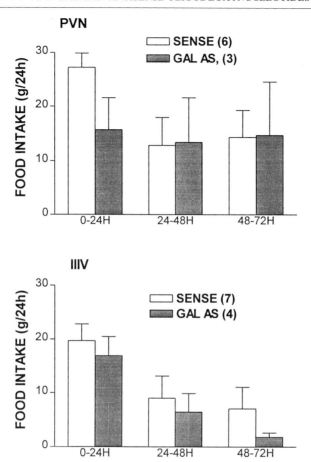

FIG. 4. Effects of a galanin (GAL) antisense or sense oligodeoxynucleotide on food intake in COL-injected hyperphagic rats. The oligodeoxynucleotides were injected once daily either into the third cerebroventricle (IIIV) or bilaterally into the paraventricular nuclei (PVN).

Blockade of Neuropeptide Y Receptors

In the absence of specific receptor antagonists, discerning the functions of the NPY cloned receptor subtypes in order to identify the "feeding receptor" has been problematic. With the use of antisense oligodeoxynucleotides the Y_1 receptor (Y_1R), the first NPY receptor subtype to be cloned, has been shown to transduce NPY signals for anxiety,[43,44] hypothermia,[45] and vasoconstriction.[46] In one study with antisense oligodeoxynucleotides, these receptors were reported to mediate the spontaneous feeding elicited presumably by endogenous NPY[45]; however, other studies have reported the opposite behavioral effects. Feeding was enhanced after treatment with two different Y_1 antisense oligodeoxynucleotides, including one that was earlier shown to reduce Y_1R density.[43,44,47] These demonstrations of increased food intake after Y_1R antisense oligos treatment contradicted the inhibition of feeding seen after pharmacologic blockade of the Y_1R[48,49] and led investigators to speculate on the existence of an additional receptor subtype that mediates feeding.[44] Indeed, the cloning of a Y_1-related receptor subtype, termed Y_5, lends credence to this possibility.

The role of this Y_5R subtype in transducing NPY-induced appetite signals was established with antisense oligodeoxynucleotide technology.[40] Two 20-mer antisense oligodeoxynucleotides, encompassing the start and stop codons (bases 19–38 and 1377–1396) were phosphorothioate end protected. Repeated icv injections of these Y_5 antisense oligodeoxynucleotides over a 2-day period decreased both NPY- and fasting-induced food intake in rats. Because no decrease in receptor transcript was detected, it was suggested that the Y_5 antisense oligodeoxynucleotide reduced receptor protein availability by an action at the posttranscriptional level.[40] No untoward behavioral effects, indicative of toxicity, were reported after 3 days of treatment with these terminally protected oligodeoxynucleotides. The specificity of action was indicated by the ineffectiveness of the corresponding missense and sense sequences. The participation of Y_5R in regulation of feeding was confirmed in another study that used an unmodified 20-mer sequence 20 bp downstream from the start codon (Table III). Repeated injections into the lateral ventricle suppressed both spontaneous and NPY-induced feeding.[50] Although suppression of spontaneous food intake oc-

[43] C. Wahlestedt, E. M. Pich, G. F. Koob, F. Yee, and M. Heilig, *Science* **259,** 528 (1993).

[44] M. Heilig, *Regul. Pept.* **59,** 201 (1995).

[45] F. J. Lopez-Valpuesta, J. W. Nyce, T. A. Griffin-Biggs, J. C. Ice, and R. D. Myers, *Proc. R. Soc. Lond. B. Biol. Sci.* **263,** 881 (1996).

[46] D. Erlinge, L. Edvinsson, J. Brunkwall, F. Yee, and C. Wahlestedt, *Eur. J. Pharmacol.* **240,** 77 (1993).

[47] A. O. Schaffhauser, S. Whitebread, R. Haener, K. G. Hofbauer, and A. Stricker-Krongrad, *Regul. Pept.* **75,** 417 (1998).

[48] P. S. Kalra, M. G. Dube, B. Xu, W. G. Farmerie, and S. P. Kalra, *J. Neuroendocrinol.* **10,** 43 (1998).

[49] A. Kanatani, A. Ishihara, S. Asahi, T. Tanaka, S. Ozaki, and M. Ihara, *Endocrinology* **137,** 3177 (1996).

[50] M. Tang-Christensen, P. Kristensen, C. E. Stidsen, C. L. Brand, and P. J. Larsen, *J. Endocrinol.* **159,** 307 (1998).

TABLE III
NEUROPEPTIDE Y_1 AND Y_5 RECEPTOR SUBTYPE ANTISENSE OLIGONUCLEOTIDE[a]

Receptor	Bases	Sequence	Modifi-cation	Effect	Ref.
Y_1R	71–88	5'-GGA GAA CAG AGT TGA ATT-3'	—	$Y_1R \downarrow$; anxiety $\uparrow$	43
			—	FI $\uparrow$	44
			S_t	FI $\leftrightarrow$; fasting-induced FI $\uparrow$	47
	157–173	5'-GCC AGG TTT CCA GAG ACC CC-3'	—	FI $\downarrow$	45
			S_t	$Y_1R \downarrow$; NPY-induced FI $\leftrightarrow$	47
			S_t	Hyperphagia $\leftrightarrow$	??
Y_5R	19–38	5'-AGA GGA CGT CCA TTA GCA GC-3'	S_t	NPY and fasting-induced FI $\downarrow$	40
			S_t	Hyperphagia $\leftrightarrow$	This report
	1377–1396	5'-TCA TGA CAT GTG TAG GCA GT-3'	S_t	NPY and fasting-induced FI $\downarrow$	40
			S_t	Hyperphagia $\leftrightarrow$	This report
	24–46	5'-GTG GAA GAA GAG GAC GTC CAT-3'	—	FI $\downarrow$; NPY-induced FI $\downarrow$	50

[a] S_t, terminal phosphorothioated; —, unmodified phosphodiester; FI, food intake; $\downarrow$, decrease; $\uparrow$, increase, $\leftrightarrow$, no change.

curred rapidly it lasted only after the first three injections given at 12-hr intervals, and there was no decrease after the fourth injection, suggesting that a compensatory increase in other orexigenic signals in response to reduced NPY function may account for this loss of efficacy of the antisense oligodeoxynucleotide. It is also likely that with the continued loss of Y_5R, the other putative orexigenic Y_1 receptor subtype, may transduce the NPY signal. In our studies of the hypothalamic factors responsible for the hyperphagia induced by COL injection into the VMN, gene expression of both the Y_1 and Y_5R subtypes was enhanced.[48] We were unable to suppress this hyperphagia by blockade of Y_5R transcripts with the antisense oligodeoxynucleotide sequences used successfully by Schaffhauser et al.[40] Daily food intake was similar after repeated injections of the two Y_5R antisense oligodeoxynucleotides or of their corresponding missense sequences (our unpublished observations, 1998). However, pharmacological blockade of the Y_1R with an icv injection of the antagonist 1229U91 partially suppressed food intake,[30] suggesting a role for the Y_1R in this hyperphagia.

Thus, antisense oligodeoxynucleotides used to block the Y_1R transcript have yielded conflicting results.[44,45,47] However, this technology has been useful to establish a role for Y_5R in transducing the NPY orexigenic signal for spontaneous and fasting-induced feeding, but not in experimentally induced hyperphagia. Blockade of the Y_1R transcript with antisense oligodeoxynucleotides has yielded conflicting results. Collectively, results of the pharmacologic and antisense experiments suggest that both Y_1 and Y_5 receptor subtypes mediate the orexigenic actions of NPY; their participation may vary under different conditions. Further investigations are warranted to unravel the role of each receptor subtype during various conditions of energy demand.

Summary and Conclusions

The effects of neurotransmitters and neuromodulators can be interrupted by either blockade or diminution in the amount of release by curtailing the availability of the neuropeptides in the nerve terminals. Theoretically, antisense oligodeoxynucleotides decrease the availability of signals by blocking the transcription process, thus offering an opportunity to dissect the relative roles of neurotransmitters that elicit similar biological responses. Both NPY and GAL stimulate feeding and LHRH secretion, but antisense oligodeoxynucleotides behaved differently in interrupting these two responses. Centrally administered antisense oligodeoxynucleotides were effective in blocking the stimulatory effects of NPY on LH release, thereby demonstrating that neuronal permeability, degradation, and toxicity of oligodeoxynucleotides are not limiting factors. Thus, for short-term studies the unmodified phosphodiester sequences can be successfully used. Because the attempts to block the behavioral effects of NPY yielded equivocal results, it is clear that newly synthesized NPY, critical for LH release, is relatively insignificant for feeding. Blockade of behavioral effects requires a longer period of effectiveness of oligodeoxynucleotides necessitating that the rate of oligodeoxynucleotide degradation be retarded. Effective protection from degradation *in vivo* can be achieved by phosphorothioating one or two terminal bases. This modification, unlike the earlier practice of phosphorothioate protection of each base, causes no toxicity and is well tolerated after central administration. Adequate controls, including vehicle and similarly modified missense or scrambled sequences, are essential to confirm specificity and to exclude toxicity. The site of administration is another important factor to be considered in the experimental design. Whereas icv injections (lateral ventricle, or IIIrd ventricle) have been largely effective in allowing access to multiple hypothalamic sites, direct injection into relevant hypothalamic nuclei may provide surgical precision to effect concentrated blockade at the site of synthesis. Earlier studies with centrally administered oligodeoxynucleotides were plagued by these limitations, resulting in inconsistent and equivocal results. However, more recent investigations, designed with these caveats in mind, have successfully used antisense oligodeoxynucleotides as exemplified by the studies to establish the role of the Y_5R subtype in transducing the orexigenic NPY signal.

Acknowledgments

Original work reported here was supported by grants (DK37273 and NS32727) from the NIH. Thanks are due to Mrs. Dawn Stewart for secretarial assistance.

[14] Blockade of Neuropathic Pain by Antisense Targeting of Tetrodotoxin-Resistant Sodium Channels in Sensory Neurons

By Josephine Lai, John C. Hunter, Michael H. Ossipov, and Frank Porreca

One of the most significant health problems in our country is the inadequate treatment of pain. The impact of pain places great burden in economic terms (approximately $100 billion annually; *NIH Guide*, Vol. 24, 1995) as well as in human suffering. Estimates suggest that as many as one-third of all Americans suffer from some form of chronic pain, and that one-third of these have pain that is resistant to the treatment efforts of the medical community. Neuropathic pain is one of the most difficult pains to treat and our understanding of this condition is currently limited. Research into mechanisms of neuropathic pain has been greatly influenced by the development of several animal models of peripheral nerve injury that reflect some aspects of neuropathic pain in humans.[1-5]

Neuropathophysiology of Peripheral Nerve Injury

Peripheral nerve injury results in a number of significant neuroanatomical and neurophysiological changes to the injured nerve. An important consequence of nerve injury is the development of spontaneous neuronal activity. The injured nerve begins to fire spontaneously and constantly at frequencies above the normal firing rate, giving rise to spontaneous pain. Normally quiescent high-threshold nociceptors also alter their electrophysiologic activity so that constantly active nociceptors lead to spontaneous pain and hyperalgesia. It is thought that this constant afferent drive leads to the development of central sensitization, which drives the development of the manifestations of neuropathic pain. Early studies in rats and rabbits have shown that chronic nerve injury produces spontaneous firing of sensory

[1] G. J. Bennett and Y.-K. Xie, *Pain* **33,** 87 (1988).
[2] Z. Seltzer, R. Dubner, and Y. Shir, *Pain* **43,** 205 (1990).
[3] S. H. Kim and J. M. Chung, *Pain* **50,** 355 (1992).
[4] J. A. DeLeo, D. W. Coombs, S. Willenbring, R. W. Colburn, C. Fromm, R. Wagner, and B. B. Twitchell, *Pain* **56,** 9 (1994).
[5] B. P. Vos, A. M. Strassman, and R. J. Maciewicz, *J. Neurosci.* **14,** 2708 (1994).

nerves.[6,7] The primary sites of this abnormal, ectopic firing was determined to be the site of injury itself and the dorsal root ganglion (DRG) of the injured nerve.[8,9] The ectopic sites are also believed to be associated with the generation of discharges that can be initiated by noxious or by normally nonnoxious mechanical or chemical stimuli.[8,10] The sensitivity to chemical stimuli includes circulating catecholamines, and may provide a basis for sympathetically maintained pain, sometimes associated with neuropathy. Ectopic firing contributes to neuropathic pain directly, by eliciting abnormal sensations (paresthesias from increased firing of large fibers) and dysesthesias and pain (from increased firing of small fibers), as well as indirectly by the resulting sensitization of neurons in the spinal cord.[10] Damaged peripheral nerves may also form peripheral collateral sprouts and ectopic foci, resulting in spontaneous discharges that are perceived as nociception[11] as well as sprouting in central connections. Thus, large fiber terminals sprout into the superficial lamina of the spinal dorsal horn to form physiopathic synapses that may be the basis whereby normally innocuous inputs now drive nociceptive transmission cells.[12,13] It has been demonstrated that the degree of ectopic discharge caused by lumbar vertebra 5/6 (L5/L6) spinal nerve ligation correlates well with behavioral signs of neuropathic pain.[14] This peripheral neuropathy, a model that was developed by Kim and Chung,[3] was assessed for neuropathic pain behavior and for ectopic discharges by electrophysiological recordings from the L5 dorsal root. It was found that the number of units showing ectopic discharge and behavior indicative of neuropathic pain both reached a peak at 1 week, and were reduced considerably at 10 weeks.[14]

Role of Voltage-Gated Sodium Channels in Neuropathic Pain

The generation of spontaneous, ectopic discharges after nerve injury strongly implicates voltage-gated sodium channels in neuropathic pain. These channels are located in the plasma membrane and permit entry of

[6] P. D. Wall and M. Gutnick, *Nature (London)* **248,** 740 (1974).

[7] E. J. Kirk, *J. Comp. Neurol.* **155,** 165 (1974).

[8] M. Devor, *in* "Textbook of Pain" (P. D. Wall and R. Melzack, eds.), p. 79. Churchill Livingstone, Edinburgh, 1994.

[9] K. C. Kajander, S. Wakisaka, and G. J. Bennett, *Neurosci. Lett.* **138,** 225 (1992).

[10] H. L. Fields, M. C. Rowbotham, and M. Devor, *in* "Handbook of Experimental Pharmacology" (A. Dickenson and J. M. Besson, eds.), p. 93. Springer-Verlag, Berlin, 1997.

[11] K. C. Kajander and G. J. Bennett, *J. Neurophysiol.* **68,** 734 (1992).

[12] C. J. Woolf, P. Shortland, and R. E. Coggeshall, *Nature (London)* **355,** 75 (1992).

[13] C. J. Woolf and T. P. Doubell, *Curr. Opin. Neurobiol.* **4,** 525 (1994).

[14] H. C. Han, D. H. Lee and J. M. Chung, *Soc. Neurosci. Abstr.* **21,** 896 (1995).

sodium ions into the cell, causing depolarization and generation of the action potential.[15,16] There is an increase in the concentrations of sodium channels at the site of injury and in the DRG of injured nerves. The importance of sodium channel activity in clinical neuropathic pain is underscored by the key observation that essentially all of the drug categories that are clinically useful exhibit a significant degree of sodium channel-blocking activity. Categories of drugs that are clinically important against neuropathic pain include the antiseizure medications, local anesthetics, tricyclic antidepressants, and orally active antiarrhythmic agents. Importantly, a common theme of all of these compounds is their demonstrated ability to block tetrodotoxin (TTX)-sensitivie and TTX-resistant (see below) sodium channels nonselectively.[17–19] A crucial aspect of these medications is that the relief of pain that is produced usually occurs at doses that are close to the production of toxic side effects (i.e., low therapeutic index). In addition, adequate pain relief is often obtained only by combinations of several medications administered at the same time (as many as three or more), which also result in increased, and often barely tolerable, side effects and decreased quality of life for patients. It is clear that the nonselective blockade of different classes of sodium channels by these agents contributes heavily to the side effects, and that pain relief might be accomplished with minimal side effects if the specific sodium channels types driving the firing of the injured nerve could be selectively blocked (see below).

Classification of Voltage-Gated Sodium Channels

A number of voltage-gated sodium channel subtypes have been identified on the basis of biophysical properties and sensitivity to tetrodotoxin (TTX). Membrane excitability in different tissues is predominantly controlled by the tissue-dependent expression of distinct genes encoding individual voltage-gated sodium channel subtypes that have been distinguished on the basis of primary structure but that can also be differentiated by their biophysical properties and sensitivity to TTX. Most voltage-gated sodium channels are characterized by rapid inactivation kinetics and sensitivity to low, nanomolar concentrations of TTX. These include rat brain types I, IIA, and III, skeletal muscle type I, [15,16] sodium channel protein 6

[15] W. A. Catterall, *Physiol. Rev.* **72,** S15 (1992).
[16] R. G. Kallen, S. A. Cohen, and R. L. Barchi, *Mol. Neurobiol.* **7,** 383 (1993).
[17] D. L. Tanelian and W. G. Brose, *Anesthesiology* **74,** 949 (1991).
[18] M. F. Jett, J. McGuirk, D. Waligora, and J. C. Hunter, *Pain* **69,** 161 (1997).
[19] E. Delpon, C. Valenzuela, O. Perez, and J. Tamargo, *J. Cardiovasc. Pharmacol.* **21,** 13 (1993).

(SCP6)[20] and its closely related peripheral nerve 4 protein (PN4)[21] and peripheral nerve 1 protein (PN1),[22,23] of which the latter appears to be the rodent ortholog of the human neuroendocrine channel. However, more persistent, slowly inactivating sodium currents have also been described, specifically in heart and sensory ganglia. The cardiac channel (H1)[24] exhibits low, micromolar ($1-5$ μM) sensitivity to TTX. More recently, a novel, TTX-resistant sodium channel has been cloned from rat DRG. This TTX-resistant channel is expressed predominantly in small-diameter, unmyelinated peripheral nerves (C fibers), and is not found in other peripheral or central neurons or nonneuronal tissues.[25] This TTX-resistant channel is identified as either peripheral nerve 3 (PN3)[26] or sensory nerve specific (SNS).[25] An additional sodium channel subtype, termed NaN, which appears to be expressed preferentially in C fibers in DRG and in trigeminal neurons, has been reported.[27] This channel subtype has a sequence that may represent a distinct subfamily of sodium channels; although its sequence characteristics predict that it may be TTX resistant, one communication shows that this channel, also known as SNS2, exhibits intermediate TTX sensitivity and its kinetics resemble that of TTX-sensitive sodium channels.[28]

Targeting individual voltage-gated sodium channel subtypes that may be either specific to sensory, nociceptive neurons[25,26] or selectively regulated[29] in response to a peripheral nerve injury may serve as a unique and novel approach to the treatment of neuropathic pain. In general, the neurons of the DRG express two main types of sodium currents: a rapidly

---

[20] K. L. Schaller, D. M. Krzemien, P. J. Yarowsky, B. K. Krueger, and J. H. Caldwell, *J. Neurosci.* **15,** 3231 (1995).

[21] P. S. Dietrich, J. G. McGivern, S. G. Delgado, B. D. Koch, R. M. Eglen, J. C. Hunter, and L. Sangameswaran, *J. Neurochem.* **70,** 2262 (1998).

[22] L. Sangameswaran, L. M. Fish, B. D. Koch, D. K. Rabert, S. G. Delgado, M. Ilnicka, L. B. Jakeman, S. Novakovic, K. Wong, P. Sze, E. Tzoumaka, G. R. Stewart, R. C. Herman, H. Chan, R. M. Eglen, and J. C. Hunter, *J. Biol. Chem.* **272,** 14805 (1997).

[23] J. J. Toledo-Aral, B. L. Moss, Z.-J. He, A. G. Koszowski, T. Whisenand, S. R. Levinson, J. J. Wolf, I. Silos-Santiago, S. Halegoua, and G. Mandel, *Proc. Natl. Acad. Sci. U.S.A.* **94,** 1527 (1997).

[24] R. B. Rogart, L. L. Cribbs, L. K. Muglia, D. D. Kephart, and M. W. Kaiser, *Proc. Natl. Acad. Sci. U.S.A.* **86,** 8170 (1989).

[25] A. N. Akopian, L. Sivilotti, and J. N. Wood, *Nature (London)* **379,** 257 (1996).

[26] L. Sangameswaran, S. G. Delgado, L. M. Fish, B. D. Koch, L. B. Jakeman, G. R. Stewart, P. Sze, J. C. Hunter, R. M. Eglen, and R. C. Herman, *J. Biol. Chem.* **271,** 5953 (1996).

[27] S. D. Dib-Hajj, L. Tyrrell, J. A. Black, and S. G. Waxman, *Proc. Natl. Acad. Sci. U.S.A.* **95,** 8963 (1998).

[28] S. Tate, S. Benn, C. Hick, D. Trezise, V. John, R. J. Mannion, M. Costigan, C. Plumpton, D. Grose, Z. Gladwell, G. Kendall, K. Dale, C. Bountra, and C. J. Woolf, *Nature Neurosci.* **1,** 653 (1998).

[29] S. G. Waxman, J. D. Kocsis, and J. A. Black, *J. Neurophysiol.* **72,** 466 (1994).

inactivating, TTX-sensitive (TTX-S I_{Na}) and a slowly inactivating, TTX-resistant (TTX-R I_{Na}) sodium current.[30–33] The relative proportions of these sodium currents in individual DRG neurons determine the wide range of firing behaviors and, consequently, functional properties of the various classes of DRG neurons. TTX-S currents are the predominant I_{Na} in all types of DRG cells at all stages of development and represent the main I_{Na} associated with the large-diameter, fast-conducting, myelinated Aβ fibers.[30,34] In the adult DRG, at least two types of TTX-S I_{Na} have been recorded from most categories of DRG neurons.[31,34] Whole-cell patch-clamp studies have indicated that large neurons of the DRG may have TTX-S, kinetically fast sodium currents and a combination of TTX-S fast and slow currents, indicating a heterogeneity of channels responsible for these currents.[34] In contrast, the TTX-R I_{Na} normally has a distribution restricted to the small-diameter, unmyelinated, capsaicin-sensitive nociceptors (C fibers).[30,32,35] The slow inactivation and rapid repriming properties of TTX-R I_{Na} appear particularly well suited to sustained firing at the depolarized potentials characteristic of an injured peripheral nerve.[32,36] Several sodium channel proteins have been identified from DRG tissue with inactivation kinetics and sensitivity to TTX suggesting mediation of the TTX-S I_{Na}. These channels include brain types I, IIA, and III,[29] SCP6 or PN4,[20,21] and PN1.[22,23] On the other hand, PN3 (or SNS), and the more recently characterized NaN/SNS2, appear to be suitable channel subtypes for the mediation of the TTX-R I_{Na} in small sensory neurons, and may contribute to the TTX-R sodium current that is believed to be associated with central sensitization in chronic neuropathic pain states.[37]

Tetrodotoxin-Resistant Peripheral Nerve 3 Sodium Channels in
  Neuropathic Pain

The preceding speculation regarding the potential pathophysiological role of PN3, based on its anatomical distribution and biophysical properties, suggests that a selective functional inhibition of PN3 may be useful in

[30] M. L. Roy and T. Narahashi, *J. Neurosci.* **12,** 2104 (1992).
[31] J. M. Caffrey, D. L. Eng, J. A. Black, S. G. Waxman, and J. D. Kocsis, *Brain Res.* **592,** 283 (1992).
[32] A. A. Elliott and J. R. Elliott, *J. Physiol.* **463,** 39 (1993).
[33] N. Ogata and H. Tatebayashi, *J. Physiol.* **466,** 9 (1993).
[34] M. A. Rizzo, J. D. Kocsis, and S. G. Waxman, *J. Neurophysiol.* **72,** 2796 (1994).
[35] J. B. Arbuckle and R. J. Docherty, *Neurosci. Lett.* **185,** 70 (1995).
[36] J. R. Elliott, *Brain. Res.* **754,** 221 (1997).
[37] J. C. Hunter, *in* "Novel Aspects of Pain Management: Opioids and Beyond" (J. Sawynok and A. Cowan, eds.). John Wiley & Sons, New York, 1999 (in press).

defining the role of this sodium channel subtype in neuropathic pain states. The concept of PN3 as a target is also supported by the observation that the expression of this channel is altered after nerve injury. A study[38] employed electrophysiological, *in situ* hybridization and immunohistochemical methods to monitor changes in the TTX-R sodium current and the distribution of PN3 in normal and peripheral nerve-injured rats. It was found that there were no significant changes in the TTX-R and TTX-S sodium current densities of small DRG neurons and no change in the expression of PN3 mRNA in the DRG up to 14 days after chronic constriction injury (CCI) of the sciatic nerve. However, the intensity of PN3 immunolabeling decreased in small DRG neurons and increased in sciatic nerve axons at the site of injury. The alteration in immunolabeling was attributed to a translocation of presynthesized, intracellularly located PN3 protein from neuronal somata to peripheral axons, with subsequent accumulation at the site of injury. The specific subcellular redistribution of PN3 after peripheral nerve injury may thus be an important factor in establishing peripheral nerve hyperexcitability and resultant neuropathic pain.[38]

In the L5/L6 spinal nerve ligation model, tight ligation of the L5/L6 spinal nerves leads to a reduction in the immunoreactivity of PN3 in the L5 and L6 DRGs. Interestingly, significantly more large cells are found to be immunopositive for PN3 in the L4 DRG 7 days after nerve ligation injury.[39] This increase in PN3 expression in the large cells after nerve ligation injury (as opposed to CCI) is an important observation because evidence suggests that large fiber input may be responsible for mediating tactile allodynia, which is clearly evident in the L5/L6 spinal nerve ligation model.

Antisense Targeting of Peripheral Nerve 3 Protein in Model of Neuropathic Pain

An antisense strategy that would selectively "knock down" PN3 in the peripheral nervous system presents a novel and unique approach to the study of sodium channel function and the contribution of selective channel subtypes to neuropathic pain because there are as yet no drugs that distinguish between sodium channel subtypes. On the basis of the cloned cDNA sequence of rat PN3, we have selected two oligomeric sequences that are unique to PN3 for the design of antisense (AS) oligodeoxynucleotides

[38] S. D. Novakovic, E. Tzoumaka, J. G. McGivern, M. Haraguchi, L. Sangameswaran, K. R. Gogas, R. M. Eglen, and J. C. Hunter, *J. Neurosci.* **18,** 2174 (1998).

[39] J. C. Hunter, D. Bian, R. M. Eglen, L. Kassotakis, J. Lai, M. H. Ossipov, S. Novakovic, L. Sangameswaran, and F. Porreca, submitted (1999).

PN3 AS sequence 1   5'-TCC TCT GTG CTT GGT TCT GGC CT-3'

PN3 MM sequence 1 5'-TCC TTC GTG CTG TGT TCG TGC CT-3'

PN3 AS sequence 2   5'-CCC ACG GAC GCA AAG GGG AGC T-3'

PN3 MM sequence 2 5'-CCC AGC GAC GAC AAG GGA GGC T-3'

FIG. 1. Antisense (AS) sequences and the corresponding mismatch (MM) control for targeting PN3. Antisense sequence 1 and sequence 2 are complementary to nucleotides 107–129 and nucleotides 5–26 of the coding region of rat PN3, respectively. The corresponding mismatch control sequences are derived from the antisense sequences by shuffling three nucleotide pairs evenly among the sequence. These ODNs were synthesized as unmodified phosphodiester analogs for intrathoracic administration.

(ODNs), and their corresponding mismatch (MM) control (Fig. 1). These molecules were prepared as phosphodiester ODNs by solid-phase synthesis (Midland Certified, Midland, TX) and made up to a final concentration of 9 μg/μl in nuclease-free, deionized water. These ODN samples were prepared immediately prior to the first injection and stored in aliquots at 4° throughout the experiment. All ODN samples were administered intrathecally at a dose of 45 μg per injection twice daily via an intrathecal catheter. Two experimental paradigms were used to test the effect of the PN3 AS ODN: (1) the animals would receive a daily dose of ODN prior to ligation injury and continue receiving the ODN after surgery, while the animals were monitored for tactile allodynia and thermal hyperalgesia daily (see below) throughout the ODN treatment. This paradigm evaluates whether PN3 AS ODN would prevent the onset of neuropathic pain due to L5/L6 nerve ligation; (2) rats would be subjected to L5/L6 nerve ligation injury or sham operation and monitored for tactile allodynia and thermal hyperalgesia. Once neuropathic pain is established after nerve injury in these animals, they would receive PN3 AS ODN daily as described above and continue to be monitored. This paradigm evaluates whether PN3 AS ODN would reverse the neuropathic pain that results from nerve injury.

For the animal subjects, adult male Sprague-Dawley rats (250–350 g) were first prepared for intrathecal ODN administration according to the method of Yaksh and Rudy.[40] Rats were anesthetized with halothane. An 8-cm length of PE-10 tubing was inserted through an incision made in the atlantooccipital membrane to the level of the lumbar enlargement of the rat. The catheter was secured to the musculature at the site of incision,

[40] T. L. Yaksh and T. A. Rudy, *Physiol. Behav.* **17**, 1031 (1976).

which was then closed. The rats received of gentamicin (4.4 mg/kg, intramuscular) as a prophylactic precaution and were allowed to recover for a minimum of 5 days before experimentation. Rats that exhibited any sign of motor deficiency were euthanized. These rats were then subjected to L5/L6 spinal nerve ligation injury according to the method of Kim and Chung,[3] or sham surgery. For peripheral nerve injury, rats were anesthetized with 4% halothane in 100% O_2 (flow rate, 2 liters/min) and maintained in the anesthetized state with 1.5% halothane in O_2. The skin over the lumbar vertebrae was shaved and sterilized with povidone-iodine (Betadine) followed with ethanol. An incision was made over the lumbar vertebrae and the left paraspinal muscles were carefully separated from the vertebral processes at the L4 to S2 levels. Removal of the L6 transverse process is completed with a small rongeur in order to allow for visual identification of the L4 to L6 spinal nerves. The L5 and L6 spinal nerves were exposed, carefully isolated, and tightly ligated with 4-0 silk suture distal to the DRG. The wounds were then sutured, and the animals allowed to recover in separate cages. Sham-operated rats were prepared in an identical fashion except that the L5/L6 nerves were not ligated. Any rats exhibiting motor deficiency were euthanized.

L5/L6 nerve-ligated rats and the sham-operated controls were then monitored for signs of neuropathic pain in the left hind paw daily after surgery. Two types of abnormal nociception were evaluated: a nociceptive response to innocuous mechanical stimuli (tactile allodynia) and a hypersensitive response to noxious thermal stimuli (thermal hyperalgesia). Tactile allodynia was determined according to the method described by Chaplan *et al.*[41] This method measures the paw withdrawal threshold of a rat in response to probing with calibrated von Frey filaments in increments equivalent to force required to displace 0.41, 0.71, 1.20, 2.00, 3.63, 5.50, 8.5, and 15.1 g. The von Frey filament is applied perpendicularly to the plantar surface of the affected paw of the rat until it buckles slightly, and is held for 3 sec. A positive response is indicated by a sharp withdrawal of the paw. The 50% paw withdrawal threshold is determined by the nonparametric method of Dixon.[42] An initial probe equivalent to 2.00 g is applied and if the response is negative the stimulus is increased one increment, otherwise it is decreased one increment. When continuous positive responses occur below the minimal stimulus (0.41 g) or negative responses above the maximal stimulus (i.e., 15.1 g), then arbitrary cutoff values of 0.20 and 15, respectively, are assigned. The stimulus is incrementally increased until a

[41] S. R. Chaplan, F. W. Bach, J. W. Pogrel, J. M. Chung, and T. L. Yaksh, *J. Neurosci. Methods* **53,** 55 (1994).
[42] W. J. Dixon, *Annu. Rev. Pharmacol. Toxicol.* **20,** 441 (1980).

positive response is obtained, then decreased until a negative result is observed. This "up–down" method is repeated until three changes in behavior are determined. The pattern of positive and negative responses is tabulated. The 50% paw withdrawal threshold is determined as $[10^{(Xf+k\partial)}]/10,000$, where Xf is the value of the last von Frey filament employed, k is the Dixon value for the positive/negative pattern, and ∂ is the mean (log) difference between stimuli. Thermal hyperalgesia was determined according to the method of Hargreaves et al.,[43] using a timer-activated radiant heat source that was focused onto the plantar surface of the rat hind paw. When the animal withdrew its paw, a photodetection device would terminate the heat source and timer. A maximal cut-off of 40 sec (about 2.5 times baseline) was used to prevent tissue damage.

Peripheral Nerve 3 Protein Antisense Oligodeoxynucleotide to Prevent as Well as to Reverse Neuropathic Pain in Lumbar Vertebra 5/6 Nerve-Ligated Rats

L5/L6 nerve ligation injury induces signs of neuropathic pain 3 to 5 days after surgery. Sham-operated rats are typically used as controls; these animals do not differ from naive rats in their nociceptive threshold. Animals that received PN3 AS ODN 3 days prior to L5/L6 nerve ligation surgery and continued to receive the ODN for 5 days did not develop tactile allodynia or thermal hyperalgesia; animals that received PN3 MM ODN and received L5/L6 ligation surgery, on the other hand, developed tactile allodynia and thermal hyperalgesia.[39] Sham-operated rats that received PN3 AS, or the MM control ODN, did not develop tactile allodynia or thermal hyperalgesia, and the paw withdrawal latency to radiant heat did not differ from that of saline-treated sham-operated rats. These data demonstrate that PN3 AS ODN prevents the development of neuropathic pain resulting from peripheral nerve injury, and the effect of this ODN is sequence specific. The administration of ODN, whether the PN3 AS or the MM control, has no effect on the basal level of nociception in the rats. The animals also did not display any behavioral toxicity due to ODN administration throughout the experiment.

PN3 AS ODN administration also reversed the tactile allodynia and thermal hyperalgesia in animals that had received L5/L6 ligation injury (Fig. 2). In these animals, the baseline 50% paw withdrawal threshold to von Frey filaments was 1.8 ± 0.38 g (Fig. 2A). The daily administration of PN3 AS ODN, but not saline or MM control, resulted in a significant increase in the 50% paw withdrawal threshold by day 2 of ODN administra-

[43] K. Hargreaves, R. Dubner, F. Brown, C. Flores, and J. Joris, *Pain* **32,** 77 (1988).

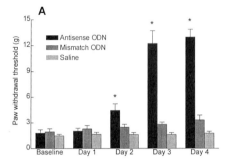

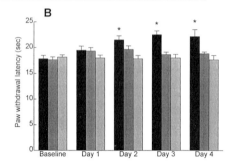

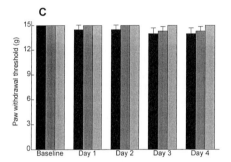

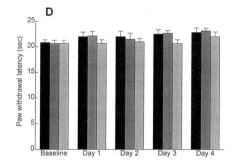

FIG. 2. PN3 AS ODN, but not MM or saline control, reversed tactile allodynia (A) and thermal hyperalgesia (B) in L5/L6 nerve-ligated rats (A and B); neither ASn or MM ODN had any effect on the baseline mechanical sensitivity (C) or baseline thermal paw withdrawal latency (D) in sham-operated rats (C and D). AS and MM sequence 2 were used in this study (see Fig.1). Day number refers to the time since first injection of ODN. Asterisks denote a significant difference from baseline as determined by ANOVA followed by least significant difference test ($p < 0.05$). Groups of six rats were used for each of the ODN or saline treatments, and were monitored daily for tactile and thermal nociceptive responses.

tion and achieved its full effect by day 3. Thus the administration of the PN3 AS ODN in nerve-injured rats leads to a significant antiallodynic effect, which is rapid in onset (evident after five injections) and highly effective [a 50% threshold of 13 ± 0.9 g (day 4) compared with 13.9 ± 0.8 g in sham-operated control; Fig. 2A and C]. Similarly, PN3 AS ODN, but not the MM control or saline, reversed the thermal hyperalgesia in L5/L6 nerve-ligated rats. The baseline paw withdrawal latency exhibited by these animals prior to ODN treatment was 17.8 ± 0.66 sec; treatment with the PN3 AS ODN resulted in a paw withdrawal latency of 22.5 ± 0.79 sec in these animals by day 3, and was not significantly different from that of sham-operated control (Fig. 2B and D). The time course of the antihyperalgesic effect and the antiallodynic effect of the PN3 AS ODN was similar (Fig. 2A and B). Furthermore, neither PN3 AS ODN nor the MM control altered the baseline paw withdrawal threshold to tactile stimulus or the baseline paw withdrawal latency to thermal stimulus, suggesting that neither ODN treatment, nor antisense to the PN3, had any effect on the nociceptive threshold of sensory input (Fig. 2C and D).

Specificity of Peripheral Nerve 3 Protein Antisense Oligodeoxynucleotide

Our data thus demonstrate that local, intrathecal treatment of L5/L6 nerve-ligated rats with a PN3 AS ODN results in an effective blockade of tactile allodynia and hyperalgesia that arise from injury to the peripheral nerve. Furthermore, several lines of evidence demonstrate that the observed antiallodynic and antihyperalgesic effects of the PN3 AS ODN is attributed to a specific knock-down of the PN3. First, the effect of the PN3 AS ODN is sequence specific, because treatment with a homologous mismatch control ODN did not produce any antiallodynic or antihyperalgesic effect. Second, two independent target sequences derived from the coding region of PN3 have similar effects in the behavioral assays, reducing the likelihood that the effects of the PN3 AS ODN could result from a cross-reactivity of the antisense with other transcripts. In this regard, it is also important to note that PN3 AS ODN and MM control had no effect on baseline response to noxious, or nonnoxious, stimuli in sham-operated animals, which argues against a sequence-independent, nonspecific effect of ODN treatment. Third, the antiallodynic and antihyperalgesic effects of the PN3 AS ODN are reversible. As seen in Fig. 2, the onset of antiallodynia and antihyperalgesia after AS ODN injection was fairly rapid, and entirely dependent on the continual administration of the ODN. On cessation of the AS ODN treatment, tactile allodynia and thermal hyperalgesia were fully recovered by 48 hr after the last ODN injection, and resuming PN3 AS ODN injection in these same animals would again result in a blockade of tactile allodynia and thermal hyperalgesia.[39] Fourth, the onset of PN3 AS ODN-mediated antiallodynia and antihyperalgesia in the L5/L6 nerve-ligated rats was concurrent with a significant reduction in the immunoreactivity of PN3 in both large and small cells in the L4 DRG, whereas the MM ODN had no effect.[39] Similarly, PN3 AS ODN, but not the MM ODN, also reduced the immunoreactivity of PN3 in the small cells of the DRG (L4,L5) in sham-operated or naive rats. This PN3 AS ODN-mediated knock-down of the PN3 protein was highly specific: PN3 AS ODN did not affect the immunoreactivity of other sodium channel subtypes, PN1 and PN4, in the DRG cells.[39]

Pathophysiological Implications of Peripheral Nerve 3 Protein in
  Neuropathic Pain

The specificity of the PN3 AS ODN in the knock-down of PN3 in
the DRG and the corresponding antiallodynia and antihyperalgesia in the
nerve-injured rats strongly implicate a critical role of this sodium channel
subtype in the manifestation of neuropathic pain on injury to the nerve.
Consequently, a selective blockade of this channel, whether by antisense
strategy or by other pharmacological agents that are highly selective for
PN3, may represent a new modality in the treatment of the neuropathic
pain state. It is interesting to note that interfering with the expression of
PN3 does not alter the baseline response to noxious or nonnoxious input
under normal physiological conditions, but acts effectively in modulating
the hypersensitivity to sensory input resulted from nerve injury. The function of PN3 therefore does not appear to be sufficient in maintaining normal
nociceptive function, but becomes a major component in mediating the
exaggerated or spontaneous conductivity of the injured nerve. This is consistent with the postulation that a TTX-R current may be important in driving
central sensitization in neuropathic pain states. On the basis of previous
observations that nerve injury is associated with a redistribution of PN3 to
the site of injury, it may be speculated that the antiallodynic and antihyperalgesic effect of PN3 AS ODN is due to limiting the accumulation of this
channel protein at the site of injury by attenuating the biosynthesis of the
protein. The significant increase in the expression of PN3 in the large cells
of the adjacent L4 DRG after the L5/L6 ligation described above may be
correlated with a previous finding that uninjured primary afferents in an
adjacent segment of the sciatic nerve may contribute to certain types of
neuropathic pain behavior.[44] In this regard, the increase in PN3 level/
function in the large cells, which normally mediate nonnoxious sensory
input, may be related to the observed tactile allodynia because PN3 AS
ODN reversed tactile allodynia with a concomitant knock-down of PN3 in
these cells. The mechanism that enhances the expression of PN3 in the
large cells of the L4 DRG remains to be established.

The antisense targeting of PN3 thus provides the first evidence that a
selective attenuation of the level/activity of PN3 is sufficient to prevent
and reverse the primary symptoms of neuropathic pain. These data promote
the concept that neuropathic pain states may be blocked by selective interference with the function of a subtype of sodium channel, allowing for pain
relief without the side effects associated with nonselective blockade of all
sodium channels. The validation of PN3 as a therapeutic target in the

---

[44] Y. W. Yoon, H. S. Na, and J. M. Chung, *Pain* **64,** 27 (1996).

treatment of chronic pain states, however, demands further elucidation of the PN3 function in relation to the onset and offset of neuropathic pain states. What our data have demonstrated is that the expression of PN3 somehow is critical for the manifestation of nerve injury-induced pain, presumably through a change in the number and/or distribution of functional PN3 channels in the injured nerve. To establish a causal relationship between PN3 function and neuropathic pain states, it is necessary to determine a correlation between TTX-R current density and changes in the expression of the PN3 protein, and the relationship between ectopic discharge and the function of PN3 in injured nerves. These studies are also of significance in that they will set the stage for the possible employment of antisense oligodeoxynucleotides to PN3 as a treatment for neuropathic pain states in humans, perhaps ultimately by a viral delivery system. As an extension of the significance of these studies, it is reasonable to suggest that PN3 expression may also be important in chronic pain due to other factors such as diabetes and acquired immunodeficiency syndrome (AIDS), as well as chronic inflammatory states.

[15] Antisense Approach for Study of Cell Adhesion Molecules in Central Nervous System

By RADMILA MILEUSNIC

Introduction

All cell types so far studied use multiple molecular mechanisms to adhere to other cells and to the extracellular matrix. Thus the specificity of cell–cell and cell–matrix adhesion must result from the integration of a number of different adhesion systems, some of which are associated with specialized cell junctions while others are not. The adhesive potential of these glycoproteins determines their general designation as cell adhesion molecules (CAMs) and substrate adhesion molecules (SAMs). However, these operational designations do not always accurately reflect their biological functions. An increasing body of evidence suggests that the biological activities of several CAMs may be accounted for by a direct activation of second-messenger systems, probably by their participation in signal transduction. Although their conventional name "cell adhesion molecules" does not accurately describe their complex biological functions, a more appropriate name, such as "recognition molecules," was never fully accepted by the research community. Hence, the term CAM is used in this chapter.

0076-6879/99 $30.00

Neural Cell Adhesion Molecule

The neural cell adhesion molecule (NCAM) was the first molecule mediating cell adhesion to be identified on the basis of functional criteria.[1,2] NCAM is a complex of three immunologically related membrane proteins of 180, 140, and 120 kDa molecular mass. They have identical N-terminal domains and differ primarily in their region of membrane association. These isoforms are primary translation products that arise through the alternative splicing of a single gene. NCAM appears during early embryonic development on derivates of all three germ layers, but after birth becomes essentially restricted to neural tissue and is expressed on both neuronal and glial cells.[3] Within the central nervous system, NCAM 180 and 140 appear to be characteristic of neurons, whereas NCAM 140 and 120 are expressed predominantly by glial cells.[4,5] Unique to NCAM are the posttranslational modifications associated with the glycan structures.[6] Studies on cell adhesion molecules suggest that an individual NCAM can function both to promote synaptic plasticity and maintain the structure of the synapse.[7] It has been shown that NCAM mediates cell–cell interactions by homophilic and heterophilic binding mechanisms,[5–8] and may act as well in synergy with other CAMs as well as with elements of cytoskeleton, and transduces extracellular signals probably by interaction with fibroblast growth factor (FGF) receptor.[9] NCAM is believed to be critical for the formation of neuronal connections during development, and for the long-lasting synaptic remodeling that occurs during memory formation.[10]

In contrast to NCAM, amyloid precursor protein (βAPP) mediates cell–matrix adhesions. Because of its association with Alzheimer's disease, the functions of βAPP are under intense investigation. βAPP is a family of eight isoforms of transmembrane glycoproteins, containing 770, 752, 751, 733, 714, 695, 695, and 677 amino acids generated by alternative splicing.[11] Full-length APP resembles a glycosylated call surface receptor.[12] After

[1] J. P. Thiery, R. Brackenburg, U. Rutishauser, and G. M. Edelman, J. Biol. Chem. 252, 6841 (1977).

[2] G. M. Edelman, Biochemistry 27, 3533 (1988).

[3] C. Goridis and J.-F. Brunet, Semin. Cell Biol. 3, 189 (1992).

[4] U. Rutishauser and T. M. Jessel, Physiol. Rev. 68, 819 (1988).

[5] G. E. Polleberg, K. Buridge, K. E. Krebs, S. R. Goodman, and M. Schachner, Cell Tissue Res. 250, 227 (1987).

[6] G. Rougon, Eur. J. Cell Biol. 61, 197 (1993).

[7] P. Doherty and F. S. Walsh, Curr. Opin. Neurobiol. 2, 595 (1992).

[8] G. Kadmon, A. Kowitz, P. Altevogt, and M. Schachner, J. Cell Biol. 110, 193 (1990).

[9] P. Doherty, J. Cohen, and F. S. Walsh, Neuron 5, 209 (1990).

[10] S. P. R. Rose, J. Physiol. 90, 387 (1996).

[11] R. Sandbrink, C. L. Masters, and K. Beyreuther, Biol. Chem. 269, 1510 (1990).

[12] J. Kang, H. G. Lemaire, A. Unterbeck, J. M. Salbaum, C. L. Masters, K. H. Grzeschlik, G. Multhaup, K. Beyreuther, and B. Müller-Hill, Nature (London) 325, 733 (1987).

synthesis and posttranslational modifications,[13–15] it may be cleaved in a secretory pathway, at α or β sites, to release a large amino-terminal ectodomain (sAPP), or at a γ site by an γ-secretase within the transmembrane domain, to release amyloidogenic peptide (βA4) involved in formation of extracellular amyloid plaques in the cerebral and limbic cortices.[16,17] The remaining carboxy-terminal fragment undergoes further processing. The high degree of evolutionary conservation of the extracellular and cytoplasmic domains of βAPP, its abundance in neurons and glia, as well as its tightly regulated differential expression during development and aging at both tissue and cellular levels suggest important functions of βAPP in normal brain tissue.[18] In neurons, βAPP is expressed in axons and dendrites,[19,20] and its localization in postsynaptic densities might suggest its role in synaptogenesis.[21] Several studies have shown that APP promotes cellular adhesion[22,23] and neurite outgrowth.[24–27] A possible role for βAPP in learning and memory formation is suggested from several studies.[28–34]

The methods reviewed here describe the application of the antisense

[13] J. Ghiso, A. Rostagno, J. E. Gardella, L. Liem, P. D. Gorevic, and B. Frangione, *Biochem. J.* **288,** 1053 (1992).

[14] A. Weidemann, G. König, D. Bunke, P. Fischer, J. M. Salbaum, C. L. Masters, and K. Beyreuther, *Cell* **5,** 115 (1989).

[15] T. Oltersdorf, P. J. Ward, T. Henriksson, E. C. Beattie, R. Neve, I. Leiberburg, and L. C. Fritz, *J. Biol. Chem.* **265,** 4492 (1990).

[16] G. Evin, K. Beyreuther, and C. L. Masters, *Int. J. Exp. Clin. Invest.* **1,** 263 (1994).

[17] F. J. Checler, *Neurochemistry* **517,** 522 (1995).

[18] D. J. Selkoe, *Annu. Rev. Neurosci.* **17,** 489 (1994).

[19] E. H. Koo, L. Park, and D. J. Selkoe, *Proc. Natl. Acad. Sci. U.S.A.* **90,** 4748 (1993).

[20] E. Storey, K. Beyreuther, and C. L. Masters, *Brain Res.* **735,** 217 (1966).

[21] K. Shigematsu, P. L. McGeer, and E. G. McGeer, *Brain Res.* **592,** 353 (1992).

[22] D. Schubert, L. W. Jin, T. Saitoh, and G. Cole, *Neuron* **3,** 689 (1989).

[23] D. Beher, L. Hesse, C. L. Masters, and K. Beyreuther, *Biol. Chem.* **271,** 1613 (1996).

[24] A. C. LeBlanc, D. M. Kovacs, H. Y. Chen, F. F. Villaré, M. Tyckocinski, L. Autilio-Gambetti, and P. Gambetti, *J. Neurosci. Res.* **31,** 635 (1992).

[25] D. Schubert and C. Behl, *Brain Res.* **629,** 275 (1993).

[26] L. W. Jin, H. Ninomiya, J. M. Roch, D. Schubert, E. Masliah, D. A. Otero, and T. Saitoh, *J. Neurosci.* **14,** 5461 (1994).

[27] W. O. Qui, A. Ferreira, C. Miler, E. H. Koo, and D. J. Selkoe, *J. Neurosci.* **15,** 2157 (1995).

[28] E. Doyle, P. Nolan, R. Bell, and C. M. Regan, *J. Neurosci. Res.* **31,** 513 (1992).

[29] J. F. Flood, J. E. Morley, and E. Roberts, *Proc. Natl. Acad. Sci. U.S.A.* **88,** 3363 (1991).

[30] G. Huber, Y. Bailly, J. R. Martin, J. Mariani, and B. Brugg, *Neuroscience* **80,** 313 (1993).

[31] U. Müller, N. Cristina, Z.-W. Li, D. P. Wolfer, H.-P. Lipp, T. Rütlicke, T. Brander, A. Aguzzi, and C. Weissmann, *Cell* **79,** 755 (1994).

[32] T. Maurice, B. P. Lockhart, and A. Privat, *Brain Res.* **706,** 181 (1996).

[33] H. Zheng, H. Jiang, M. Trumbauer, R. Hopkis, D. Sirinathsinghji, K. A. Stevens, M. W. Conner, H. H. Slunt, S. S. Sisodia, H. Y. Chen, and L. H. T. VanderPloeg, *Ann. N.Y. Acad. Sci.* **777,** 421 (1996).

[34] H. Meziane, J.-C. Dodart, C. Mathis, S. Litle, J Clemens, S. M. Paul, and A. Ungerer, *Proc. Natl. Acad. Sci. U.S.A.* **95,** 12683 (1998).

approach to the study of the role of CAMs in the central nervous system in long-lasting synaptic remodeling that is believed to occur during the process of memory formation. However, before focusing on these studies a short account of the animal model in which all studies were performed will be described.

Animal Model

On the morning after hatching, chicks are placed in pairs in aluminum pens (20 × 25 × 20 cm) with scattered chick crumbs. The pens are maintained at 28–30° and illuminated by an overhead red light and allowed to equilibrate for 1 hr before pretraining by two 10-sec presentations of a small (2.5-mm) white bead. Birds are scored as either "peck" or "nonpeck." Ten minutes after the last pretraining trial, each bird is trained by a single presentation of a 4-mm bright chrome bead coated with methyl anthranilate (MeA). In this training trial, the behavior of each bird is scored as "peck," "peck and shake" (the disgust response to methyl anthranilate), or "nonpeck." Birds that fail to peck twice during pretraining, or fail to peck during training (<15% in all), are not used for further analysis. At different time points posttraining, each chick is tested by being offered a dry chrome bead for 30 sec, and its response ("peck" or "avoid") is noted. Each bird is trained and tested only once. Learning and retention are represented by avoiding the bead and therefore calculated as an avoidance score on testing. Forgetting, or amnesia, is indicated by trained birds pecking the dry bead on testing. Scoring during both training and testing is done by an experimenter blind as to prior treatment. Retention scores between groups are compared by G-test, using a χ^2 table. Memory for the association lasts for several days and its establishment involves a well-characterized molecular and cellular cascade in a particular forebrain region, the intermediate medial hyperstriatum ventrale (IMHV) and lobus parolfactorius (LPO).[10]

The chick system is useful for exploring "synaptic plasticity issues," because the learning task is precise and sharply timed, and also permits one to be sure that any observed effect of an injected substance is specific to retention and not either to acquisition or to concomitant processes such as visual acuity, arousal, or motor activity. Further, the role of other cell adhesion molecules in the cascade of molecular and biochemical changes leading to synaptic modulation has been well mapped, so that the effects of either blocking or attempting to rescue functional CAM activity can be set into an established context.

Intracerebral Injections

Bilateral intracerebral injections of oligodeoxynucleotides (ODNs) or vehicle (2 μl) are made using a 5-μl Hamilton syringe fitted with a plastic

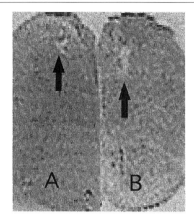

FIG. 1. Toluidine blue staining of chicken brain near the site of ODN injection. (A) Chick that received one injection of ODN; (B) chick that received two injections of ODN. Arrows point toward the lesion induced by ODN injection. Original magnification: ×20.

sleeve to allow penetration to 3.5 mm. Correct placement of injections into the IMHV is ensured by using a custom-built head holder and routinely monitored visually postmortem. These injections are rapid (<30 sec per bird) and cause no observable distress to the chicks. To detect any gross morphological changes that can be caused by single and multiple injections of ODNs into the IMHV, brains are routinely examined by staining with toluidine blue (Fig. 1).

Oligodeoxynucleotides and Their Purification

There is a body of evidence showing that fully protected ODNs are effective in the study of immediate-early genes if injected only in a single dose.[35,36] However, our studies of 24-hr-old chicks revealed that a single injection of more than 1 μg of ODNs exerts a toxic effect on behavior and damages the brain. These effects are observed in animals treated with both antisense and scrambled ODNs and are likely to be the consequence of the known phosphorothioate interaction with basic proteins.[37,38] In contrast to fully protected ODNs, unprotected ODNs appear to be well tolerated and effective in the study of inducible genes such as immediate-early genes,[39] but

[35] B. J. Chiasson, M. G. L. Hong, and H. A. Robertson, *Neurochem. Int.* **31,** 459 (1997).
[36] S. Ogawa and D. W. Pfaff, *Chem. Sense* **23,** 249 (1998).
[37] M. K. Gosh, K. Gosh, O. Dahl, and J. S. Cohen, *Nucleic Acids Res.* **21,** 5761 (1993).
[38] J. F. Milligan, R. J. Jones, B. C. Froehler, and M. D. Matteucci, *Ann. N.Y. Acad. Sci.* **13,** 228 (1994).
[39] R. Mileusnic, K. Anokhin, and S. P. R. Rose, *Neuroreport* **7,** 1269 (1996).

their short half-life makes them inefficient in the study of gene products that are constitutively expressed or have much longer half-lives.

End-protected ODNs have proved to be a much better approach for studies of constitutively expressed proteins such as CAMs. End-protected ODNs contain greatly reduced sulfur content, when compared with fully protected ODNs, hence a shorter half-life and reduced toxicity. Additional purification by use of NAP-5 columns (Pharmacia, Piscataway, NJ) has been shown to be efficient in elimination of trace amounts of low molecular weight impurities that can compromise some of the ODN effects.

Purification. The purification method described here allows repeated injections of ODNs without detrimental effects to behavior. High-performance liquid chromatography (HPLC)-purified end-protected phosphorothioate ODNs [400–500 μg in Tris–borate–EDTA (TBE) buffer] can be applied to an NAP-5 column, prewashed with 5–7 ml of phosphate-buffered saline (PBS). ODNs are eluted with 3 ml of PBS. Almost 90% of ODNs are eluted in fractions 4, 5, and 6 (4–5 drops/fraction, ~200 μl). The ODN concentration in each fraction is measured as absorbance at 260 nm (A_{260}). Fractions with a high concentration of ODNs are kept at $-40°$. Low molecular weight toxic impurities can be eluted by further addition of PBS.

Effect of Oligodeoxynucleotides on Behavior

Phosphorothioate oligodeoxynucleotides are synthesized at Kings College (Molecular Medicine Unit, London, UK).

Neural Cell Adhesion Molecule Oligodeoxynucleotides. For behavioral studies three different antisense (AS) ODNs corresponding to positions 190 (15-mer), 207 (18-mer), and 332 (18-mer) are used (exon 1; source: chicken, cDNA to mRNA, clones pEC[254, 265]; 2 μl/hemisphere, 0.30 μg/μl).[40] Scrambled (SC) controls have the same base composition as the AS ODNs but in a random sequence that is not complementary to any other mRNA sequence in the species.

Comments. Memory retention is significantly reduced only in the group of animals injected with the AS ODNs corresponding to positions 207–223 of the Ig1 domain, while two other tested ODNs have no behavioral effect. Intracerebral injection of AS ODNs affects memory consolidation and produces amnesia 3 hr posttraining only if given immediately after hatching and 12 hr before training. If the entire dose of ODNs is applied in one injection, 12 hr before training, chicks show serious behavioral deficits,

[40] B. A. Cunningham, J. J. Hemperly, B. A. Murray, E. A. Prediger, and R. Brackenbury, *Science* **236**, 799 (1987).

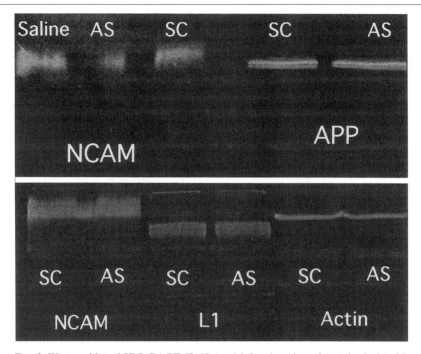

FIG. 2. Western blot of SDS–PAGE (5–12%, w/v) fractionation of proteins isolated from brains of ODN-treated animals. Brains were homogenized in TBS (1:10, w/v). Membrane and soluble fractions can be obtained by high-speed centrifugation. Proteins were separated by SDS–PAGE under reducing conditions on a 5–12% polyacrylamide gel and transferred to nitrocellulose. Blots were routinely checked by posttransfer staining with Ponceau S. The nitrocellulose was incubated in blocking buffer [Tris-buffered saline (TBS, pH 7.5) containing 5% defatted milk powder and 0.05% Tween 20]. Antibodies were diluted in blocking buffer according to the instructions of the manufacturer and the blot was incubated overnight at 4°. After three 10-min washes, peroxidase-conjugated secondary antibody was added at a dilution of 1:500 in blocking buffer and the blot incubated overnight at 4°. Immunoreactive bands were detected with diaminobenzidine (DAB). *Top:* Western blots of membrane fraction isolated from animals injected twice with NCAM antisense and APP antisense-treated animals. *Bottom:* Western blots of membrane fraction isolated from NCAM antisense-treated animals (single injection) probed with anti-NCAM, anti-L1, and anti-actin antibodies.

most probably owing to accumulation of degradation products of ODNs.[41] Thus AS ODNs affect memory consolidation and produce amnesia only if the injections are separated by approximately 10–12 hr, the final injection being 11–12 hr before training. The apparent slight reduction in memory retention after injection of SC ODNs, when compared with saline-injected

[41] M. Dragunow, P. A. Lawlor, B. J. Chiasson, and H. A. Robertson, *Neuroreport* **5,** 305 (1993).

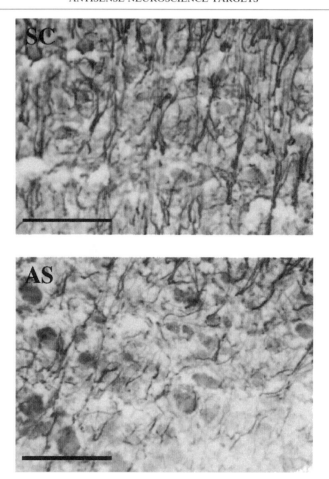

Fɪɢ. 3. Immunocytochemistry of APP in APP antisense-treated animals. Chicks were killed immediately after testing and forebrains were removed, frozen slowly in precooled isopentane, and stored at $-40°$. Frozen 15-μm cryostat sections were collected at $-18°$ onto Super Frost Gold Plus slides (Merck), air dried for 30 min at room temperature, and stored at $-40°$ until processed. Sections were equilibrated to room temperature, fixed with cold 4% paraformaldehyde in TBS (pH 7.5) for 8 min, and washed once with TBS for 5 min. After fixing and washing, slides were drained and 50 μl of anti-APP probing antibody (MAb 22C11; Boehringer) was applied per section. Sections were incubated in a moist box for 1.5 hr at room temperature and then washed three times for 5 min each with TBS. After draining, 50 μl per section of the anti-APP probing antibody was applied again and the above-described procedure repeated. After incubation with primary antibody, sections were probed with secondary antibodies diluted with TBS (pH 7.5 or 9.0) containing 0.1% Nonidet P-40 and 2% bovine serum albumin. FITC-conjugated anti-mouse IgG (Vector Laboratories), diluted 1 : 75 at pH 9, was applied. After the final wash, sections were preserved with a fluorescent antifade mountant (Dako). To check that the injections were correctly placed, sections adjacent to the fluorescently

controls, is not significant. There is no effect of any of the AS ODNs or their SC counterparts on behavior during pretraining or training; the proportions of chicks pecking at the beads, and showing the disgust response on training, are the same for all groups.

Amyloid Precursor Protein Oligodeoxynucleotides. For behavioral studies we use 16-mer AS ODNs designed to correspond to the transcription start site (position -146) and to AUG_{1786} of βAPP mRNA, immediately upstream of a ribozyme-binding site. SC controls have the same base composition as the AS ODNs, but in a random sequence that is not complementary to any other mRNA sequence in the species.

Comments. Increasing amounts of antisense oligodeoxynucleotides (AS ODNs) (0.6–1 μg/hemisphere) are injected intracerebrally at 12 hr pretraining and chicks are tested at different times posttraining. Controls are treated with scrambled (SC) ODNs or saline and trained and tested similarly. Memory retention is significantly reduced in animals injected with 0.8 or 1 μg of antisense ODNs. The slight retention deficit resulting from injection of scrambled ODNs is not significant when compared with saline-injected controls. Memory consolidation is affected only if AS ODNs are injected 6–12 hr before training, and amnesia is evident by 30 min posttraining and lasts for at least the subsequent 24 hr.

Specificity of Oligodeoxynucleotide Effects

Many antisense ODNs have been found to have non-sequence-specific effects.[41–43] These findings emphasize the importance of using proper controls, such as (1) histological examination for ODN-induced neurotoxic damage and, more importantly, (2) measurement of the inhibition of targeted gene expression.

The issue of NCAM antisense specificity was addressed by examining its effects on the expression of NCAM itself, on its close congener L1, and

[41] M. Dragunow, P. A. Lawlor, B. J. Chiasson, and H. A. Robertson, *Neuroreport* **5**, 305 (1993).

[42] C. A. Stein and Y.-C. Cheng, *Science* **261**, 10004 (1993).

[43] C. A. Stein and A. M. Krieg, *Antisense Res. Dev.* **4**, 309 (1994).

FIG. 3. (*continued*) probed sections were fixed and washed as described above, then stained with 0.25% aqueous toluidine blue, dehydrated by passing them through 80–100% ethanol, and preserved with DPX mountant. In the same sections, counterstaining with propidium iodide (PI) was done to check the cell numbers. After three 5-min washes with TBS, sections were counterstained with PI (5 μg/ml) for 5 min at room temperature, washed three times (5 min each) with TBS, and preserved with a fluorescent antifade mountant. Sections were examined on a Zeiss Axiophot. Bar: 100 μm.

on actin. L1 (NgCAM), like NCAM and actin, belongs to the superfamily of immunoglobulins. It shares many functional and structural similarities with NCAM. The extracellular domains of both CAMs consist of Ig-like domains followed by fibronectin type III-like domains. As Fig. 2 shows, antisense treatment results in a significant reduction of NCAM protein in double-injected animals (Fig. 2, top) but had no effect on either L1, a structurally similar member of the CAM family, or on a functionally and structurally unrelated protein such as actin (Fig. 2, bottom). Although this is indicative of a degree of specificity, the results must still be viewed with caution, as a variety of unexpected sequence-nonspecific effects of ODNs have been observed,[43,44] including preliminary suggestions that NCAM AS ODN may affect levels of expression of other proteins, such as SNAP and glyceraldehyde-3-phosphate dehydrogenase (G3PDH).[45]

Western blotting and immunocytochemistry addresses the question of APP antisense specificity. By scanning and computing the peak area of Western blots (Fig. 2), it is shown that animals treated with βAPP AS ODNs show a 30–40% decline (4.379 ± 0.402%; mean ± SD) in APP levels compared with scrambled ODNs (6.993 ± 0.364%; mean ± SD).

The specificity of APP antisense may in addition be examined by immunocytochemical analysis. Figure 3 shows the distribution of immunofluorescence in brain sections taken around the injection site in the IMHV. The predominant membrane staining of nerve cell bodies and their dendrites in the SC ODN-treated animals changes to a less pronounced surface pattern in AS ODN-treated animals, accompanied by neurite retraction and cell deattachment from the matrix.

Conclusions

There has been a remarkable convergence of evidence pointing to a key role for cell adhesion molecules in the process of maintenance of cellular integrity and synaptic plasticity. The antisense method has proved to be a powerful tool with which to study the role of CAMs in the CNS, in the processes of transient or long-lasting synaptic remodeling. However, there is a growing body of literature reporting problems, pitfalls, and unexpected nonantisense effects.

Problem of application: The distribution of uptake can be circumvented by direct intracerebral injections of purified end-capped ODNs to the targeted brain structures.

[44] M. A. Guvakova, L. A. Yakubov, I. Vlodavsky, J. L. Tonkinson, and C. A. Stein, *J. Biol. Chem.* **270,** 2620 (1995).
[45] L. A. Roberts, R. Mileusnic, S. P. R. Rose, and M. Schachner, *Eur. J. Neurosci.* **10**(Suppl. 10), 122 (1998).

Problem of efficiency: Not all antisense sequences work well, nor are their effects expressed at the same time and to the same degree. Single injection of ODNs has been shown to be efficient in studies targeting inducible gene products. To deplete constitutively expressed proteins or proteins with a low turnover rate, repeated injections have been shown to be efficient.

Problem of specificity: Determination of specificity is one of the crucial problems in antisense research. Therefore, Western blot analysis and immunocytochemical analysis of functionally and structurally similar proteins should be routinely performed as controls for non-sequence-specific antisense effects.

Acknowledgments

Much of the work in this chapter is based on collaboration with Christine Lancashire. This work was supported by an MRC-Foresight LINK grant to S.P.R.R.

[16] Sequence Design and Practical Implementation of Antisense Oligonucleotides in Neuroendocrinology

By Inga D. Neumann and Nicola Toschi

Introduction

In neurobiological research focused on the relation between neuroendocrine and behavioral parameters it is a challenge to reveal the functional involvement of endogenous neuropeptides of the brain. Highly specific antagonists, suited to reveal receptor-mediated actions of neuroactive substances, are not always available, and often the use of antisense oligodeoxynuclcotide (ODN) strategies is the only alternative. When such techniques came to life, the neurobiologists' dream of highly selective, fast, and reversible inhibition of the expression of any gene product potentially involved in brain functions seemed to have come true. Still, the discovery of several unexpected effects of antisense molecules as well as the present lack of clarity with respect to their mechanisms of actions should somewhat temper enthusiasm. However, antisense techniques have been growing in status as a powerful research tool, especially for the study of the molecular basis of brain functions, for several reasons. Compared with other systems, ODNs are taken up by neurons and glia relatively easily, limiting the necessity to employ other delivery systems (cationic lipids, viral vectors). Further, whereas in many systems unmodified ODN are rapidly degraded before being biologically effective, a higher stability of ODNs in brain tissue has been reported.[1] This could be due to the nonproliferating nature of nervous system cells. Because of the spatial and temporal specificity afforded by antisense techniques (i.e., the possibility to infuse antisense ODN into the brain region of interest at a particular time point) and the reversibility of their action they are sometimes considered superior to germ line null mutation (knockout) strategies (e.g., irreversible developmental defects[2] and compensation[3]). Further, provided prior knowledge of the targeted sequence is given, antisense ODN strategies allow the functional study of any particular neuropeptide in, e.g., behavioral or neuroendocrine regulation on any laboratory animal, including mice and rats.

Antisense targeting requires the introduction of short ODNs into cells, which can be achieved by two different approaches. The first entails the use of DNA vectors that direct the expression of complementary mRNA from large inverted gene fragments. This can be efficiently achieved in cell lines, but requires complex vectors (e.g., viruses) *in vivo*. Therefore, the more commonly used technique is based on the administration of short, single-stranded DNA sequences complementary to the sense strand of the targeted mRNA. Gene expression is hence inhibited through partial sequence-specific hybridization with the targeted mRNA encoding a particular neuropeptide or its receptor protein. Enormous advances have been made in the characterization of the several different antisense mechanisms of action as well as their limitations. As a consequence, the "state of the art" procedure is shifting away from the simple steps of (1) designing a single oligonucleotide for hybridization to a target gene, (2) ordering the antisense ODN from the DNA synthesis laboratory, and (3) administering the antisense sequence as well as respective controls to the brain in order to cause translational arrest and selective reduction in neuropeptide (or receptor) expression. Nowadays it is evident that the successful downregulation of neuropeptide or receptor availability through antisense ODN application is dependent on a variety of factors, including (1) target mRNA accessibility, (2) specificity of binding, (3) mechanisms of cellular uptake, (4) stability of the binding and enzymatic degradation of the antisense ODN, (5) the time scale of intracerebral antisense administration and reactions, and (6) turnover of the targeted gene. The optimal parameters (anti-

[1] M. McCarthy, *in* "Applied Antisense Oligonucleotide Technology" (S. A. Stein and A. M. Krieg, eds.), p. 83. Wiley-Liss, New York, 1998.

[2] S. G. N. Grant, T. J. O'Dell, K. A. Karl, P. L. Stein, P. Soriano, and E. R. Kandel, *Science* **258,** 1903 (1992).

[3] E. Hummler, T. J. Cole., J. A. Blendy, R. Ganss, A. Aguzzi, W. Schmid, F. Beermann, and G. Schutz, *Proc. Natl. Acad. Sci. U.S.A.* **91,** 5647 (1994).

sense ODN sequences, dosages, time scales, mode of administration, etc.) are mainly determined empirically to date, giving rise to great uncertainty and mixed feelings whenever antisense is drawn into the picture.

In this chapter we briefly discuss the methodological issues that should be considered when selecting a particular antisense ODN sequence and experimental design. Further, we review the methodological approaches used in our laboratory in order to study the physiological significance of various neuropeptides, including vasopressin, corticotropin-releasing hormone, and oxytocin, in behavioral and neuroendocrine regulation.

Theoretical Considerations for Selection of Antisense Oligodeoxynucleotide Sequence

It is evident that most, if not all, possible antisense-mediated effects rely on the successful binding of the ODN to its target mRNA portion. The reported efficiencies of different ODNs are widely varying,[4] and the causes for this as well as the wealth of possible antisense actions and reaction kinetics are not well understood to date.[5] This ignorance results in a severe lack of guidelines in target selection for ODN design. Currently, the most widely used strategy is a "shotgun" approach. One or several short ODNs complementary to sequences distributed throughout a target RNA are more or less randomly selected (a popular approach has been to target regions flanking the initiation codon AUG or the splicing sites of introns) and screened for inhibitory effects. This approach is clearly less than exhaustive, expensive, and of limited efficiency. We now outline the main issues in antisense technology and discuss possible approaches for tackling them.

Target mRNA Accessibility

It is well known that, unlike its DNA sibling, RNA rarely exists in a long uninterrupted double helix in nature.[6,7] Instead, unique architectures are formed by combining short helical regions with noncanonical base pairs (e.g., "wobble" GU, AA, and UU pairs) and unpaired bases, giving rise to a wealth of substructures such as hairpin loops, internal bulges, tetraloops, multiple junctions, and pseudoknots (Figs. 1 and 2).[7,8] The higher order structure of an RNA molecule is crucial in many of its biological functions, where specific shapes are adapted for catalysis, molecular recognition, and

---

[4] J. J. Zhao and G. Lemke, *Mol. Cell. Neurosci.* **11**(1–2), 92 (1998).
[5] A. Calogero, G. A. P. Hospers, and N. H. Mulder, *Pharmacy World Sci.* **19**(6), 264 (1997).
[6] N. Breton, C. Jacob, and P. Daegelen, *J. Biomol. Struct. Dynam.* **14**(6), 727 (1997).
[7] O. C. Uhlenbeck, A. Pardi, and J. Feigon, *Cell* **90**, 833 (1997).
[8] J. A. Doudna and E. A. Doherty, (1998). *Folding Design* **2**(5), R65 (1998).

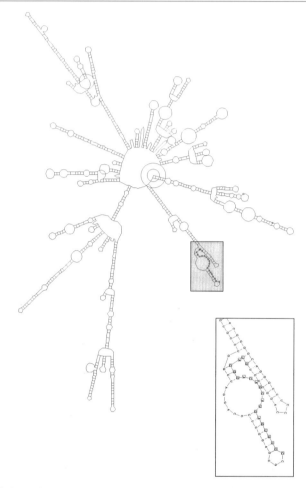

Fig. 1. Minimum free energy structure of mRNA encoding the rat vasopressin V1 receptor as predicted by the *mfold* package.[9] The enlarged area shows the segment complementary to the antisense ODN used in Ref. 12. The 5' end of the molecule is marked by two concentric circles. It appears that the region selected for antisense targeting contains part of an open loop consisting of approximately 10 bases, thereby facilitating ODN–mRNA hybridization and subsequent cleavage by RNase H.

so on. The effect this can have on ODN efficiency is evident, as the selected target region must be open to the base-pairing reactions that lead to the formation of the heteroduplex necessary for triggering the antisense effects. Targeting a sterically inaccessible region will almost certainly drastically

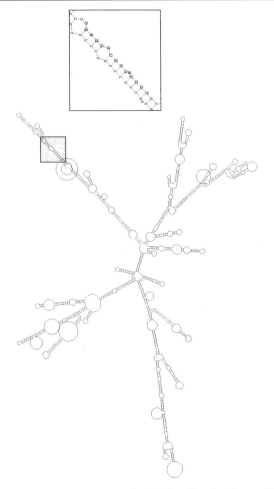

FIG. 2. Minimum free energy structure of mRNA coding for the rat CRH1 receptor as predicted by the *mfold* package.[9] The enlarged area shows the segment complementary to the antisense ODN used in Ref. 13. The 5' end of the molecule is marked by two concentric circles. This particular secondary structure prediction suggests that the area selected for targeting is involved in the formation of a significant amount of self-structure, thereby disfavoring efficient ODN–mRNA hybridization. A reduction in local receptor expression could not be observed in the original study (see text).

reduce the efficiency of the ODN, however well the overall experimental framework might be designed.

A promising approach to this fundamental issue is computer-based modeling of the molecular structure of the target mRNA. It is possible to predict the base-pairing pattern within an RNA molecule [i.e., its secondary structure, which seems to account for most of the free energy of folding (Figs. 1 and 2)[8]] by a variety of different methods and use this information as a guideline for determining the regions of the molecule that might be more accessible to specifically designed ODNs. The major paradigms that can be applied to secondary structure prediction include the minimization of the free energy of folding,[9] kinetic simulations of the folding process,[6] simulation of the evolution toward the biological active molecular structure through genetic algorithms,[10] and secondary structure prediction by phylogenetic comparison based on sequence alignment.[11]

It has already been shown that segments involved in the formation of long double-stranded regions should be avoided as targets as they are less accessible to external binding,[4] and that targeting sequences within the first several hundred nucleotides of a large RNA molecule is more effective (probably owing to the higher reliability of the modeling results for these regions). Thus, antisense-relevant information can be extracted from a reliable secondary structure prediction (both of the target molecule and of the ODN), and structural knowledge about the target molecule and the antisense ODN allows informed selection of the sequence to be targeted on the basis of steric and energetic accessibility. The affinity of the ODN with the mRNA can be optimized as a function of the position of the target sequence within the molecule, and at the same time control ODNs can be built with homogeneously low affinity to most portions of the target molecule. It is conceivable to reduce nonspecific binding by taking into account the affinities with bystander mRNAs, and nonstandard structures such as base triplexes, which are thought to determine the catalytic activity of RNA (e.g., in ribozymes) could also be accounted for.

Figures 1 and 2 show minimum free energy secondary structure predictions for the mRNA encoding the rat vasopressin (V1) receptor and the corticotropin-releasing hormone 1 (CRH1) receptor, respectively, highlighting the areas that have been selected for antisense targeting in our laboratory.[12,13] In these studies the sequences were empirically selected without taking secondary structure information into account. It appears that, especially in the case of the CRH receptor mRNA, the antisense ODN sequence is likely to be complementary to mRNA structures involved in the

[9] M. Zuker and D. Sankoff, *Bull. Math. Biol.* **46,** 591 (1984).
[10] F. H. D. Van Batenburg, A. P. Gultyaev, and C. W. A. Pleij, *J. Theor. Biol.* **174** (1995).
[11] R. Lück, G. Steger, and D. Riesner, *J. Mol. Biol.* **258,** 3 (1996).
[12] R. Landgraf, R. Gerstberger, A. Montkowski, J. C. Probst, C. T. Wotjak, F. Holsboer, and M. Engelmann, *J. Neurosci.* **15**(5), 4250 (1995).
[13] G. Liebsch, R. Landgraf, R. Gerstberger, J. C. Probst, C. T. Wotjak, M. Engelmann, F. Holsboer, and A. Montkowski, *Regul. Pept.* **59,** 229 (1995).

formation of relatively stable helical regions, thereby disfavoring efficient antisense ODN–mRNA binding (Fig. 2). This could provide an explanation for the fact that antisense-induced downregulation of local CRH receptor expression could not be found to accompany antisense-induced behavioral effects.[13] In contrast, when targeting the vasopressin V1 receptor (Fig. 1), a significant reduction in septal receptor expression accompanying effects on cognitive behavior was described.[12]

Stability to Degradation and Reaction Kinetics

Among the major unresolved issues of antisense technology is the problem of understanding the kinetics involved in the reactions triggered by ODN administration *in vivo*. Unmodified ODNs are polyanions with a phosphodiester backbone, and are rapidly degraded under physiological conditions by single-stranded nucleases, primarily 3′-exonucleases, drastically reducing the antisense effects. Because of this, ODN modifications (phosphorothioate and methyl phosphonate modifications[14]) have been designed to retard degradation. To reduce nonspecific protein-binding effects and (limited) toxicity that can be caused, for example, by phosphorothioate ODNs,[14,15] second-generation ODNs have been introduced,[16] including the creation of end-capped versions and chimeric molecules that have chemically modified "wings" around a central phosphorothioate "window."

It is clear that improving binding ability and reducing toxicity are major goals in antisense research. The dynamics of degradation (and more generally the time evolution of the concentrations) both of the ODN and of the DNA:RNA heteroduplex, coupled with the "noise" caused by variable cellular penetration ability, need to be thoroughly understood in order to maximize experimental and therapeutic efficiency. The above-described molecular modeling efforts can represent a significant aid in constructing more stable and efficient molecules. Further, it is conceivable to model the reaction kinetics by nonlinear differential or delay equations based on the law of mass action.[17,18] The analysis of the resulting system could, if successful, yield insight into a variety of different aspects of antisense action, such as the exact onset delay caused by intracellular peptide stores, the tentative characterization of the early (i.e., transient) effects of antisense actions, and the isolation of so-called rebound phenomena (e.g., in the

---

[14] B. Weiss, G. Davidkova, and S. P. Zhang, *Neurochem. Int.* **31**(3), 321 (1998).
[15] I. Neumann, *Neurochem. Int.* **31**, 363 (1997).
[16] J. F. Milligan, M. D. Matteucci, and J. C. Martin, *Biochem. Soc. Trans.* **24**, 630 (1997).
[17] A. K. Dutt and A. Datta, *J. Phys. Chem.* **102**(41), 7981 (1998).
[18] D. A. Fennell, *Antisense Nucleic Acid Drug Dev.* **6**, 49 (1997).

concentration of mRNA) that could be caused by antisense ODN administration (see below).

Specificity of Antisense Action

A celebrated feature of antisense techniques is their specificity. It is generally stated that longer ODNs (with an upper bound posed by the necessity to be able to penetrate the cell membrane) will achieve greater specificity and ensure selective binding to the target sequence. However, because on average RNase H requires less than 10 base pairs between ODN and target to initiate cleavage, increasing the length of the ODN after a certain threshold will not only reduce cellular penetration, but also reduce specificity, because a long ODN will contain many 10-mer sequences that can have undesired interactions with RNAs other than the target, thereby initiating their cleavage and triggering undesired effects. Further, increasing the lengths of the ODN in order to increase specificity might hamper efficiency, as longer ODNs tend to exhibit a significant amount of self-structure,[4] requiring higher energy to bind stably to their target sites. These are just examples of the several unexpected mechanisms of action that have been observed with antisense ODNs. Because the different targeted products and other molecules involved in their physiological functions might have widely varying turnovers and intracellular stores, the time scales of several antisense mechanisms of action could differ significantly (see below). A possible approach to these issues is the above-mentioned nonlinear dynamic modeling of the reactions involved. Albeit uncertain, the outcome of such a modeling effort could provide orientative information about the major parameters, time scales, and response curves involved in the reactions.

Methodological Approaches for *in Vivo* Use of Antisense
    Oligodeoxynucleotide

Clearly, the first step is the design of a suitable antisense ODN sequence(s) (with a particular chemical modification) as well as of appropriate control sequences. Depending on the hypothesis to be tested, various alternative are given regarding the time scale and localization of intracerebral administration. We were able to distinguish two different types of antisense ODN action: *acute effects* on neuronal activity of, for example, oxytocinergic or vasopressinergic neurons (reviewed in Ref. 19) can be observed within

[19] I. D. Neumann and Q. J. Pittman, *in* "Modulating Gene Expression by Antisense Oligonucleotides to Understand Neuronal Functioning" (M. M. McCarthy, ed.), p. 43. Kluwer Publishing, Boston, 1999.

a few (up to 24) hours after intracerebral application; *long-term effects* are achieved after repeated or chronic infusion into the brain (over 2 to 6 days). Albeit highly sequence specific, the acute effects were shown not to be due to classic antisense mechanisms of action (translational arrest and consequent neuropeptide depletion), as they were observed at a time when neuronal neuropeptide stores were still unaffected.[20,21] In contrast, after prolonged treatment behavioral or neuroendocrine effects were described to be due mainly to a reduction in target neuropeptide/receptor protein expression.[12,13,22–25]

Although in numerous studies antisense ODNs have been administered into the cerebral ventricles, this procedure bears a variety of disadvantages. Efficient uptake has been shown to occur only in areas close to the ventricular system (for review see Ref. 1); to increase net uptake, relatively high doses (up to 50 μg of chemically modified antisense ODN) must be administered, rendering an impact on the well-being of the animals more than likely. Signs of motor disturbances, increase in body temperature, suppression of food and fluid intake (see also results in mice[26]), and expression of interleukin 6 were observed after a single infusion of antisense ODN regardless of sequence and type of chemical modification (phosphorothioated and 3′–3′ inverted internucleotide linkages). Interleukin expression clearly indicates ongoing inflammatory processes; these could not be reduced by additional purification of the commercially purchased ODN by ion-exchange chromatography.[27] Although thorough studies are missing, signs of sickness behavior such as impairment of cognitive parameters [social recognition[12] and anxiety-related,[13] stress-coping (our unpublished observation, 1998), suckling,[20] and sexual behaviors[22]] have never been observed after acute, repeated, or chronic antisense ODN infusion directly into a distinct brain region. Further, the local specificity of neuronal ODN uptake can be enhanced by the

[20] I. D. Neumann, D. W. F. Porter, R. Landgraf, and Q. J. Pittman, *Am. J. Physiol.* **267**, R852 (1994).

[21] I. Neumann, P. Kremarik, and Q. J. Pittman, *Neuroscience* **69**(4), 997 (1995).

[22] M. McCarthy, S. P. Kleopoulos, C. V. Mobbs, and D. W. Pfaff, *Neuroendocrinology* **59**, 2 (1994).

[23] A. Akabayashi, C. Wahlestedt, J. T. Alexander, and S. F. Leibowitz, *Mol. Brain Res.* **21**, 55 (1994).

[24] C. Wahlestedt, E. Merlo Pich, G. F. Koob, F. Yee, and M. Heilig, *Science* **259**, 528 (1993).

[25] K. M. Standifer, C. C. Chien, C. Wahlestedt, G. P. Brown, and G. W. Pasternak, *Neuron* **12**, 805 (1994).

[26] U. M. Sarmiento, J. R. Perez, J. M. Becker, and R. Narayanan, *Antisense Res. Dev.* **4**, 99 (1994).

[27] B. Schöbitz, G. Pezeshki, J. C. Probst, J. M. H. N. Reul, T. Skutella, T. Stöhr, F. Holsboer, and R. Spanagel, *Europ. J. Pharm.* **331** (1997).

use of small infusion volumes (0.5–1.0 μl). We therefore summarize some methodological approaches for the local infusion of antisense ODN directly into a specific brain region.

Acute, Local Infusion of Antisense Oligodeoxynucleotide into the Hypothalamic Supraoptic or Paraventricular Nuclei of Conscious Rats

Acute, short-term antisense effects could be observed as soon as 4 hr after local infusion of 2.5 μg/0.5 μl of end-capped phosphorothioated antisense ODNs directed against the mRNA sequence encoding the signal peptide of the oxytocin precursor (5′–3′ sequence: CAG GCC ATG GCG TTG GTG) and the vasopressin precursor (CAT GGC GAG CAT AGG TGG). Vehicle as well as mixed bases (ATC GAC TGT CGA AGG TTC) were used as controls. We were able to demonstrate antisense-induced effects on the electrophysiological activity of hypothalamic oxytocin and vasopressin neurons, stimulation of Fos expression in oxytocin neurons of the supraoptic nucleus, as well as neuropeptide release from neurohypo-physial terminals into the blood.[19–21] The device used for acute infusion is described in Fig. 3: two guide cannulas (22 gauge, 15 mm) are implanted with their tips resting 2 mm above the left and right supraoptic nucleus (SON) and closed by stylettes 3 days before the experiment. The upper ends of the stylettes (25-gauge cannula) are sealed with glue and a 2-mm piece of silicon tubing is fitted, ensuring proper fit to the guide cannula. Twice daily (after surgery) the animals are carefully handled and the stylettes removed, cleaned, and replaced in order to allow adaptation of the animals to these procedures. The infusion device used on the day of the experiment consists of a piece of polyethylene (PE) tubing (i.d. 0.75 mm; o.d. 1.22 mm; length, ~ 20–30 cm) filled with the appropriate infusion solution and its upper end connected to the 5-μl microsyringe filled with distilled water (allows expensive antisense solution to be saved and keeps the syringe clean). The two solutions are separated by a tiny air bubble at the upper end of the tubing. Additional control over the infused volume is attained by setting colored marks at a distance corresponding to 0.5 μl ahead of the bubble. A piece of fused silica tubing (o.d. 130 μm; length 22 mm) is glued (using two-component glue) onto the lower end of the PE tubing. Owing to its small dimensions and the sensitive material it consists of (fused silica fiber), it can easily get stuck during tissue passage or break with any slight movement of the conscious animal. To obviate this problem we employ an additional protection cannula (insertion cannula, 27 gauge; Fig. 3). Having inserted the silica tubing (which should be kept retracted at this stage), the insertion cannula is lowered into the guide cannula until the former extends 1 mm. The final depth is determined by a piece of PE-

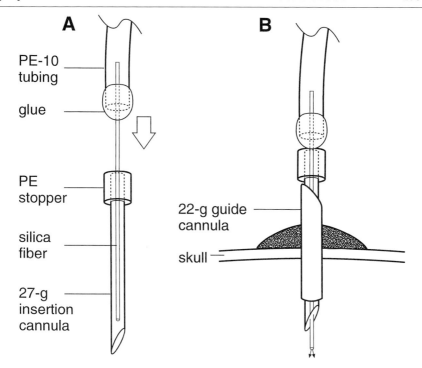

FIG. 3. Device for acute antisense ODN infusion directly into the targeted brain region. The device consists of a chronically implanted 22-gauge guide cannula, a 27-gauge insertion cannula, and a fused silica fiber. (A) Infusion system before lowering into the guide cannula. The silica fiber is retracted for greater protection as described in text. (B) Final arrangement, with the silica fiber lowered into the insertion cannula until the tip reaches the brain region of interest.

50 stopper at its upper end. In this way, the silica tubing is protected and can be finally inserted, reaching the target region and allowing local infusion. The device should be kept in place for another 10 to 20 sec to ensure optimal distribution of the infused solution.

A similar device was designed for studying the acute effects of antisense ODN on local oxytocin or vasopressin release within the supraoptic nucleus as measured by microdialysis (Fig. 4). This release, which is thought to occur mainly from dendrites and perikarya,[28] can take place

[28] J. F. Morris, D. V. Pow, H. A. Sokol, and A. Ward, in "Vasopressin" (P. Gross, D. Richter, and G. L. Robertson, eds.), p. 171 John Libbey, London, 1993.

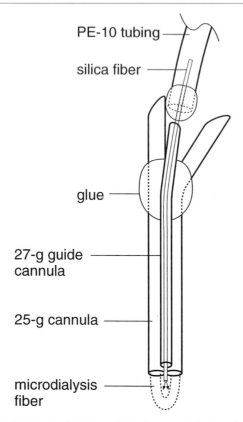

PE-10 tubing

silica fiber

glue

27-g guide cannula

25-g cannula

microdialysis fiber

FIG. 4. Device for local microdialysis and simultaneous infusion of antisense ODN into the dialyzed region. The device consists of a U-shaped microdialysis probe attached to a 27-gauge guide cannula for insertion of the silica fiber infusion cannula. The latter is attached to a microliter syringe via PE-10 tubing.

independently of neurohypophysial secretion[29,30] and can be stimulated by a variety of pharmacological and physiological stimuli (for review see Refs. 31 and 32). We routinely use U-shaped microdialysis probes with a 10-kDa molecular cutoff membrane. As described in Fig. 4, a guide

[29] I. Neumann, M. Ludwig, M. Engelmann, Q. P. Pittman, and R. Landgraf, *Neuroendocrinology* **58,** 637 (1993).

[30] C. T. Wotjak, J. Ganster, G. Kohl, F. Holsboer, R. Landgraf, and M. Engelmann, *Neuroscience* **85**(4), 1209 (1998).

[31] R. Landgraf, *J. Neuroendocrinol.* **7,** 243 (1995).

[32] I. Neumann, Q. J. Pittman, and R. Landgraf, *in* "Oxytocin" (R. Ivell and J. A. Russell, eds.), p. 173. Plenum Press, New York, 1995.

cannula (27 gauge) can be glued between the side-by-side 25-gauge cannulas of the microdialysis probe, with its lower tip resting above the U-shaped probe formed by the microdialysis membrane. Such experiments were mainly performed under urethane anesthesia, allowing the simple insertion of a silica fiber attached to a microsyringe into the guide cannula and subsequent infusion of the vehicle, antisense, or mixed base solution into the microdialyzed area.

However, we have not yet been able to detect convincing, sequence-specific effects of either oxytocin antisense ODN on local, suckling-stimulated oxytocin release or vasopressin antisense ODN on osmotically stimulated vasopressin release within the supraoptic nucleus.[15,19] Whereas in microdialysates sampled from the jugular vein stimulated oxytocin and vasopressin levels were blunted only in the antisense-treated groups, a reduction in hypothalamic neuropeptide release was found in both anti-sense- and mixed base-treated groups. This indicates that end-capped phos-phorothioated oligonucleotides unspecifically affected intrahypothalamic, but not peripheral, neuropeptide release; this effect was observed even after administration of a fifth of the original dose directly into the dialyzed area (0.4 μg/0.5 μl).

Repeated or Chronic Infusion of Antisense Oligodeoxynucleotide Directly into Brain Region of Interest

To reveal "true" antisense effects, i.e., neuroendocrine or behavioral changes due to reduced neuropeptide availability or expression of their respective receptor, the antisense ODN needs to be applied over several days either by repeated acute or continuous infusion. The duration of the treatment should depend on the turnover of the targeted neuropeptide/ receptor protein and has been reported to vary between 2 and 6 days.[12,13,24,25,33] It is advisable to measure the physiological/behavioral consequences of reduced neuropeptide availability in a functional context rather than simply the reduction in neuropeptide concentration. Examples are the antisense-induced deficits in vasopressin–receptor interactions, which became evident as reduced cognitive abilities[12]; the blockade of neuropeptide Y synthesis, which resulted in decreased feeding behav-ior[23]; and the inhibition of oxytocin receptor expression, which caused not only reduced local oxytocin binding but also a reduction in relevant behaviors.[22]

To study the selective involvement of CRH, vasopressin, and oxytocin

---

[33] D. A. Clemett, M. I. Cockett, C. A. Marsden, and K. C. F. Fone, *J. Neurochem.* **71,** 1271 (1998).

in the regulation of basal as well as stress-induced adenohypophysial corticotropin adrenocorticotropic hormone (ACTH) secretion, appropriate antisense ODNs were repeatedly infused into the hypothalamic paraventricular nucleus (PVN), using the silica tubing infusion device described above (Fig. 3). This tiny brain region would have been severely damaged by (bilateral) infusion via osmotic minipump—an elegant technique that has required relatively large infusion cannulas to date (see below). Five days prior to the experiment the animals were fitted with guide cannulas resting above the left and right PVN as well as a chronic jugular vein catheter. As a first approach, a cocktail containing antisense ODNs (800 ng each) directed against the mRNAs encoding CRH, vasopressin, and oxytocin was infused three times at 12-hr intervals into both PVN. This was expected to block the synthesis of all three putative ACTH secretagogues, thereby avoiding conterregulatory effects. As a result, swim stress-induced secretion of ACTH into blood was found to be reduced in animals treated with antisense, but not in animals treated

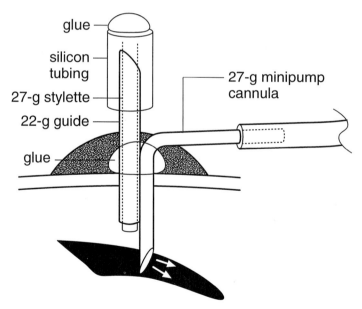

FIG. 5. Device for chronic intracerebral infusion of antisense ODN directed against a neuropeptide receptor via a 27-gauge cannula attached to an osmotic minipump, and acute infusion of the respective neuropeptide via a 22-gauge guide cannula attached to the minipump cannula (see text). The guide cannula is closed by a stylette consisting of a 27-gauge cannula and a piece of silicon fiber sealed with glue at its upper end to ensure proper fit.

with vehicle or scrambled sequence or in untreated animals.[34] The finding that *basal* secretion of ACTH was unaffected by the repeated antisense treatment supports the idea that neuroendocrine effects of antisense treatment are more easily detectable in a physiologically challenged system. The comparison between untreated and other control animals (vehicle, mixed bases) validated the infusion method used. Repeated infusion of the respective control solution into both the left and right PVN did not alter ACTH secretion, which is thought to be strongly dependent on the activity of parvocellular, especially CRH, neurons of the PVN. Hence our infusion device can also be used for repeated infusion of the antisense ODN into relatively small brain regions without dramatic tissue irritation or damage.

Still, we prefer to infuse antisense ODN chronically into the brain region of interest, whenever it is large enough for local infusion, via a 27-gauge cannula connected to Alzet osmotic minipumps (Alza, Palo Alto, CA). These pumps are available in varying sizes corresponding to total infusion volumes between 100 μl and 2 ml and can be used for between 1 and 28 days (depending on the size). It should be mentioned that the absolute infusion duration could be increased by partly covering the semipermeable surface of the minipump with glue—a procedure that also reduces the infusion volume per time. In our laboratory, however, we use the "unmodified" Alzet model 1007D (nominal infusion rate, 0.5 μl/hr), for instance, for intraseptal infusion of antisense ODN against the mRNA encoding the vasopressin V1 receptor, thus demonstrating its involvement in the ability of male rats to recognize conspecific juveniles (social discrimination[12]). For this experiment, septal vasopressin receptor downregulation was assessed not only by receptor autoradiography but also in a functional context. Synthetic vasopressin was infused intracerebroventricularly on day 4 of antisense ODN treatment via an additionally implanted guide cannula above the contralateral lateral ventricle, resulting in an improved social memory in control (scrambled sequences, vehicle), but not antisense ODN-treated, animals. This was due to the antisense-induced lack of V1 receptor proteins in the behaviorally relevant brain region.

We have developed a technical alternative for both chronic (antisense ODN) and acute (synthetic peptide) infusions directly into the antisense-targeted region as described in Fig. 5. All parts of this combination device can be prepared before implantation. In this way, we can avoid the implantation of cannulas on both sites of the skull, which is technically difficult owing to the relatively large space necessary for fixation of the minipump device to the skull.

---

[34] R. Landgraf, T. Naruo, M. Vecsernyes, and I. Neumann, *Eur. J. Endocrinol.* **137**, 326 (1997).

Using this experimental design it would be possible to study the involvement of, for example, brain prolactin receptors in anxiety-related behavior as well as neuroendocrine regulation; the study would rely on antisense techniques, as a prolactin receptor antagonist is not available. Intracerebral infusion of exogenous prolactin reduced anxiety behavior in the elevated plus-maze. In animals treated with the antisense ODN directed against the mRNA encoding the neuropeptide receptor infusion of the exogenous neuropeptide should not result in behaviorally relevant effects.

To summarize, according to the neurobiological hypothesis to be tested the experimental approach should be thoroughly selected within a variety of available alternatives. This will also hold true for the selection of target mRNA sequences, where the secondary and possibly tertiary molecular structures of the antisense and mRNA molecules should be explored by techniques such as minimization of the free energy of folding or kinetic folding simulations and examined according to stability and specificity criteria, to name a few. The differentiation and characterization of all of the effects antisense targeting can produce are essential for the correct and fruitful implementation of such techniques, which, without thorough knowledge of their effects, can produce misleading or, even worse, harmful results (in the case of therapeutic use).

Acknowledgment

Supported by Deutsche Forschungsgemeinschaft (NE 465/3-1, 4-1).

[17] Localization of Oligonucleotides in Brain by in Situ Hybridization

By Katsumi Yufu, Yutaka Yaida, and Thaddeus S. Nowak, Jr.

Introduction

Localization of administered agents within tissue is a critical component of in vivo antisense studies, particularly in heterogeneous tissues such as brain. Oligonucleotides prelabeled in various ways have been used to monitor tissue distribution with reasonable success.[1–6] A potential problem with

[1] L. Whitesell, D. Geselowitz, C. Chavany, B. Fahmy, S. Walbridge, J. R. Alger, and L. M. Neckers, Proc. Natl. Acad. Sci. U.S.A. 90, 4665 (1993).

[2] F. Yee, H. Ericson, D. J. Reis, and C. Wahlestedt, Cell. Mol. Neurobiol. 14, 475 (1994).

[3] D. Wielbo, C. Sernia, R. Gyurko, and M. I. Phillips, Hypertension 25, 314 (1995).

this approach is the sustained presence of the labeling moiety beyond the lifetime of the parent molecule, resulting in dissociations between the distribution of intact oligonucleotide and that of cleaved label. An alternative approach is the use of *in situ* hybridization with a labeled complementary sequence to detect the administered oligonucleotide.[7,8] Conditions are easily chosen that require essentially a full-length sequence to be present to form stable hybrids, allowing this method to specifically detect intact oligonucleotide. Procedures are described for the use of both radiolabeled and fluorescein-labeled probes to detect antisense oligonucleotides in sections of fresh-frozen and perfusion-fixed brain. While the methods detailed here address the detection of phosphorothioate (PS) and phosphodiester (PO) oligonucleotides, the general approach should be readily applicable to other constructs.

Methods of Procedure

Oligonucleotides and Routes of Administration

The antisense oligonucleotide employed in these studies is an all-phosphorothioate (PS) 20-mer with the sequence 5'-CCTTCATCTTGGTCAG-CACC-3', which is complementary to a region of conserved 70-kDa heat shock protein-coding sequence reported to have antisense effects.[9] Gel-purified preparations are obtained from Gene Link (Thornwood, NY). The oligonucleotide is dissolved at a concentration of 1 mM in artificial cerebrospinal fluid consisting of 20 mM HEPES (pH 7.4), 1.3 mM CaCl$_2$, 1 mM MgCl$_2$, 112 mM NaCl, 3.1 mM KCl, 5.5 mM glucose, and 2 mM sodium lactate. Wistar rats (250–350 g; Hilltop Lab Animals, Scottdale, PA) are anesthetized with 1–2% halothane administered in 70% N$_2$–30% O$_2$ and placed in a stereotaxic frame, after which the skull is exposed and burr holes drilled. For intraventricular injections polyethylene tubing connected to a 30-gauge needle is filled with oligonucleotide solution and inserted into the lateral ventricle at the level of the septum (coordinates: 3.5 mm deep, 1.5 mm lateral, and 0.5 mm posterior to bregma). Oligonucleo-

[4] R. R. Sakai, L. Y. Ma, P. F. He, and S. J. Fluharty, *Regul. Pept.* **59**, 183 (1995).
[5] W. C. Broaddus, S. S. Prabhu, G. T. Gillies, J. Neal, W. S. Conrad, Z.-J. Chen, H. Fillmore, and H. F. Young, *J. Neurosurg.* **88**, 734 (1998).
[6] R. Grzanna, J. R. Dubin, G. W. Dent, Z. Ji, W. Zhang, S. P. Ho, and P. R. Hartig, *Mol. Brain Res.* **63**, 35 (1998).
[7] Y. Yaida and T. S. Nowak, Jr., *Regul. Pept.* **59**, 193 (1995).
[8] A. Szklarczyk and L. Kaczmarek, *J. Neurosci. Methods* **60**, 181 (1995).
[9] K. Sato, H. Saito, and N. Matsuki, *Brain Res.* **740**, 117 (1996).

tide is infused at a rate of 0.5 μl/min, administering up to 20 nmol in 10 μl. For direct tissue injection glass capillaries with tip diameters of ~50 μm are drawn, filled with oligonucleotide solution, fitted with polyethylene tubing attached to a syringe pump, and positioned in the hippocampus (2.5 mm deep, 2.0 mm lateral, and 4.0 mm posterior to bregma). A total of 2 nmol of oligonucleotide is administered in 2 μl over 10 min. The infusion apparatus is withdrawn, the scalp is sutured and the animals are returned to their cages.

Preparation of Tissue Samples

Frozen Sections. Animals are briefly anesthetized and decapitated, after which the brain is removed, frozen in hexane at $-40°$, and stored at $-70°$. Sections (16 μm) are cut on a cryostat, thaw mounted on polylysine- or gelatin-coated slides, briefly air dried, and stored desiccated at $-70°$.

Vibratome Sections. Animals are perfused transcardially with 100 mM sodium phosphate (pH 7.4) followed by 4% paraformaldehyde in the same buffer. Brains are removed, postfixed for several hours, and transferred through several changes of 10 mM sodium phosphate-buffered saline (PBS) overnight. Sections (50 μm) are cut on a vibrating tissue slicer and stored refrigerated in PBS containing 0.02% sodium azide.

In Situ Hybridization

Probe Labeling. The hybridization probe is a 20-mer phosphodiester sense sequence complementary to the injected oligonucleotide that is 3′ end labeled with terminal deoxynucleotide transferase,[10] using enzyme and reagents provided in a kit from Boehringer Mannheim (Indianapolis, IN). The reaction is performed in a microcentrifuge tube in a volume of 50 μl containing 0.2 M potassium cacodylate, 25 mM Tris-HCl (pH 6.6), bovine serum albumin (0.25 mg/ml), 2.5 mM CoCl$_2$, 10 pmol of probe, 25 units of terminal transferase, and either 62.5 μCi (approximately 50 pmol) of α-[^{35}S]thio-dATP (NEN Life Science Products, Boston, MA) or 1 nmol of 11-fluorescein-dUTP (Boehringer Mannheim). Labeling is accomplished at 37° for 40–60 min, after which 40 μg of glycogen in a volume of 2 μl is added as a carrier, and the mixture is extracted with an equal volume of phenol–chloroform–isoamyl alcohol (25:25:1). The aqueous phase is collected after centrifugation, the organic phase is extracted with 50 μl of 10 mM Tris-HCl (pH 8.0), 1 mM EDTA, 1% sodium dodecyl sulfate (TE-SDS) and recentrifuged, and the combined aqueous phases are precipitated at $-70°$ after the addition of 10 μl of 3 M sodium acetate and 350 μl of

[10] R. Roychoudhury and R. Wu, *Methods Enzymol.* **65,** 43 (1980).

ethanol. The pellet is collected by microcentrifugation for 15 min at maximum speed, rinsed in 80% ethanol, air dried, and dissolved in 50 μl of TE-SDS. Radiolabeled probes are subjected to further purification on Sephadex G-25 spin columns (Quick Spin; Boehringer Mannheim) equilibrated with the same buffer.

Hybridization Mix. The hybridization reagent consists of 25% formamide, 2× SSC, sonicated salmon DNA (0.5 mg/ml), yeast tRNA (0.25 mg/ml), 1× Denhardt's solution (1% each Ficoll, polyvinylpyrrolidone, and bovine serum albumin), 100 mM dithiothreitol, and 10% dextran sulfate in addition to labeled probe. Components for 10 ml of hybridization mix are typically prepared in a volume of 9 ml and stored frozen in 450-μl aliquots at $-70°$, to which a 50-μl preparation of labeled probe is added just prior to use.

Processing of Frozen Sections. Slide-mounted frozen sections are processed for hybridization essentially as previously described,[11] using glass slide racks and trays accommodating 200 ml of the solutions to be employed, and containing a magnetic stirrer for agitation. Sections are fixed for 5 min in 4% paraformaldehyde in 100 mM sodium phosphate (pH 7.4), rinsed in two changes of 70% ethanol, and twice in 2× SSC (0.3 M sodium chloride, 0.03 M sodium citrate, pH 7.0), transferred to 0.1 M triethanolamine hydrochloride (pH 8.0), and immediately acetylated by the dropwise addition of acetic anhydride to 0.25% with rapid and continuous stirring for 10 min. Sections are then transferred to 0.2 M Tris, 0.1 M glycine for 30 min, rinsed in 2× SSC, dehydrated through 70, 80, 90, and 100% ethanol, and extracted in chloroform for 5 min. To each section is applied 10 μl of hybridization mix containing 0.2 pmol of either ^{35}S- or fluorescein-labeled probe, followed by a coverslip. Hybridization takes place for 3 hr at room temperature, followed by a brief wash in 2× SSC and several 20-min washes in 2× SSC, 25% formamide. These conditions are chosen to be approximately 25° below the estimated melting temperature (T_m) for PO–PO hybrids (46° for this pair of oligonucleotides), taking into account the lower T_m expected for PO–PS hybrids,[12] and the desirability of carrying out hybridization and stringency washes at as much as 10° below the T_m.[13] Sections hybridized with radiolabeled probes are then rinsed briefly in 2× SSC, dehydrated through graded alcohol, and placed against Kodak (Rochester, NY) SB5 film to obtain the autoradiographic image. Fluorescein label is detected by immunocytochemistry. Briefly, sections are rinsed in 2× SSC and placed

[11] T. S. Nowak, Jr., *J. Cereb. Blood Flow Metab.* **11**, 432 (1991).
[12] C. A. Stein, C. Subasinghe, K. Shinozuka, and J. S. Cohen, *Nucleic Acids Res.* **16**, 3209 (1988).
[13] J. Sambrook, E. F. Fritsch, and T. Maniatis, "Molecular Cloning: A Laboratory Manual," 2nd Ed. Cold Spring Harbor Press, Plainview, New York, 1989.

for 30 min in 0.5% blocking reagent (Boehringer Mannheim) prepared in PBS. Slides are then transferred to mouse anti-fluorescein monoclonal antibody (diluted 1 : 1000 in blocking reagent; Sigma, St. Louis, MO) and incubated at 4° overnight. This takes place in small polyethylene slide boxes holding approximately 5 ml, and the antibody solution is typically reused several times. Slides are washed several washes in blocking reagent. To each section is then applied 50 μl of a 1 : 200 dilution of biotin-labeled goat anti-mouse IgG (Kirkegaard & Perry Laboratories, Gaithersburg, MD) and slides are incubated for 1 hr at room temperature in a humidified chamber, washed several times, treated with 3% H_2O_2, 10% methanol in PBS for 20 min to quench endogenous peroxidase, and again washed several times in PBS and blocking reagent. Sections are then incubated for 1 hr in 50 μl of peroxidase-conjugated streptavidin (diluted 1 : 200 in blocking reagent; Kirkegaard & Perry), washed in PBS containing 0.05% Tween 20, and transferred to 100 mM Tris-HCl (pH 7.6). Peroxidase is detected with diaminobenzidine (1 mg/ml), $NiSO_4$ (1 mg/ml), 0.0015% H_2O_2 in the same buffer, after which the slides are rinsed, dehydrated through graded alcohols and xylene, and coverslipped in Permount.

Processing of Vibratome Sections. In initial studies up to two floating sections were placed in 7-ml vials and subjected to a brief pretreatment involving incubations in successive 1-ml volumes of 50 mM NH_4Cl in PBS and 0.3% Triton in PBS for 30 min each, as typically employed to permeabilize sections for immunocytochemistry,[14] after which sections were washed in 2× SSC and hybridized as described below. Enhanced detection is obtained with a sequence of steps carried out essentially as described above for frozen sections on slides, using volumes of 1 ml. Sections are washed twice in 2× SSC, transferred to triethanolamine and acetylated with acetic anhydride, incubated in Tris–glycine, rinsed in 2× SSC, dehydrated through graded alcohol, and extracted in chloroform. Chloroform is removed and the sections rehydrated by reversing the graded alcohol sequence. Sections are then washed twice for 5 min in 2× SSC and covered with 100 μl of hybridization mix containing fluorescein-labeled probe. After 3 hr of hybridization at room temperature under continuous gentle agitation sections are washed briefly in 2× SSC, followed by several changes of 2× SSC, 25% formamide over 1 hr. The hybridized probe is then detected by immunocytochemistry, using procedures comparable to those employed for frozen sections as detailed above. Antibody and streptavidin–peroxidase dilutions and incubation times are identical, but 200-μl volumes are used to ensure adequate coverage of the sections, and the various washes employ

[14] K. Vass, W. J. Welch, and T. S. Nowak, Jr., *Acta Neuropathol.* **77,** 128 (1988).

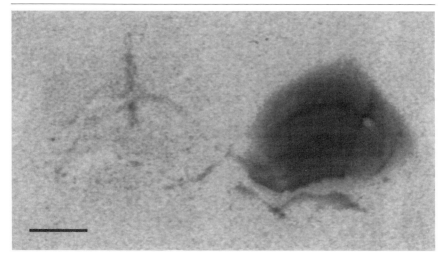

FIG. 1. Localization of injected oligonucleotide with radioactive probe. PS oligonucleotide was injected into hippocampi of the same animal at intervals of 48 hr (left) or 1 hr (right) prior to killing, and frozen sections were processed for *in situ* hybridization with ^{35}S-labeled probe. At the early postinjection interval an intense focus of hybridization signal is readily detected throughout the hippocampus, extending as well into adjacent cortex. A nearly complete loss of signal is evident on the side injected 48 hr earlier, although noticeable signal remains along the injection track as well as diffusely throughout the region of initial distribution. Bar: 1 mm.

volumes of 1 ml each. The sections are finally mounted on gelatin-coated slides and coverslipped.

Results

Hybridization with radiolabeled probes in frozen sections allows the straightforward localization of injected oligonucleotide in brain (Fig. 1). Under these injection conditions a region of tissue approximately 3 mm in diameter is targeted, as evident when hybridization is performed in the initial hours after infusion. There is a significant loss of signal by 48 hr after injection, even in the case of relatively stable PS oligonucleotides. Previous studies have shown that this method detects a comparable distribution of PO oligonucleotide at early intervals, with, as expected, a more rapid loss of signal.[7]

The cellular localization of oligonucleotide after intrahippocampal injection can be evaluated by hybridization with nonradioactive detection meth-

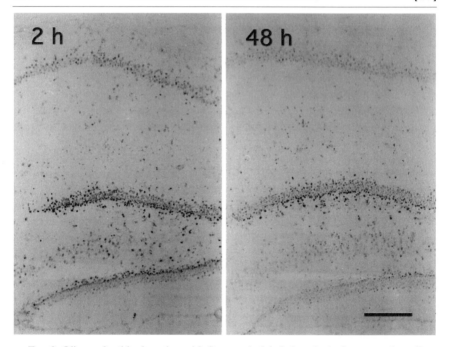

FIG. 2. Oligonucleotide detection with fluorescein-labeled probe in frozen sections. Rats received intrahippocampal injections of 2 nmol of PS oligonucleotide and brains were fresh frozen 2 or 48 hr postinjection. Sections were hybridized with fluorescein-labeled probe, followed by immunoperoxidase detection. Frozen sections preferentially reveal early and persistent nuclear localization, readily detected in major neuron populations. Bar: 200 μm.

ods, as illustrated in Figs. 2 and 3, although distinct compartments in specific cell types show differential sensitivity to detection depending on the processing of the tissue. Frozen sections allow the identification of an early component of cellular uptake that appears to be consistent with a nuclear localization, readily detected in major neuron populations but likely including other cell types (Fig. 2). A comparable distribution is evident 2 or 48 hr after injection, with a decrease in signal intensity during this interval. Conversely, Vibratome sections show a diffuse signal at early time points that is consistent with the expected general distribution of injected oligonucleotide as detected with radioactive probes in frozen sections, but with little evidence of discrete cellular localization (Fig. 3). By 48 hr there is a prominent detection of the hybridization signal in uniformly scattered cells throughout the territory of the injection, with evidence of an apparent vesicular localization at higher magnification. This compartment of oligonucleotide distribution, not detected in fresh frozen sections, appears to involve a glial cell type.

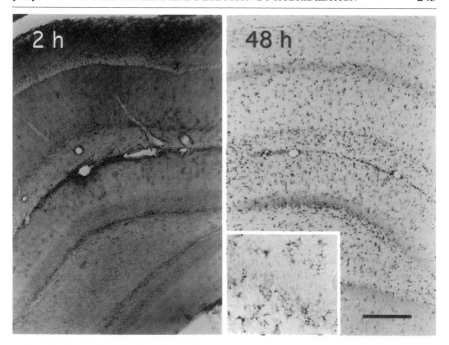

FIG. 3. Oligonucleotide detection with fluorescein-labeled probe in perfusion-fixed brain. Rats received intrahippocampal injections of 2 nmol of PS oligonucleotide and brains were perfusion fixed 2 or 48 hr postinjection. Vibratome sections obtained at early intervals exhibit diffuse labeling corresponding to the oligonucleotide distribution detected with radiolabeled probe (see Fig. 1). In contrast, a distinct glial uptake is evident at late intervals. Localization to an apparently vesicular compartment is seen at higher magnification (inset). Bar: 200 μm.

Discussion

The preceding studies demonstrate the feasibility of using *in situ* hybridization approaches to detect oligonucleotides in brain. An overall regional localization is most easily detected using ^{35}S-labeled probes in frozen sections, while immunodetection of non-radioactive probe is necessary for cellular localization. Autoradiographic approaches are less useful in Vibratome sections owing to increased section thickness and high background, but non-radioactive methods can be applied to either fresh frozen tissue or Vibratome sections of fixed material. Hybridization and wash stringency can be readily adjusted as appropriate for the structure, base composition, and length of the target and probe. Results obtained with this method[7,8] are consistent with studies using other approaches to document the relative instability of PO versus PS oligonucleotides.[1] These results also indicate the relatively limited tissue distribution of oligonucleotides reported in most studies, whether administered by intraparenchymal or intraventricular routes,[3,4] although the penetration of injected oligonucleotide is clearly dose dependent.[6] A study examining the distribution of PS oligonucleotide prelabeled with ^{35}S demonstrated a generally comparable decrease in overall signal for 48 hr after intraparenchymal injection, but this was accompanied by a several fold increase in the volume of distribution over time.[5] This may reflect in part the 1000-fold larger load of oligonucleotide (2 μmol) delivered by the high-flow infusion technique employed in the latter study, perhaps resulting in a saturation of local binding and cellular uptake,[15] but could also include a component arising from metabolism and redistribution of label in nonnucleic acid molecules. The hybridization method has yet to be applied to examine the more extensive distributions of PS oligonucleotides reported after prolonged infusion.[1]

An important consideration is the differential detection of oligonucleotide in distinct cellular elements depending on the method of tissue processing. A nuclear component of the signal, prominent in major neuron populations, is evident only in frozen sections, whereas the hybridization in a vesicular glial compartment is detected only in Vibratome sections. This early preferential uptake of oligonucleotides into neurons has been described in several studies employing prelabeled constructs.[16,17] Several reports have also described a progressive redistribution of fluorescence to a vesicular localization in neurons after the initial nuclear uptake.[17,18] We have found no evidence of such a neuronal compartment in the hybridization studies, although identification of the cell type accounting for the late hybridization signal in Vibratome sections remains uncertain. Preliminary double-labeling results (not shown) indicate that these cells are negative for both glial fibrillary acidic protein and isolectin B4, ruling out both astrocytes and microglia, and may represent a glial precursor positive for NG2 proteoglycan that also displays activation in response to diverse insults.[19] Persistent retention of fluorescein in various cell types has been reported in brain after injection of labeled oligonucleotide.[1,20] The extent to which discrepancies in results obtained with these approaches reflect a localization of fluorescence independent of intact oligonucleotide or rather

[15] S. L. Loke, C. A. Stein, X. H. Zhang, K. Mori, M. Nakanishi, C. Subasinghe, J. S. Cohen, and L. M. Neckers, *Proc. Natl. Acad. Sci. U.S.A.* **86**, 3474 (1989).

[16] S. Ogawa, H. E. Brown, H. J. Okano, and D. W. Pfaff, *Regul. Pept.* **59**, 143 (1995).

[17] W. Sommer, X. Cui, B. Erdmann, L. Wiklund, G. Bricca, M. Heilig, and K. Fuxe, *Antisense Nucleic Acid Drug Dev.* **8**, 75 (1998).

[18] S. P. Zhang, L. W. Zhou, M. Morabito, R. C. Lin, and B. Weiss, *J. Mol. Neurosci.* **7**, 13 (1996).

[19] A. Nishiyama, M. Yu, J. A. Drazba, and V. K. Tuohy, *J. Neurosci. Res.* **48**, 299 (1997).

[20] M. Bannai, M. Ichikawa, M. Nishihara, and M. Takahashi, *Brain Res.* **784**, 305 (1998).

provide further examples of differential detection by the hybridization method, remains to be determined.

Regarding the factors responsive for this differential detection, it may be speculated that probes have greater access to the nuclear compartment in the briefly fixed frozen sections than in more strongly fixed tissue from perfused brain. Conversely, oligonucleotide in a more labile glial compartment may be lost from the frozen sections. A related issue is the relative absence of discrete cellular uptake detected with the use of radioactive probes in frozen sections. This undoubtedly reflects in part a preferential detection of superficial hybridization signal, as somewhat greater quenching would be expected for a radioactive probe localized within the nuclear compartment. In addition, there may be reduced penetration of radioactive vs. non-radioactive probes due to size differences arising during the labeling reactions. When ^{35}S- and fluorescein-labeled probes were 5′-end-labeled with ^{32}P and compared on sequencing gels it was evident that fluorescein-labeled probes were significantly smaller than those that were radiolabeled (unpublished observation). These results suggest that the two methods of tissue processing must both be applied to obtain a complete characterization of oligonucleotide distribution in brain. It remains to be determined whether this may also be a consideration in other tissues. From a practical perspective the use of frozen sections may be most relevant in the majority of studies, in which the general localization of antisense delivery and successful targeting of cells at early intervals may be of primary interest. The later redistribution to a vesicular glial compartment likely reflects a step in the clearing of stable PS oligos, and has not been detected following injection of PO oligos. It will be of interest to determine whether other stable antisense structures are similarly processed.

Acknowledgment

This work was supported in part by USPHS grant NS32344 and T.S.N.

[18] Use of Antisense Oligonucleotides in Human Neuronal and Astrocytic Cultures

By Valeria Sogos, Monica Curto, Mariella Setzu, Isabella Mussini, Maria Grazia Ennas, and Fulvia Gremo

Antisense oligodeoxynucleotides can potentially block the expression of a single gene in living cells, by binding to a complementary mRNA. These molecules can enter cultured cells, recognize a specific gene sequence, and inhibit its expression.[1-3] For this reason, antisense oligodeoxynucleotide strategies are largely used to understand specific gene functions in biological processes and during development.[4] Several factors must be considered when treating cells with antisense and appropriate controls, which must be chosen to demonstrate that the effects of treatment are due to specific gene inhibition; indeed, there are many nonspecific effects that are not due to antisense mechanisms alone.

In this chapter we describe standard protocols for antisense treatment of human fetal brain cultures and for controls that are required to interpret results.

Choice of Oligonucleotide Antisense

An antisense molecule can be considered specific if it determines a significant decrease in target RNA and related protein levels and if there is no strong loss of cell viability. The interaction between antisense and RNA requires a minimum level of affinity. This can be reached, in human cells, with an oligonucleotide 17–24 nucleotides long. This size should ensure both specificity and sufficient uptake into cells. In fact, a sequence at least 17 nucleotides long has a high probability of being unique in the human genome and consequently of being specific for the target RNA.[5] On the other hand, oligonucleotides that are too long might be unable to cross the cellular membrane efficiently.

The use of several oligonucleotides at the same time is not recommended, because it increases nonspecific effects while poorly enhancing their activity. Standard computer programs for choosing sequences for PCR (polymerase chain reaction) primers or DNA probes can be used to select the optimal target for antisense oligonucleotides, which must be examined for analogies with other sequences in genome databanks. Most regions of a mRNA molecule are available for binding to antisense compound, but a strong secondary structure of mRNA can diminish the portion of the molecule accessible to oligonucleotides.[6] In addition, numerous sequences can unspecifically bind to small molecules[7] and proteins.[8,9] As a conse-

[1] J. F. Milligan, M. D. Matteucci, and J. C. Martin, *J. Med. Chem.* **36,** 1923 (1993).
[2] C. A. Stein and C. Cheng, *Science* **261,** 1004 (1993).
[3] C. Helene and J. J. Toulme, *Biochim. Biophys. Acta* **1049,** 99 (1990).
[4] K. Augustine, *Mutat. Res.* **396,** 175 (1997).
[5] A. D. Branch, *Trends Biochem. Sci.* **23,** 45 (1998).
[6] N. Milner, K. U. Mir, and E. M. Southern, *Nature Biotechnol.* **15,** 537 (1997).
[7] A. D. Ellington and J. W. Szotastak, *Nature (London)* **346,** 818 (1990).
[8] L. C. Bock, L. C. Griffin, J. A. Latham, E. H. Vermaas, and J. J. Toole, *Nature (London)* **355,** 564 (1992).
[9] L. Yakubov, Z. Khaled, L. M. Zhang, A. Truneh, V. Vlasson, and C. A. Stein, *J. Biol. Chem.* **268,** 18818 (1993).

quence, oligonucleotides can cause many biological effects that are not due to antisense mechanisms alone. For example, they can affect cell differentiation[10] and proliferation[11] and can cause partial destruction of other RNAs.[5] For these reasons several oligonucleotides, each directed against a different site in the same target RNA, should be tested in order to find the most suitable one.

Oligonucleotides are rapidly degraded in most cells by nucleases. Several oligonucleotide modifications have been suggested[12] in order to increase their nuclease resistance, cellular uptake, and specificity of binding to target mRNA. Phosphorothioate oligonucleotides (S-ODNs) are the most popular antisenses used in culture. Because of their nuclease resistance, they can be used at a lower concentration compared with unmodified ODNs.

Multiple controls are essential in antisense treatment, first to demonstrate the efficiency of antisense gene inhibition and second to demonstrate the absence (or minimal presence) of nonspecific effects.

Cell Treatment

Antisense Administration

Procedure. Suspend lyophilized antisense oligonucleotides in Dulbecco's modified Eagle's medium (DMEM) at a 1 mM concentration (stock solution) and store at $-20°$.

Cells are grown on polylysine-coated dishes, in DMEM plus 10% fetal calf serum (FCS).[13] Before treatment, wash cells twice with DMEM to remove serum, which can inhibit the antisense effect, and incubate with oligonucleotide solution in serum-free DMEM. The optimal concentration should be obtained by testing different oligonucleotide concentrations. In our model (human fetal neuronal enriched cultures[13]) 3 μM S-ODN antisense for dystrophin and CD4 were sufficient to inhibit gene expression, while higher concentrations determined nonspecific toxic effects (see Fig. 1). The use of liposomes (Lipofectin) did not increase the uptake of oligonucleotides, but had a toxic effect on neurons and astrocytes (our unpublished observations, 1998). After 2 hr, add heat-inactivated serum (10%). At different times (from a few hours to several days, depending on the turnover of

[10] H. Kamano, T. Tanaka, Y. Yamaji, K. Ikeda, Y. Hata, T. Shiotani, T. Ishida, J. Takahara, and S. Irino, *Biochem. Int.* **26,** 537 (1992).

[11] A. M. Krieg, A. K. Yi, S. Matson, T. J. Waldschmidt, G. A. Bishop, R. Teasdale, G. A. Koretzky, and D. M. Klinman, *Nature (London)* **374,** 546 (1995).

[12] A. J. Baertschi, *Mol. Cell. Endocrinol.* **101,** R15 (1996).

[13] V. Sogos, M. G. Ennas, I. Mussini, and F. Gremo, *Neurochem. Int.* **31,** 447 (1997).

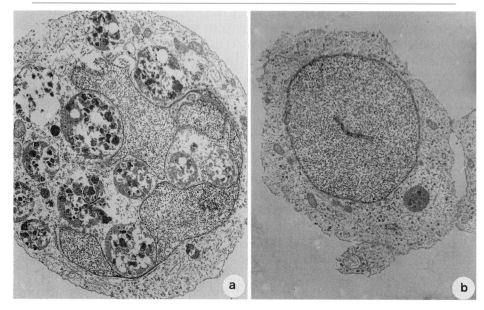

FIG. 1. TEM micrograph of human neuronal cultures treated with antisense oligonucleotides against dystrophin for 4 days. When a too-high concentration of the antisense was used (in this case, 8–10 μM), ultrastructural modifications could be observed in treated cells (a) versus controls (b): nuclear morphology was altered and large inclusions, with heterogeneous content, were abundant in the cytoplasm. Original magnification: ×18,700. [From V. Sogos, M. G. Ennas, I. Mussini, and F. Gremo, *Neurochem. Int.* **31,** 447 (1997).]

the protein) cells are fixed or harvested. Two different controls must be performed: (1) cell incubation with a mismatched oligonucleotide, which should differ from the antisense sequence in a few bases; and (2) cell incubation with DMEM alone. Compare oligonucleotide- and mismatch-treated cells for nonspecific effects.

Uptake

Fluorochrome-conjugated (Fig. 2) or biotin-conjugated oligonucleotides can be used to observe their uptake and localization in cells.[13] When both astrocytes and neurons are present in the culture, an immunostaining for specific markers of the two populations, such as neurofilament (NF) for neurons and glial fibrillary acidic protein (GFAP) for astrocytes, must be associated in order to identify the cell type. This method is not suitable when sequences that aspecifically bind to the substrate are used, resulting in a high fluorescent background.

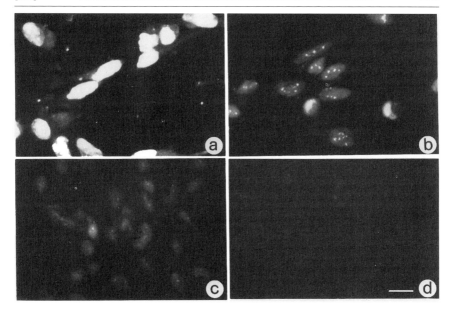

FIG. 2. Antisense uptake by human fetal brain cultures. A fluorescent oligonucleotide against brain dystrophin (3 μM) was added to the medium. After 30 min (a) antisense accumulated in the nuclei, whereas the cytoplasm was negative. Staining significantly decreased after 1 hr (b) and it was very low after 15 hr (c). Controls incubated with free fluorescein were negative (d). Bar: 10 μm. [From V. Sogos, M. G. Ennas, I. Mussini, and F. Gremo, *Neurochem. Int.* **31,** 447 (1997).]

Procedure. Cells are grown on polylysine-coated glass coverslips.[14] After two washes with DMEM, incubate cells with 3.3 μM fluorescein isothiocyanate (FITC)-conjugated oligonucleotide in serum-free DMEM. At different times (15 min, 30 min, 1 hr, 2 hr, and 24 hr) gently rinse cells with warm (37°) phosphate-buffered saline (PBS) and fix in cold methanol for 4 min at $-20°$ (for 24-hr treatments, add 10% heat-inactivated serum after 2 hr). For negative controls incubate samples with free fluorescein. Stain cells for NF or GFAP (see Immunocytochemistry) with a TRITC-conjugated secondary antibody.

Alternatively, cells can be incubated with biotin-conjugated oligonucleotides according to the same protocol as for FITC–oligonucleotides. After fixation, incubate coverslips with horseradish peroxidase (HRP)-conjugated biotin (1 : 800; Vector, Burlingame, CA) for 30 min, wash three times with PBS, and reveal with diaminobenzidine (DAB) (1 mg/ml in PBS, plus 10 μl of 30% H_2O_2). Staining for NF or GFAP must be performed according

[14] V. Sogos, F. Bussolino, E. Pilia, S. Torelli, and F. Gremo, *J. Neurosci. Res.* **27,** 706 (1990).

to the alkaline phosphatase–anti-alkaline phosphatase (APAAP) method (see Immunocytochemistry).

Controls for Specific Gene Inhibition

Antisense-specific gene inhibition can be demonstrated by measuring the target RNA and protein levels in treated cells versus controls. Several controls must be performed, treating cells with different oligonucleotides. The best control sequences are mismatched oligonucleotides, which maintain the same base composition as the antisense. A sense oligonucleotide complementary to antisense can also be used, as well as an antisense for a protein that is never expressed in the examined cell population.

Reverse transcription (RT)-PCR is a useful method for a quantitative analysis of the transcript. As an internal control an independent mRNA, such as glyceraldehyde-3-phosphate dehydrogenase (GAPDH), should also be tested.

Biochemically measuring antisense effects on protein level is more complicated, because it depends on protein turnover.

Alternatively, treatment efficiency can be assayed by immunocytochemistry, which is not a quantitative method, but allows observation of antigen localization. Thus, using specific antibodies, treated and untreated cells can be stained for the target protein and results compared. Antisense effect on a specific cell type in a mixed culture can be demonstrated by double immunostaining.

Depending on the target protein, other controls for specificity of gene inhibition can be performed. For example, we treated human fetal brain cultures (containing neurons and astrocytes) with antisense for dystrophin. By alternative splicing, three different isoforms can be synthesized from the same gene: (1) brain (expressed by neurons but not by astrocytes), (2) muscle (expressed by astrocytes, but not by neurons), and (3) cerebellum Purkinje cell specific. Different antisenses, specific for the gene portion differing in the protein isoforms, were added to the cultures. Using RT-PCR, we demonstrated that each population had a decrease in dystrophin expression only when the specific antisense was used, while no effect was observed after treatment with any of the other antisenses.[13]

Immunocytochemistry

There are several protocols for *in vitro* immunostaining. Either monoclonal antibodies (MAbs) or polyclonal antibodies can be used. MAbs work well in cell staining and give a low background, but usually recognize only a small epitope. Polyclonals are a mixture of antibodies directed against different epitopes of the same antigen, so that they can label a truncated protein not recognized by a monoclonal. When cultures are not pure, a double staining for NF or GFAP and target protein is necessary to identify the cell type. Antigen–antibody binding can be detected by different methods. Here we describe indirect immunofluorescence and immunoenzymatic methods such as avidin–biotin–peroxidase and alkaline phosphatase–anti-alkaline phosphatase (APAAP) techniques. Immunofluorescence gives a high resolution of subcellular structures and is the best method for double labeling, but its staining decreases with time. Immunoenzymatic methods give permanent labeling and have a higher sensitivity, but have lower resolution than fluorescent labels. Moreover, they cannot be used for double labeling when antigens have the same localization.

For immunocytochemistry, cells are grown on polylysine-coated coverslips (see above). At various intervals after treatment, remove coverslips from dishes, fix, and continue immunostaining for the target protein.

Fixation

Fixation represents a crucial step in immunocytochemistry. Effective fixation must preserve cell structure and not interfere with specific antigen–antibody binding. Different fixatives are commonly used and the right choice depends on the nature of the antigen, the antibody used, and mode of detection.

Two kinds of fixatives are generally used: organic solvents and cross-linking reagents. Organic solvents (alcohol, acetone) determine lipid extraction and cell dehydration. Cross-linking reagents, such as glutaraldehyde and paraformaldehyde, fix cells by forming intermolecular bridges. There are no general rules for choosing the most suitable fixation, but it is important to know the advantages and disadvantages of both methods. Glutaraldehyde and paraformaldehyde preserve cell structure better than organic solvent, and therefore they are commonly used for electron microscopy (EM) observations. Moreover, cross-linking reagents create a network of cell proteins that can mask the antigen. In addition, paraformaldehyde often provides fixed cells with an intense background in immunofluorescence. On the other hand, alcohol and acetone can destroy certain elements of cell architecture: by extracting lipids from plasma membrane, they permeabilize the cells, allowing access of the antibody to the antigen. Thus, they are not suggested for immunostaining of membrane proteins. We routinely fix cultured cells, for light microscopy immunocytochemistry, either with absolute methanol or 4% paraformaldehyde, depending on the antigen.

Methanol Fixation. Rinse cells once with warm PBS and then incubate with ice-cold methanol for 4 min at $-20°$. Remove methanol and let samples dry. Store at $-20°$. Cells must be rehydrated for 5 min in PBS before processing for immunocytochemistry.

Paraformaldehyde Fixation. A 4% solution of paraformaldehyde should be prepared fresh: add 800 mg to 10 ml of water and heat to 60° (no more than 65°, because formic acid may form). Add a few drops of 1 N NaOH to help dissolve paraformaldehyde, let the solution cool, and then add 10 ml of 2× PBS. Rinse cells once with warm PBS and then incubate with warm paraformaldehyde for 15 min at 37° and for 45 min at room temperature. Wash the cells three times with PBS. Fixed cells can be frozen with sucrose after treatment to preserve morphology. Incubate the cells (three times, 5 min each) with increasing concentrations (3.5, 7, 15, and 25%) of sucrose in PBS. Remove the solution and store at $-20°$. Carefully wash the cells with PBS before performing immunocytochemistry.

Pretreatment

To permeabilize cells, incubate them with 0.1% Triton X-100 in PBS for 5 min at room temperature. If the antigen is localized on plasma membrane this step must be omitted; nor may Triton X-100 be used in the following steps. In this case, cells can be permeabilized before the second labeling (NF or GFAP).

Incubate cells in 10% normal serum for 15 min, to block unspecific binding sites. The serum must be from the same species in which the secondary antibody is made.

Immunofluorescence

1. Incubate cells with primary antibody (against the antisense target protein) in PBS with 0.1% Triton X-100 (about 50 μl over each coverslip) in a humid chamber, for 1 hr at room temperature or overnight at 4°. Antibody dilutions must be tested. Generally a concentration ranging between 0.1 and 10 μg/ml is used. Incubate negative controls with nonimmune serum.

2. Wash three times in PBS–Triton for 5 min each time.

3. Incubate with an FITC-conjugated secondary antibody against the host species of the primary antibody for 30 min at room temperature. Dilutions must be titrated between 1:10 and 1:300.

4. Wash three times in PBS–Triton for 5 min each time.

5. For double labeling, incubate cells with anti-NF or anti-GFAP antibody. In this step, the antibody must be made in a different species from the primary antibody in step 1.

6. Wash three times in PBS–Triton for 5 min each time.

7. Incubate with an FITC-conjugated secondary antibody against the host species of the primary antibody for 30 min at room temperature.

8. Wash three times in PBS–Triton for 5 minutes each time.

9. Mount each coverslip on a slide with PBS–glycerol (1 : 1) and observe under a fluorescence microscope.

Immunoenzymatic Methods

1. Incubate with primary antibody as described in the preceding section.

2. Wash three times in PBS–Triton for 5 min each time.

3. Incubate with a biotinylated secondary antibody against the host species of the primary antibody for 30 min at room temperature. Dilutions must be titrated between 1 : 100 and 1 : 1000.

4. Wash three times in PBS–Triton for 5 min each time.

5. Incubate with horseradish peroxidase (HRP)-conjugated avidin (diluted 1 : 250 to 1 : 3000) for 30 min at room temperature.

6. Wash three times in PBS–Triton for 5 min each time.

7. Incubate in diaminobenzidine solution (dissolve 1 mg/ml in PBS; add 10 μl of 30% H_2O_2) for 10 min.

8. Block the reaction by washing with water.

9. For double labeling, incubate cells with anti-NF or anti-GFAP antibody. In this step, the antibody must be made in a different species from that in which the primary antibody was made in step 1.

10. Wash three times in PBS–Triton for 5 min each time.

11. Incubate cells with the link antibody from the APAAP kit (Dako, Carpinteria, CA) for 30 min at room temperature.

12. Wash three times in PBS–Triton for 5 min each time.

13. Incubate cells with alkaline phosphatase–anti-alkaline phosphatase immunocomplex (APAAP; Dako) for 30 min at room temperature.

14. Wash three times in PBS–Triton for 5 min each time.

15. Incubate cells with BCIP/NBT solution for 10 min to overnight.

16. Block the reaction by washing with water.

17. Mount each coverslip on a slide with gelatin.

Reverse Transcriptase-Polymerase Chain Reaction

Reverse transcriptase-polymerase chain reaction (RT-PCR) is a sensitive technique for detecting mRNA in biological samples. Total RNA can be extracted from cell cultures using commercial reagents, such as Trizol (GIBCO, Grand Island, NY), by following an easy protocol:

1. After treatment, rinse cells with serum-free DMEM.

2. Lyse cells, adding 1 ml of Trizol per 10 mm dish.

3. Pass the cell lysate several times through a Pasteur pipette.

4. Add 0.2 ml of chloroform, shake, and incubate for 5 min at room temperature.

5. Centrifuge for 15 min at 4° at 12,000g.

6. Transfer the upper phase to a new tube with 0.5 ml of 2-propanol, incubate for 10 min at room temperature, and centrifuge for 10 min at 4° at 12,000g.

7. Wash the pellet with 1 ml of 75% ethanol, vortex, and centrifuge for 5 min at 7500g at 4°.

8. Dissolve RNA in RNase-free water.

9. Measure RNA concentration by absorbance at 260 nm in a spectrophotometer.

10. Check RNA integrity by running an agarose gel (see below).

Identical amounts of RNA are then reverse transcribed into cDNA. Several kits for reverse transcription, with their own protocols, are available. It is convenient to use an oligo(dT) primer, preparing total cDNA, which can be used to study expression of several mRNAs. Alternatively, a gene-specific primer can be used, transcribing an mRNA only.

cDNA can now be amplified by PCR. PCR protocols should be optimized for every gene assayed. Primers must be carefully selected. Several computer programs can help in finding the more suitable pair of primers. Primers should generate a cDNA fragment that can be resolved on a 1.5% agarose gel (150–500 bp long). Self-hybridization and interprimer hybridization, which decrease the effective concentration of primers, must be avoided. It is absolutely necessary that every primer be checked for similarities with other sequences present in databases.

Primers are incubated with samples in a mixture containing DNA polymerase and $MgCl_2$ in a thermal cycler. Cycling conditions must be optimized for each amplified gene sequence by testing a positive control. Standard PCR consists of an initial denaturation step (94–95° for 1–3 min), followed by a number of cycles (25–40) of amplification: denaturation (95–96° for 30–60 sec), annealing (primer-specific temperature for 60 sec), and extension (72–74° for 60–70 sec), with a final elongation step (72° for 7 min). The optimal annealing and extension temperature should be determined empirically by testing increasing temperatures to obtain the maximum of specificity and amount of product. For quantitative PCR with a nonradioactive label, incorporate digoxigenin-11-dUTP (DIG) during the PCR. Run a 2% agarose gel with an aliquot of the PCR mixture in Tris–borate–EDTA buffer (0.09 M Tris–borate, 0.002 M EDTA, pH 8). Incubate the gel in 2 M NaOH, 5 M NaCl for 30 min. Transfer cDNA to a nylon membrane by blotting for 18–24 hr in 10× SSC (87.7 g of NaCl, 44.1 g of sodium citrate, per liter, pH 7.0). Rinse the membrane in washing buffer (0.1 M maleic acid, 0.15 M NaCl, 0.3% Tween 20) and incubate for 30 min in blocking buffer (1% blocking reagent in washing buffer) at room temperature. Incubate with anti-digoxigenin IgG conjugated to alkaline phosphatase diluted in blocking buffer. Wash the membrane and incubate in CSPD (Boehringer Mannheim, Indianpolis, IN), the chemiluminescent substrate for alkaline phosphatase, 1 : 100 in 0.1 M Tris-HCl, 0.1 M NaCl, pH 9.5, for 5 min. Seal membranes in a hybridation bag and incubate for 5–10 min at 37° to enhance the luminescent reaction. Expose the membrane to X-ray film for 15–30 min at room temperature. Measure the signal corresponding to the target mRNA with a densitometer. Normalize the amount of mRNA by comparison with levels of an independent mRNA, such as actin or GAPDH.

Controls for Biological Effects

After antisense treatment, modifications in cell viability and morphology must be examined. Controls must be included in order to demonstrate that the biological effects are a consequence of specific gene inhibition. In fact, too-high concentrations of antisense can cause nonspecific cell toxicity and death. Moreover, modifications in cell ultrastructure can be observed as consequence of antisense degradation products.[15] Ultrastructural modification can be observed by transmission electron microscope (TEM). To detect cell toxicity, cell viability should be examined. The easiest method is trypan blue staining. By using an image analyzer associated with immunocytochemistry, the percentage of dead cells within a population can be evaluated.

Cell Counting

Images are captured by a charge coupled device (CCD) camera connected to a PC-hosted frame grabber and saved on a hard disk. No particular CCD sensitivity or linearity is required. Image size should be set to the largest possible format (512 × 512 pixels or higher). Choose the lowest microscope magnification assuring the clear recognition of single cells. Usually, a ×10 objective fits this purpose. Set the microscope lamp light intensity to a level that does not saturate the signal. Background pixels should never exceed the value of 250.

Cells are sampled from 10 different fields of each of 10 different slides.

---

[15] R. Wagner, *Nature* (*London*) **372,** 333 (1994).

Images should be preprocessed to minimize positional and random noise (apply background subtraction and Despeckle filter). Cell counting is based on the ratio between the total cell area (assuming cells are in a monolayer) and the mean area of the cell. Because the interest is in the relative frequency or percentage of neurons and glial cells, the absolute number of cells in each slide or field is irrelevant. The mean cell area is easily obtained by contouring a number of well-defined profiles of neurons and glial cells. Process a number of cells sufficient to make the standard deviation of the mean cell area less than 10% of the same mean. Differential segmentation of neurons from glial cells is based on their different staining intensity: fuchsin-stained neurons are lighter than Mayer's hematoxylin-stained glial cells. For each slide, sum the number of neurons and glial cells found in the different fields and divide the total by the mean cell area. Note that the number of cells should be large, on the order of thousands. From the number of neurons and glial cells, calculate the percentage of neurons for each slide.

To apply statistical tests (i.e., Student t test) percentages should be transformed to arcsine values. This transform is required to correct in part the skewness of percentage distributions, which hinders the application of statistical methods that assume data are normally distributed.

Reliability of the method may be assessed by correlating cell counts obtained as described above with cell counts performed manually by an expert operator. A good agreement between automated and subjective data is demonstrated when the regression slope and the correlation coefficient are both between 0.9 and 1.0. However, to avoid any influence between the two methods, data should be evaluated only at the end of the session.

Electron Microscopy

1. Wash monolayer cells once with medium and once with PBS.
2. Remove PBS and replace with 1% gluteraldehyde in 100 mM sodium cacodylate (pH 7.2) for 30 min at room temperature.
3. Wash cells (3 × 10 min) in 100 mM sodium cacodylate (pH 7.2).
4. Dry samples and put them in a humid chamber.
5. Postfix in 0.66 M osmium tetroxide (OsO$_4$) in 0.1 M sodium phosphate buffer (1 hr at room temperature). *Caution:* Be careful with osmium tetroxide; even the fumes are dangerous.
6. Wash in 0.1 M phosphate buffer (3 × 5 min).
7. Dehydrate at 4° in 50% ethanol (2 × 5 min), in 70% ethanol (2 × 10 min), and in 95% ethanol (1 × 5 min); transfer the samples at room temperature and incubate in 95% ethanol (1 × 5 min) and in 100% ethanol (3 × 10 min, and then overnight at 4°).

8. Wash twice in 100% ethanol at room temperature.

9. Remove ethanol and incubate in methyl cyanide acetonitrile (3 × 15 min).

10. Incubate the samples in a mix of two-thirds acetonitrile and one-third Epon resin for 20 min.

11. Incubate in acetonitrile and Epon (1 : 1) for 20 min and then in one-third acetonitrile plus two-thirds Epon resin for 20 min.

12. Replace with fresh Epon resin for 20 min in a slow-turning wheel.

An alternative method is to substitute propylene oxide (2 × 5 min) for acetonitrile, but it is necessary to quickly remove the cells from culture dishes; otherwise propylene oxide will dissolve the plastic and the cell layer will float off.

13. Incubate dishes at 60° overnight for polymerization. Heat silica gel in an oven in order to achieve low humidity. After one night in the oven, the resin should be hard enough to trim and section.

14. Remove Epon-embedded cells from the plastic surface of the dish.

Biological materials can be examined by TEM, after cutting ultrathin sections (60 nm) with an ultramicrotome, using glass or diamond knives. Sections are floated on water and transferred to specimen support grids (copper grids).

Positive Staining with Uranyl Acetate/Reynolds' Lead Citrate

To improve membrane contrast, sections are contrasted with uranyl acetate and lead citrate. It is known that uranyl ions react strongly with phosphate and amino groups, so that nucleic acids and certain proteins are highly stained. Lead ions bind negatively charged components such as hydroxyl groups and osmium-reacted areas.

Uranyl Acetate Staining

1. Dilute a saturated solution of uranyl acetate 1 : 1 with double-distilled water.

2. Put a few drops of this solution on a clean square of Parafilm in a petri dish.

3. Float each grid with mounted sections (section side down) on a drop of the uranyl acetate for 5–15 min.

4. Rinse each grid in boiled distilled water and keep in a covered petri dish for subsequent staining with lead.

Uranyl salts are photolabile and must be protected from bright light.

Caution: Salts of uranium are toxic and may contain radioactive isotopes.

Take precaution to avoid breathing the dust when weighing the powder: weigh the powder in a fume hood or wear a dust mask or filter.

Preparation of Reynolds' Lead Citrate

1. In a clean 50-ml volumetric flask combine 1.33 g of lead nitrate, 1.76 g of sodium citrate, and 30 ml of CO_2-free double-distilled water.

2. Shake the solution for several minutes, and then six or seven times over a 30-min period. A milky white suspension of lead stain with no large particles will generate. If not, continue shaking until the particles are dissociated.

3. Add 6 ml of 1 N NaOH freshly prepared in CO_2-free water and swirl to mix the solutions. The milky solution should turn clear; if not, add a few more drops of NaOH, up to 8 ml. Adjust to pH 12.0 $\pm$ 0.1 by adding NaOH. A proper pH is extremely important because otherwise poor staining or precipitation will occur.

4. Add CO_2-free double-distilled water to the volumetric flask to bring the solution to a final volume of 50 ml.

5. Store the solution at 4° in a tightly closed flask; this prolongs shelf life (up to 6 months).

Caution: Lead salts are extremely toxic and may contain radioactive isotopes. Take care in weighing the powder: use gloves, a fume hood, or a filter over the face and mouth.

Use of Reynolds' Lead Citrate

Before staining the grids with uranyl acetate and then with lead citrate, prepare a small quantity of fresh NaOH pellets and place at one side of the covered petri dish to produce a CO_2-free atmosphere.

1. Centrifuge lead stain before use at 15,000g for 5 min.

2. Place a few drops of this solution on a clean square of Parafilm in a petri dish.

3. Float each grid with mounted sections (section side down) on a drop of lead citrate for 30 sec to 15 min.

4. Rinse each grid in double-distilled water and dry on a piece of Whatman (Clifton, NJ) 3MM paper.

Sections are now ready to be examined by transmission electron microscopy.

[19] Pharmacokinetic Properties of Oligonucleotides in Brain

By WOLFGANG SOMMER, MATTHEW O. HEBB, and MARKUS HEILIG

Introduction

Antisense oligonucleotides can be used to suppress translation and protein expression in neurons and glial cells and have become a popular technology with which to study gene function in the brain. While researchers continue to report the successful use of antisense oligonucleotides in the brain, few attempts have been made to systematically study the pharmacokinetics of these compounds within brain tissue. The ability of oligonucleotides to enter a cell and specifically hybridize to the target mRNA is a key process that is necessary for suppression of gene expression. The pharmacokinetic profile of a particular antisense oligonucleotide will confer the selectivity, specificity, and localization of its antisense effects. Therefore, knowing the fate of the oligonucleotides after intracerebral administration is essential for both an effective experimental design and accurate interpretation of the results.

Important pharmacokinetic determinants of antisense efficacy in the brain include (1) the accurate delivery of the oligonucleotides to the desired intracerebral locus (loci), (2) the extent to which the oligonucleotides penetrate the targeted cell population, and (3) the resistance of the oligonucleotides to nuclease-mediated degradation in the cellular environment. When applying antisense technology to a specific physiological system, the determination of the temporal sequence of oligonucleotide distribution, uptake, and degradation is essential for predicting the period of maximal functional effect. While human antisense therapy is still in its infancy, these parameters are also crucial to the design of any such pharmaceuticals. Thorough evaluation of oligonucleotide pharmacokinetics in the central nervous system (CNS) will provide answers to questions such as the following: What areas of the brain are accessible to oligonucleotides? What metabolites are produced on degradation, and what are their effects? Are oligonucleotides stored within the brain? Which kind of cells exhibit the most efficient uptake of oligonucleotides? Does the method of delivery influence oligonucleotide pharmacokinetics?

The focus of this chapter is on methodologies that are effective for studying pharmacokinetics of oligonucleotides within brain tissue. These techniques are designed to determine the fate of centrally administered

METHODS IN ENZYMOLOGY, VOL. 314

oligonucleotides and to establish methodological parameters that maximize the cellular uptake and physiological resistance of these molecules. However, there are several variables that can alter the effects of antisense oligonucleotides *in vivo*. A comprehensive review of basic oligonucleotide pharmacology is beyond the scope of this chapter and can be found elsewhere.[1]

Use of Phosphorothioate-Modified Oligonucleotides in Brain

In the brain, short polymers of deoxyribonucleic acid are regarded as foreign substances or cellular debris, and are readily digested by endogenous nucleases. Unmodified, phosphodiester oligonucleotides are rapidly degraded *in vivo* by endo- and exonucleases. A common modification that is used to increase oligonucleotide resistance is phosphorothioate substitution, in which an oxygen atom of the phosphate linkage group is replaced with a sulfur atom. Although the molecular charge of phosphodiester oligonucleotides is conserved by phosphorothioate modification, the sulfur atoms hinder nuclease activity, possibly by increasing the dissociation time of the enzyme–oligonucleotide complex (reviewed in Stein and Cohen, 1989[2]). The sulfur atom is, however, much larger and more polarizable than the oxygen atom, increasing the ionic affinity of sulfur-modified oligonucleotides for other charged cellular components and, therefore, contributing to their nonspecific effects and impeding their internalization. Phosphorothioate oligonucleotides have been shown to bind to several cell surface proteins, and, in the brain, to various heparin-binding factors, including basic fibroblast growth factor (bFGF).[3] Thus, phosphorothioate oligonucleotides tend to have a decreased cellular uptake, increased potential for toxicity, and altered bioavailability.

The use of partially modified or end-capped oligonucleotides, instead of fully modified oligonucleotides, can reduce the toxic potential of these compounds. These chimeric molecules have a reduced number of phosphate groups that possess the phosphorothioate substitution, thus reducing their sulfur content while maintaining a portion of the nuclease resistance of fully modified molecules. It has been reported that repeated infusions of phosphorothioate oligonucleotides into the amygdala produced marked

[1] W. Sommer, D. W. Pfaff and S. Ogawa, *in* "Modulating Gene Expression by Antisense Oligonucleotides to Understand Neural Functioning" (M. M. McCarthy, ed.), pp. 9–26. Kluwer Academic Publishers, Norwell, 1998.

[2] C. A. Stein and J. S. Cohen, *in* "Antisense Inhibitors of Gene Expression" (J. S. Cohen, ed.), pp. 97–117, 1989.

[3] M. A. Guvakova, L. A. Yakubov, I. Vlodavsky, J. L. Tonkinson and C. A. Stein, *J. Biol. Chem.* **270,** 2620 (1995).

cellular damage, whereas the same treatment with partially modified oligo-nucleotides did not.[4] As the degree of nuclease resistance generally corre-lates with the extent of phosphorothioate modification, partially modified oligonucleotides may differ in terms of pharmacodynamics from fully modi-fied molecules. For example, many groups have used the same antisense sequence in the striatum to target the mRNA of the immediate-early gene c-*fos*. The optimal period of suppression for this oligonucleotide varied between 1 hr (with end-capped oligonucleotides) and 10 hr (with fully modified oligonucleotides), depending on the degree of phosphorothioate substitution.[5–9]

Determination of Oligonucleotide Pharmacokinetics: Some Pitfalls

The fate of intracerebrally administered oligonucleotides can be moni-tored by tagging the molecules with various markers that can facilitate their identification, either in brain sections or by gel electrophoresis. While most radioactive isotopes that are used to label oligonucleotides (^{32}P, ^{33}P, or ^{35}S) cause only minor changes in the chemical structure of the nucleic acids, the addition of fluorescence or biotin residues, usually done via a carbon spacer, may have greater impact. However, it has been shown that oligonu-cleotides that were labeled at various positions within the molecule, using fluorescent markers, biotin, digoxigenin, 5-bromodeoxyuridine, or radioac-tive markers, generally had similar distribution patterns in the brain.[10–14]

Previous studies have used tissue extracts to examine the stability of labeled oligonucleotides within the brain. As described above, catabolic enzymes that are present within cellular environments can reduce the integ-rity of the labeled oligonucleotide, possibly liberating free marker mole-cules. Because of their increased resistance to nuclease-mediated degrada-

[4] B. J. Chiasson, M. G. Hong and H. A. Robertson, *Mol. Brain Res.* **57,** 248 (1998).

[5] B. J. Chiasson, J. N. Armstrong, M. L. Hooper, P. R. Murphy and H. A. Robertson, *Cell. Mol. Neurobiol.* **14,** 507 (1994).

[6] M. O. Hebb and H. A. Robertson, *Mol. Brain Res.* **47,** 223 (1997).

[7] M. L. Hooper, B. J. Chiasson and H. A. Robertson, *Neuroscience* **63,** 917 (1994).

[8] W. Sommer, B. Bjelke, D. Ganten and K. Fuxe, *Neuroreport* **5,** 277 (1993).

[9] W. Sommer, R. Rimondini, W. O'Connor, A. C. Hansson, U. Ungerstedt and K. Fuxe, *Proc. Natl. Acad. Sci. U.S.A.* **93,** 14134 (1996).

[10] F. Yee, H. Ericson, D. J. Reis and C. Wahlestedt, *Cell. Mol. Neurobiol.* **14,** 475 (1994).

[11] W. Sommer, X. Cui, B. Erdmann *et al., Antisense Nucleic Acid Res. Dev.* **8,** 75 (1998).

[12] K. H. Schlingensiepen and M. Heilig, *in* "Antisense—From Technology to Therapy" (R. Schlingensiepen, W. Brysch, and K. H. Schlingensiepen, eds.), pp. 186–223. Blackwell Science, Berlin, 1997.

[13] R. R. Sakai, L. Y. Ma, D. M. Zhang, B. S. McEwen and S. J. Fluharty, *Neuroendocrinology* **64,** 425 (1996).

[14] S. Ogawa, H. E. Brown, H. J. Okano and D. W. Pfaff, *Regul. Pept.* **59,** 143 (1995).

tion, phosphorothioate oligonucleotides are considered to be more stable than unmodified oligonucleotides. In fact, it was demonstrated that intact, fully modified phosphorothioate oligonucleotides, labeled with either radioactive or fluorescence markers, could be recovered from brain tissue extracts after as long as 2 days.[3,11,15] In contrast, unmodified oligonucleotides, labeled with biotin or digoxigenin, were partially degraded within 4 hr after their administration into the brain.[10]

The method of tissue preparation and the means by which the oligonucleotide label is visualized (i.e., autoradiography, fluorescence microscopy) are potential sources of variability that can alter apparent the results of pharmacokinetic studies. There are both benefits and drawbacks to using freshly frozen or fixed brain tissue. For example, the use of radioactive markers allows for rapid tissue processing because sections of freshly frozen brain can be directly exposed to autoradiographic film for localization of the oligonucleotide signal. However, this method produces limited resolution of intracellular localization, particularly near the administration site, where the oligonucleotide signal is intense. To examine intracellular localization, fluorescence labeling tends to be more suitable. The appearance of fluorescence signal was similar in brains that were directly frozen after decapitation or prepared after transcardial perfusion with fixative. This is an important issue because, when using fixed tissue, potential postmortem changes in the distribution of the oligonucleotide (i.e., by liquid-phase diffusion) must be excluded by the experimental protocol. We have found that stable fixation that was useful for histology, including immunohistochemistry, could be achieved by the addition of glutaraldehyde to 4% paraformaldehyde fixative, and by performing all necessary incubations and washing steps at 4°. Thus, with appropriate precautions, common oligonucleotide-labeling techniques such as 5' or 3' end labeling with radioactive (i.e., ^{32}P, ^{33}P, ^{35}S) or fluorescent [i.e., fluorescein isothiocyanate (FITC), tetramethyl-6-carboxyrhodamine (TAMRA)] markers can be useful for studying the pharmacokinetics of oligonucleotides within the brain.

Biodistribution and Pharmacokinetics of Oligonucleotides in Striatum

Stability of Oligonucleotides after Infusion into Striatum

We were interested in determining the fate of end-capped, partially modified, and fully modified phosphorothioate oligonucleotides after their infusion into the striatum of rats. Oligonucleotides are labeled at the 3'

---

[15] P. K. Liu, A. Salminen, Y. Y. He *et al.*, *Ann. Neurol.* **36,** 566 (1994).

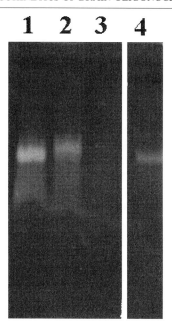

FIG. 1. Recovery of 3′-FITC-labeled phosphorothioate oligonucleotides from striatal tissue 20 min (lane 1), 4 hr (lane 2), and 24 hr (lane 3) after the infusion. Animals were injected with 1 nmol of oligonucleotide into the striatum. Tissue extracts from striatum were analyzed on a 12% polyacrylamide gel. A single band of the expected size is clearly seen at the 20-min and 4-hr time points, demonstrating that intact oligonucleotides could be recovered. The absence of the signal at 24 hr is most likely due to lack of sensitivity of detection method. Lane 4, markers.

terminal with FITC and/or at the 5′ terminal with either $[^{35}S]ATP\gamma S$ or $[\alpha-^{33}P]dATP$ and purified by either gel electrophoresis or high-performance liquid chromatography (HPLC). Radiolabeled oligonucleotides (5×10^5 cpm) are added to a solution of unlabeled oligonucleotides (1.0 nmol) and injected into the striatum (as described below). Tissue is obtained at various time points after the infusion. Radioactive, partially and fully modified oligonucleotides are recovered 24 hr after their infusion into the striatum. The presence of intact, end-capped oligonucleotides can demonstrated up to 4 hr after infusion. Similar results are obtained with FITC-labeled oligonucleotides (Fig. 1). The intensity of the single band of fluorescent or radioactive signal that is obtained by polyacrylamide gel electrophoresis (PAGE) appears to decrease with time. In fact, only 20 min after the infusion of fully modified, radiolabeled phosphorothioate oligonucleotides, a large portion of the injected radioactivity has disappeared from the injection site. This phenomenon may reflect a rapid clearance of the oligonucleotides from the parenchyma into the cerebrospinal fluid or the transport of oligonucleotides to other brain regions (discussed below).

Distribution of Oligonucleotides after Infusion into Striatum

As described above, both FITC and radiolabeled oligonucleotides are injected into the midrostral region of the striatum [coordinates from bregma: anteroposterior (AP), + 1.0 mm; dorsoventral (DV), −6.0 mm; lateral (LAT), ±3.0 mm].[16] Within 1 hr of infusion, the oligonucleotide signal is observed throughout the striatum, apparently restricted from the cortex by the corpus callosum (Fig. 2A and B). Both the temporal and spatial patterns of distribution are similar for either fluorescently or radioactively labeled oligonucleotides, regardless of the degree of phosphorothioate modification. Oligonucleotide distribution is found to extend throughout the bulk of the rostrocaudal axis, disappearing at the level of the central globus pallidus. By 24 hr postinfusion, the oligonucleotide signal is restricted to the hemisphere into which it was infused, demonstrating that generalized diffusion throughout the brain does not occur. Unexpectedly, there is considerable signal intensity emanating from two major projection nuclei of the striatum, the globus pallidus (Fig. 2C and D) and the substantia nigra.[11,17] Both fluorescent and radioactive oligonucleotide signals are observed with the same temporal latency in projection nuclei, suggesting that the intact molecule is being transported. The oligonucleotide signal can be detected in the globus pallidus as early as 30 min after infusion into the striatum, whereas the signal in the substantia nigra is typically not seen until 4 hr postinfusion. This difference in transport times is consistent with the relative distances from the striatum to the globus pallidus and to the substantia nigra (Fig. 2). These findings have significant implications concerning the behavioral effects of oligonucleotides that are infused within specific brain regions. Interpretation of experimental results must take into account the possibility of oligonucleotide transport and physiological consequences in alternate regions of the brain.

Uptake and Intracellular Localization of Oligonucleotides in Striatum

Phosphorothioate oligonucleotides, like their phosphodiester analogs, are large anionic molecules that do not readily pass through hydrophobic cellular membranes. Internalization of phosphorothioate oligonucleotides

---

[16] G. Paxinos and C. Watson, "The Rat Brain in Stereotaxic Coordinates." Academic Press, San Diego, California, 1997.

[17] M. O. Hebb, Experimental Investigations into the Pathophysiology of Huntington's Disease. Ph.D. thesis. Dalhousie University, Halifax, Canada, 1998.

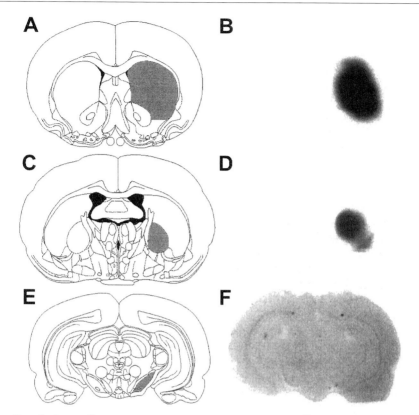

Fig. 2. Autoradiographs demonstrating the distribution of ^{35}S 5'-radiolabeled "end-capped" phosphorothioate-modified oligonucleotides 1 hr after the infusion (1 μl) into the striatum. Extensive diffusion is evident throughout the caudate putamen (B) and a robust signal emanates from the globus pallidus (D). No radioactivity was detected in the substantia nigra at this time point (F). The section in (F) has been overexposed to reveal the tissue morphology. (A), (C), and (E) show schematically the region of interest in (B), (D), and (F), respectively. The areas shaded in gray correspond to the regions shown on the right-hand side. Diagrams have been modified from the atlas of Paxinos and Watson.[16]

is believed to occur through active pinocytotic mechanisms, although there have also been reports of receptor-mediated oligonucleotide transport into cells.[18,19] We have found that 30 min after infusion into the striatum, phosphorothioate oligonucleotides are taken up and accumulate within a large

[18] S. L. Loke, C. A. Stein, X. H. Zhang et al., Proc. Natl. Acad. Sci. U.S.A. **86,** 3474 (1989).
[19] L. A. Yakubov, E. A. Deeva, V. F. Zarytova et al., Proc. Natl. Acad. Sci. U.S.A. **86,** 6454 (1989).

number of medium-sized neurons and in cells with a perivascular location (pericytes). At this early time point (30 min), intense oligonucleotide signal emanates from the cytoplasmic, nuclear, and dendritic regions of neurons (Fig. 3). There is also a diffuse signal in the interstitial space that disappears within a few hours after infusion. At 6 hr postinfusion, the diffuse cytoplasmic and nuclear labeling that was observed at early time points has changed to a punctate cytoplasmic signal. This punctate pattern is more pronounced at 24 hr postinfusion. Electron microscopy confirms that the punctate oligonucleotide signal is truly of an intracellular origin and that the oligonucleotides have been localized to cytoplasmic vesicles.[9]

Uptake of Oligonucleotides: Neuronal Versus Glial

In addition to modulating gene expression in neurons, researchers continue to report the successful use of antisense oligonucleotides to suppress gene expression in glial cells. However, investigations into oligonucleotide distribution and pharmacokinetics in glial cells have been scarce. We have demonstrated that FITC-labeled oligonucleotides that are infused into the striatum are rapidly taken up by neurons, but not glial cells. However, at 24 hr postinfusion a small but substantial number of astroglia and microglia exhibit punctate, cytoplasmic oligonucleotide labeling (Fig. 4).[11] While the punctate appearance of these signals suggests that the oligonucleotides are encapsulated in intracellular vesicles, the physiological effect on gene expression in this system remains unclear. Thus, the interaction between neuronal and glial activity in the CNS represents another means by which gene expression in either cell types may be altered. Determination of the effects of antisense oligonucleotides in each specific experimental system is important to elucidate changes that are due to direct, antisense-mediated suppression, or indirectly through alterations in neighboring cellular physiology.

Summary

The application of antisense oligonucleotides is becoming a popular technique with which to study the function of genes in the brain. These compounds can serve as a useful alternative when conventional pharmacological tools lack adequate specificity or are unavailable. Theoretically, the versatility of this technology is enormous. However, as with any genetic methodology, theoretical success is hindered by numerous practical difficulties. In this chapter we address several potential pitfalls that arise when using antisense oligonucleotides. Importantly, these include the biodistribution of the molecules within the region of delivery, as well as their potential

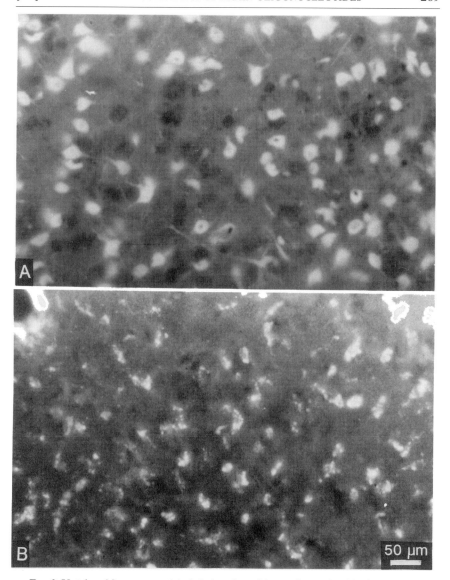

FIG. 3. Uptake of fluorescence-labeled phosphorothioate oligonucleotides is shown 30 min (A) and 6 hr (B) after a single infusion (1 μl) into the striatum. The diffuse staining of cytoplasm, nuclei of nerve cell bodies, and their dendrites early after the infusion changes to a granular appearance of the fluorescence at the 6-hr time point. [From Sommer *et al.* (1996).[9]]

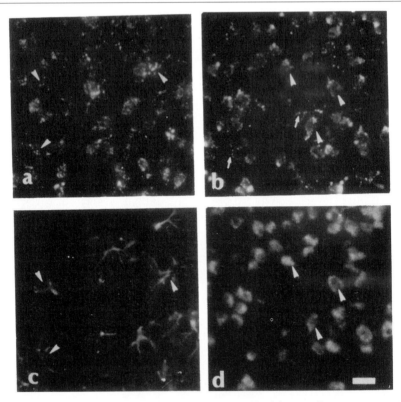

Fɪɢ. 4. Simultaneous localization of astroglia-specific (c) as well as neuron-specific (d) antigens and FITC-labeled phosphorothioate oligonucleotide (a and b) 24 hr after the infusion (1 nmol/1 μl) into the striatum. Examples of double-labeled cells are marked by arrowheads. Note that the number of double-labeled neurons greatly exceeds the number of astrocytes. The astrocytes were identified by staining with a rabbit antiserum to glial fibrillary acidic protein (GFAP), followed by detection with a rhodamine-labeled anti-rabbit antibody (c). Neurons were identified by staining with the mouse monoclonal antibody to neuron-specific nuclear protein, NeuN, which was detected with a rhodamine-conjugated anti-mouse antibody (d). Bar: 20 μm. [Adapted from Sommer et al. (1998).[11]]

transport to one or more projection nuclei. Also, determination of oligonucleotide pharmacokinetics is essential for effective experimental design and assessing the interval of maximal antisense effects. Furthermore, the stability of the oligonucleotides, which can be altered by biochemical modification (i.e., phosphorothioate substitution), may impact on the temporal development and duration of antisense effects. Finally, it may be important to assess the neuronal–glial interactions in a particular region and the effects of antisense oligonucleotides on gene expression in both cell types.

Experimental Protocols

Preparation of Oligonucleotides for Central Nervous System Infusion

Oligonucleotides are synthesized with various degrees of phosphoro-thioate modification, i.e., no substitutions (phosphodiester oligonucleo-tides), modification solely at the terminal nucleotides ("end-capped" oligo-nucleotides), substitution at every second phosphate group (partially modified oligonucleotides), or modification at all phosphate groups (fully modified oligonucleotides). For labeling with FITC or TAMRA, oligonucle-otides are synthesized with either a 5′ or 3′ amino-linked six-carbon spacer. Unincorporated dye is removed by gel filtration, followed by HPLC. Cus-tom synthesis of all oligodeoxynucleotides is performed by Biotez (Berlin, Germany) or Genosys (The Woodlands, TX).

5′ Radiolabeling of Oligonucleotides

1. Add 0.5 nmol of cold oligonucleotide to 5.0 μl of [^{35}S]ATPγS (10 mCi/ml, 600 Ci/mmol; Amersham, Arlington Heights, IL) in a 30-μl reac-tion volume, together with buffer and T4 polynucleotide kinase (according to the supplier protocol).

2. Incubate at 37° for 1–3 hr.

Note: When using phosphorothioate oligonucleotides, we have found that extended incubation times (up to 3 hr at 37°) results in a substantially greater proportion of labeled oligonucleotides than is obtained with shorter incubation times.

3. Oligonucleotide purification: Load the entire reaction volume on a 12% nondenaturing polyacrylamide (vertical) minigel and isolate the la-beled oligonucleotides by electrophoresis. A migration distance of approxi-mately 3 cm into the gel is usually sufficient for oligonucleotide separation. Bromphenol blue is useful as a tracking dye in the samples because, in a 12% polyacrylamide gel, it comigrates with 15-base oligonucleotides.

4. Expose the gel to X-ray film for 2–5 hr. Excise the desired band and elute the oligonucleotides in 400 μl of sterile water at 60° for 10–16 hr.

5. Transfer the solution to a fresh Eppendorf tube, using 1.0 μl to measure the amount of incorporated radioactivity. Store at −20° until use. Typically, this protocol yields approximately 12×10^6 cpm of labeled phosphorothioate oligonucleotides per reaction.

Preparation of Radioactively Labeled Oligonucleotides for Infusion into Brain

1. Calculate 5×10^5 cpm of radiolabeled oligonucleotides per infusion; desiccate the respective aliquots under vacuum centrifugation.

2. Resuspend the labeled oligonucleotides in the appropriate infusion vehicle [i.e., artificial cerebrospinal fluid (CSF), 0.9% saline, Ringer acetate] containing unlabeled oligonucleotides at the concentration that is typically used in functional studies.

3. Inject the oligonucleotides into the desired brain region according to the protocol used for the functional experiments. Spare an aliquot of the injected solution for subsequent polyacrylamide gel electrophoresis and standardization of recovered radioactivity from tissue extracts.

Infusion of Antisense Oligonucleotides into Freely Moving Animals

Required equipment (when using rats as subjects): surgical stereotaxic apparatus (i.e., Kopf), stainless steel guide cannula (23 gauge), stainless steel stylet (25 gauge), stainless steel infusion cannula (25 gauge), polyethylene tubing (minimum, 1 m in length), CMA/100 microinjection pump (Carnegie-Medicin), and 10-μl microsyringe (model 801; Hamilton, Reno, NV)

1. Cannula implantation: Using standard procedures for stereotaxic surgery, implant and secure a guide cannula into the desired cerebral locus. Maintain cannula patency during the recovery period by inserting a stylet through the guide cannula. To ensure that the stylet does not become dislodged, it is convenient to use a guide cannula with a threaded end, into which screw-topped stylets or infusion cannulas can be firmly secured.

2. Animal care and handling: Allow a minimum of 1 week for recovery before oligonucleotide infusion, making certain to handle the animals regularly during the recovery period. Inconsistent handling can produce variability in animal behavior and stress responses. It is especially important to habituate the animals to brief periods of physical restraint that are required during the insertion of the infusion cannula. Ensure that cannula caps screw on easily and remain secure.

3. Preparation for infusion: Attach the infusion cannula to one end of the polyethylene tubing (at least 1 m in length). Rinse the infusion cannula, tubing, and microsyringe with sterile water. Fill the microsyringe with sterile water and then, after filling the cannula and tubing, attach the free end of the tubing to the microsyringe. Care must be taken to ensure that no air has been pulled into the tubing or microsyringe. Using the plunger of the microsyringe, expel a volume of water from the cannula that is approximately 0.3 μl greater than the volume of oligonucleotide solution to be infused. Pull up the equivalent of 0.3 μl of air, followed by the appropriate volume of oligonucleotides. This should leave a 1- to 2-mm air bubble between the oligonucleotide solution and the tubing water. (*Note:* Initially securing the microsyringe into the injection pump greatly facilitates loading

the of the cannula.) Before inserting the cannula into the brain, advance the microsyringe plunger by 0.5–1 μl to force any inadvertent air bubbles from the tip of the cannula. Ensure that the oligonucleotide solution is being extruded efficiently by the microsyringe. To assess the delivery efficiency of the system, it may be valuable to measure a test volume of solution with a pipette, prior to oligonucleotide infusion.

4. While gently holding the animal, remove the stylet and insert the infusion cannula through the guide implanted in the skull. Secure the cannula in place (we use self-adapted dust caps that screw into place). Infuse the oligonucleotide solution at 0.25 μl/min for infusions into brain parenchyma and at 10.0 μl/min for intracerebroventricular (icv) infusions. During the infusion period, observe the animals for signs of pathology and ensure that the tubing does not become tangled. After infusion, leave the cannula in place for an additional 2 min to allow oligonucleotide diffusion and to minimize the amount of solution drawn up into the guide cannula on retraction. Remove the infusion cannula and replace the stylet.

5. Harvest brain tissue 1–12 hr after oligonucleotide infusion and analyze for antisense effects and oligonucleotide distribution.

Recovery of Injected Oligonucleotides from Brain Tissue

1. Animals should be swiftly decapitated and the desired brain regions excised, snap frozen in liquid nitrogen, and stored at $-70°$ until use.

2. Homogenize the tissue in 10 vol (w/v) of extraction buffer [50 mM Tris (pH 8.8), 1 mM EDTA, 0.5% Tween 20, proteinase K (0.5 mg/ml)] by intermittent gentle agitation over a 3-hr period at 50°. Initially dissociate the tissue by aspirating the suspension several times through a pipette tip.

3. Extract the lysate three times with 1 volume of phenol–chloroform–isoamyl alcohol (25/24/1, v/v; pH 8.0).

4. Precipitate the supernatant with 3 volumes of ethanol in the presence of 0.3 M NaCl, 10 mM MgCl$_2$, and 10 μg of yeast tRNA.

5. Resuspend the precipitate in 50 μl of sterile water.

6. Load aliquots of an appropriate volume onto a 12% nondenaturing polyacrylamide gel.

7. Blot the gel on filter paper (3MM; Whatman, Clifton, NJ), dry it, and expose it to X-ray film for 3–10 days, depending on signal intensity.

Histological Evaluation of Fluorescence-Labeled Oligonucleotides

1. To establish temporal effects on oligonucleotide distribution, animals should be decapitated at various time points following delivery. Following extraction, the brains should be frozen on dry ice and stored at $-70°$ until use.

2. Using a freezing microtome, cut 14 μm-thick sections through the regions of interest and mount on uncoated slides (Superfrost Plus; Boehringer Mannheim, Indianapolis, IN).

3. For fluorescence-labeled oligonucleotides, apply glass coverslips with a solution of glycerol and phosphate-buffered saline (PBS; 3:1, v/v). Sections can be immediately viewed by standard fluorescence microscopy.

4. For radioactively labeled slides, allow the sections to dry at room temperature, then expose to autoradiographic film for 2 days (Hyperfilm-3H; Amersham).

Note: It is our experience that, for visualizing fluorescently labeled oligonucleotides, animals may alternatively be transcardially perfused with fixative to yield results of similar quality (see below).

Colocalization of Immunolabeled Cellular Antigens with Fluorescence-Labeled Oligonucleotides in Brain

1. Tissue preparation: After oligonucleotide infusion and recovery, animals are given a terminal dose of anesthetic (i.e., sodium pentobarbital, >100 mg/kg) and transcardially perfused through the left ventricle with 60 ml of saline followed by 120 ml of 4% paraformaldehyde–0.25% glutaraldehyde (v/v) in 0.1 M phosphate-buffered saline (pH 7.4). Brains should be postfixed in the same fixative for 2 hr, immersed in 10% phosphate-buffered sucrose until the tissue sinks (1–2 days), and then frozen. For the use of unfrozen tissue in these protocols, perfused brains should be postfixed overnight, then transferred to a 0.12 M phosphate buffer solution containing sodium azide until further processing.

2. Tissue sectioning: Cryoprotected brains can be cut into 14-μm-thick coronal sections, using a freezing microtome or cryostat. Sections should be thaw-mounted on uncoated slides (Superfrost Plus). Thicker ($\sim$50 μm) sections may be obtained from unfrozen, fixed brains that are cut on a Vibratome and collected into a 0.12 M phosphate buffer solution containing sodium azide. It is recommended that the time interval between sectioning the unfrozen brains and completing tissue processing be minimized to prevent leaching of oligonucleotide within the tissue.

3. Primary antibodies: Incubate sections with antiserum to the cellular antigen of interest (i.e., glial fibrillary acidic protein, neuron-specific nuclear protein) as specified by the manufacturer protocol for the individual sera or monoclonal antibodies. Perform all necessary incubations at 4°.

4. Wash several times with cold phosphate-buffered saline.

5. Secondary antibodies: It is recommended that a fluorescence-labeled secondary antibody be used to visualize the antigen, instead of other standard methods (i.e., peroxidase, avidin–biotin–diaminobenzidine). Second-

ary antibodies conjugated with FITC, rhodamine, and CY2 or CY3 fluoro-chromes have worked successfully in our laboratory. It is important to choose fluorescent markers for the oligonucleotide and secondary antibodies that have nonoverlapping or minimally overlapping emission wavelengths (i.e., FITC and rhodamine).

Coverslip. Glass coverslips should be applied with a glycerol-based mounting medium (i.e., Citifluor, Marivac). It may be necessary to use a mounting medium that contains an antifade agent, as some media can facilitate fading and leaching of fluorescent signals from cells within 24 hr. We have also obtained quality micrographs from brain sections that were not coverslipped. Noncoverslipped sections tend to desiccate rapidly, however.

[20] Antisense Oligonucleotides: Preparation and *in Vivo* Application to Rat Brain

By WOLFGANG TISCHMEYER

Introduction

Long-term plastic changes in the brain, including those supporting memory formation, are assumed to depend on permanent functional alterations in neuronal cells that require reprogramming of gene expression. Molecular genetic techniques, interfering with the expression of specific genes, have been used in combination with neurobiology at the systems level to analyze the significance of defined gene products for behavioral plasticity. In this context, current knockout techniques bear some limitations, such as molecular compensation, developmental defects, background genotypes, and the restriction to the mouse species. Antisense oligodeoxynucleotides (ODNs), designed to inhibit the synthesis of specific proteins by hybridization to the respective mRNA, may provide an alternative to circumvent some of these problems. As required for studies of learning and memory formation, intracerebral microinjection of antisense ODNs exerts locally and temporally restricted effects. Reports have demonstrated the usefulness of antisense approaches to study the role of specific proteins in the nervous system in a behavioral context, and specifically during learning and memory formation (for review, see Refs. 1–4 and this volume).

[1] C. Wahlestedt, *Trends Pharmacol. Sci* **15**, 42 (1994).
[2] W. Sommer and K. Fuxe, *Neurochem. Int.* **31**, 425 (1997).

In our laboratory, intracerebral application of 15- to 20-mer antisense ODNs to rats has been used to address the question of the functional significance of inducible transcription factors, such as c-Jun, Jun-B, and c-Fos, for processes of learning and memory formation. As one of the major problems for *in vivo* use of natural, phosphodiester ODNs (O-ODNs) is their rapid degradation by nucleolytic enzymes, phosphorothioate-modified ODNs (S-ODNs)[5] were generated. Solid-phase automated synthesis of ODNs has been performed using cyanoethyl phosphoramidite chemistry.[6-8] To minimize the risk of batch differences as a potential source of artifacts, several batches of antisense ODNs and of the corresponding control ODNs for a given experiment were alternately synthesized using a 1-μmol scale, instead of bulk production of each ODN using a larger scale synthesis range. During the subsequent steps of deprotection and purification, these batches of antisense ODNs and control ODNs were processed in parallel. Reversed-phase high-performance liquid chromatography (HPLC)[9,10] has been used to separate full-length ODNs, containing the large, hydrophobic dimethoxytrityl (DMT) protection group, from failure sequences in the crude reaction mix by differences in hydrophobicity. Subsequently, the DMT group has been cleaved from the purified ODNs under acidic conditions. Precipitation, desalting, and extraction of the detritylated ODNs with diethyl ether have been performed to remove residual by-products and to change the ammonium salt form of the ODN to the more soluble and natural sodium salt form. Before use in animal studies, the homogeneity of each ODN has been analyzed electrophoretically.[11] For distribution studies, S-ODNs were radiolabeled at the 5' end via an amino linker[12] and purified electrophoretically.

Theoretical and practical considerations of general methods for synthesis, purification, modification, functionalization, and labeling of ODNs have

---

[3] B. J. Chiasson, M. G. Hong, and H. A. Robertson, *Neurochem. Int.* **31,** 459 (1997).

[4] W. Tischmeyer and R. Grimm, *Cell. Mol. Life Sci.* **55,** 564 (1999).

[5] H. Vu and B. L. Hirschbein, *Tetrahedron Lett.* **32,** 3005 (1991).

[6] S. L. Beaucage and M. H. Caruthers, *Tetrahedron Lett.* **22,** 1859 (1981).

[7] M. D. Matteucci and M. H. Caruthers, *J. Am. Chem. Soc.* **103,** 3185 (1981).

[8] S. P. Adams, K. S. Kavka, E. J. Wykes, S. B. Holder, and G. R. Galluppi, *J. Am. Chem. Soc.* **105,** 661 (1983).

[9] H.-J. Fritz, R. Belagaje, L. E. Brown, R. H. Fritz, R. A. Jones, R. G. Lees, and H. G. Khorana, *Biochemistry* **17,** 1257 (1978).

[10] G. Zon and J. A. Thompson, *BioChromatography* **1,** 22 (1986).

[11] R. Frank and H. Koster, *Nucleic Acids Res.* **6,** 2069 (1979).

[12] B. C. F. Chu, G. M. Wahl, and L. E. Orgel, *Nucleic Acids Res.* **11,** 6513 (1983).

been described in detail.[13–17] The protocols given below have been routinely used in our laboratory.

Automated Synthesis of Oligodeoxynucleotides

Materials. Reagents for ODN synthesis are purchased from Applied Biosystems (Foster City, CA)

Acetonitrile (ACN): <30 ppm water

Controlled pore glass (CPG) DNA synthesis solid supports (columns for 1-μmol synthesis scale)

β-Cyanoethyl-DNA phosphoramidites: 0.1 M in anhydrous ACN

Tetrazole–ACN

Acetic anhydride–lutidine–tetrahydrofuran

1-Methylimidazole–tetrahydrofuran

Trichloroacetic acid–dichloromethane

Iodine–water–pyridine–tetrahydrofuran, 0.02 M

Tetraethylthiuram disulfide (TETD)–ACN

Argon 4.8 (Linde, Höllviegelskreuth, Germany)

Procedures

Using a PCR-MATE EP model 391 DNA synthesizer (Applied Biosystems), solid-phase automated synthesis of O-ODNs and S-ODNs has been performed according to the manufacturer recommendations.

For synthesis of phosphorothioate-containing ODNs, the sulfurizing reagent TETD–ACN replaces the iodine reagent, with all other reagents used for standard phosphoramidite chemistry remaining unchanged. Stepwise sulfurization is achieved by conversion of internucleotide cyanoethyl phosphite to the phosphorothioate triester, which is subsequently converted to phosphorothioate diester during deprotection using concentrated ammonia after the completion of synthesis. The synthesis cycle is largely identical to that employed for the synthesis of O-ODNs, except that capping is performed after the sulfurization step. Cycle times are about 15 min longer

[13] R. W. A. Oliver, "HPLC of Macromolecules: A Practical Approach." IRL Press, Oxford, 1989.

[14] D. Rickwood and B. D. Hames, "Gel Electrophoresis of Nucleic Acids: A Practical Approach." IRL Press, Oxford, 1990.

[15] S. Agrawal, "Protocols for Oligonucleotides and Analogs." *Methods Mol. Biol.* **20** (1993).

[16] S. Agrawal, "Protocols for Oligonucleotide Conjugates." *Methods Mol. Biol.* **26** (1994).

[17] S. T. Crooke, "Antisense Research and Application." Springer-Verlag, Berlin, 1998.

than the corresponding, normal O-ODN cycles. Thus, synthesis of a 20-mer S-ODN using TETD–ACN will take about 7 hr.

For subsequent purification of ODNs by reversed-phase HPLC, the large, hydrophobic dimethoxytrityl (DMT) protection group is left on the 5′ end of the ODNs, i.e., the synthesis of the ODNs is performed in the "Trityl ON" mode.

After completion of synthesis, the solid support is washed with ACN for approximately 1 min and dried with argon for another minute before the column is removed from the synthesizer.

Notes

Some of the synthesis reagents tend to precipitate and might obstruct delivery lines and inhibit reagent flow during the synthesis. It is thus advisable to monitor the use of reagents after completion of each ODN synthesis. For the same reason, it is advisable to wash all chemicals out of the delivery lines before shutting down the synthesizer for a longer period. This can be efficiently done by performing a "sham synthesis" using ACN instead of synthesis reagents.

Cleavage and Deprotection of Oligodeoxynucleotides

Materials

Ammonium hydroxide solution, ~30% (Applied Biosystems)
Triethylamine (TEA; Merck, Rahway, NJ)
Triethylamine acetate, 2 *M* (TEAA; Applied Biosystems)

Procedures

1. Remove the solid support from the column and transfer it into a 2-ml safelock tube. Add approximately 2 ml of fresh concentrated ammonium hydroxide solution, seal the tube, and shake thoroughly at room temperature for at least 1 hr to cleave the ODN from the solid support and to deprotect the phosphate groups.

2. Spin down the solid support, and cool to 0°.

3. Remove the supernatant and transfer aliquots of about 1 ml to screw-capped Eppendorf tubes.

4. Rinse the solid support with 1 ml of fresh ammonium hydroxide solution and combine the wash solution with the previous supernatant.

5. Seal the tubes tightly and incubate at 55° for 8 hr to deprotect the exocyclic amines.

6. Cool to 0°.

7. Concentrate to dryness by vacuum centrifugation at room temperature.

8. Dissolve and combine the ODN residues in a total volume of 50 μl of 0.1 M TEAA, adjusted to pH 10 by addition of TEA to prevent detritylation. Store at $-20°$ until purification by HPLC.

Notes

The use of concentrated ($\sim 30\%$) ammonia is essential. After use, store the ammonia tightly sealed at $4°$

After cleavage, store the solid support at $4°$ until the effectiveness of ODN cleavage has been proved. If cleavage has been incomplete, repeat it using fresh, concentrated ammonia solution.

Heating and acidic pH should be avoided to minimize detritylation during evaporation. As a precaution to maintain basic pH, portions of about 5 μl of TEA can be added to the deprotection mix before and periodically during evaporation. However, in our experience, the presence of TEA during evaportion has no critical influence on the product yield after reversed-phase HPLC.

Purification of Oligodeoxynucleotides by Reversed-Phase
  High-Performance Liquid Chromatography

Materials

Equipment: The following LiChroGraph HPLC system (Hitachi, Tokyo, Japan) has been used:
LC-Organizer with rheodyne type 7125 manual injector and 50-μl sample loop
UV detector (L-4000) with a flow cell of 2-mm path length
Intelligent pump (L-6200A)
Fraction collector (L-5200)
Chromato-integrator (D-2500)
Four-channel online degasser (Knauer, Berlin, Germany)
HPLC columns
LiChroCART 250-4 HPLC cartridge (Merck), prepacked with LiChrosorb RP-18 (7 μm)
LiChroCART 4-4 guard column (Merck), prepacked with LiChrospher 100 RP-18 (5 μm)
Eluants
A: 0.1 M TEAA, pH 7.0
B: ACN, HPLC grade (Merck)

Procedures

Sample Loading

1. Centrifuge the ODN sample (dissolved in 50 μl of 0.1 M TEAA, as described above) for 5 min at room temperature to remove particulate matter.

2. Inject the supernatant and start the chromatography. HPLC conditions are as follows. Gradient:

Time (min)	Eluant A (%)	Eluant B (%)
0	92	8
20	72	28
30	60	40
40	50	50
50	50	50

Adjust the flow rate to 1 ml/min.

3. Detection at 254 nm is appropriate for ODN amounts resulting from a 1-μmol synthesis (depending on the capacity of the column used and the amount of loaded sample, longer wavelengths, e.g., 280–300 nm, may be suitable).

4. Record at 2.56 ODU AUFS (i.e., plot attenuation, 10); chart speed, 10 mm/min.

5. Perform sample collection: 18–28 min from start time; 30 sec (i.e., 0.5 ml) per fraction.

After Completion of Chromatography

1. Briefly vortex the fractions corresponding to the product peak.

2. Dilute an aliquot of 10 μl of each fraction by addition of 500 μl of water and determine the absorbance at 260 nm (A_{260}) spectrophotometrically against a water blank.

3. Collect the fractions representing the central region of the product peak. Discard fractions from the leading and trailing ends of the peak, i.e., fractions with A_{260} below 50% of peak maximum.

4. Add 1 ml of sterile, deionized water per fraction, mix, and evaporate to dryness under vacuum. Repeat this procedure to remove volatile eluants.

5. Dissolve and combine the residues in sterile, deionized water and transfer equal aliquots of the ODN solution to two 2-ml safelock tubes for subsequent detritylation. Concentrate to dryness.

Notes

To protect the HPLC column from material that is irreversibly adsorbed to the packing material, the use of a guard column is advisable. Replace the guard column periodically when the back pressure of the HPLC system becomes too high.

To avoid the formation of gas bubbles in the mixing chamber and at the column, which may interfere with solvent delivery and detection, eluants should be degassed either by flushing with helium or by use of a commercially available degasser.

Retention times for the more hydrophobic S-ODNs will be slightly longer than those for O-ODNs owing to the more hydrophobic phosphorothioate linkage. Under the HPLC conditions used, tritylated 15- to 20-mer O-ODNs and S-ODNs will elute approximately 18–28 min from the start time. However, because base composition and sequence length affect the mobility retention time, it is advisable to examine the proper period for sample collection for a given ODN in an analytical run using about 1–2% of the sample and HPLC conditions as described above, except that the recorder range must be appropriately adjusted.

Product peaks of S-ODNs are often seen as a doublet because of the stereochemistry of the phosphorothioate linkage.

Detritylation and Precipitation of Oligodeoxynucleotides

Materials

Acetic acid, 80% (v/v)
Sodium acetate, pH 5 (3 *M*).
2-Propanol
Ethanol

Procedures

1. Dissolve the dried ODN residues by addition of 440 μl of 80% acetic acid per tube and vortex. O-ODNs will dissolve immediately; S-ODNs may require a brief heating to attain solution. Alternatively, S-ODNs can be dissolved in a minimum amount of sterile, deionized water before addition of acetic acid.

2. Incubate for 30 min at room temperature.

3. Add 50 μl of 3 *M* sodium acetate, pH 5, per tube and vortex.

4. Add 1400 μl of 2-propanol per tube and mix thoroughly.

5. Incubate at −20° for 1 hr.

6. Centrifuge at 14,000 rpm, 2° for 15 min.

7. Decant or remove the supernatants with a pipette and discard them. Be careful not to lose the precipitates.

8. Wash the precipitates with 300 μl of prechilled ethanol.

9. Centrifuge again. Remove and discard the supernatants.

10. Remove residual ethanol from the precipitates by brief evaporation under vacuum.

Desalting of Oligodeoxynucleotides

Materials

NAP-10 disposable columns (Pharmacia, Piscataway, NJ) containing Sephadex G-25
Sterile, deionized water

Procedures

1. Equilibrate the column with three 5-ml volumes of sterile, deionized water. Dissolve and combine the ODN precipitates of a given synthesis batch in a total volume of 500 μl of sterile, deionized water and add the sample to the gel.
2. Elute with six 0.5-ml volumes of sterile, deionized water and collect fractions of 0.5 ml.
3. Dilute an aliquot of 10 μl of each fraction by addition of 500 μl of water and determine the A_{260} spectrophotometrically against a water blank.
4. Combine the peak fractions (usually fractions 3–5) and evaporate to dryness *in vacuo*.

Ether Extraction of Oligodeoxynucleotides

Caution: Ether is extremely flammable. Use a well-ventilated, explosion-proof fume hood.

Materials

Diethyl ether
Sterile, deionized water

Procedures

1. Dissolve the ODN residue in 0.5 ml of sterile, deionized water.
2. Mix diethyl ether with an equal volume of sterile, deionized water. Vortex thoroughly and allow phases to separate.
3. Add an equal volume of water-saturated ether (upper phase) to the ODN solution and vortex.
4. Allow phases to separate, or centrifuge briefly.
5. Remove and discard the ether (upper layer) from the sample.

6. Repeat the extraction twice.

7. Remove residual ether from the sample by evaporation with a water pump.

8. Concentrate to dryness.

Determination of Product Yield

1. Dissolve the purified ODN in 500 μl of sterile, deionized water.

2. Dilute an aliquot of 5 μl by addition of 500 μl of water and determine the A_{260} spectrophotometrically against a water blank.

3. To calculate the total number of A_{260} units of purified ODN, multiply the determined absorbance by 50. One A_{260} unit corresponds to approximately 33 μg of ODN. The approximate yield of purified ODN expressed in micromoles can be calculated by dividing the total number of A_{260} units by 10 times the number of nucleotides of the given ODN.

4. Remove an aliquot of about 1 A_{260} unit for subsequent electrophoretic analysis (see below) and concentrate to dryness.

5. Concentrate the bulk of the ODN to dryness and store at $-20°$ until use.

5'-End Radiolabeling of Oligodeoxynucleotides

Caution: Before ordering or using radioisotopes, consult the radiation protection adviser at your institution.

Materials

Amino linker-phosphoramidite (commercially available from different suppliers), 0.1 M in anhydrous ACN

Sodium borate, pH 8.5 (0.1 M)

Glycine, 0.2 M in 0.1 M sodium borate, pH 8.5

^{35}S labeling reagent (^{35}SLR; Amersham, Arlington Heights, IL)

Procedures

1. For functionalization of ODNs at the 5' end, place the amino linker solution on an extra base position of the DNA synthesizer and assign this position to the 5' end of the ODNs sequences to be synthesized. Synthesize the ODNs in the "Trityl ON" mode, using the coupling cycles for O-ODNs or for S-ODNs as described above (in our laboratory, this protocol has been used for functionalization and subsequent labeling of S-ODNs).

2. Cleave and deprotect with ammonia as described above.

3. Purify the crude product by HPLC as described above. Owing to

the lower degree of hydrophobicity of the functionalized ODNs compared with DMT-bearing ODNs, modify the gradient conditions as follows:

Time (min)	Eluant A (%)	Eluant B (%)
0	92	8
30	72	28
40	60	40
50	50	50

The ODN elutes approximately 15–25 min from the start time.

4. Perform 2-propanol precipitation, ether extraction, and desalting as described above, except that detritylation with acetic acid is not necessary.

5. Dissolve the ODN (100 to 150 nmol) in 50 μl of 0.1 M sodium borate, pH 8.5.

6. Transfer 250 μCi of [35]SLR, brought to room temperature, to an Eppendorf tube. Remove the solvent by directing a gentle stream of dry nitrogen onto the surface.

7. Cool the tube containing the dried [35]SLR residue on ice and add the ODN solution.

8. Incubate on ice for 30 min.

9. Terminate the reaction with 100 μl of 0.2 M glycine in 0.1 M sodium borate, pH 8.5.

10. Desalt the ODN solution by precipitation and by passage through a NAP-10 column as described above.

11. Combine the ODN-containing fractions and concentrate to dryness.

12. To avoid radioactive contamination of the HPLC system, it is advisable to separate radiolabeled ODNs from unincorporated [35]SLR by polyacrylamide gel electrophoresis as described below.

Denaturing Polyacrylamide Gel Electrophoresis
  of Oligodeoxynucleotides

Caution: Acrylamide is a potent neurotoxin. Ethidium bromide is a powerful mutagen. Ultraviolet radiation is dangerous to the eyes.

Materials

Slab gel polyacrylamide gel electrophoresis (PAGE) apparatus, gel length ≥20 cm (e.g., PROTEAN II xi slab cell; Bio Rad, Hercules, CA)

Solution of 38% acrylamide and 2% N,N'-methylenebisacrylamide (commercially available)

10× TBE (0.89 M Tris–borate, 20 mM EDTA)
Ammonium persulfate (APS), 10%
N,N,N',N'-Tetramethylethylenediamine (TEMED)
Urea
Formamide (deionized)
Ethidium bromide (saturated solution in water)
Bromphenol blue
Xylene cyanol
Fluorescent thin-layer chromatography (TLC) plate
Millex HV 0.45-μm pore size filter (Millipore, Bedford, MA)

Procedures

Gel Preparation

1. To prepare a gel of 19% polyacrylamide concentration (recommended for ODNs of ≤25 nucleotides), dissolve 31.5 g of urea in a mixture of 37.5 ml of acrylamide solution, 7.5 ml of 10 × TBE, and 7.5 ml of water.
2. Add 450 μl of 10% APS and mix.
3. Degas under vacuum, using a water pump, for several minutes.
4. Add 35 μl of TEMED and mix gently.
5. Pour a gel of 1.5 mm thickness, insert the comb (slots of 1- and 2-cm width have been used for analytical and preparative runs, respectively), and allow the gel to polymerize for about 2 hr.
6. Place the gel into the electrophoresis chamber, add upper and lower buffers (1 × TBE), remove the comb, and flush the slots with buffer to remove urea and residual acrylamide.
7. Prerun the gel for 1 hr at 20–25 mA.

Analysis of Oligodeoxynucleotides by Polyacrylamide Gel Electrophoresis

1. Dissolve up to 1 A_{260} unit of ODN in 5 μl of water.
2. Add 5 μl of a mixture of formamide–ethidium bromide (50 : 1, v/v) to the sample, vortex, and spin down the sample from the tube walls.
3. Denature at 55° for 5 min, then place on ice immediately.
4. After the prerun of the gel, flush the slots again with buffer.
5. Load the samples.
6. Loading of tracking dyes, e.g., bromphenol blue and xylene cyanol, onto an unused adjacent slot helps to assess the position of the ODN in the gel during and after the run (in a 19% polyacrylamide gel, bromphenol blue and xylene cyanol comigrate with ODNs of approximate lengths of 6 and 22 nucleotides, respectively).

7. Run the gel at 20–25 mA until the bromphenol blue reaches the bottom of the gel.

8. On completion of electrophoresis, transfer the gel onto a piece of transparent plastic wrap.

9. Photograph the gel on a transilluminator, using short-wavelength (302 or 254 nm) ultraviolet light.

Purification of Radiolabeled Oligodeoxynucleotides by Polyacrylamide Gel Electrophoresis

1. Dissolve the ODN in 40 μl of sterile, deionized water.

2. Add an equal volume of formamide, vortex, and spin down the sample from the tube walls.

3. Denature at 55° for 5 min, then place on ice immediately.

4. After the prerun of the gel, flush the slots again with buffer.

5. Depending on the ODN amount in the sample, load 20–40 μl per slot (of 2-cm width) onto the gel.

6. Loading of tracking dyes, of unlabeled ODN, and of unreacted [35]SLR onto adjacent slots may be helpful in identifying the labeled ODN.

7. Run the gel at 20–25 mA until the bromphenol blue reaches the bottom of the gel.

8. On completion of electrophoresis, transfer the gel onto a fluorescent TLC plate covered with a transparent plastic wrap.

9. Illuminate from above, using an ultraviolet lamp in a dark room. To minimize nicking of the ODN, the use of ultraviolet light of longer wavelengths (e.g., 366 nm) is advisable. The ODN appears as a dark band against the fluorescent background.

10. Excise the ODN band, which is usually the slowest migrating band. (Note that the labeled ODN may migrate slightly slower than the unlabeled ODN. Separation from unincorporated [35]SLR as well as the efficiency of excision of radiolabelled ODN from the gel can be monitored by autoradiography before and subsequent to band excision.)

11. Transfer the ODN-containing gel pieces into 2-ml Eppendorf tubes.

12. Chop the gel pieces with a disposable pipette tip.

13. Add 1 ml of sterile, deionized water to each tube and incubate overnight at 37° in a shaker incubator.

14. Centrifuge for 5 min at room temperature.

15. Collect and combine the supernatants.

16. Vortex the remaining gel pellets with 1 ml of sterile, deionized water, centrifuge, collect the supernatants, and combine them with the previous ones.

17. Pass the combined supernatants through a Millex HV 0.45-μm pore size filter.

18. Concentrate the effluent and desalt by precipitation and by passage through a NAP-10 column as described above.

Application of Oligodeoxynucleotides to Rat Brain

Animals

Male 7-week-old Wistar rats housed on a 12-hr light/dark cycle (light on at 6 A.M.) and given free access to standard laboratory chow and tap water

Materials

Stereotaxic apparatus (Stoelting, Wood Dale, IL)

Stainless-steel guide cannulas for chronical implantation, 0.9-mm o.d., 0.6-mm i.d., 5 mm in length (suitable for targeting both the dorsolateral hippocampus and the lateral ventricle): The upper end of each cannula is sealed by means of a socket filled with silicon rubber.

Blunt-tipped stainless steel injector cannulas, 0.5-mm o.d., 0.3-mm i.d.: The length of the injector cannulas is adjusted by means of a socket such that their tips are located 0.1 mm below the tips of guide cannulas during injection. This helps to push away possible closures and to minimize accumulation of injection solution inside the lumen of the guide cannulas during injection

Polyethylene catheter tubing: 0.8-mm o.d., 0.4-mm i.d. (Reichelt, Heidelberg, Germany)

Microsyringe (CR-700-50; Hamilton, Reno, NV): 50 μl

Stainless steel anchor screws (1.2 $\times$ 3 mm)

Sterile swabs (Sugi steril; Kettenbach, Eschenburg, Germany)

Dental cement (Paladur; Kulzer, Wehrheim, Germany)

Isotonic (0.9%) saline (Braun, Melsungen, Germany)

Ethanol, 70%

Hydrogen peroxide, 4%

Pentobarbital sodium (400 mg dissolved in 50 ml of propylene glycol–50 ml of sterile, deionized water)

Surgical Procedures

1. Before implantation, soak the guide cannulas in 70% ethanol. Evaporate under vacuum to remove all ethanol from the surface and the interior of the cannulas.

2. Anesthetize a rat with an intraperitoneal injection of pentobarbital sodium (40–45 mg/kg).

3. Shave the scalp and fix the rat in the stereotaxic apparatus.

4. Make an incision slightly behind the eyes and open the scalp.

5. Remove connective tissue from the exposed skull area and swab it dry.

6. To visualize the bone suture junctions lambda and bregma, bleach the skull with a sterile swab wetted with 4% hydrogen peroxide.

7. Adjust the incisor bar of the stereotaxic apparatus until the height of the lambda skull point is 1 mm ventral to the height of the bregma skull point.

8. Using stereotaxic coordinates according to Paxinos and Watson,[18] determine the location for bilateral placement of cannulas (we used the following: posterior, -2.8 mm from bregma; lateral, ±2.2 mm, corresponding to the CA1 region in the dorsal hippocampus). Mark the correct sites with drawing ink.

9. Drill holes through the skull at the marked positions by means of a dentist's drill. Using a sterile needle, incise the endocranium before implantation of cannulas.

10. Using the electrode manipulator arm of the stereotaxic apparatus, insert the first guide cannula to the desired depth (e.g., 3.2 mm from the skull for intrahippocampal implantation). Fix it with a small amount of dental cement.

11. Insert and fix the second guide cannula as described.

12. Drill a hole 3 mm posterior to one of the cannulas and insert the anchor screw. Take care not to go entirely through the skull.

13. Dry the skull. Fix the cannulas to the skull and the anchor screw by covering with dental cement.

14. Allow polymerization of the cement for several minutes, treat the scalp wounds with antibacterial powder, and remove the animal from the stereotaxic apparatus.

15. During a 1-week recovery period before beginning the experiments, handle the rats daily to habituate them to the injection procedure.

Intrahippocampal Injections

1. Dissolve the ODN of interest to the desired concentration in sterile, deionized water.

2. Fill a 50-μl mycrosyringe completely with isotonic saline. Avoid the formation of air bubbles.

3. Attach the catheter tubing to the mycrosyringe, inject nearly all of the saline from the syringe into the tubing, and cut the tubing at the meniscus

---

[18] G. Paxinos and C. Watson, "The Rat Brain in Stereotaxic Coordinates," 2nd Ed. Academic Press, Sydney, 1986.

(the length of the tubing that remains attached to the syringe is suitable to allow injection into freely moving rats).

4. Connect the injector cannula with the end of the tubing opposite to the syringe, and fill the injector with saline from the tubing.

5. To separate the saline inside the injector and tubing from the ODN solution to be delivered, first draw 1 μl of air before drawing a few microliters of the ODN solution. Mark the position of the air bubble to allow monitoring of its migration during and after the injection procedure.

6. Inject freely moving rats in an uncovered cage. Place the injector gently into the guide cannula, and inject 1 μl of the solution at a rate of approximately 0.5 μl/min. Subsequently, leave the injector in place for another minute.

7. For bilateral application, repeat the injection procedure in the opposite hemisphere.

8. On completion of all experiments, verify the correct location of the guide cannulas histologically.

Notes

Animal studies must comply with national and state legislation.

In our studies,[19,20] 1 μl of a 2 mM solution of (either antisense or control) S-ODNs was injected twice into each hippocampus. The first injection was given 10 hr prior to the behavioral experiments, considering reports of an 8-hr onset latency for antisense S-ODNs.[21,22] To maintain efficient local concentrations of ODNs during the behavioral studies, a second injection was given 2 hr before the beginning of experiments. However, our studies were performed to suppress the expression of inducible transcription factors, i.e., rapidly and transiently induced, short-lived proteins. To efficiently suppress constitutively expressed proteins with longer half-lives, the treatment must be adjusted accordingly.

Autoradiographic and immunohistochemical studies revealed that, using the preceding protocol of intrahippocampal injections, the distribution of S-ODNs was chiefly restricted to brain regions surrounding the injection

[19] W. Tischmeyer, R. Grimm, H. Schicknick, W. Brysch, and K.-H. Schlingensiepen, *Neuroreport* **5,** 1501 (1994).

[20] R. Grimm, H. Schicknick, I. Riede, E. D. Gundelfinger, T. Herdegen, W. Zuschratter, and W. Tischmeyer, *Learning Memory* **3,** 402 (1997).

[21] B. J. Chiasson, M. L. Hooper, P. R. Murphy, and H. A. Robertson, *Eur. J. Pharmacol.* **227,** 451 (1992).

[22] S. Suzuki, P. Pilowsky, J. Minson, L. Arnolda, I. J. Llewellyn-Smith, and J. Chalmers, *Am. J. Physiol.* **266,** R1418 (1994).

sites, i.e., the hippocampal formation, adjacent cortical areas, and callosal fibers.[20,23,24]

After intrahippocampal application of c-*fos* antisense S-ODN to rats as described above, approximately 70% inhibition of induced c-Fos expression was observed in the dentate gyrus.[20] However, a few scattered cells still expressed high levels of c-Fos, possibly owing to cell type-specific differences in the uptake of ODNs. Thus, inhibition of gene expression by antisense ODNs is not always complete.

Acknowledgment

Work in the author's laboratory has been supported by the Kultusministerium of the Land Sachsen-Anhalt.

[23] W. Brysch, A. Rifai, W. Tischmeyer, and K.-H. Schlingensiepen, *in* "Methods in Molecular Medicine: Antisense Therapeutics" (S. Agrawal, ed.), p. 159. Humana Press, Totowa, New Jersey, 1996.
[24] W. Tischmeyer, R. Grimm, K. Lohmann, H. Schicknick, and E. D. Gundelfinger, *in* "Neurochemistry: Cellular, Molecular and Clinical Aspects" (A. Teelken and J. Korf, eds.), p. 1117. Plenum, New York, 1997.

[21] Application of Antisense Techniques to Characterize Neuronal Ion Channels *in Vitro*

By STEPHEN G. VOLSEN, REGIS C. LAMBERT, YVES MAULET, MICHEL DE WAARD, SAMANTHA GILLARD, PETER J. CRAIG, RUTH BEATTIE, and ANNE FELTZ

Introduction

Current gene cloning and genomic initiatives provide the neuroscientist with an exponentially increasing bank of genetic sequence data that details the molecular identity of novel brain proteins.[1] Often in the initial absence of selective ligands, the functional properties of the more recently cloned brain proteins remain unclear.[2] The antisense approach conceptually offers a solution to this problem.[3]

Multiple mechanisms have been proposed that describe the molecular

[1] M. R. Brownstein, J. M. Trent, and M. S. Boguski, *Trends Neurosci. Suppl.* **21,** 27 (1998).
[2] D. B. Kell, *Trends Biotechnol.* **16,** 491 (1998).
[3] B. Weiss, G. Davidkova, and S. P. Zhang, *Neurochem Int.* **31,** 321 (1997).

events that precipitate gene-specific knockdown by antisense oligonucleo-tides. These include inhibition of transcription, translational arrest, disrup-tion of RNA processing, and RNase H-mediated transcript degradation. Despite the many questions that remain unanswered, antisense oligonucleo-tides have been applied successfully in CNS research. We focus our attention here on their *in vitro* applications and discuss both the strategies and methods that we have developed and applied successfully to studies of neuronal voltage-dependent ion channels. Wherever possible, the effects of antisense treatment were examined at multiple levels, i.e., transcription transduction, mRNA, protein and/or functionally by electrophysiological measurements. Each level of analysis affords complementary data that when taken together, greatly facilitate the final interpretation of results.

Antisense Reagents and Transfection Procedures

Antisense Oligonucleotide Design

One of the first, and perhaps most important, considerations when embarking on an antisense study is the design of the oligonucleotides. As a general rule, it is advantageous to demonstrate antisense effects using two reagents targeted to different sequences on the transcript of interest. This reduces significantly the opportunity of confusing nonspecific, se-quence-independent effects with true antisense knockdown.

The exact site on the transcript against which to target an antisense oligonucleotide has been a matter that historically has been approached empirically and remains of concern today.[4] Many of the first-generation reagents spanned the initiation codon,[5] and although such oligonucleotides have on occasion proved active, many have not. Indeed, effective antisense reagents targeted to regions either deep in the coding sequence and/or in untranslated domains have also been described. The choice of the sequence is, in fact, determined by the nature of the study. The knockdown of a complete family of proteins necessitates the definition of a motif conserved among the different mRNAs encoding each member of the group; targeting a single protein of the family involves the selection of a gene-specific sequence. This consideration often restricts the choice to a small number of possible sequences. In our studies, antisense oligonucleotides were always targeted to domains in coding sequence of the mRNA far from the initiation codon, and we assume that their efficacy is primarily dependent on RNase H activity. This helps provide a few guidelines in the choice of the oligonu-

[4] F. Eckstein, *Nature Biotechnol.* **15,** 519 (1997).
[5] B. Weiss, L. Zhou, S. P. Zhang, and Z. H. Quin, *Neuroscience* **55,** 607 (1993).

cleotide. It should show no significant homology (fewer than 6 contiguous nucleotides) to any other sequence. On a theoretical basis, the sequence of a 20-mer should not occur more than once in the genome and we currently use oligonucleotides of this general size in our studies. It should be noted that as few as five duplexed DNA–RNA base pairs can be cleaved by RNase H[6] and that partial annealing to unrelated mRNAs may alter the required specificity of action of the antisense oligonucleotide. This risk is, however, minimized by the fact that such hybrids are much less stable and require higher antisense oligonucleotide concentrations. To facilitate some of the general design features of antisense oligonucleotides, various software programs are available that are predominantly used for efficient primer preparation. It should be noted that we have also had some success in using commercial sources to provide first-generation, "ready to go" reagents.[7] This can save considerable time and effort, although initial costs are not insignificant.

In addition, elegant data cast new light on the problem of selecting effective antisense reagents. It is clear that the interaction of an oligonucleotide with an mRNA transcript is largely dependent on the molecular conformation of both species.[8] Accessible sites of potential interaction are, however, inefficiently predicted using current molecular modeling and thermodynamic methods. Taking these factors into account, combinatorial oligonucleotide array techniques have been developed. These methods allow the simultaneous assessment of binding of all oligonucleotides within a defined RNA locus, so comprehensively highlighting accessible sites against which to target antisense.[9] Such novel techniques are proving effective in a number of systems and we are currently testing reagents so generated to knock down calcium channels in native neurons. Although not available today as a routine laboratory technique, the oligonucleotide array technology marks a clear breakthrough for rational antisense design. This will undoubtedly yield, as preliminary data suggest, second-generation reagents of greater potency and efficacy. In addition, they provide what is probably the most meaningful negative control reagents in these experiments, namely an antisense oligonucleotide raised against an area determined by the array analysis to be inaccessible to duplex formation. It is clear that the final design of antisense oligonucleotides will become more scientifically based and efficient in the future.

[6] S. Shibahara, S. Mukai, T. Nishihara, H. Inoue, E. Ohtsuka, and H. Morisawa, *Nucleic Acids Res.* **15,** 4403 (1987).
[7] S. E. Gillard, S. G. Volsen, W. Smith, R. E. Beattie, D. Bleakman, and D. Lodge, *Neuropharmacology* **36**(3), 405 (1997).
[8] W. F. Lima, B. Monia, D. J. Ecker, and S. M. Freier, *Biochemistry,* **31,** 12055 (1992).
[9] N. Milner, K. U. Mir, and E. M. Southern, *Nature Biotechnol.* **15,** 537 (1997).

Antisense Oligonucleotide Preparation and Chemistry

Having selected the site against which to prepare antisense oligonucleotides, one is next faced with the choice of backbone chemistry. We have consistently used fully phosphorothioated oligonucleotides in our experiments.[7,10,11] Their inherent stability to both endo- and exonuclease degradation is of considerable merit when studying, as in the case of calcium channels, molecular targets whose turnover and half-life are measured in numbers of days as opposed to hours.[10-13] Whereas other reagents, including methyl phosphonates, are also active in increasing oligonucleotide half-life, only phosphorothioate modification supports RNase H activity. When purified to single peak purity, as assessed by high-performance liquid chromatography (HPLC), and used in concentration ranges up to 1 μM, we have not observed problems of cellular toxicity with these reagents in either central or peripheral neurons *in vitro*.

The addition of a fluorescent label to the oligonucleotide is essential to assess both the overall transfection efficiency and, of critical importance in electrophysiological experiments, to differentiate transfected from untransfected cells[10] (see Electrophysiology, below). It is worth noting that aging of this solution can induce hydrolysis of the fluorescein label, as exemplified by the appearance of two bands on polyacrylamide gel electrophoresis in 8 M urea. In such cases, the oligonucleotide solution should then be discarded.

Transfection Procedures

The most demanding procedure to assess the efficacy of an antisense knockdown of an ionic channel is the electrophysiological approach. The experimental aim is to assess the effects of either antisense or control oligonucleotides on target currents as a function of time posttransfection. To minimize changes appearing with development, this approach can be carried out only on postmitotic neurons. Embryonic neurons are known to incorporate oligonucleotide in the absence of any vector. For example, it has been shown[7] that the selective knockdown of the pore-forming subunit of P-type calcium channels in rat cerebellar Purkinje cells can be achieved

[10] R. C. Lambert, Y. Maulet, J. Mouton, R. Beattie, S. G. Volsen, M. de Waard, and A. Feltz, *J. Neurosci.* **17,** 6621 (1997).

[11] R. C. Lambert, Y. Maulet, J. Dupont, S. Mykita, P. Craig, S. G. Volsen, and A. Feltz, *Mol. Cell. Neurosci.* **7,** 239 (1996).

[12] M. Passafarro, F. Clementi, and E. Sher, *J. Neurosci.* **12,** 3372 (1992).

[13] R. C. Lambert, F. McKenna, Y. Maulet, E. M. Talley, D. A. Bayliss, L. L. Cribbs, J. H. Lee, E. Perez-Reyes, and A. Feltz, *J. Neurosci.* **18,** 8605 (1998).

by continually exposing the cultured cells to serum-free medium containing 1 μM antisense oligonucleotide and recording 8 days later. However, in our hands, postnatal neurons show virtually no spontaneous uptake of antisense molecules. Therefore, we have used and refined formulations of antisense oligonucleotides with specific cationic vectors, as developed by other groups.[14,15] The complexes formed are taken up by postmitotic neurons and render such cells amenable to electrophysiological analysis.

Vector Choice

To obtain sufficient numbers of identifiable target cells for analysis, the antisense transfection rate must be optimized. In a culture dish containing between 10^4 and 10^6 neurons, transfection rates as low as 1% can be exploited electrophysiologically. However, in our hands it has been difficult to improve on this level because the general understanding of the critical steps governing transfection efficiency is poor. Because the general understanding of the critical steps governing transfection efficiency is poor, current protocols using antisense oligonucleotides simply aim at facilitating the passage of a large number of the antisense molecules across the cell membrane and through the lysosomal compartment, so as to be delivered finally to the cytoplasm or nucleus.

Because our primary aim was to knock down channel activity and perform detailed electrophysiological analyses, only vectors that did not disturb the electrical properties (e.g., minimal leak current) of the cells could be used. On this basis we excluded the use of first-generation cationic compounds, including calcium phosphate and diethylaminoethyldextran, although these have proved valuable in the study of numerous channels in various expression systems. Since the late 1980s, two main groups of synthetic vectors have been developed for gene transfer: cationic lipids[16,17] and cationic polymers. In working with neurons, we initially found that lipospermine (Transfectam) was a reasonable vector, inducing less damage than other available reagents, but subsequent patch-clamp analysis proved highly problematic.[18] The vector we now use is the polycationic polymer

[14] J. P. Behr and B. Demeneix, *Mol. Therapeut. Gene Ther. Oligonucleotides* **1**, 5 (1997).

[15] J. S. Remy, D. Goula, A. M. Steffan, M. A. Zanta, O. Boussif, J. P. Behr, and B. Demeneix, "Self-Assembling Complexes for Gene Delivery: From Laboratory to Clinical Trial" (A. V. Kabanov, P. L. Felgner, and L. W. Seymour, eds.), pp. 135–148. John Wiley & Sons, New York, 1998.

[16] F. Lezoualch, I. Seugnet, A. L. Monnier, J. Ghysdael, J. P. Behr, and B. A. Demeneix, *J. Biol. Chem.* **270**, 12100 (1995).

[17] B. Schwartz, C. Benoist, B. Abdallah, D. Scherman, J. P. Behr, and B. Demeneix, *Hum. Gene Ther.* **6**, 1515 (1995).

[18] F. Barthel, P. Feltz, J. P. Behr, P. Sassone-Corsi, and A. Feltz, *J. Neurochem.* **54**, 1812 (1990).

polyethyleneimine (PEI, 50 kDa; Sigma, St. Louis, MO).[19] This reagent provides an acceptable transfection rate and good electrical properties, and induces no toxicity when used at a concentration below 180 μM for a 1- to 4-hr exposure.[11] Formulation of a fully phosphorothioated antisense oligonucleotide with PEI in NaCl, 150 mM, as we routinely use, appears to produce submicrometric particles which are especially small ($<$500 nm by electron microscopic analysis; J. S. Remy, personal communication, 1998). The way the complexes are formulated has a dramatic effect on their charge and size. For example, plasmid DNA in a mixture of linear polymers (with repeated CH_2–CH_2–NH sequences) with a mean molecular mass of 22 kDa (Exgene 500; Euromedex, Souffelweyersheim, France) exhibits different sizes depending on whether the formulation medium contains NaCl or glucose. Preliminary experiments carried out with differently substituted oligonucleotides (phosphothioates versus phosphodiesters) in PEI also show that the oligonucleotide nature and formulation (NaCl or glucose)[20] dramatically affect the size and charge of the complexes formed (D. Goula, personal communication, 1998). No doubt much pertinent information on the biophysical characteristics of the efficient transfection complexes will be obtained from quasielastic light scattering (QUELS) measurements.

Transfection Procedure

In brief, the required amount of PEI diluted in 10 μl (per 3.5-cm-diameter petri dish to be transfected) of a 9% NaCl solution at pH 7.1 is added to the DNA solution prepared in the same volume of saline solution. No improvement was observed when formulating oligonucleotide in a glucose solution. The best transfection rates were obtained when mixing the PEI and DNA such that the amine nitrogen and phosphate are present in a molar ratio of 10 to 1, respectively. At first, for electrophysiological recordings to be carried out 1–3 days after transfection, 500 nM oligonucleotide solutions were prepared. However, we have reduced the concentration to 300 nM as this reduces the amount of PEI necessary to condense the oligonucleotide and thus reduces potential toxicity (see Table I). For expression experiments, plasmid DNA (usually at 0.5–1 μg/μl) was condensed with PEI as described above.

[19] O. Boussif, F. Lezoualch, M. A. Zanta, M. D. Mergny, D. Scherman, B. Demeneix, and J. P. Behr, *Proc. Natl. Acad. Sci. U.S.A.* **92,** 7297 (1995).

[20] D. Goula, J. S. Remy, P. Erbacher, M. Wasowicz, G. Levi, B. Abdallah, and B. A. Demeneix, *Gene Ther.* **5,** 712 (1998).

TABLE I
POLYETHYLENEIMINE TOXICITY MEASURED
IN CEREBELLAR GRANULE CELLS

Concentration of PEI (μM)	Cell survival percent + SEM (n)
60, 90, 120, 150	100
150	63 ± 8 (5)
240	44 ± 9 (4)
360	25 ± 6 (3)

Preparation of Stock Solutions

Polyethyleneimine Solution. We routinely used PEI (50 kDa; Sigma) for which we have tested toxicity and transfection rates. A stock solution of 100 mM monomers ($-CH_2-CH_2-NH$, one nitrogen interspaced by two carbon residues) of PEI, brought to pH 7.0 by HCl addition, is kept in 1-ml aliquots at 4° for up to 6 months.

Oligonucleotide Solution. Oligonucleotide aliquots (50 μM) are prepared and stored at $-20°$. An approximately 50 μM solution is first prepared, using the optical density (OD) value given by the supplier. A further OD measurement is carried out after dilution to obtain a more accurate estimate of DNA concentration. The latter nominal oligonucleotide concentration is used in the preparation of the transfection medium. This calibration must be systematically performed so as to achieve the correct formulation of fluorescent oligonucleotide in PEI.

NaCl Solution. A 150 mM NaCl solution is filtered with a 0.22-μm pore size Millipore (Bedford, MA) filter.

Transfection Medium. Proteins have been shown to aggregate strongly with the PEI/DNA particles, thus serum-free medium is used during the period of transfection.

Antimitotic Treatment. Importantly, it should be noted that transfection is efficient when highly enriched neuronal populations are directly exposed to the transfection medium. Therefore, division of glia should be minimized by an antimitotic treatment 1 day prior to the transfection. In our hands, fluorodeoxyuridine has fewer long-term toxic effects than cytosine arabinoside. One day after cell dissociation, fluorodeoxyuridine (4×10^{-4} M) is added to the culture medium for 4 hr and then washed.

Preparation of Polyethyleneimine/DNA Complex in a 10-to-1 Ratio ($Z = 10$). It is to be remembered that in PEI there is one nitrogen per monomer, so that 3 nmol of PEI brings in the same amount of nitrogen and that 1 μg of DNA contains 3 nmol of phosphate. As a result, a practical rule is to add 0.3 μl of 100 mM PEI to condense 1 μg of DNA with $Z = 10$.

Transfection Protocol. At the time of transfection, for each petri dish to be transfected, 10 μl of a 150 mM NaCl solution containing PEI is gently added to 10 μl of a 150 mM NaCl oligonucleotide solution and agitated by gentle movement of the pipette tip (four to six movements). Condensation occurs rapidly and is visualized by the change in fluorescein emission of the oligonucleotide from its characteristic yellow to a faint pink color. The solution is gently added to 0.8 ml of serum-free transfection medium. This volume of transfection medium is the minimum volume necessary to cover the neuronal cells when cultured in a 3.5-cm-diameter petri dish, thus maximizing contact with PEI/DNA particles. For a culture dish of different diameter, the volume of transfection medium would need to be adjusted appropriately. When using larger volumes of transfection medium, transfection rates are somewhat decreased unless the culture dishes are gently centrifuged. Neurons are exposed for 3 hr to the transfection medium at 35° in an incubator before being returned to the serum-containing culture medium.

Experimental Techniques to Assess Antisense Knockdown

Inhibition of Protein Synthesis by Antisense Oligonucleotides: Use of Cell-Free Reticulocyte Lysate System

It may at first impression seem curious to use the reticulocyte lysate system to inhibit the expression of a particular protein, as the main function of this expression system is precisely to express the protein in question. However, it should be noted that not all antisense oligonucleotides are equally efficient in inhibiting the expression of particular genes. The reasons for this difficulty are multiple, including (1) the importance of a stable duplex formation, (2) the specificity of antisense oligonucleotide action, (3) the ease of transfection, and (4) the localization of the duplex on the mRNA transcript. Some of these difficulties can be preempted by an *in vitro* analysis, using the reticulocyte lysate system. The reticulocyte lysate system has several immediate advantages relative to cell transfection because (1) this system has the ability to radiolabel a specific protein selectively, which eases the analysis of antisense oligonucleotide action, (2) the delivery of the antisense oligonucleotide to the translation machinery of the cell is highly simplified, as the antisense oligonucleotide can be readily added to the reaction mixture *in vitro,* (3) the optimization of various parameters, including time of incubation and concentration of antisense oligonucleotide, is considerably eased, and (4) the analysis of the results of antisense oligonucleotide action is straightforward. Finally, this system provides a unique opportunity to test the effects of a single antisense oligonucleotide on a series of mRNAs if, for example, the reagent is designed to knock down the expression of a family of proteins.[10] Thus, this system allows the optimization of an antisense oligonucleotide sequence that best prevents the expression of a particular mRNA.

Plasmid Selection. In vitro translation requires either cDNA or cRNA but in both cases a prokaryotic phage RNA polymerase promoter is used for the initiation of transcription. However, not all vectors are equivalent. As for cRNA expression in *Xenopus* oocytes, there are several pitfalls that should be avoided in order to obtain good protein expression. The efficiency of translation is mostly dependent on structural features of the 5' and 3' ends. For example, the RNA polymerase promoter should not be located too far away from the start of translation and false starts and stops should be avoided. The insertion of 36 bp from the 5' untranslated region of the alfalfa mosaic virus increases the yield of translated product and diminishes the amount of incorrectly initiated translation products. A well-defined Kozak consensus sequence (A/GCC AUG G) should be used to define the first methionine of the protein. Finally, an efficient polyadenylation signal should be included downstream of the stop codon for greater RNA stability. We have found that vectors derived from the pcDNA3 backbone fulfill these requirements and present the additional advantage that they possess a eukaryotic promoter, which is useful for cell transfection. As a rule of thumb, we found that plasmids that provided good protein expression in *in vitro* translation also work well in the *Xenopus* oocyte expression system.

In Vitro Transcription/Translation. Several commercial kits are now available for *in vitro* transcription and translation. The two processes, transcription and translation, can either be performed separately or coupled within the same microcentrifuge tube. We have successfully used the TNT system (Promega, Madison, WI). These reagents allow an efficient coupled *in vitro* transcription/translation using a rabbit reticulocyte lysate system and one radioactively labeled amino acid, such as [^{35}S]methionine, [^{35}S]cysteine, or [^{3}H]/[^{14}C]leucine. It does not require the usual RNA synthesis of standard reticulocyte lysate translations.

Protocol. As for *in vitro* transcription, great care must be taken to reduce RNase contamination. This can be achieved by using RNase-free microcentrifuge tubes and pipette tips, and the additional use of a ribonuclease inhibitor is strongly recommended. A typical 50-μl *in vitro* translation reaction using [^{35}S]methionine as radioactive amino acid is set up in the following way:

Rabbit reticulocyte lysate	25 μl
Reaction buffer	2 μl
RNA polymerase (T7, T3, or SP6)	1 μl
1 mM Amino acid mixture minus methionine	1 μl
[^{35}S]Methionine (1000 Ci/mmol) at 10 mCi/ml	4 μl (translational grade)
RNase ribonuclease inhibitor (40 U/μl)	1 μl
DNA template (1 μg/μl)	1 μl
Nuclease-free H$_2$O	15 μl

The reaction should be incubated at 30° for 1 to 2 hr.

In vitro translation has several limitations, including the size of the protein that can be generated. We found that proteins as large as 200 kDa may be difficult to synthesize, particularly if the number of hydrophobic regions is high. In addition, the yield of the protein synthesized may be low, indeed below purification level. Protein concentration in the lysate can be determined by trichloroacetic acid precipitation in the presence of 2% casamino acids and we determined a concentration of 1–2 nM for the voltage-dependent calcium channel.[21] Higher concentrations (up to 50 nM) have been reported by other authors.[22]

Antisense Oligonucleotide-Induced Inhibition of Translation. As normal translation is carried out in the presence of an RNase inhibitor, it is necessary to omit it from reactions in which the effects of antisense oligonucleotide are to be tested. The reticulocyte lysate may by itself contain some RNase H required for the antisense oligonucleotide-induced degradation of the mRNA produced from the cDNA added. It is, however, a good idea to supplement the basic translation reaction with exogenous RNase H. We add 2 μl of recombinant RNase H (Promega) at 1.5 units/μl in the 50-μl translation reaction. The addition of RNase H does not affect the efficiency of protein translation.[10] We have found that antisense knockdown in selective translation can be achieved by the addition of 300 nM oligonucleotide in the translation reaction in addition to the RNase H (see Fig. 1).

Analysis. Generally, it is not necessary to immunoprecipitate the translated protein, as background translation is minimal. It is satisfactory to analyze the protein translated by running about 2–5 μl of the translational mixture on a sodium dodecyl sulfate (SDS)–polyacrylamide gel followed by autoradiography. Exposure time should vary depending on the nature of the radioactive amino acid incorporated, the specific activity of this

[21] M. De Waard, D. R. Witcher, M. Pragnell, H. Liu, and K. P. Campbell, *J. Biol. Chem.* **270,** 12056 (1995).

[22] E. J. Neer, B. M. Denker, T. C. Thomas, and C. J. Schmidt, *Methods Enzymol.* **237,** 226 (1994).

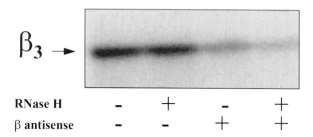

A

$\beta_3 \rightarrow$

RNase H   −   +   −   +

β antisense   −   −   +   +

B

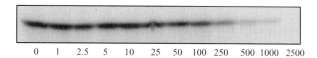

0   1   2.5   5   10   25   50   100   250   500   1000   2500

β antisense (nM)

C

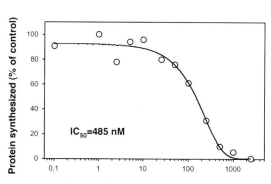

$IC_{50} = 485$ nM

Antisense oligonucleotide concentration (nM)

Fig. 1. Synthesis of the ^{35}S-labeled β_3 subunit is blocked by a selective antisense oligonucleotide. (A) Effect of β antisense (500 nM) and exogenous RNase H (1 unit) addition on *in vitro* translation reactions of ^{35}S-labeled β_3. The β antisense used here, and subsequently, was identical to the oligonucleotide used by Lambert *et al.*[10] and contains two mismatches. The GenBank nucleotide accession number for β_3 is LO2315. Lanes 3 and 4 illustrate that addition of exogenous RNase H is not required for inhibition of protein synthesis but does produce greater inhibition. (B) Effect of increasing concentrations of β-antisense oligonucleotide on β_3 synthesis. The concentration of antisense was varied between 1 and 2500 nM in

amino acid, the number of radioactive amino acids incorporated, the size of the protein, and the quality of the translation. The latter depends mostly on the cDNA construct being used. In our hands, as an indication, we found that under optimal conditions, using [^{35}S]methionine as amino acid marker and for a protein of about 60 kDa in size containing about 10 methionines, an exposure time of 5–6 hr should be sufficient. If it turns out to be much longer under these conditions (2–4 days), then it is advisable to subclone the insert of interest into a new vector more suitable for such analysis. Another indication of poor translation quality is provided by the background radioactivity that can be induced by the translation of endogenous proteins. If this background is too high, then the specific translation of the protein under study may be difficult to detect. If quantifying the effect of antisense oligonucleotide is required, then it may be useful to use a PhosphorImager (Bio-Rad, Hercules, CA) to scan the gel and appropriate software (Molecular Analysis) to quantify the decrease in radioactivity. Alternatively, if antibodies directed against the protein translated are available, then the decrease in translation efficiency induced by the antisense oligonucleotide can be quantified by direct immunoprecipitation of the labeled protein and quantifying the change in radioactivity by β or γ counting. Another approach, although more exacting than those described above, is to quantify the decrease in protein expression by changes in binding to a defined ligand. This procedure was used to determine the decrease in calcium channel β-subunit translation by measuring β-subunit binding to its α_1 site.[10]

Posttransfection Analysis of Target Protein by Immunocytochemistry and mRNA Hybridization

After cell transfection, analyses to evaluate the levels of target protein and/or mRNA are of considerable importance in the assessment of antisense efficacy. Quantitative mRNA analysis on single cells or on small subpopulations of transfected cells within a heterogeneous culture is highly exacting and we have found no simple solution to this issue. In tissue

---

FIG. 1. (*continued*) the presence of 1 unit of RNase H and compared with a control translation in the absence of antisense. Complete inhibition is evident at 2.5 μM. In (A) and (B), the translation reactions were 25 μl each. Five microliters of each reaction was used for protein separation on a 9% SDS–polyacrylamide gel. The gel was then dried and radioactivity quantified by phosphoimaging (Bio-Rad). (C) The intensities of bands shown in (B) were quantified with the PhosphorImager, normalized to the maximum intensity obtained, and shown as a function of oligonucleotide antisense concentration. The data obtained were fitted according to a sigmoidal equation: $f(x) - IC_{50}/\{1 + \exp[-(x - x_0)/b]\}$, where $IC_{50} = 485.8$ nM, $b = -203.3$, and $x_0 = -294.8$.

section, *in situ* mRNA hybridization is our favored method for the analysis for both mRNA knockdown and oligonucleotide penetration.[23] However, despite the fact that these methods can be used on cell smears or cultured neurons grown on coverslips, if the transfection efficiency is low then such experiments have in our hands proved highly impractical because one first needs to identify transfected cells and then perform *in situ* mRNA hybridization. Single-cell polymerase chain reaction (PCR) does not overcome this problem, nor do Northern/RNase protection assays.

Target-specific antibodies are critical tools in antisense experiments. We have generated an extensive panel of polyclonal and monoclonal reagents[24] both to define the basal neuronal level of the calcium channel protein under study and also assess its knockdown after antisense treatment.[7] A significant difficulty in assessing protein knockdown is encountered when differentiating between functionally active protein and inactive quiescent subunits, which, in the case of ion channels, may reside deep in the cytoplasm. The synthesis, processing, and deposition of ion channels into the membrane is an extremely inefficient process and it has been estimated that only ~30% of the protein produced may end up in an active form in the membrane.[25] If antibodies specifically label the active protein, confocal microscopy can help to evaluate the level of membrane-inserted protein and therefore the oligonucleotide efficacy.[26] However, because an antibody will generally bind to its epitope in both active and inactive forms, demonstrating protein knockdown may be difficult. To help overcome these problems we have developed and used antibodies against the selective toxins that define calcium channels. For example, by labeling oligonucleotide-treated Purkinje cells with ω-agatoxin IVA and/or ω-conotoxin GVIA initially and subsequently localizing extracellularly bound toxin with monoclonal antibodies, we were able to demonstrate and quantify antisense effects that correlated well with electrophysiological and pharmacological parameters.[7] The technical protocol for this staining method and quantification technique is given below.

Protocol. Glass-adherent neurons are fixed in prewarmed 4% (w/v) paraformaldehyde in phosphate-buffered saline (PBS) for 20 min at 37°, washed in PBS, and incubated with either 100 μl of ω-agatoxin IVA (50 nM), ω-conotoxin GVIA (1 μM), or PBS for 10 min at room tempera-

[23] P. Craig, A. McCainsh, A. McCormack, W. Smith, R. Beattie, J. Priestley, J. Yip, S. Averill, R. Lonbottom, and S. G. Volsen, *J. Comp. Neurol.* **397,** 251 (1998).

[24] R. E. Beattie, S. G. Volsen, D. Smith, A. McCormack, S. E. Gillard, J. P. Burnett, A. Gillespie, M. M. Harpold, and W. Smith, *Brain Res. Prot.* **1,** 307 (1997).

[25] J. D. Clemens, *Trends Neurochem. Sci.* **19,** 163 (1996).

[26] N. S. Berrow, V. Campbell, E. M. Fitzgerald, K. Brickley, and A. Dolphin, *J. Physiol.* **482,** 481 (1995).

ture. After washing, cells are incubated for a further 30 min with anti-toxin monoclonal antibodies (10 μg/ml). The Purkinje cell phenotype is also characterized by immunohistochemical analysis with antibodies specific for calbindin 28 kDa, γ-aminobutyric acid (GABA), and synaptophysin. Cultures are then incubated with fluorescein isothiocyanate (FITC)-conjugated goat anti-mouse IgG (20 μg/ml) for 30 min at room temperature and washed repeatedly. Coverslips are mounted on glass microscope slides in an antifade mountant (Citifluor) to reduce fluorescence quenching. Cells are photographed on a Leica (Bensheim, Germany) DM IRB UV microscope and the relative fluorescence intensities are measured with an Astrocam digital camera operating with Optimas image analysis software (see Fig. 2a and b; see color insert).

Inhibition of Protein Synthesis in Xenopus Oocytes: A Cell Model for Exogenous Protein Expression

Unlike cell-free systems, the *Xenopus* oocyte can correctly carry out posttranslational modifications including precursor processing, phosphorylation, and glycosylation of several proteins of exogenous origin. It will also direct proteins toward the correct intracellular compartment in an active form. The *Xenopus* oocyte expression system also supports a facilitated delivery of the antisense oligonucleotide by direct cytoplasmic injection of the molecule. Thus in *Xenopus* oocytes, antisense oligonucleotides have been useful in multiple applications.[27] Indeed, several studies reveal that the inhibition of mRNA levels by oligonucleotide antisense treatment can be close to 100% in this system.

Xenopus Oocyte Preparation and Maintenance. Xenopus laevis frogs are the preparation of choice because of the ability of females of this species to produce oocytes all year round. The animals are maintained under a 12 hr light–12 hr dark cycle at 16° and fed twice a week. Albino *Xenopus* frogs can be used if nuclear injections are to be performed, the absence of plasma membrane pigmentation favoring the nucleus localization. To harvest oocytes, the frogs need to be anesthetized with 0.03% ethyl *p*-aminobenzoate (Sigma). Care must be taken that frogs are not fully covered with the anesthetic solution, to avoid drowning the animals. Ovaries are surgically removed by creating two incisions within the skin and muscle layer. The incisions should be sutured immediately after removal of the oocytes and the animals returned to the tank after recovery from the anesthetics. Frogs are allowed to recover for at least 2 months before being reused. Follicle membranes from isolated batches of oocytes are enzymatically digested for

[27] P. J. Green, O. Pines, and M. Inouye, *Annu. Rev. Biochem.* **55,** 569 (1986).

2 hr with collagenase type IA (2 mg/ml; Sigma) in Ca^{2+}-free Barth's solution [in mM: NaCl 88, KCl 1, $MgSO_4$ 0.82, $NaHCO_3$ 2.4, HEPES 15, NaOH (pH 7.4)]. After defolliculation, oocytes at stage V and VI are isolated under a stereomicroscope from their less mature precursors and washed several times with standard Barth's solution [in mM: NaCl 88, KCl 1, $MgSO_4$ 0.82, $Ca(NO_3)_2$ 0.33, $CaCl_2$ 0.41, $NaHCO_3$ 2.4, HEPES 15, NaOH (pH 7.4)]. The oocytes are then maintained in defined nutrient oocyte medium (DNOM), as developed by Eppig and Dumont,[28] supplemented with gentamicin (50 μg/ml) for 1–2 days prior to antisense oligonucleotide injection.

Antisense Injection into Xenopus Oocytes. Oligonucleotides are dissolved in water at a final concentration of 100 μM (stock solution). Care should be taken to ensure that the solution is free of solid contaminating particles by a rapid microcentrifugation in order to avoid electrode clogging during the aspiration or injection procedures. Five to 6 μl of this solution is loaded into the injection electrode by vacuum aspiration. *Xenopus* oocytes are injected with the antisense solution after placing them into a microchamber designed for cell immobilization and containing Barth's medium. The volume of injection should not exceed 50 nl. Usually, 20 nl is an ideal volume as it does not exceed one-fiftieth of the cell volume, which on average is 1 μl. The *Xenopus* oocyte system has the advantage that the concentration and time of injection of the antisense oligonucleotide can be controlled accurately. These parameters should be determined empirically and depend on the nature of the experiment (mRNA or protein), the transcript targeted (endogenous or expressed), the half-life of the protein, and the stability of the oligonucleotide.

The time of antisense oligonucleotide injection is a crucial parameter. It will vary depending on whether mRNA degradation or protein knockdown is to be studied. If one wants to analyze the degradation of an endogenous *Xenopus* mRNA, then the time of injection of antisense oligonucleotide is of little relevance. However, if the degradation of exogenous transcripts is of interest after cDNA injection into the nucleus, then there should be a delay between injections of 12 to 24 hr. Various techniques facilitate the study of mRNA degradation after antisense oligonucleotide injection. Northern blot as well as RNase protection assays have been used successfully. It should be mentioned that the kinetics of mRNA degradation after antisense oligonucleotide injection have been analyzed in oocytes and that the maximum cleavage level of transcripts occurs within 60 min, a time period that is shorter than the lifetime of the oligonucleotide.[29] Therefore, it is not necessary to delay between antisense oligonucleotide injection and mRNA level determination. If, however, the antisense oligonucleotide

[28] J. J. Eppig and J. N. Dumont, *In Vitro* **12,** 418 (1976).
[29] J. Shuttleworth and A. Colman, *EMBO J.* **7,** 427 (1988).

a

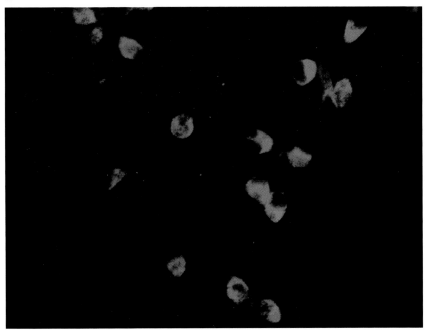

b

FIG. 2. Immunocytochemical characterization of oligonucleotide-treated Purkinje cells after 8 days in culture. Photomicrographs show fluorescence after exposure to an anti-AgaIVA antibody after treatment with an $\alpha 1 A$ mismatch oligonucleotide (a), and with an $\alpha 1 A$ antisense oligonucleotide (b). Note the reduced signal in antisense-treated cells. [From S. E. Gillard *et al.*, *Neuropharmacology* **36,** 405 (1997) with permission.]

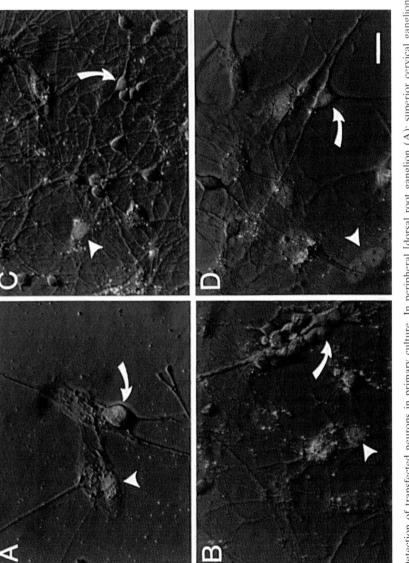

FIG. 3. Detection of transfected neurons in primary culture. In peripheral [dorsal root ganglion (A); superior cervical ganglion (B)] and central [granule cells of the cerebellum (C); hippocampus (D)] neurons transfected with 5'-fluorescein-conjugated oligonucleotides, nuclei of some neurons (arrows) and glia (arrowheads) presented clear fluorescence. Cells were visualized with a modulation optic system using both bright-field and standard fluorescein epifluorescence illumination. Note that the apparent staining intensity observed here is enhanced by the long exposure time used. Bar: 25 μm. [From R. C. Lambert et al., Mol. Cell Neurosci. **7**, 239 (1996) with permission.]

injection is aimed at inhibiting the translation of a protein, then it is advisable to inject it 2 days after oocyte preparation, either alone (in the case of an endogenous protein) or along with the mRNA of exogenous origin (when an exogenous protein is studied).[30] The success of hybrid arrest of translation is determined by the half-life of the protein itself, as the antisense oligonucleotide-induced transcript degradation is not the rate-limiting step. Therefore, biochemical and functional studies aimed at determining protein levels or the effects of protein depletion after antisense oligonucleotide injection should be performed after an interval of time longer than the half-life of the protein studied. This can be as short as a few hours or as long as a few days and needs to be determined empirically.

In contrast, the concentration of antisense oligonucleotide injected seems to be of less importance than the time of injection. It is recommended to inject an $\sim 2 \ \mu M$ concentration of antisense oligonucleotide. For a 20-mer antisense, this corresponds to approximately 12.5 ng of antisense oligonucleotide. This concentration can be varied within the range of 10 to 50 ng per oocyte.[29] Such a concentration of antisense oligonucleotide (e.g., 50 ng) is in great molar excess compared with the concentration of the transcript in the oocyte. A concentration of endogenous transcript of about 100 pg is current in oocytes, which corresponds to about 5000 times less mRNA molecules than antisense oligonucleotide (assuming an mRNA sequence only 10 times longer than the antisense oligonucleotide). However, such a molar excess appears to be important for efficient mRNA degradation. This molar ratio also holds in the case of exogenous poly(A) mRNA injection into oocytes and antisense oligonucleotide injection has revealed a potent strategy to degrade the mRNA resulting from the transcription of a single cDNA injected into the nucleus.[29] The point of injection (animal versus vegetal pole) in the oocyte seems of little importance, as diffusion of the antisense oligonucleotide does not appear to be restricted.

Analysis of Effects of Antisense Oligonucleotide Injection into Xenopus Oocytes. The translation products of an injected oocyte and the effect of antisense oligonucleotide can be measured by several biochemical, pharmacological, and electrophysiological assays based on the expected bioactivities of the desired proteins. For protein studies, *Xenopus* oocytes have been successfully used for secretion, uptake, ligand binding, immunochemical, and electrophysiological studies demonstrating the wide array of antisense oligonucleotide applications.[30]

Electrophysiology

When using an antisense strategy to study ionic channels, it is likely that electrophysiological techniques will be the major analytical tools. We

---

[30] J. Alder, B. Lu, F. Valtorta, P. Greengard, and M. Poo, *Science* **257,** 657 (1992).

have used antisense oligonucleotides to knock out the expression of calcium channel subunits in primary cultured neurons. The effects of the subunit depletion were measured by recording currents in the whole-cell patch-clamp configuration. Because electrophysiological techniques do not generally result in the gathering of a large amount of data either quickly or easily, such studies are both exacting and time consuming. When possible, the efficacy of the antisense oligonucleotide should be ascertained by other techniques before commencing electrophysiology (see above). The main problem is that in antisense studies, one must wait for the turnover of the target protein and this precludes the use of an internal control as in pharmacological studies. There is no way to record the current before, during, and after the treatment. Therefore, the effect on the channel activity cannot be quantified in the same cell, and is assessed by significant differences between mean current properties measured in antisense oligonucleotide-treated and control cells. Considering both that channel activity may be dependent on many physiological or experimental parameters (including phosphorylation state, G protein regulation, intracellular dialysis) and that electrophysiological techniques independently deal with individual cells, which thus limits the size of the studied population, one must be careful in reducing every source of variability to obtain accurate statistics.

A major point in antisense studies of neuronal primary cultures is the accurate detection of positively transfected cells. To achieve this, we have used a fluorescein group conjugated in the 5' position with the phosphorothioate oligonucleotides (see Antisense Oligonucleotide Design, above). Under this condition transfected cells exhibit a green fluorescent labeling of their nucleus with limited photobleaching when observed with standard fluorescein epifluorescence illumination [Nikon (Garden City, NY) epifluorescence filter DM510 B2A] (see Fig. 3; see color insert). In addition, depending on the cell type and the culture, a variable number of cells display weak staining of their cytoplasm and/or a strong punctate staining either in the intracellular compartment or on the cell membrane. However, the antisense oligonucleotides seem to be efficient only in neurons with a stained nucleus, and the study should be restricted to this cell population. In some cases (antisense oligonucleotides directed against the Ca_vT α_1-subunit family[13]) the efficiency of the antisense oligonucleotide could be related to the intensity of the nuclear labeling, but the relationship is not systematic and depends on the antisense oligonucleotide used. Generally, the intensity may be variable from transfection to transfection and nuclear staining is often faint. However, under our conditions (using phosphorothioate oligonucleotides), the fluorescein labeling was strong enough to allow visualization of the transfected cells using modulation optics combined with simultaneous bright-field and epifluorescence illumination (Fig. 3; see color insert). Therefore, no modification in the optical set-up was necessary between the cell detection process and patch-clamp analysis. No doubt the use of an image enhancement system would greatly facilitate the detection process.

To reduce variability, transfection methods with short incubation times (see Transfection Procedure, above) are preferred to allow the effect of the antisense oligonucleotide on protein synthesis to be synchronized. In addition, it is preferable to record in both control and antisense oligonucleotide-transfected neurons from the same batch of culture in order to reduce culture-to-culture and transfection-to-transfection discrepancies. Finally, if the expected effect of the antisense treatment is a reduction in the current amplitude, improvement in the significance of the mean effect can be obtained by considering current densities rather than current amplitudes.

Another critical aspect of antisense oligonucleotide studies is the choice of controls. Because only a few stained neurons are observed in each petri dish, the other neurons may constitute easy controls. However, if a cell displaying a clear staining of its nucleus can be considered as positively transfected, the contrary is not straightforward. In addition, a close observation of the neurons generally helps to choose the more favorable cells for patch clamp. If unstained cells are used as controls, the selection criteria may be more rigorous for the large control cell population than for the small transfected cell population, inducing unexpected bias in the statistics. Therefore, ideal controls are cells from the same culture transfected with oligonucleotides that theoretically do not induce any effect. Such oligonucleotides should be similar in composition to the antisense, which implies the use of sense or scrambled sequences. In some cases, the sense oligonucleotide seems to weakly induce the antisense effect (possibly by an action during the mRNA transcription process), and therefore multiple assays with different scrambled oligonucleotides are preferred (see Fig. 4A).

Theoretically, antisense oligonucleotides are highly specific tools but past studies have clearly showed that this point must be determined with much care. When considering antisense oligonucleotides designed to suppress the expression of a specific channel subunit, its absence of effect on other channel populations must be tested. Ideally, the unaffected currents can be recorded at the same time as the affected current. For example, when studying low voltage-activated calcium channels in sensory neurons, we could record simultaneously both low and high voltage-activated calcium currents. Therefore, we had an internal control because the specificity of the effect was checked in every recorded cell: anti-β-subunit oligonucleotides affected only high voltage-activated currents[10] (see Fig. 5A) and, in contrast, anti-Ca$_v$T α_1-subunit oligonucleotides suppressed specifically low voltage-

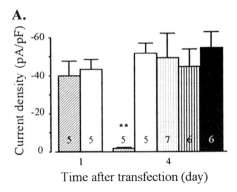

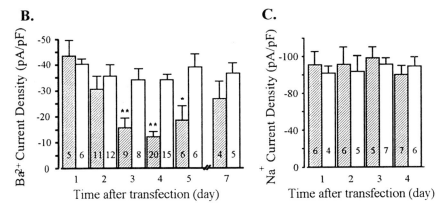

FIG. 4. Transfections with a generic calcium channel β-subunit antisense oligonucleotide specifically affect calcium channel activity. Histogram in (A) represents the mean ($\pm$ SEM) high voltage-activated calcium current density evaluated in dorsal root ganglion neurons transfected either with a calcium channel β-subunit antisense oligonucleotide (shaded columns) or three different scrambled oligonucleotides (white columns and columns filled with vertical lines), or with the sense oligonucleotide (black column). Currents were recorded 1 and 4 days after transfection. Four days after transfection, the current density is significantly decreased in cells treated with the antisense oligonucleotide, but the current densities measured in scrambled and sense oligonucleotide-transfected cells remained similar and constant. (B) Histogram representing the mean ($\pm$ SEM) Ba^{2+} current density recorded in granule cells of the cerebellum transfected, respectively, with a β-subunit antisense oligonucleotide (shaded columns) and a scrambled oligonucleotide (white columns) on different days after transfection. The current density in scrambled oligonucleotide-transfected cells remained constant, but it progressively decreased in antisense oligonucleotide-transfected cells, reaching a minimum 4 days after transfection, which corresponds to the turnover of the protein. The histogram in (C) represents the mean ($\pm$ SEM) Na^+ current density recorded in granule cells of the cerebellum transfected, respectively, with a calcium channel β-subunit antisense oligonucleotide (shaded columns) and a scrambled oligonucleotide (white columns). The lack of effect of the antisense oligonucleotide on the Na^+ current demonstrates the selectivity of the antisense oligonucleotide for voltage-dependent calcium channels. In (A), (B), and (C) numbers of cells used to calculate the means are indicated inside the columns. $**p < 0.001$; $*p < 0.05$; Student t test. [From R. C. Lambert et al., Mol. Cell Neurosci. **7**, 239 (1996) with permission.]

A.

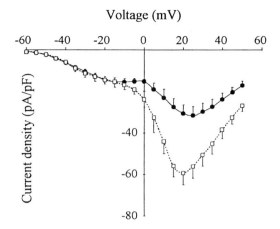

B.

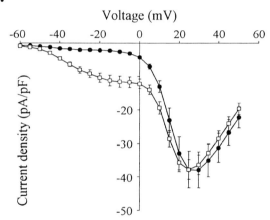

FIG. 5. Use of an antisense strategy to clarify the molecular structure of low voltage-activated calcium channels. (A) Mean calcium current densities measured at various voltages in nodosus sensory neurons transfected either with a calcium channel β-subunit antisense oligonucleotide (●; $n = 14$) or a scrambled oligonucleotide (□; $n = 11$). Four days after transfection, these $I-V$ curves clearly indicate that, in the same cell, depletion of β subunits specifically affects the high voltage-activated calcium current but does not modify the low voltage-activated calcium current. [From R. C. Lambert *et al., J. Neurosci.* **17,** 6621 (1997) with permission.] A reverse image is obtained in (B), where sensory neurons were transfected either with an antisense oligonucleotide designed to block the Ca_vT α_1-subunit expression (●; $n = 11$) or a scrambled oligonucleotide (□; $n = 11$). The specific suppression of the low voltage-activated calcium current obtained 3 days after transfection demonstrates that members of the Ca_vT subunit family encode this current in sensory neurons. [From R. C. Lambert *et al., J. Neurosci.* **18,** 8605 (1998) with permission.]

activated currents[13] (see Fig. 5B). However, this cannot always be achieved and then a theoretically unaffected current can be studied independently in the same way as the target current[11] (see Fig. 4B and C).

Finally, as the antisense oligonucleotide effect is based on the turnover of the protein, this effect develops slowly. Its kinetics can be estimated by a systematic study of the channel activity in antisense and scrambled oligonucleotide-transfected cells according to days/hours after transfection (see Fig. 4B). This part of the study checks that the observed modifications are compatible with an antisense effect and highlights the turnover of the knockdown protein. This information is not only of particular interest for our knowledge of the cell biology of ion channels, but is also practical, because the duration of the turnover compared with the viability of the cell culture is the main limitation in applying the antisense strategy to the study of ionic channel function.

Conclusion

The antisense strategy is a potent tool with which to identify the function of a specific neuronal gene. It provides a powerful approach to link the enormous amount of data generated by molecular cloning to functional studies in the native environment. It will undoubtedly cast new light on both the role of precise subunits in receptor channels or voltage-dependent channels and ancillary subunits linking channels to other protein complexes such as the cytoskeleton and exocytosis machinery. However, reliable conclusions from an antisense approach can be attained only by robust statistical studies. At present, this may perhaps limit the use of this strategy to certain endeavors. Doubtless, further developments improving both transfection rates and oligonucleotide chemistry, and in predicting the action of the antisense oligonucleotides, are to be expected and will increase the range of application of the antisense oligonucleotide strategy significantly.

Section III

Antisense in Nonneuronal Tissues

[22] Antisense Inhibition of Sodium–Calcium Exchanger

By MARTIN K. SLODZINSKI and MAGDALENA JUHASZOVA

A critical problem in assessing the physiological role of the sodium–calcium exchanger (NCX) in arterial and cardiac myocytes (and other types of cells) has been the absence of a selective and specific blocker.[1–4] We have successfully tested antisense oligodeoxynucleotides as a means of inhibiting expression of the NCX.[5–7]

Antisense Oligodeoxynucleotides

Antisense oligodeoxynucleotides (AS-ODNs) have great promise for the inhibition of specific gene expression in cells. In theory, AS-ODNs inhibit gene expression by Watson–Crick base pair binding to the complementary RNA sequences, thereby preventing translation of the mRNA.[8–11] Inhibition of gene expression, which results from hydrogen bonding between nucleic acid bases from the sense (mRNA) and antisense (AS-ODN) sequence should, presumably, be highly specific, given a long enough AS-ODN. The haploid human genome contains about 3×10^9 bases. A random sequence of 17 nucleotides long or longer would have a low probability of occurring more than once, and therefore a high probability of being unique. Thus, antisense strategies have the potential for sensitive and selective inhibition of the NCX expression.

[1] C. Gatto, C. C. Hale, W. Xu, and M. A. Milanick, *Biochemistry* **34,** 965 (1995).

[2] G. J. Kaczorowski, R. S. Slaughter, V. F. King, and M. L. Garcia, *Biochim. Biophys. Acta* **988,** 287 (1989).

[3] Z. Li, D. A. Nicoll, A. Collins, D. W. Hilgemann, A. G. Filoteo, J. T. Penniston, J. N. Weiss, J. M. Tomich, and K. D. Philipson, *J. Biol. Chem.* **266,** 1014 (1991).

[4] J. B. Smith, R.-M. Lyu, and L. Smith, *Biochem. Pharmacol.* **41,** 601 (1991).

[5] M. K. Slodzinski and M. P. Blaustein, *Am. J. Physiol.* **275,** C251 (1998).

[6] M. K. Slodzinski and M. P. Blaustein, *Am. J. Physiol.* **275,** C459 (1998).

[7] M. K. Slodzinski, M. Juhaszova, and M. P. Blaustein, *Am. J. Physiol.* **269,** C1340 (1995).

[8] M. M. McCarthy, P. J. Brooks, J. G. Pfaus, H. E. Brown, L. M. Flanagan, S. Schwartz-Giblin, and D. W. Pfaff, *Neuroprotocols* **2,** 67 (1993).

[9] L. Neckers, A. Rosolen, and L. Whitesell, *J. Immunother.* **12,** 162 (1992).

[10] J.-J. Toulme, "Antisense RNA and DNA" (J. H. A. Murray, ed.), p. 175. Wiley-Liss, New York, 1992.

[11] R. W. Wagner, *Nature (London)* **472,** 333 (1994).

Antisense inhibition of translation has four major theoretical mechanisms of gene expression inhibition. First, AS-ODNs targeted upstream of the initiation codon (AUG) can block the initiation complex through steric forces that cause physical arrest of translation initiation. Second, AS-ODNs complementary to the coding region can prevent polypeptide chain elongation if the RNA–DNA duplex is cleaved by RNase H or base modified by double-stranded RNA (dsRNA) adenosine deaminase. Third, AS-ODNs can sterically prevent the transport of mRNA (e.g., out of the nucleus) or impede translocation of the mRNA along the ribosome, pre-mRNA processing (5' cap addition, splicing, or polyadenylation), ribosomal binding, and translation. Fourth, AS-ODNs linked with bioactive groups (e.g., heavy metals or photosensitizers) can cleave the mRNA. Overall, the preceding are all putative AS-ODN mechanisms; one or more of these mechanisms may function in individual cases, but the exact mechanism has not been completely described.[8–12]

Concerns about the use of AS-ODNs include problems with the stability of the ODNs and problems with cellular uptake. Rapid degradation of AS-ODNs by endogenous and exogenous (e.g., serum) DNases is a particular concern. Replacement of the negatively charged oxygen on the phosphate with a sulfur atom (phosphorothioate) or methyl group (methylphosphonate) confers resistance to degradation. Furthermore, phosphorothioated AS-ODNs are better transported into cells than unmodified AS-ODNs. This uptake process is saturable, and is therefore likely to be a receptor-mediated function.[13]

An antisense molecule can be considered "specific" if no gross loss of cell viability is observed and if the levels of the target RNA, and/or its associated protein, fall below the control levels.[14] However, AS-ODNs can have "nonantisense" effects by acting on some molecule other than its intended target. It may be impossible to prevent destruction of some unintended nontargeted mRNA. The ratio of intended to unintended mRNA destruction depends on a combination of factors. First, the AS-ODN and target mRNA must colocalize.[12] Second, the AS-ODN must have access to the complementary site on the mRNA, which could be buried by proteins and intramolecular bonding. Third, AS-ODNs are charged molecules and can potentially bind nonspecifically to charged proteins, thereby producing a biologically significant, antisense-independent effect.

When an ODN has a sequence of four repeated guanine (G) residues,

[12] G. M. Arndt and G. H. Rank, *Genome* **40**, 785 (1997).
[13] Q. Zhoa, S. Matson, C. J. Herrera, E. Fisher, H. Yu, and A. M. Krieg, *Antisense Res. Dev.* **3**, 53 (1993).
[14] A. D. Branch, *Trends Biochem. Sci.* **23**, 45 (1998).

the affinity of the ODN is altered. Such ODNs can, for example, activate the Sp1 transcription factor nonspecifically.[15] This factor is a "general" transcription factor that interacts with G–C boxes in the promoter region of many viral and cellular genes.[16] The phosphorothioated oligodeoxynucleotides used in these experiments contained four or more tendem guanine residues with a potential of three hydrogen-bonding sites per residue. To combat these pitfalls of AS-ODNs, the base composition must be carefully designed. Even if the AS-ODNs does not contain a G quartet, the sense and scrambled (or mismatched) controls must also be designed so that they do not contain four or more tandem Gs.

A convincing demonstration of antisense gene inhibition requires a clear distinction between antisense mechanism and nonspecific effects of the AS-ODNs. First, an assessment of the cell viability should be demonstrated. Second, the targeted protein or RNA level must be measured and compared with external and internal controls. Third, the target sequence must be screened for homologies among other protein. Fourth, the AS-ODN design should prevent secondary and tertiary structures from forming, minimize tandem guanine residues, and be of the highest chemical purity. Fifth, mismatched controls, containing the same base composition in a scrambled order, must be used to control for sequence-independent effects of AS-ODNs. Sixth, nuclease-resistant AS-ODNs should be used to minimize the effects of DNase activity. Seventh, the lowest dose of AS-ODNs should be used to minimize nonspecific effects and to maximize the sensitivity. All of these considerations must be taken into account for the proper and careful use of AS-ODNs.[8,10,11,14]

Design of Antisense Oligodeoxynucleotides Targeted to NCX1 mRNA

The AS-ODNs were designed to target the region upstream (5') to, and encompassing the ATG start codon of the NCX1 mRNA (Table I). This area is highly conserved among the different, alternatively spliced NCX1 isoforms.[17] No consensus on appropriate design of AS-ODNs has been widely accepted. Our design sought to maximize the probability of steric interference with ribosomal attachment and translocation while maintaining the possibility of RNase H activity.

Chimeric phosphorothioated oligodeoxynucleotides were used. This design has natural deoxynucleotides flanked by four phosphorothioate-modi-

[15] J. R. Perez, L. Yuling, C. A. Stein, S. Majumder, A. van Oorschot, and R. Narayanan, *Proc. Natl. Acad. Sci. U.S.A.* **91**, 5957 (1994).
[16] B. F. Pugh and R. Tjian, *Cell* **61**, 1187 (1990).
[17] P. Kofuji, W. J. Lederer, and D. H. Schulze, *Am. J. Physiol.* **263**, C1241 (1992).

TABLE I
SEQUENCES OF CHIMERIC ANTISENSE, SCRAMBLED, AND SENSE OLIGONUCLEOTIDES TARGETED
TO NCX1 mRNA

Oligonucleotide	Sequence	Oligonucleotide	Sequence
NCX1′ [a]	5′-AACAATTGGAAGTCTCA-3′	NCX1″	5′-TTGTACAACATGCTT-3′
CHAS′ [b]	5′-*TGAG*ACTTCCAAT*TGTT*-3′	CHAS″	5′-*AAGC*ATGTTGTA*CAA*-3′
CHNS′ [b]	5′-*TAGT*ACCTTCTAT*GAGT*-3′	CHNS″	5′-*CAGA*TATATCA*GATG*-3′
CHSS′ [b]	5′-*AACA*ATTGGAAGT*CTCA*-3′	CHSS″	5′-*TTGT*ACAAC*ATGCTT*-3′

[a] The original rat heart NCX1 oligodeoxynucleotide sequence NCX1′ runs from base −26 to base −10; NCX1″ runs from base −9 to base +6. The underlined ATG in sequence NCX1″ indicates the start codon.
[b] CHAS′ and CHAS″ are the chimeric (mixed natural and phosphorothioated deoxynucleotides) antisense oligonucleotides, CHSS′ and CHSS″ are the chimeric (mixed natural and phosphorothioated deoxynucleotides) sense oligonucleotides, and CHNS′ and CHNS″ are the respective chimeric scrambled (nonsense) oligonucleotides corresponding to the normal sequences (NCX1′ and NCX1″) shown just above them. The four 5′ and four 3′ phosphorothioated deoxynucleotides in each of the antisense and scrambled oligonucleotides are indicated by boldface italics.

fied deoxynucleic acids at the 3′ and 5′ ends. Phosphorothioated deoxynucleic acids are resistant to nuclease degradation. Preliminary studies indicated that in cardiac myocytes, chimeric phosphorothioated ODNs were less toxic than fully phosphorothioated ODNs. Chimeric phosphorothioated ODNs are stable (exonuclease resistant) and are transported into cells better than unmodified ODNs.[13]

Custom-synthesized chimeric phosphorothioated ODNs can be obtained from Oligos Etc. (Wilsonville, OR). A pair of chimeric AS-ODNs (Table I) was targeted to a contiguous region of the mRNA around the start codon (underlined ATG in the normal rat heart NCX1″ sequence shown in Table I).[14,17–19] Each AS-ODN was designed with 15 or 17 bases. The "optimal length" results from a balance between the rapid uptake of shorter ODNs and the greater specificity of longer ODNs.[20] Two tandem ODNs were used to improve specificity and efficacy.[20] The nonspecific effects of tandem guanine bases were minimized by the selection of targeted sequences with low guanine and cytosine content.

In parallel cultures, a second set of ODNs containing the same base composition, but with scrambled (nonsense) sequence (Table I) was used to control for nonspecific or toxic effects of the ODNs. To ensure that the

[18] L. V. Hyrshko, D. A. Nicoll, J. N. Weiss, and K. D. Philipson, *Biochim. Biophys. Acta* **1151,** 35 (1995).
[19] D. H. Nicoll, S. Longoni, and K. D. Philipson, *Science* **250,** 562 (1990).
[20] L. J. Maher III and B. J. Dolnick, *Nucleic Acids Res.* **16,** 3341 (1988).

antisense knockdown of the NCX was dependent on sequence and not on non-sequence-dependent structure, a sense (S-ODN) sequence with base pair composition complementary to that of our antisense probe (Table I) served as an additional control in some experiments. All sequences were compared with known sequences in GenBank and EMBL, using the Wisconsin Package sequence analysis program (Genetics Computer Group, Inc., Madison, WI); no significant homologies to other sequences were found. All ODNs were gel purified and screened for secondary structure.

Control cells were grown without ODNs. In parallel, cells were grown in media containing the AS-ODN pair, S-ODN pair, or the nonsense (NS)-ODN pair. The final concentration of the ODNs was 0.5 μM, continuously from the time of initial plating.

Strong effort was made to avoid artifacts. First, the cell viability was determined. AS-ODNs, at a low concentration of 0.5 μM, were not lethal in cardiac myocytes and arterial myocytes, and appeared to have negligible effects on arterial myocyte cell proliferation. Second, biochemical evidence of NCX knockdown was obtained by immunoblotting. AS-ODNs reduced NCX protein expression by about 50–60% in cells treated for 7 days. These data are consistent with the physiologic results both temporally and quantitatively.

Primary Neonatal Cardiac Myocyte Cell Culture

The product of the NCX1 gene is expressed at high density in the plasma membrane of cardiac cells.[21,22] To conduct AS-ODN experiments in cardiac myocytes, a cell-culturing method is needed that minimizes fibroblast contamination, serum nuclease activity, and cell–cell interactions, while maintaining cell viability.

A cell culture technique[5] that is based on previously published methods is employed.[23,24] Neonatal rat ventricular myocytes are prepared from 1- to 2-day-old Sprague-Dawley rats. Hearts are removed and placed in ice-cold digestion buffer [(in mM) 10 N-2-hydroxyethylpiperazine-N'-2-ethanesulfonic acid (HEPES), 10 pyruvate (Na$^+$ salt), 5 L-glutamine, 1 nicotinamide, 0.4 L-ascorbate, 1 adenosine, 1 D-ribose, 1 MgCl$_2$, 1 taurine,

[21] M. Juhaszova, A. Ambesi, G. E. Lindenmayer, R. J. Bloch, and M. P. Blaustein, *Am. J. Physiol.* **266**, C234 (1994).

[22] M. Juhaszova, H. Shimizu, M. Borin, R. K. Yip, E. M. Santiago, G. E. Lindenmayer, and M. P. Blaustein, *Ann. N.Y. Acad. Sci.* **779**, 318 (1996).

[23] G. L. Engelmann, C. McTiernam, R. G. Gerrity, and A. M. Samarel, *Technique* **2**, 279 (1990).

[24] A. Lokuta, M. S. Kirby, S. T. Gaa, W. J. Lederer, and T. B. Rogers, *J. Cardiovasc. Electro.* **5**, 50 (1994).

2 DL-carnitine, 26 KH_2CO_3, and gentamicin (10 $\mu g/ml$) in Jocklick minimal essential medium (GIBCO-BRL, Grand Island, NY)]. The great vessels, atria, and pericardium are dissected away and discarded. The ventricles are minced and washed in ice-cold digestion buffer to remove blood. The minced ventricles are incubated (1 g/10 ml) in digestion buffer containing collagenase (0.5 mg/ml; Worthington Biochemical, Freehold, NJ) and $CaCl_2$ (50 μM) for 15 min at 37°. After gentle trituration, the supernatants from the first three digestions are discarded. The next two supernatants are retained and centrifuged at 500g. Cells are collected from the resulting pellets, and are plated at 10^5 cells/well in Dulbecco's modified Eagle medium (DMEM; GIBCO-BRL) containing heat-inactivated 10% fetal bovine serum (FBS; HyClone, Logan, UT) (55° for 4 hr to minimize exonuclease activity[25]) and appropriate ODNs (see below). After 20–24 hr, the cells are exposed to 4000 rad of γ radiation from a ^{137}Cs radiation source (J. L. Shepherd and Associates, Glendale, CA) to prevent fibroblast overgrowth.[24] The medium is changed immediately after the irradiation, and thrice weekly thereafter.

The purity of the cardiac myocyte cultures is determined immediately after physiologic experiments. Cells are fixed in 95% cold ethanol and labeled with a nuclear stain, 4,6-diamidino-2-phenylindole (DAPI), followed by mouse anti-rat troponin T antibodies (Sigma, St. Louis, MO). Rhodamine-conjugated donkey anti-mouse antibodies (Jackson Immuno-Research, West Grove, PA) are used as the secondary stain. Rat heart cells are also labeled with polyclonal antibodies raised against the canine cardiac NCX1 protein, which cross-reacts with the rat antigen.[21] Fluorescein isothiocyanate (FITC)-conjugated goat anti-rabbit antibodies (Jackson Immuno-Research) are used as a secondary stain. Cells in the cultures are identified as cardiac myocytes on the basis of positive staining for troponin T and NCX1; in our experience, almost all cells in these cultures should meet these criteria.

Primary Arterial Myocyte Cell Culture

In contrast to cardiac myocytes, arterial myocytes express relatively little NCX in the plasma membrane.[21,22,26,27] Nevertheless, in order to conduct AS-ODN experiments in arterial myocytes, as in the case of the cardiac cells, a cell-culturing method is required that minimizes fibroblast contamination and serum nuclease activity while maintaining cell viability and that enables these cells to proliferate normally.

---

[25] J.-P. Shaw, K. Kent, J. Bird, J. Fishback, and B. Rroehler, *Nucleic Acids Res.* **19,** 747 (1991).
[26] M. P. Blaustein, *Am. J. Physiol.* (*Cell Physiol.* 33) **264,** C1367 (1993).
[27] A. P. Somlyo, R. Broderick, and A. V. Somlyo, *Ann. N.Y. Acad. Sci.* **488,** 228 (1986).

Primary cultured arterial myocytes are prepared from rat mesenteric arteries. The arteries are excised under aseptic conditions, and placed in Hanks' balanced salt solution (HBSS; Sigma) containing penicillin (100 units/ml) and streptomycin sulfate (100 μg/ml). After removal of connective tissue, the blood vessels are incubated for 35 min at 37° in HBSS with collagenase (2 mg/ml). The adventitia is then separated from the arterial media, and the latter is incubated overnight in DMEM containing 10% fetal calf serum (air/5% CO_2, 37°). The next day the arterial media is placed in HBSS containing collagenase (2 mg/ml) and elastase (0.5 mg/ml) and incubated for 40 min (air/5% CO_2, 37°). The cells are triturated with a Pasteur pipette and resuspended in DMEM containing the appropriate ODNs and heat-inactivated 10% fetal calf serum (55° for 4 hr to minimize exonuclease activity[25]). The suspended cells are plated onto glass coverslips on 35-mm culture plates. The medium is changed thrice weekly.

Arterial myocyte culture purity is determined immediately after physiologic experiments, as well as in parallel cells grown on gridded coverslips (Eppendorf, Madison, WI). Cells are fixed in 95% cold ethanol. The myocytes are identified by immunochemical methods with mouse anti-rat smooth muscle α-actin antibodies; FITC-conjugated goat anti-mouse antibodies are used as the secondary stain. DAPI is used to label nuclei in cells grown on gridded coverslips for cell counting.

Cells exposed to antisense ODNs are indistinguishable, in terms of mortality (cardiac and arterial myocytes) and proliferation rate (only arterial myocytes proliferate) when compared with control (no ODNs present), scramble-, or sense-treated cells.[7]

[^{35}S]Methionine Labeling of Cells

To estimate the kinetics of AS-ODN knockdown of the cardiac myocyte NCX protein, it is important to know the half-life of the NCX protein. The high level of NCX protein expression in cardiac myocytes helps maximize the signal-to-noise ratio for these determinations.

Pulse–chase labeling of proteins can be used to estimate the average protein half-life of all the proteins in cells. We employed [^{35}S]methionine to label all cellular proteins in the cardiac myocytes, and subsequently determined the half-life of the NCX protein.

After 3 days in culture, cardiac myocytes are incubated in methionine-free DMEM with 10% FBS containing 5 μCi of [^{35}S]methionine per milliliter for 24 hr. The cells are washed and incubated in DMEM containing 10% FBS (without labeled methionine) for up to 7 days. The cells are harvested as for immunoblotting (see below), at various times, and the NCX is immunoprecipitated. The immunoprecipitated proteins are solubilized in sodium dodecyl sulfate (SDS) sample buffer, separated by SDS–polyacrylamide gel electrophoresis (PAGE), and immunoblotted. The [^{35}S]methionine-labeled proteins harvested at different times after the pulse label are identified and quantified by densitometry of the nitrocellulose membrane. We performed the densitometric scan with a PhosphorImager, using ImageQuant software (Molecular Dynamics, Sunnyvale, CA). To calculate a half-life for the NCX protein, the [^{35}S]methionine activity in each band, at each time point, is normalized to the density of the respective 170-kDa NCX band (equivalent to a measurement of microcuries of ^{35}S per milligram of protein).

Immunoprecipitation

Pulse–chase labeling of cellular proteins can be used to estimate the average protein half-life of all proteins in a cell. Then, with immunoprecipitation, the [^{35}S]methionine signal of a specific protein can be followed over time to estimate its half-life. The cardiac myocytes are washed with phosphate-buffered saline (PBS; in mM: 120 NaCl and 2.7 KCl in 10 sodium phosphate buffer, pH 7.4) and harvested (at appropriate time points, e.g., 0, 6, 12, 24, 48, 96, and 144 hr) into PBS containing protease inhibitor cocktail (in μM): 1 pepstatin A, 100 leupeptin, and 50 phenylmethylsulfonyl fluoride (PMSF). Protein concentrations are determined by the bicinchoninic acid (BCA) reagent assay (Pierce, Rockford, IL), standardized with bovine serum albumin. The proteins (100 μg) are then solubilized in a solubilization buffer: 150 mM NaCl, 50 mM tris(hydroxymethyl)aminomethane, 1% Triton X-100, 0.5% deoxycholate, 0.1% SDS, 0.01% sodium azide, and protease inhibitor cocktail at pH 8.0. After solubilization, immunoprecipitation buffer (150 mM NaCl, 25 mM HEPES, 0.01% sodium azide, at pH 7.4) is added in a 2:1 ratio to a total volume of 1 ml; 50 μl of protein A–Sepharose beads is added for 3–4 hr and maintained at 0°. This "preclearing step" is used to remove proteins that bind nonspecifically to the protein A–Sepharose beads. The protein A–bead suspension is centrifuged at 5000 rpm for 3 min, the beads are discarded, and the supernatant is incubated overnight with 5 μl of anti-NCX antibody (or with preimmune serum). The next day, 50 μl of protein A–Sepharose beads is added to the supernatant (containing antibodies) for 3 hr; after centrifugation, as above, the supernatant is discarded. The protein A–Sepharose bead pellet is resuspended in immunoprecipitation buffer and centrifuged; the resulting supernatant is discarded. This "washing" step is repeated three times to remove proteins that are not bound specifically to the protein A–Sepharose beads. To elute the antibodies and immunoprecipitated proteins from the protein A–Sepharose beads, the beads are incubated in 100 μl of 3× concentrated SDS sample buffer with 1% 2-mercaptoethanol at 70° for 20 min. After centrifugation, as above, the final supernatant, containing NCX protein and anti-NCX IgG, is analyzed by SDS–PAGE. The immunoprecipitated NCX protein is detected by immunoblotting. [^{35}S]Methionine labeling of immunoprecipitated protein is determined by phosphoimaging.

Immunoblotting

Immunoblotting is used to determine the decay of the [^{35}S]methionine signal relative to the amount of the NCX protein at each time point, in order to estimate the NCX protein half-life. Furthermore, immunoblotting can also provide direct biochemical evidence of AS-ODN knockdown of the NCX protein. This is most reliable when the level of protein expression is high, as in cardiac myocytes. The reliability of this method may be greatly decreased when the level of relevant protein expression is low (as in the case of arterial myocytes).

Cells are grown on 100-mm tissue culture dishes because of the large numbers of cells needed for these biochemical experiments. The cells are washed with PBS and harvested into PBS containing protease inhibitor cocktail. Protein concentrations are determined by the BCA assay. The proteins are solubilized in SDS sample buffer containing 1% 2-mercaptoethanol, and separated by SDS–PAGE [8% gel; calibrated with prestained protein molecular mass markers from Bio-Rad (Hercules, CA)]. The proteins are transferred to NitroBind nitrocellulose membrane (Micron Separations, Westboro, MA) at 100 V for 1.5 hr in an ice bath to avoid overheating. The extent of protein transfer is verified by Ponceau-S staining. Membranes are blocked with 5% nonfat dry milk in TBST (Tris-buffered saline with 0.1% Tween 20) at room temperature. Blots are incubated overnight with either preimmune serum or rabbit antiserum raised against the NCX.[21] After washing, the membranes are incubated in a solution containing anti-rabbit IgG conjugated to horseradish peroxidase (Amersham, Arlington Heights, IL). Enhanced chemiluminescence (ECL; Amersham) of the luminol oxidized by horseradish peroxidase is used to detect NCX labeling.

Cytoplasmic Ca^{2+} Determination Using Fura-2

The presence and effects of AS-ODN knockdown of NCX protein must be evaluated functionally. In cardiac and arterial myocytes, knockdown of function is evaluated by ratiometric fluorescence imaging methods in cells loaded with the Ca^{2+}-sensitive dye, Fura-2.[7] Full details of the imaging methods have been published.[5–7] These methods enable the investigator to evaluate the viability and function of the NCX in individual cells.

Results and Discussion

Half-Life of Sodium–Calcium Exchanger Protein

At least in theory, antisense inhibition of proteins relies on prevention of mRNA translation into protein. This putative model of antisense mechanism suggests that the activity of protein with a long half-life would take a long time to knock down. Correlation of the time course of antisense inhibition with the protein half-life supports this model.

To determine the $t_{1/2}$ of the NCX in cardiac myocytes, a pulse chase with [^{35}S]methionine was used to label the proteins.[5] This method of labeling does not directly interfere with the synthesis of proteins. Nevertheless, densitometric methods enabled us to identify the ^{35}S-labeled band that comigrated with the NCX (i.e., the 170-kDa band), which was identified by immunolabeling. The ^{35}S label in the NCX bands decayed with a half-life ($t_{1/2}$) of 33 hr.

When the cardiac myocytes were exposed to AS-ODNs (0.5 μM) targeted to NCX1, no detectable effect on the protein (measured by immunoblotting) was observed after 2 days of treatment. After treatment for 7 days, however, the NCX1 protein was reduced by 50%.[5]

Knockdown of Sodium–Calcium Exchanger Function by Antisense Oligodeoxynucleotides in Cardiac and Arterial Myocytes

NCX can mediate both Ca^{2+} influx and Ca^{2+} efflux. Thus AS-ODN-induced knockdown of NCX should inhibit both modes of net Ca^{2+} transport. Digital imaging with Fura-2 was therefore used to obtain physiological evidence of knockdown of the NCX in primary cultured neonatal rat cardiac myocytes[6] and arterial myocytes.[6,7] An important advantage of this method is the direct visualization of single cells. Moreover, this method is used to measure the key parameter in NCX function: the dynamic changes in the cytosolic Ca^{2+} concentration ([Ca^{2+}]$_{CYT}$). To assess the action of AS-ODNs on NCX-mediated Ca^{2+} influx, the effect of external Na^+ removal on [Ca^{2+}]$_{CYT}$ was determined. To test for effects of AS-ODNs on Ca^{2+} efflux via NCX, cyclopiazonic acid (CPA) and caffeine (CAF) were added to the physiologic salt solution (PSS) to inhibit sarcoplasmic reticulum (SR) Ca^{2+} buffering and unload the SR Ca^{2+} stores. The prolonged elevation of

$[Ca^{2+}]_{CYT}$ after introduction of CPA + CAF, and the absence of a significant rise in $[Ca^{2+}]_{CYT}$ after external Na^+ depletion, indicate that NCX-mediated Ca^{2+} efflux and influx are both knocked down in As-ODN-treated cardiac and arterial myocytes.[5–7]

Knockdown of NCX activity in cardiac myocytes was first observed after 96 hr of incubation with the AS-ODNs. The NCX activity was not knocked down in all of the AS-ODN-treated cells: only about one-half to two-thirds of the cells treated with AS-ODNs for 4–7 days exhibited NCX knockdown effects. This heterogeneity in the response to AS-ODNs is not surprising. Similar behavior has also been reported in other preparations, and has been attributed to heterogeneity in ODN uptake.[8]

In cardiac myocytes, AS-ODNs did not seem to interfere with other cell functions such as high external K^+-induced Ca^{2+} entry, maintenance of resting $[Ca^{2+}]_{CYT}$ levels, and maintenance of CPA- and CAF-sensitive stored Ca^{2+}.[5] In arterial myocytes, cell proliferation as well as the activation by the vasoconstrictor serotonin (5-HT) were maintained in AS-ODN-treated cells.[6] This latter activity, as well as the individual cell analysis, enabled us to distinguish between NCX knockdown and nonresponsive cells.

The AS-ODNs apparently knocked down cardiac myocyte NCX activity, as was demonstrated by the functional and biochemical assays. In arterial myocytes, AS-ODNs apparently knocked down cell NCX function completely. Relative to arterial myocytes, cardiac myocytes express substantially more NCX protein than do arterial myocytes.[6,21]

The knockdown of NCX activity by AS-ODNs was reversible.[6] This recovery, which was observed at 5 days, but not at 1 or 2 days, is a further indication that the effect of the AS-ODNs was specific.

Taken together, our results in cardiac and arterial myocytes demonstrate that carefully designed AS-ODNs targeted to NCX1 can selectively and effectively inhibit expression of their target.

Acknowledgments

We thank Dr. M. P. Blaustein for comments and suggestions. This work was supported by an SRIS Award from the University of Maryland, Baltimore (to M.J.), and by NIH Grant HL-45215 and NS-16106 (to M. P. Blaustein) and Training Grant GM-02521580.

[23] Targeted Delivery of Antisense Oligonucleotides to Parenchymal Liver Cells *in Vivo*

By E. A. L. Biessen, H. Vietsch, E. T. Rump, K. Fluiter,
M. K. Bijsterbosch, and T. J. C. van Berkel

Oligodeoxynucleotides (ODNs) have been shown to inhibit gene expression at various levels both *in vitro* and *in vivo*.[1-5] *In vivo,* the efficacy of ODN-induced regulation of genes in specific cell types may be suboptimal owing to poor accumulation of ODNs in these cells. In addition, untimely elimination of ODNs via renal clearance, degradation, and scavenger receptor-mediated uptake[6] may further impair their therapeutic activity. These hurdles can be at least partly overcome by targeted delivery of the ODNs to the desired site of action. A number of approaches have been suggested to facilitate the entry of polyanionic ODNs into the aimed target cell[7-11] (for a review see Ref. 12). Neutral and cationic liposomes are considered to be attractive ODN carriers because they markedly enhance cellular uptake under *in vitro* conditions. Like native ODNs, however,[13] liposomally formulated ODNs are mainly captured by cells of the reticuloendothelial system in lungs, spleen, and liver,[13-15] as a result of which the ODN concentration in the target cell will be suboptimal. After local delivery of ODNs

[1] C. A. Stein and Y. C. Cheng, *Science* **261,** 1004 (1993).

[2] R. W. Wagner, *Nature (London)* **372,** 333 (1994).

[3] J. F. Milligan, M. D. Matteucci, and J. C. Martin, *J. Med. Chem.* **36,** 1923 (1993).

[4] R. W. Wagner and W. M. Flanagan, *Mol. Med. Today* **3,** 31 (1997).

[5] D. E. Szymkowski, *Drug Discov. Today* **1,** 415 (1996).

[6] E. A. L. Biessen, H. Vietsch, J. Kuiper, M. K. Bijsterbosch, and T. J. C. van Berkel, *Mol. Pharmacol.* **53,** 262 (1998).

[7] P. L. Felgner, T. R. Gadek, M. Holm, R. Roman, H. W. Chan, M. Wenz, J. P. Northrop, G. M. Ringold, and M. Danielsen, *Proc. Natl. Acad. Sci. U.S.A.* **84,** 7413 (1993).

[8] R. Leventis and J. R. Silvius, *Biochim. Biophys. Acta* **1023,** 124 (1990).

[9] C. F. Bennett, M. Y. Chiang, H. Chan, J. E. Shoemaker, and C. K. Mirabelli, *Mol. Pharm.* **41,** 1023 (1992).

[10] A. M. Tari, S. D. Tucker, A. Deisseroth, and G. Lopez-Berestein, *Blood* **80,** 601 (1994).

[11] J. G. Lewis, K. Y. Lin, A. Kothavale, W. M. Flanagan, M. D. Matteucci, R. B. DePrince, R. A. Mook, R. W. Hendren, and R. W. Wagner, *Proc. Natl. Acad. Sci. U.S.A.* **93,** 3176 (1996).

[12] S. T. Crook, *J. Drug Targeting* **3,** 185 (1995).

[13] H. Sands, L. J. Gorey-Feret, A. J. Cocazza, F. W. Hobbs, D. Chidester, and G. L. Trainer, *Mol. Pharmacol.* **45,** 932 (1994).

[14] M. Inagaki, K. Togawa, B. I. Carr, K. Ghosh, and J. S. Cohen, *Transplant. Proc.* **24,** 2971 (1992).

[15] J. Frese, C. H. Wu, and G. Y. Wu, *Adv. Drug Deliv. Rev.* **14,** 137 (1994).

encapsulated into virus capsid-coated liposomes, Morishita *et al.*[16] could enhance ODN uptake by vascular endothelial cells, leading to cell-specific antisense effects. Nevertheless, this approach is not feasible for specific delivery of ODNs to most other cell types, such as the parenchymal liver cell (PC). The PC is the target for antisense-mediated downregulation of the expression of a number of clinically relevant target genes, including that of the atherogenic apolipoprotein A,[17] cholesterol ester transfer protein,[18,19] and viral proteins from hepatitis B/C virus.[20–22] A promising way to enhance the local bioavailability of ODNs in this cell type involves conjugation of the ODNs to a ligand for a receptor uniquely expressed by PCs, such as the asialoglycoprotein receptor.[23] *In vitro,* glycotargeted delivery to this receptor has been shown after noncovalent complexation of the ODNs to a conjugate of poly (L-lysine) and asialoorosomucoid,[24–27] while a study by Lu *et al.*[24] also indicated an altered biodistribution of glycotargeted ODNs *in vivo.* In comparison with the rather bulky ODN carriers based on poly(L-lysine)/asialoorosomucoid conjugates, covalent attachment of ODNs to a low molecular weight ligand for a cell-specific receptor confers the advantage that it is synthetically more accessible, less laborious to make, and pharmaceutically applicable. Hangeland *et al.*[28] have reported enhanced uptake of a 7-mer heptathymidinylate (T_7) methyl phosphonate by HepG2 cells after conjugation to a *N*-acetylgalactosamine-terminated glycopeptide. Subsequent *in vivo* studies of this glycoconjugate in mice also showed that hepatic uptake of this conjugate was increased.[29] As short (and in particular uncharged) oligothymidinylates are poor substrates for hepatic scavenger receptors, which are responsible for the rapid elimination of ODNs by

[16] R. Morishita, G. H. Gibbons, Y. Kaneda, T. Ogihara, and V. J. Dzau, *Gene* **149,** 13 (1994).
[17] R. Morishita, J. Higaku, I. Kida, M. Aoki, A. Moriguchi, R. Lawn, Y. Kaneda, and T. Ogihara, *Circulation* **94,** I-39 (1996).
[18] M. Sugano and N. Makino, *J. Biol. Chem.* **271,** 19080 (1996).
[19] M. Sugano, N. Makino, S. Sawada, S. Otsuka, M. Watanabe, H. Okamoto, M. Kamada, and A. Mizushima, *J. Biol. Chem.* **273,** 5033 (1998).
[20] T. Mizutani, N. Kato, M. Hirota, K. Sugiyama, A. Murakami, and K. Shimotohno, *Biochem. Biophys. Res. Commun.* **212,** 906 (1995).
[21] T. Wakita and J. R. Wands, *J. Biol. Chem.* **269,** 14205 (1994).
[22] K. Nakazono, Y. Ito, C. H. Wu, and G. Y. Wu, *Hepatology* **23,** 1297 (1996).
[23] G. Y. Wu and C. H. Wu, *J. Biol. Chem.* **267,** 12436 (1992).
[24] X. M. Lu, A. J. Fischman, S. L. Jyawook, K. Hendricks, R. G. Tompkins, and M. L. Yarmusch, *J. Nucl. Med.* **35,** 269 (1994).
[25] B. A. Bunnell, F. A. Askari, and J. M. Wilson, *Somatic Cell Mol. Genet.* **18,** 559 (1992).
[26] M. Reinis, M. Damkova, and E. Korec, *J. Virol. Methods* **42,** 99 (1993).
[27] G. G. Ashwell and J. Harford, *Annu. Rev. Biochem.* **51,** 531 (1982).
[28] J. J. Hangeland, J. T. Levis, Y. C. Lee, and P. O. P. Ts'o, *Bioconjugate Chem.* **6,** 695 (1995).
[29] J. J. Hangeland, J. E. Flesher, S. F. Deamond, Y. C. Lee, P. O. P. Ts'o, and J. J. Frost, *Antisense Nucleic Acid Drug Dev.* **7,** 141 (1997).

cells of the reticuloendothelial system, the latter results leave unanswered whether longer, charged, and miscellaneous ODN sequences can also be redirected to the aimed target cell *in vivo*.[6,30] In this respect, it is crucial to analyze the tissue distribution and to identify the cellular uptake sites within the liver, using full-length antisense sequences. In this study *in vivo* evidence is provided that untimely elimination of a miscellaneous 20-mer ODN by the preceding scavenger pathways can be circumvented and, concomitantly, accumulation by parenchymal liver cells can be enhanced after derivatization with a small-sized synthetic galactoside with high affinity for the asialoglycoprotein receptor.

Synthesis of L_3G_4

p-Aminophenyl β-D-galactopyranose is converted into the phenyl isothiocyanate derivative by reacting *p*-aminophenyl β-D-galactopyranose (244 mg, 0.9 mmol) with thiophosgene (0.52 ml, 5.1 mmol) in ethanol–H_2O (4:1, v/v; 50 ml) for 2 hr at room temperature. After reaction excess thiophosgene is removed by perspiration and the mixture concentrated under reduced pressure. The residue is dissolved in a small volume of distilled water (1–2 ml), the solvent neutralized, and after concentration under reduced pressure chromatographed over a Kieselgel 60 column (40 ml) using CH_2Cl_2–methanol (4:1, v/v) as eluent. p-(β-D-Galactopyranosyloxy)phenyl isothiocyanate can be isolated as a white crystalline powder in almost 100% yield. Subsequently, the activated glycoside (0.10 mmol; 32 mg) and lysyllysylysine tetraacetate (0.03 mmol; 20 mg) are dissolved in a mixture of 0.1 *M* sodium hydrogen carbonate (pH 8.5) and dimethylformamide (DMF) (1:1, v/v; 4 ml), and the solution is stirred for 18 hr in the dark at room temperature. Progression of the reaction is monitored by thin-layer chromatography (TLC). The reaction mixture is concentrated *in vacuo,* and the residue is chromatographed over Kieselgel 60 using acetronitrile–H_2O (4:1, v/v) as eluent. The pure glycopeptide (L_3G_4) can be isolated in 67% yield and the chemical identity of the compound is established by mass spectroscopy and nuclear magnetic resonance (NMR). The high affinity of L_3G_4 (see below) makes it an attractive homing device for specific delivery of antisense ODN to hepatocytes.

Conjugation of ODN-AS5′ with L_3G_4

Conjugation of L_3G_4 to the antisense ODN is accomplished via a two-step procedure. In the first step, L_3G_4 is linked to the 5′ end of the antisense

[30] A. M. Pearson, A. Rich, and M. Krieger, *J. Biol. Chem.* **268**, 3546 (1993).

ODN for apolipoprotein A (nucleotides 83–92; TGA CTT TAG T; ODN-AS5'). In the second step, this ODN is ligated to the phosphorylated and 3'-capped ODN-AS3' (nucleotides 93–102 of the apo A gene; CGT CGT GGA T-cap; cap, 3'-amino-2-hydroxypropyl). This target sequence (nucleotides 83–102) was selected because it is specific for apo A (and not plasminogen), because it is highly accessible (being not involved in hairpin-like structures), and because it is close to the translation initiation site. 5'-Aminothymidinyl-derivatized ODN-AS5' is conjugated to L_3G_4 at the 5'-NH_2 of the terminal thymidine as follows. L_3G_4 (2 μmol, 3.3 mg) is dissolved in 75 μl of DMF–H_2O (2 : 1, v/v) and 0.5 μl of *N,N*-diisopropylethylamine (2 μmol), 0.79 mg of 2-(1*H*-benzotriazole-1-yl)-1,1,3,3-tetramethyluronium (HBTU; 2 μmol), and 0.27 mg of hydroxybenzotriazole (HOBT; 2 μmol) are added. The reaction is started by addition of 70 μg (20 nmol) of ODN-AS5' (as a Li^+ salt) in 25 μl of H_2O, and the mixture is incubated for 18 hr at 37°. Next, the ODN is precipitated by addition of acetone (80%, v/v), and the ODN pellet is lyophilized, resolved in water, and precipitated again with acetone (80%, v/v). After lyophilization, the conjugate is purified by gel electrophoresis over 19% polyacrylamide under denaturing conditions in TBE buffer (90 m*M* Tris–borate, 0.1 m*M* EDTA, pH 8.4) containing 7 *M* urea. The conjugate can be isolated from the gel after visualization by UV shadowing [see Fig. 1; the relative electrophoretic mobility (R_f) is 1.02; the R_f of underivatized ODN-AS5' is 0.40; R_f values for xylene cyanol and bromphenol blue are 1.0 and 0.0, respectively] and is desalted over Sephadex G-25, yielding 29 μg of L_3G_4-ODN-AS5 (31%; 7.26 × 10^7 AU/mol · cm). Mass spectrometry is in agreement with the presumed chemical structure (see Scheme 1). Attempts to increase the coupling yield by changes in solvent composition or coupling reagents were not successful. The pursued procedure leads to markedly higher yields than do reactions catalyzed by DCC, DCC–HOBT, EDC, or EEDQ. Furthermore, DMF–H_2O appears to be superior to DMA–H_2O, acetonitrile–H_2O mixtures, or pure H_2O. Efforts to couple L_3G_4 to the 5' amino-derivatized ODN, using a Heinzer base according to the procedure of Oberhauser and Wagner,[31] failed for phosphodiester ODNs longer than 10 bases, and gave only low yields for smaller ODN sequences. Although coupling yields may seem moderate, solution-phase acylation reactions between negatively charged ODNs and carboxylic groups are considered notoriously difficult.[28,29,31,32] Similar yields are reported for solution-phase coupling by Oberhauser and Wagner,[31] while the two-step procedure used by Hangeland *et al.* to conjugate a heptathymidine to a triantennary glycopeptide in solution, using

[31] B. Oberhauser and E. Wagner, *Nucleic Acids Res.* **20,** 533 (1992).
[32] O. Zelphati, E. Wagner, and L. Leserman, *Antiviral Res.* **25,** 13 (1994).

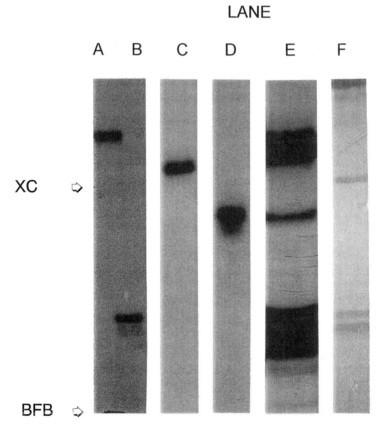

FIG. 1. Analysis of the antisense ODN derivatives by gel electrophoresis on 19% polyacrylamide under denaturing conditions. Lane A, L_3G_4-[^{32}P]ODN; lane B, [^{32}P]ODN-AS3'; lane C, [^{32}P]ODN-SE; lane D, [^{32}P]ODN; lane E, crude mixture of the ligation of L_3G_4-ODN-AS5' to [^{32}P]ODN-AS3' (the upper band is the ligation product L_3G_4-[^{32}P]ODN); lane F, crude mixture of the coupling reaction of L_3G_4 to ODN-AS5' (UV shadowing; upper band is the conjugated L_3G_4-ODN). Arrows indicate position of the mobility markers bromphenol blue (BFB) and xylene cyanol (XC).

cystamine and a thiol/amine cross-linker, resulted in an overall yield of only 14%.[28]

Preparation of [^{32}P]ODN and L_3G_4-[^{32}P]ODN

The phosphorylated ODN-AS3' (Fig. 1, lane B) is enzymatically ligated to the ligand-conjugated ODN-AS5' without prior purification. Internally labeled [^{32}P]ODN and L_3G_4-[^{32}P]ODN are prepared enzymatically from

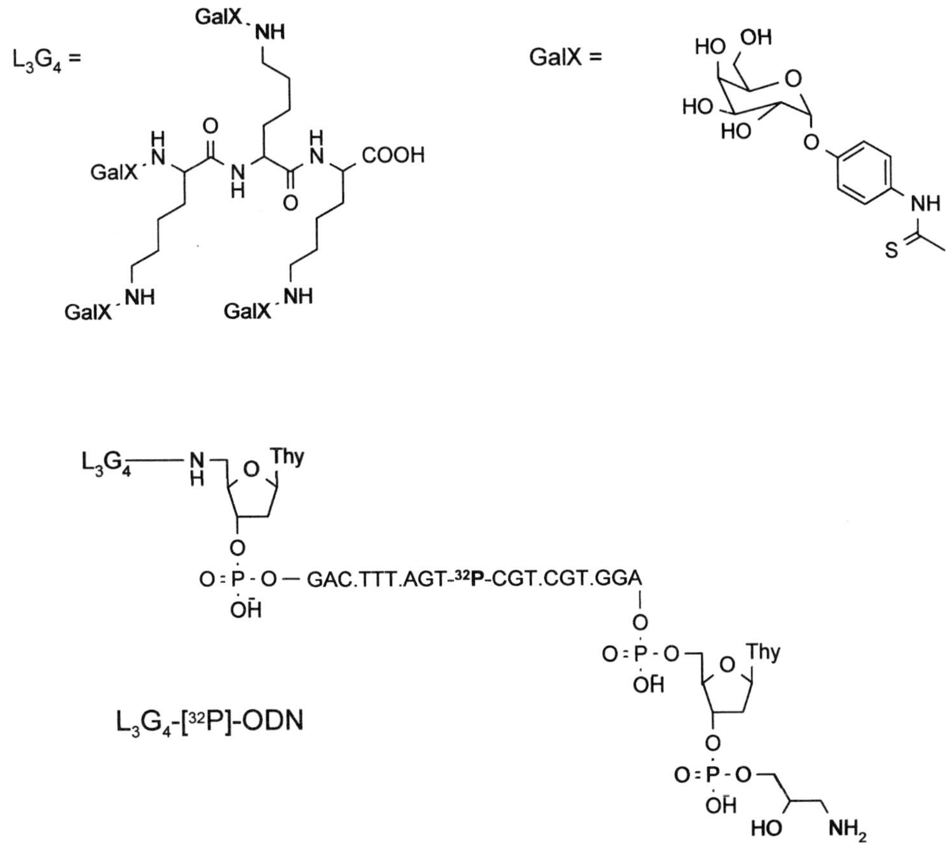

SCHEME 1. Chemical structures of L_3G_4 and L_3G_4-[^{32}P]ODN.

L_3G_4-ODN-AS5′ and ODN-AS3′. ODN-AS3′ is 5′-end phosphorylated using [γ-^{32}P]ATP and T4 polynucleotide kinase according to Sambrook *et al.*[33] The phosphorylated product is then ligated for 18 hr at 8° to ODN-AS5′ or L_3G_4-ODN-AS5′, using T4 DNA ligase in the presence of ODN-SE (5′-TTT CTG AAA TCA GCA GCA CCT GAG). The ligation products are isolated by gel electrophoresis on 19% polyacrylamide in TBE, the gel is autoradiographed, the ligation product is excised from the gel [R_f 0.88 and 1.23, respectively; R_f (xylene cyanol) = 1.0 and R_f (bromphenol blue) = 0.0]. L_3G_4-[^{32}P]ODN runs at an electrophoretic mobility compara-

---

[33] J. Sambrook, E. F. Fritsch, and T. Maniatis, "Molecular Cloning: A Laboratory Manual," 2nd Ed. Cold Spring Harbor Laboratory Press, Cold Spring Harbor, New York, 1989.

ble to that of a 27-mer ODN ($R_f = 0.0555 \, N - 0.2521$; $r^2 = 0.9731$), and considerably slower than [^{32}P]ODN-AS3′, L_3G_4-ODN-AS5′, or nonconjugated [^{32}P]ODN (see Fig. 1). Ligation yields of [^{32}P]ODN and L_3G_4-[^{32}P]ODN are approximately 50%. Typically, specific activities of [^{32}P]ODN and L_3G_4-[^{32}P]ODN are 100–250 cpm/ng. An extended cooling protocol for annealing L_3G_4-ODN-AS5′ and [^{32}P]ODN-AS3′ to the complementary ODN-SE strand appear to be critical to obtain good ligation yields.

Stability of L_3G_4-[^{32}P]ODN and [^{32}P]ODN in Presence of Serum, Parenchymal Liver Cells, and Titrosomes

To determine the stability of the ODN derivative in serum, internally labeled [^{32}P]ODN or L_3G_4-[^{32}P]ODN (63 pmol, 0.5 μg) are incubated for 180 min with 300 μl of freshly isolated rat serum (diluted 1:1 with phosphate-buffered saline, pH 7.2, 37°) or with parenchymal liver cells [2×10^6 cells/ml in Dulbecco's modified Eagle's medium (DMEM) plus 2% bovine serum albumin (BSA), 4°]. At the indicated times, 30-μl samples are taken and ODN is isolated by phenol extraction and subsequent precipitation in ice-cold isopropanol. The pellet is electrophoresed on denaturing 19% polyacrylamide at an $N/N - 1$ resolution level, and the gel is autoradiographed on X-Omat Kodak (Rochester, NY) film and analyzed on an Instant-Imager. Analysis of the stability of the 3′-capped L_3G_4-[^{32}P]ODN at 37° in the presence of serum shows that the glycoconjugate is much more stable than the underivatized ODN (Fig. 2). Only 40% of the conjugate is degraded within 3 hr of incubation. The half-life of the glycoconjugate in serum is calculated to be 200 ± 20 min, which is about 10-fold higher than that of underivatized [^{32}P]ODN (19 ± 6 min). As a measure of the stability of L_3G_4-[^{32}P]ODN during lysosomal processing, the ODN (63 pmol, 0.5 μg in 30 μl of water) is incubated for 180 min in sodium acetate buffer (0.1 M, pH 4.5) in the presence of 150 μl of titrosomes (a highly purified preparation of rat liver lysosomes with a specific acid phosphatase activity 25-fold higher than liver homogenate, prepared according to the procedure of Huisman et al.[34]). From the stability studies it may be concluded that the glycoconjugate is degraded in the lysosomal compartment with a half-life of 2 min.

In Vitro Studies of Uptake and Binding of [^{125}I]ASOR and L_3G_4-[^{32}P]ODN to Parenchymal Liver Cells

L_3G_4-[^{32}P]ODN binding to the asialoglycoprotein receptor is studied by incubating isolated rat parenchymal liver cells (1–1.5×10^6 cells) for 2

[34] W. Huisman, J. M. W. Bouma, and M. Gruber, *Biochim. Biophys. Acta* **297,** 98 (1973).

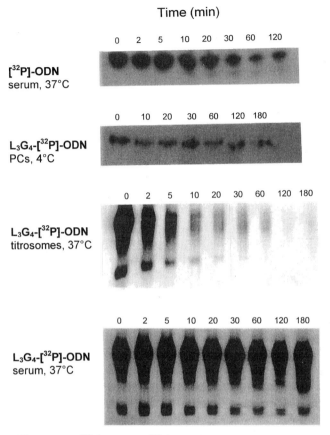

FIG. 2. Stability of L_3G_4-[^{32}P]ODN and [^{32}P]ODN in the presence of serum, parenchymal liver cells, or titrosomes. L_3G_4-[^{32}P]ODN or [^{32}P]ODN (10 nM) was incubated for 180 min in rat serum (diluted 1:1 with phosphate-buffered saline; 37°), in the presence of titrosomes (a cellular fraction highly enriched in lysosomal enzymes) (0.1 M sodium acetate buffer, pH 4.5; 37°) or in the presence of parenchymal liver cells (2×10^6/ml in DMEM + 2% BSA; 4°). At the indicated times, samples (30 μl) were taken and put on ice. The [^{32}P]ODN derivative was isolated by phenol–chloroform extraction and subsequent precipitation with ice-cold 2-propanol. The precipitates were subjected to gel electrophoresis on a 19% polyacrylamide gel under denaturing conditions (80 mM Tris–90 mM boric acid–2 mM EDTA–7 M ureum), and the gels subsequently analyzed with a PhosphorImager. The major bands have an electrophoretic mobility identical to untreated [^{32}P]ODN or L_3G_4-[^{32}P]ODN.

hr at 4° in a Lab-Shaker (150 rpm) with 0–300 nM L_3G_4-[^{32}P]ODN in DMEM + 2% BSA. The half-life of the ODN glycoconjugate in the presence of PCs at 4° is approximately 2 hr (Fig. 2). After incubation cells are washed thoroughly and cell-associated radioactivity is determined and

TABLE I

INHIBITION CONSTANTS (pK_i) OF GLYCOSIDES DERIVED FROM COMPETITION
ASSAYS OF L_3G_4-[^{32}P]ODN AND [^{125}I]ASOR BINDING[a]

Inhibitor	L_3G_4-[^{32}P]ODN binding	[^{125}I]ASOR binding
L_3G_4	—	8.19 ± 0.01
L_3G_4-ODN-AS5	—	7.63 ± 0.11
L_3G_4-ODN	6.59 ± 0.20	—
ASOR	7.94 ± 0.10	8.19 ± 0.01
GalNAc	4.88 ± 0.06	5.06 ± 0.12
GlcNAc	≪1.0	≪1.0
ODN	≪1.0	≪1.0

[a] Rat PCs (1×10^6 cells/500 μl) were incubated for 2 hr at 4° in DMEM + 2% (w/v) BSA with [^{125}I]ASOR (10 nM) in the presence of asialoorosomucoid, GalNAc, GlcNAc, L_3G_4-ODN-AS5, L_3G_4-ODN, and ODN. After incubation, cells were washed thoroughly and cell-bound radioactivity was determined and corrected for protein content. pK_i values were calculated from the competition curves, using a computerized nonlinear regression procedure.

corrected for protein content. Binding data are analyzed according to a single-site binding/competition model, using nonlinear regression (Graph-PAD; Isis Software).[35] The *in vitro* competition studies of ^{125}I-labeled asialooromucoid ([^{125}I]ASOR) binding to isolated rat hepatocytes show that free L_3G_4 is able to inhibit [^{125}I]ASOR binding to the asialoglycoprotein receptor in a competitive fashion. The affinity is 6.5 nM (pK_i 8.19 ± 0.01) (Table I). Subsequently, we have tested whether L_3G_4 is also recognized by the asialoglycoprotein receptor on conjugation to the decanucleotide ODN-AS5'. L_3G_4-ODN-AS5' inhibits [^{125}I]ASOR binding at an inhibition constant of 23 nM (pK_i 7.63 ± 0.11). Apparently, derivatization of L_3G_4 with ODN-AS5' reduces only slightly its affinity for the asialoglycoprotein receptor. In contrast, underivatized ODN-AS5' is not able to displace [^{125}I]ASOR binding at concentrations of up to 200 nM.

To assess the relative contribution of the asialoglycoprotein receptor in the association of L_3G_4-[^{32}P]ODN to hepatocytes, we studied the interaction of internally labeled L_3G_4-[^{32}P]ODN with isolated hepatocytes. First, equilibrium conditions for L_3G_4-[^{32}P]ODN binding at 4° were assessed by analyzing the kinetics of its association with hepatocytes. In the presence of parenchymal liver cells, L_3G_4-[^{32}P]ODN is stable for 2 hr at 4° (Fig. 2). Equilibrium binding at 1 nM of L_3G_4-[^{32}P]ODN is achieved within 2 hr of incubation (data not shown). Saturation binding studies of L_3G_4-[^{32}P]ODN

[35] E. A. L. Biessen, D. M. Beuting, H. Vietsch, M. K. Bijsterbosch, and T. J. C. Van Berkel, *J. Hepatol.* **21**, 806 (1994).

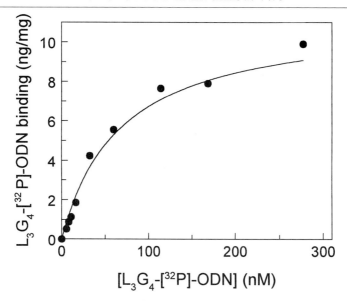

FIG. 3. Saturation binding of L_3G_4-[^{32}P]ODN to isolated rat parenchymal liver cells. PCs (1×10^6 cells/500 μl) were incubated for 2 hr at 4° in DMEM plus 2% (w/v) BSA with 0–270 nM L_3G_4-[^{32}P]ODN in the absence or presence of 100 mM GalNAc. After incubation, cells were put on ice and washed thoroughly, and membrane-bound radioactivity was determined and corrected for protein content. Specific binding, defined as the differential binding in the presence and absence of GalNAc (mean of a duplicate experiment), is plotted against the ligand concentration.

indicate that binding is monophasic, saturable ($B_{max} = 11 \pm 1$ ng/mg), and of high affinity ($K_d = 68 \pm 13$ nM) (Fig. 3). In addition, we have investigated the effect of various inhibitors of the asialoglycoprotein receptor on L_3G_4-[^{32}P]ODN binding to hepatocytes (Table I). ASOR, N-acetylgalactosamine, and unlabeled L_3G_4-ODN are all able to inhibit L_3G4-[^{32}P]ODN binding by 80–90%, whereas N-acetylglucosamine is ineffective. The inhibition constants from the L_3G_4-[^{32}P]ODN and [^{125}I]ASOR binding studies are essentially the same, suggesting that L_3G_4-[^{32}P]ODN binding could be attributed to the asialoglycoprotein receptor.

Subsequently, we have investigated whether these glycoconjugated ODNs are also efficiently and specifically taken up by parenchymal liver cells *in vitro* as follows. Freshly isolated parenchymal liver cells are incubated for 10 min at 37° with L_3G_4-[^{32}P]ODN or [^{32}P]ODN (10 nM). During incubation at 37°, L_3G_4-[^{32}P]ODN appears to be stable for 15 min, as judged by PAGE analysis. After incubation, the cell suspensions are put on ice, the cells are washed with ice-cold DMEM plus 2% (w/v) BSA, and incubated

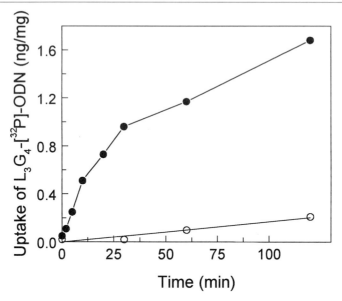

FIG. 4. Kinetics of uptake of L_3G_4-[^{32}P]ODN by isolated rat parenchymal liver cells. PCs (1×10^6 cells/500 μl) were incubated for 0–2 hr at 37° in DMEM plus 2% (w/v) BSA with 10 nM L_3G_4-[^{32}P]ODN in the absence (●) or presence (○) of 100 mM GalNAc. After incubation, cells were put on ice and washed thoroughly, and membrane-associated radioactivity was removed by an EGTA incubation step (see *In Vitro* Studies). After a subsequent cell wash, the cell-associated binding (uptake) was determined and corrected for protein content.

for 5 min at 4° in a Lab-Shaker (150 rpm) with DMEM–2% (w/v) BSA–EGTA (5 mM) to remove membrane-associated L_3G_4-[^{32}P]ODN. Finally, the cells are washed once with DMEM plus 0.2% (w/v) BSA and with DMEM, and the cell-associated radioactivity is counted and corrected for protein content. Nonspecific uptake is determined in the presence of 100 mM N-acetylgalactosamine. Uptake of L_3G_4-[^{32}P]ODN by PCs, i.e., total cell-associated radioactivity after removal of membrane-bound ligand by treatment with 5 mM EGTA, proceeds linearly in time for 10–15 min and tends to level off after 25 min of incubation (Fig. 4). Within the first 20 min, L_3G_4-[^{32}P]ODN uptake is 35-fold more rapid than nonspecific uptake in the presence of 100 mM GalNAc.

To investigate whether L_3G_4-[^{32}P]ODN uptake involves the classic pathway, we have measured the effect of various uptake inhibitors on the internalization of L_3G_4-[^{32}P]ODN by PCs in comparison with that of [^{125}I]ASOR. Cells are preincubated for 30 min at 37° with DMEM plus 2% (w/v) BSA in the absence or presence of sodium azide (10 mM), monensin (25 μM), colchicine (100 μM), or sucrose (250 mM), after which L_3G_4-

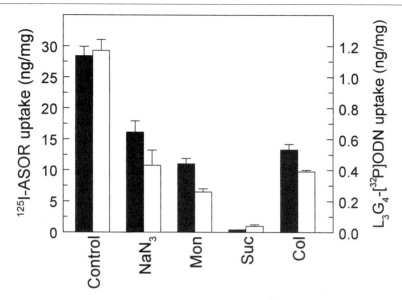

FIG. 5. Effect of various uptake inhibitors on L_3G_4-[^{32}P]ODN (open columns) and [^{125}I]ASOR uptake (filled columns) by rat PCs. Rat PCs (1×10^6 cells/500 μl) were incubated for 30 min at 37° with DMEM plus 2% BSA (control) or DMEM–2% BSA supplemented with sodium azide (NaN$_3$, 10 mM), sucrose (suc, 250 mM), monensin (mon, 10 μM), or colchicine (col, 100 μM). Subsequently, L_3G_4-[^{32}P]ODN or [^{125}I]ASOR was added to a final concentration of 100 and 10 nM, respectively. After incubation for 15 min at 37°, the cells were put on ice, washed once with DMEM plus 2% (w/v) BSA, and incubated for 10 min at 4° with EGTA [5 mM in DMEM plus 2% (w/v) BSA] to remove EGTA-releasable membrane-associated radioactivity. Finally, the cells were washed thoroughly and cell-associated radioactivity was determined and corrected for protein content.

[^{32}P]ODN (10 nM) or [^{125}I]ASOR (10 nM) is added. After incubation for 15 min at 37°, the cells are washed thoroughly. From Fig. 5 it can be derived that all of the tested agents reduce L_3G_4-[^{32}P]ODN uptake. Sodium azide (10 mM), monensin (0.025 mM), and colchicine (0.1 mM) inhibit uptake by approximately 70%, while sucrose (200 mM) almost completely prevents uptake (~96% inhibition). In agreement, [^{125}I]ASOR uptake is similarly reduced after incubation with sodium azide, monensin, colchicine, and sucrose.

Last, we have verified whether the L_3G_4-[^{32}P]ODN is dissociated from the asialoglycoprotein on acidification of the lysosomal compartment. L_3G_4-[^{32}P]ODN binding to the asialoglycoprotein receptor displays a pH dependence (Fig. 6) similar to the binding of the classic substrate [^{125}I]ASOR. Ligand binding is reduced to 50% of control values (pH 7.40) at pH 6.07

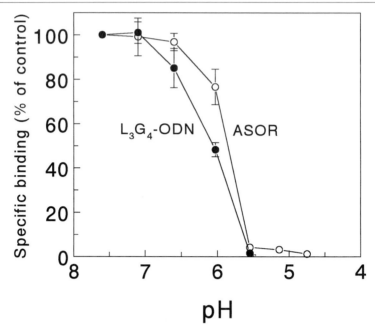

FIG. 6. Effect of pH on binding of [^{125}I]ASOR (○) or L$_3$G$_4$-[^{32}P]ODN (●) to isolated rat parenchymal liver cells. PCs (1 × 10^6 cells/500 μl) were incubated for 2 hr at 4° in Tris–acetate buffer [20 mM Tris-acetate, 150 mM NaCl, 5 mM KCl, 2 mM CaCl$_2$, 1 mM MgCl$_2$, 2% (w/v) BSA; pH 4.75–7.60] with 10 nM [^{125}I]ASOR or 100 nM L$_3$G$_4$-[^{32}P]ODN in the absence or presence of 100 mM GalNAc. After incubation, cells were put on ice and washed thoroughly, and membrane-bound radioactivity was determined and corrected for protein content. Specific binding is defined as the differential binding in the presence and absence of GalNAc (mean ± SD of a triplicate experiment).

and pH 5.91, respectively, and is undetectable below pH 5.5. This implies that L$_3$G$_4$-[^{32}P]ODN is released from the receptor during lysosomal processing.

In Vivo Fate of [^{32}P]ODN or L$_3$G$_4$-[^{32}P]ODN

From the preceding *in vitro* studies it is clear that PC uptake is greatly facilitated by derivatization of ODN with L$_3$G$_4$. To validate that this glyco-conjugation also improves specific accumulation of ODN into PCs *in vivo,* we monitored liver uptake and serum decay of intravenously injected L$_3$G$_4$-[^{32}P]ODN and [^{32}P]ODN in the rat. Male Wistar rats (250–300) are anesthetized by intraperitoneal injection of 15–20 mg of sodium pentobarbital. Four micrograms in 500 μl of PBS is injected into the inferior vena cava,

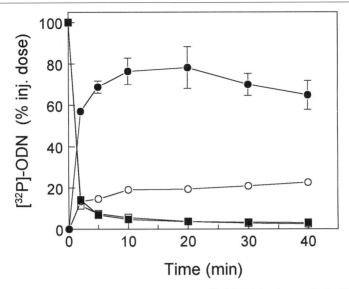

FIG. 7. Serum decay and liver uptake of L_3G_4-[^{32}P]ODN in the rat. L_3G_4-[^{32}P] ODN (●, ■), or underivatized [^{32}P]ODN (○, □) (4 μg in 500 μl of PBS), was injected intravenously into rats. At the indicated times, radioactivities in serum (□, ■) and the liver-associated radioactivities (○, ●) were determined. Values are means ± SEM of three experiments.

and at the indicated times the radioactivities in serum and liver are determined as described previously.[36] L_3G_4-conjugated and nonconjugated ODNs are cleared with equal rapidity from the bloodstream (Fig. 7): within 2 min after injection only 11.5 ± 1.5 and 14.3 ± 1.7% of the injected dose, respectively, resides in the serum. Liver uptake of [^{32}P]ODN amounts to 19.1 ± 0.6% whereas L_3G_4-[^{32}P]ODN is almost quantitatively taken up by the liver (77 ± 6% of the injected dose). Crucial for interpreting the preceding pharmacokinetic data of L_3G_4-[^{32}P]ODN is the stability of L_3G_4-[^{32}P]ODN under *in vivo* conditions. The induced liver uptake could be almost completely prevented by preinjection of *N*-acetylgalactosamine (400 mg/kg), which blocks galactose receptor-mediated uptake (Fig. 8A). Preinjection of *N*-acetylglucosamine, which does not interfere with galactose receptor-mediated substrate recognition, has no effect on liver uptake of L_3G_4-[^{32}P]ODN.

Apparently, glycoconjugation has a stimulatory effect on the hepatic uptake of the antisense drug. To verify that, within the liver, the antisense drug was indeed taken up by the correct cell type we have analyzed the

[36] T. J. C. Van Berkel, Y. B. De Rijke, and J. K. Kruijt, *J. Biol. Chem.* **266,** 2282 (1991).

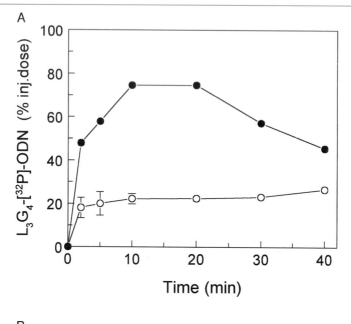

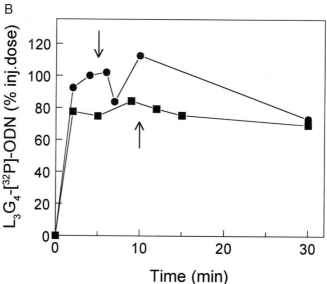

Fig. 8. (A) Effect of preinjection of N-acetylglycosamines on liver uptake of L_3G_4-[^{32}P]ODN. N-Acetylgalactosamine (○) or N-acetylglucosamine (●) (both 400 mg/kg in 250 μl of PBS) were injected intravenously in rats. At 1 min after injection, L_3G_4-[^{32}P]ODN (4 μg in 500 μl of PBS) was administered by intravenous injection into the vena cava. At the indicated times, the liver-associated radioactivities were determined. Values are means of two

contribution of the various liver cell types to the hepatic uptake of the ODN. Rats were anesthetized and injected with [32P]ODN or L_3G_4-[32P]ODN (4 μg in 500 μl of PBS), and at 10 min after injection the vena porta was cannulated and the liver was perfused with collagenase (0.01%, w/v) (8°). Parenchymyal cells (PCs), Kupffer cells (KCs), and endothelial cells (ECs) were subsequently isolated by centrifugal elutriation as previously described[36] and monitored for associated radioactivity. From the PC, EC, and KC uptake (in nanograms per milligram wet weight) the relative contribution of the cell types to the total liver uptake was calculated.[36] The underivatized ODN was primarily internalized by endothelial cells (10.5% of the total injected dose) and Kupffer cells (7.7%), and not by PCs (0.9%). In contrast, liver uptake of L_3G_4-[32P]ODN could be mainly attributed to PCs (58 ± 4% of the total injected dose), whereas endothelial and Kupffer cells played only a marginal role in the clearance of L_3G_4-[32P]ODN (11 ± 7 and 8 ± 4%, respectively; Fig. 9). Preinjection of N-acetylgalactosamine reduced PC uptake by 90% to 5 ± 2%, leaving Kupffer and endothelial cell uptake unaltered. This suggests that only the parenchymal liver cell uptake is mediated by galactose-recognizing receptors. To verify the nature of the liver-associated ODN, we have investigated whether the liver-associated ODN could be released from the liver by displacing the extracellularly bound ODN through injection of N-acetylgalactosamine (400 mg/kg) 5 and 10 min after administration of L_3G_4-[32P]ODN. It can be concluded from Fig. 8B that liver-associated radioactivity was not significantly affected by N-acetylgalactosamine injection, suggesting that the liver-associated radioactivity reflects internalized, nonreleasable rather than extracellularly bound L_3G_4-[32P]ODN.

Concluding Remarks

In vivo application of antisense ODNs for the modulation of the expression of target genes in the PC is seriously hampered because it does not markedly accumulate in this cell type. We show in the current study that this drawback can be overcome by glycoconjugating ODNs to a synthetic ligand for the asialoglycoprotein receptor. Previous studies already illus-

---

FIG. 8. (*continued*) experiments. (B) Release of liver-associated L_3G_4-[32P]ODN by intravenous injection of N-acetylgalactosamine in the rat. L_3G_4-[32P]ODN (4 μg in 500 μl of PBS) was injected intravenously into rats. At 5 min (■; indicated by the arrow) and 10 min (●) after injection of the radiolabel, N-acetylgalactosamine (400 mg/kg; in 250 μl of PBS) was administered by intravenous injection. At the indicated times, the liver-associated radioactivities were determined. Values are means of two experiments.

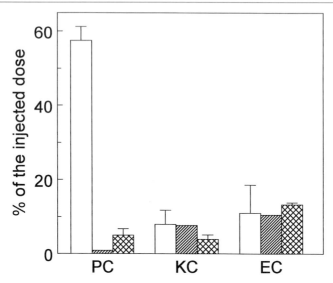

FIG. 9. Relative contribution of various cell types to the liver uptake of L_3G_4-$[^{32}P]ODN$ (open columns), L_3G_4-$[^{32}P]ODN$ after preinjection of GalNAc (hatched columns), or $[^{32}P]ODN$ (crosshatched columns). $[^{32}P]ODN$ (4 μg in 500 μl of PBS) or L_3G_4-$[^{32}P]ODN$ (4 μg in 500 μl of PBS) was injected into rats 1 min after preinjection of PBS (250 μl) or GalNAc (150 mg in 250 μl of PBS). Parenchymal, endothelial, and Kupffer cells were isolated from the liver 10 min after injection of the radiolabel and the cellular radioactivity was counted. Values (except that of GalNAc-treated rats) are means of three experiments ($\pm$ SD) and are expressed as a percentage of the total liver uptake.

trated that the bioavailability of drugs and genes at the desired site can be considerably improved through targeting.[24,26,29,35,37–41] Requisite for successful glycotargeting to the PC is the availability of a high-affinity ligand for a PC-specific receptor such as the asialoglycoprotein receptor.[23] Various research groups have designed glycopeptide mimics of multivalent N-linked oligosaccharides that display nanomolar affinities for the asialoglycoprotein receptor.[39–42] These ligands generally contain glycoside units that are attached to a small peptide scaffold. We have synthesized a tetraantennary

[37] E. Bonfils, C. Dupiereux, P. Midoux, N. T. Thuong, M. Monsigny, and A. C. Roche, *Nucleic Acids Res.* **20,** 4621 (1992).
[38] M. Chiu, T. Tamura, M. S. Wadhwa, and K. G. Rice, *J. Biol. Chem.* **269,** 16195 (1994).
[39] J. Haensler and F. C. Szoka, *Bioconjugate Chem.* **4,** 85 (1993).
[40] J. R. Merwin, G. S. Noell, W. C. Thomas, H. C. Chion, M. E. De Rome, T. D. McKee, and M. A. Findeis, *Bioconjugate Chem.* **5,** 612 (1994).
[41] C. Plank, K. Zatloukal, M. Cotten, K. Mechtler, and E. Wagner, *Bioconjugate Chem.* **3,** 533 (1992).
[42] R. T. Lee and Y. C. Lee, *Glycoconjugate J.* **4,** 317 (1987).

lysine-based galactoside (L_3G_4) with an affinity as high as that of the aforementioned glycopeptides (K_i 6.5 nM), using an accessible two-step synthetic protocol.

Competition studies of [^{125}I]ASOR binding to hepatocytes established that the affinity of L_3G_4 for the asialoglycoprotein receptor (ASGPr) was only slightly reduced after conjugation to ODN-AS5', suggesting that L_3G_4 may be an appropriate homing device for targeting of ODNs to the asialoglycoprotein receptor. Therefore, the glycoconjugated ODN-AS5' was ligated to the ^{32}P-labeled 3' end of the antisense ODN (ODN-AS3') to yield an internally labeled 20-mer, and it was assessed if hepatocyte binding and uptake of the ODNs were influenced by this derivatization. L_3G_4-[^{32}P]ODN appeared to bind to hepatocytes in a saturable fashion. L_3G_4-[^{32}P]ODN binding could be almost completely (>95%) displaced by unlabelled L_3G_4-ODN and by conventional ligands for the ASGPr (namely, N-acetylgalactosamine, ASOR), but not by N-acetylglucosamine. The inhibition constants of N-acetylgalactosamine and ASOR for displacing L_3G_4-[^{32}P]ODN were essentially similar to that for [^{125}I]ASOR. Apparently, L_3G_4-[^{32}P]ODN binding to PCs can be fully attributed to the ASGPr. *In vitro* studies at 37° demonstrated that the ASGPr not only binds glycoconjugated ODN but also mediates its uptake and processing. L_3G_4-[^{32}P]ODN uptake was also reduced by established inhibitors of lysosomal uptake, similar to [^{125}I]ASOR.

In concert with the preceding *in vitro* data, *in vivo* liver uptake of ODNs appeared to be considerably enhanced to almost 80% of the injected dose by coupling to L_3G_4. Liver uptake is close to values reported for a biantennary cluster glycoside with high affinity for the asialoglycoprotein receptor[38] and to that of ASOR itself. Because the serum degradation of the glycoconjugate proceeds at a much slower rate than hepatic uptake, because liver uptake can be inhibited by inhibitors of the asialoglycoprotein receptor, and because liver uptake of capped phosphodiester ODNs (or ODN degradation products) has been previously shown to be low,[6] we may assume that liver association of the glycoconjugate reflects uptake of the intact glycoconjugate. Analysis of the liver cell types revealed that the PC was responsible for the induced liver uptake. Accumulation of L_3G_4-derivatized ODNs in this cell type was 60-fold higher on derivatization with L_3G_4 as compared with underivatized ODNs. Hangeland *et al.*[29] have reported enhanced hepatic uptake of an uncharged methyl phosphonate dT$_7$, conjugated to a trisgalactosylated glycopeptide in the mouse. In agreement, we report in this study that elimination of a miscellaneous, charged, and full-length ODN sequence by scavenger receptors can be prevented, and that liver uptake of the ODN can be enhanced, after derivatization of the ODN to a synthetic glycopeptide tag. Moreover, we demonstrate that uptake is mediated by the targeted asialoglycoprotein receptor, and that it reflects internalized ODN. In conclusion, glycotargeting of ODN sequences using cluster galactosides is an effective way to enhance the local bioavailability of ODN in the PC under *in vivo* conditions. Apparently, the natural tendency of ODNs to be eliminated by the reticuloendothelial system (i.e., macrophages, Kupffer cells, and liver endothelial cells) and by cells that express scavenger receptor-type proteins[6,43,44] can be overcome through conjugation to a synthetic low molecular weight ligand for the asialoglyco-protein receptor.

Acknowledgment

This study was supported by Grant M93.001 of the Dutch Heart Foundation.

[43] A. Rifai, W. Brysch, K. Fadden, J. Clark, and K. H. Slingensiepen, *Am. J. Pathol.* **149**, 717 (1996).
[44] M. K. Bijsterbosch, M. Manoharan, E. T. Rump, R. L. A. De Vrueh, R. Van Veghel, K. L. Tivel, E. A. L. Biessen, C. F. Bennett, P. D. Cook, and T. J. C. Van Berkel, *Nucleic Acids Res.* **25**, 3290 (1997).

[24] Antisense Methods for Discrimination of Phenotypic Properties of Closely Related Gene Products: Jun Kinase Family

By Frédéric Bost, Robert McKay, Nicholas M. Dean, O. Potapova, and Dan Mercola

Introduction

The mitogen-activated protein kinase (MAPK) pathways are involved in the regulation of important phenotypic properties such as transformation, proliferation, differentiation, and DNA repair. Of the MAPK pathways the three that have been well characterized are the following: the extracellular-signaling regulated kinase (ERK) pathway, the Jun N-terminal kinase (JNK) pathway, and the p38MAPK pathway. These pathways mediate extracellular signals (growth factor, stress stimuli, cytokines) as well as DNA-damaging events by activating a cascade of kinases leading to the activation of transcription factors such as c-Jun, ATF-2, and Elk-1.[1-10] The

[1] F. Bost, R. McKay, N. Dean, and D. Mercola, *J. Biol. Chem.* **272**, 33422 (1997).

kinases of the cascade consist of the mitogen-activated protein kinase kinase kinases (MAPKKKs) such as PAK65, which activates the mitogen-activated protein kinase kinases (MAPKKs) such as MEKK, and the mitogen-activated protein kinases (MAPKs) such as JNK. Among these kinases several isoforms are reported; for example, at least four isoforms are described for MEKK: MEKK1, -2, -3, and -4.[11] So far, 10 different isoforms have been identified for JNK in humans,[7] four of them forming the JNK1 group, four others the JNK2 group, and the last two comprising the JNK3 group. The JNK1 and -2 isoforms are ubiquitous whereas JNK3 is expressed in only a few tissues such as brain, testis, and heart.[12,13]

"Knockout" experiments for some of the different isoforms (JNK1, -2, and -3) provide evidence of differential roles of JNKs in development. For example, it was shown that JNK3 has a role in brain development; in contrast, no defect was detected in the development of JNK1$^{-/-}$ and JNK2$^{-/-}$ mice. Although knockout experiments provide a specific way to target one isoform, they mainly furnish information about nonredundant functions. Moreover, certain biophysical properties of the JNKs such as the differing affinities of JNK1 and JNK2 for the three major substrates c-Jun, ATF2, and Elk-1,[1–10] suggest that activation of the different isoforms may lead to different phenotypic consequences. Antisense methods have the potential to discriminate among genes of homologous sequences, thereby providing a powerful potential means of dissecting the roles of JNK1, JNK2, and JNK3 and their isoforms in bringing about the various reported activities of apoptosis,[14–18] proliferation,[1–4] and DNA repair[2–19]

[2] O. Potapova, A. Haghighi, F. Bost, C. T. Liu, M. J. Birrer, R. Gjerset, and D. Mercola, *J. Biol. Chem.* **271,** 14041 (1997).

[3] T. Smeal, B. Binétruy, D. A. Mercola, M. Birrer, and M. Karin, *Nature (London)* **354,** 494 (1991).

[4] T. Smeal, B. Binétruy, D. A. Mercola, A. Bardwick-Grover, G. Heidecker, U. Rapp, and M. Karin, *Mol. Cell. Biol.* **12,** 3507 (1992).

[5] M. Hibi, A. Lin, T. Smeal, A. Minden, and M. Karin, *Genes Dev.* **7,** 2135 (1993).

[6] B. Dérijard, M. Hibi, I. H. Wu, T. Barrett, B. Su, T. Deng, M. Karin, and R. J. Davis, *Cell* **76,** 1025 (1994).

[7] S. Gupta, D. Campbel, B. Dérijard, and R. J. Davis, *Science* **267,** 389 (1995).

[8] C. Livingstone, G. Patel, and N. Jones, *EMBO J.* **14,** 1785 (1995).

[9] M. Cavigelli, F. Dolfi, F. X. Claret, and M. Karin, *EMBO J.* **14,** 5957 (1995).

[10] H. Van Dam, D. Wilhemm, I. Rher, A. Steffen, P. Herrlich, and P. Angel, *EMBO J.* **14,** 1798 (1995).

[11] G. R. Fanger, N. L. Johnson, and G. L. Johnson, *EMBO J.* **16,** 4961 (1997).

[12] T. Y. Ip, and R. J. Davis, *Opin. Cell Biol.* **10,** 205 (1998).

[13] D. D. Yang, C. Y. Kuan, A. J. Whitmarsh, M. Rincon, T. S. Zheng, R. J. Davis, P. Rakic, and R. A. Flavell, *Nature (London)* **389,** 865 (1997).

[14] Z. Xia, M. Dickens, J. Raingeaud, R. J. Davis, and M. E. Greenberg, *Science* **270,** 1326 (1995).

and other stress responses.[6,10] For example, we have observed that the JNK pathway mediates proliferation of certain human tumor lines such as T98G human glioblastoma,[20] MCF human breast carcinoma,[21] and A549 lung carcinoma cells.[1] Using antisense methods, it has been possible to show that these effects are largely due to the activation of JNK2 and not JNK1.[22,23]

Nevertheless, several controversies have been raised about studies using antisense oligonucleotides.[24] We describe here experimental procedures to select highly specific antisense oligonucleotides that work at low concentrations. These methods are illustrated using oligonucleotides that eliminate the two major groups of JNK isoforms: JNK1 and JNK2 in cancer cells. Methods to study the consequences of this elimination on the transformed phenotype are described below.

Transfection Procedures for Antisense Oligonucleotides *in Vitro*

To transfect oligonucleotides into cells it is necessary to use appropriate techniques; the two most commonly used are transfection using phosphate calcium or transfection using DNA-containing cationic liposomes (lipofection). We find that commercially available lipofection reagents are effective and work with a large variety of cells.

However, there are marked differences in the efficiency of transfection with cationic liposomes depending on the cell line used. We, therefore, recommend testing different elapsed times of lipofection when using a new cell line rather than altering the oligonucleotide concentration or oligonucleotide ratios. We have varied the elapsed time of exposure from 4.5 hr

[15] D. J. Wilson, K. A. Fortner, D. H. Lynch, R. R. Mattingly, I. G. Macara, J. A. Posada, and R. C. Budd, *Eur. J. Immunol.* **26,** 989 (1996).

[16] M. Verheij, R. Bose, X. H. Lin, B. Yao, W. D. Jarvis, S. Grant, M. J. Birrer, E. Szabo, L. I. Zon, J. M. Kyriakis, A. Haimovitz-Friedman, Z. Fuks, and R. N. Kolesnick, *Nature (London)* **380,** 75 (1996).

[17] Y. R. Chen, C. F. Meyer, and T. H. Tan, *J. Biol. Chem.* **271,** 631 (1996).

[18] N. L. Johnson, A. M. Gardner, K.M. Diener, C. A. Lange-Carter, J. M. B. GleavyJarpe, A. Minden, M. Karin, L. I. Zon, and G. L. Johnson, *J. Biol. Chem.* **271,** 3229 (1996).

[19] V. Adler, A. Polotskaya, J. Kim, L. Dolan, R. Davis, M. Pincus, and Z. Ronai, *Carcinogenesis* **9,** 2073 (1996).

[20] O. Potapova, F. Bost, N. Dean, and D. Mercola, *Proc. Am. Assoc. Cancer Res.* **39,** 251 (1998).

[21] O. Potapova, H. Fakhrai, S. Baird, and D. Mercola, *Cancer Res.* **56**(2), 280 (1996).

[22] F. Bost, O. Potapova, R. McKay, N. Dean, and D. Mercola, *Cancer Gene Ther.* **5,** S32 (1998).

[23] F. Bost, R. McKay, M. Bost, O. Potapova, N. Dean, and D. Mercola, *Mol. Cell. Biol.* **19,** 1938 (1999).

[24] C. A. Stein and A. M. Krieg, *Antisense Res. Dev.* **4,** 67 (1994).

(most cells) to 24 hr (T98G glioblastoma cells), using the protocol described here. Some cell lines such as human primary keratinocytes have the capacity to internalize oligonucleotide without any uptake enhancer, but this is unusual. Here the method is illustrated with A549 human lung carcinoma cells, which exhibit typical behavior. In the examples described here a single antisense oligonucleotide that is complementary to a sequence common to all isoforms of the JNK1 group and similarly a single oligonucleotide complementary to all isoforms of the JNK2 group are employed. Unless otherwise noted the isoform families are termed "JNK1" and "JNK2" respectively.

Assay

Day 1: A549 cells are seeded in a 150-cm^2 flask, 6-well tissue culture plates, or 24-well plates at a density of 42,000 cells/cm^2 in order to achieve ~70–80% confluency on day 2. During the first 18 hr, the cells are kept in Dulbecco's modified Eagle's medium (DMEM; GIBCO-BRL, Gaithersburg, MD) complemented with 10% fetal bovine serum (FBS, Irvine Scientific, Santa Ana, CA), L-glutamine (2 mM), penicillin G (1000 units/ml), and streptomycin sulfate (1000 μg/ml) at 37° in a 10% CO$_2$ atmosphere.

Day 2: Transfection. Prewarmed minimum essential medium (MEM) with no FBS, no antibiotics, and no L-glutamine is mixed with Lipofectin reagent (GIBCO-BRL) containing N-[1-(2,3-dioleoyloxy)-propyl] N,N,N-trimethylammonium/dioleoylphosphatidylethanolamine (DOTMA/DOPE) at a final concentration of 15 μg/ml of medium. The mix of MEM–Lipofectin is kept at room temperature for 30–45 min prior to use. A 10 μM stock solution of antisense oligonucleotide dissolved in H$_2$O is then added to the MEM–Lipofectin to reach a final concentration of 0.4 μM. It is important to vortex the mixture well and then let it stand for 15 min at room temperature. In the meantime, cells are rinsed twice with prewarmed phosphate-buffered saline (PBS) in order to remove serum and antibiotics, which interfere with transfection efficiency. It is important to avoid any dryness of the cells during the transfection, by adding a sufficient volume of transfection medium ($\geq$100 μl/cm^2). After 4.5 hr at 37° in a 10% CO$_2$ atmosphere, the transfection medium is removed from the plates. The cells are then immediately washed with prewarmed PBS, and then fed with prewarmed medium containing the appropriate amount of FBS (see above). Cell lysate is prepared at the appropriate time after lipofection for analysis. When two or more oligonucleotides are used for lipofection, the combined concentration is maintained at $\leq$0.4 μM.

For mRNA analysis, total cellular RNA is isolated by lysis in 4 M

guanidium isothiocyanate followed by cesium chloride gradient centrifugation according to the usual methods (i.e., Dean et al.[25–27]).

For protein analysis, cells are washed twice with cold PBS and the plates of cells kept on ice at all times. Total protein is extracted by adding 100 μl of whole cell extract buffer [WCE: 25 mM HEPES (pH 7.7), 0.3 M NaCl, 1.5 mM MgCl$_2$, 0.1% Triton X-100, phenylmethylsulfonyl fluoride (PMSF, 100 μg/ml), tolylsulfonyl phenylalanyl chloromethyl ketone (TPCK, 100 μg/ml), 1 μM EDTA, leupeptin (2 μg/ml), aprotinin (2 μg/ml), 20 mM β-glycerophosphate, 0.1 mM Na$_3$VO$_4$]. Cells are maintained for 5 min at 4°, harvested using a cell scraper (Costar, Cambridge, MA), and the lysate rotated slowly at 4° for 30 min. After centrifugation at 12,000g, the protein concentration of the cell extracts is determined, for example, by the Bradford dye method, following the procedure provided by the manufacturer (Bio-Rad, Hercules, CA). The same protein extract can be used for Western analysis and kinase assays.

The determination of the efficiency of transfection, i.e., the percentage of target mRNA or protein eliminated, is described below. We first optimize transfection efficiency (see Transfection Procedures, above, and Transfection Efficiency, below) by determination of the optimum elapsed time of lipofection, usually examined at multiples of 4 hr, rather than alteration of concentration or oligonucleotide/lipid ratios. High transfection efficiencies can usually be achieved by increased elapsed time, even up to 24 hr. Cells that do not tolerate prolonged exposure to the lipofection conditions, e.g., LnCAP, begin to exhibit signs of decreased attachment. The majority of established human tumor lines that we have worked with exhibit ≥80% elimination of target protein 24–48 hr after lipofection for 4–12 hr, depending on the half-life (see below) of the preformed target protein.

Transfection Efficiency

It is believed that a true antisense mechanism results from oligonucleotide–target mRNA hybrid formation in the nucleus, which stimulates the cleavage of the mRNA at one or more sites near the terminus of the hybrid complex.[28–30] Analogous cytoplasmic complexes may lead to translation

[25] N. M. Dean, R. McKay, T. P. Condon, and C. F. Bennett, *J. Biol. Chem.* **269,** 16146 (1994).
[26] N. M. Dean and R. McKay, *Proc. Natl. Acad. Sci. U.S.A.* **91,** 11762 (1994).
[27] N. M. Dean, R. McKay, L. Miraglia, R. Howard, S. Cooper, J. Giddings, P. Nicklin, L. Meister, R. Zeil, T. Geiger, M. Muller, and D. Fabbro, *Cancer Res.* **56,** 3499 (1996).

arrest.[28-30] However, owing to the lack of universal correlation between steady state mRNA levels and steady state protein product levels, the best criterion for monitoring the effects of lipofection is the fractional reduction of steady state protein target. This requires the availability of well-character- ized reagent antibodies for either Western analysis, immunoprecipitation of metabolically labeled cells, or other specific isolation methods, for exam- ple, the use of gelatin-coated beads for the isolation of fibronectin[31] or fractional extraction for the isolation of plasminogen activator inhibitor 1.[32] For JNKs, an enzymatic assay that discriminates among the various target proteins can also be used. In the case of JNKs, the major isoforms of JNK1 migrate at different molecular weights on polyacrylamide gel electrophoresis (PAGE) and can, therefore, be largely distinguished by Western analysis using a "pan-JNK" antibody such as anti-JNK1 antibodies (SC-571) available from Santa Cruz Biotechnology (Santa Cruz, CA), or can be distinguished by enzymatic assay after the so-called in-gel kinase assay as described by Hibi *et al.*[5] PAGE analysis after immunoprecipitation is another alternative.

Both oligonucleotide and protein controls are essential. The appropriate oligonucleotide controls for antisense experiments have been described in detail.[24] The full panel includes parallel lipofection with sense sequence, scrambled sequence, or random control oligonucleotides, and antisense sequences bearing one or a small number of mismatched bases, i.e., mis- matched with respect to the complementary target sequence. If comparing the effects of targeting two related genes such as JNK1 and JNK2, this would amount to the use of at least six control experiments and when combined antisense JNK1 + 2 are involved, an additional three control experiments are required, making the analysis cumbersome. As the elimina- tion of one target, say JNK1, by the respective antisense reagent does not alter significantly the steady state levels of JNK2 itself, the unaltered JNK2 level provides a demonstration of specificity. Therefore, we accept this result as an internal control and routinely omit mismatch oligonucleotides.

Protein controls consist of (1) examining the effects of antisense lipofec- tion on a "housekeeping" gene product such as β-actin, (2) including an untreated or "mock-lipofection" case, and (3) including a positive control

[28] S. T. Crooke, *Annu. Rev. Pharmacol. Toxicol.* **32,** 329 (1993).

[29] S. T. Crooke, *Antisense Nucleic Acid Drug Dev.* **8,** 133 (1998).

[30] N. M. Dean, R. McKay, L. Miraglia, T. Geiger, M. Muller, D. Fabbro, and C. F. Bennett, *RNA Interactions: Ribozymes Antisense (Biochem. Soc. Trans).* **24,** 623 (1996).

[31] C. J. Der, J. F. Ash, and E. J. Stanbridge, *J. Cell Sci.* **52,** 151 (1981).

[32] J. Cárcamo, F. M. B. Weis, F. Ventura, R. Wieser, J. L. Wrama, L. Attisano, and J. Massagué, *Mol. Cell. Biol.* **14,** 3810 (1994).

that exhibits readily detectable levels of the target protein or activity. In the case of JNK, this latter control is readily arranged by irradiation of living cells prior to harvest by UV-C at 40–100 μJ/cm^2 (see below). The results are expressed as relative protein or activity levels, i.e., the amount of signal (normalized by β-actin) after antisense treatment is divided by the signal (normalized by β-actin) of untreated cases ("signal" strength may be determined by digitalization as described; see Abbreviated Half-Life Determination Procedure, below). Typical procedures are described in the next section.

Strategy to Select Specific Antisense Oligonucleotides Targeting Major Jun N-Terminal Kinase Isoforms

Gene-Walk Screening of Jun N-Terminal Kinase Oligonucleotides

There is no rule to predict the most effective oligonucleotide able to eliminate the messenger of a target gene. It is, therefore, necessary to have a rigorous approach to check the specificity, efficacy, and affinity of several candidate antisense oligonucleotides in order to determine the best sequence for eliminating the most steady state mRNA of the gene of interest. A method that combines these features has been developed by Isis Pharmaceuticals (Carlsbad, CA) and consists of choosing one antisense oligonucleotide complementary to the target sequence every 50 to 150 nucleotides along the sequence of the targeted gene. These oligonucleotides are then synthesized on a small scale and tested for their ability to eliminate the steady state mRNA level. Phosphorothioate backbone chemistry is recommended as this formulation is nuclease resistant, readily available, and a common element of second-generation compounds.[28–30,33]

To identify effective antisense oligonucleotides capable of entering and inhibiting specifically JNK1 or JNK2 mRNA expression in A549 cells, 26 phosphorothioate oligonucleotides complementary to JNK1 and 13 phosphorothioate oligonucleotides complementary to JNK2 were prepared. In the case of JNK, the target sequences were chosen approximately every 50 nucleotides along JNK1 and every 130 nucleotides along JNK2 transcripts from the 5' start of the transcription to the 3' noncoding region.

Gene screening is performed by measuring steady state mRNA levels by Northern analysis as previously described.[25–27,34] Twenty-four hours after

---

[33] R. A. McKay, L. J. Miraglia, L. L. Cummins, S. R. S. R. Owens, H. Sasmor, and N. M. Dean, *J. Biol. Chem.*, in press (1999).

[34] R. McKay, L. L. Cummins, M. J. Graham, E. A. Lesnik, S. R. Owens, M. Winniman, and N. M. Dean, *Nucleic Acids Res.* **24,** 411 (1996).

lipofection (carried out as described; see Transfection Procedures, above), mRNA is prepared.

See Fig. 1 for an example of a screening for JNK1 and JNK2. As shown, most candidate oligonucleotides have only a small effect on the steady state level of JNK1 or JNK2 mRNA (Fig. 1A and B). However, their degree of activity greatly varies depending on the targeted region (Fig. 1A and B). Among the most potent oligonucleotides of the array complementary to JNK1 is ISIS 12539 (JNK1AS^{ISIS12539}) (5′-CTCTCTGTAGGCCCGCTTGG-3′) located at the 3′ end of the coding region. This oligonucleotide leads to a reduction of the steady state mRNA level by 95% (Fig. 1A).

ISIS 12560 (JNK2AS^{ISIS12560}) (5′-GTCCGGGCCAGGCCAAAGTC-3′), located at the 5′ end of the coding region of JNK2 genes, is the most efficient oligonucleotide in reducing JNK2 steady state mRNA levels. Lipofection with this oligonucleotide leads to a decrease in the steady state JNK2 message level by 92% compared with the control JNK2 steady state mRNA level. The oligonucleotides targeted to the regions flanking either side of the sequences for JNK1AS^{ISIS12539} and JNK2AS^{ISIS12560} also inhibit JNK1 and JNK2 steady state mRNA, but in a manner that decreases with increasing distance from the optimum sequence (Fig. 1A and B), suggesting that discrete regions of the target mRNA are accessible and sensitive to the antisense-mediated elimination of steady state mRNA as previously described.[28,29]

Cross-Inhibition Verification for Antisense Targeting Jun N-Terminal Kinase Isoforms

To determine the role of a given isoform, antisense oligonucleotides must specifically eliminate mRNA/protein of the isoform of interest, not any other isoform mRNA/protein and not unrelated mRNA/proteins. It is, therefore, necessary to check the specificity of the antisense selected from the gene walk screening. Twenty-four hours after the transfection, mRNA and protein levels are examined by Northern and Western analysis, respectively. After such analysis, when "lead" candidate antisense sequences have been chosen, it is mandatory to use control oligonucleotides such as sense and/or scrambled oligonucleotide corresponding to the candidate antisense response in a repeat Northern assay in order to confirm specificity. An antisense oligonucleotide less effective than the leading compound and selected from the messenger walk screening can also be considered as a good additional control.

Having identified specific lead antisense compounds, a cross-inhibition test is essential in order to show isoform class specificity. As shown in Fig.

A

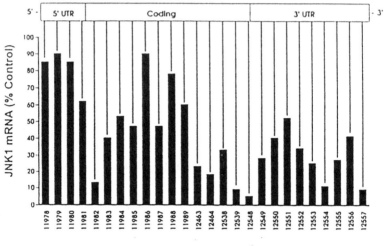

JNK1 Oligonucleotide n°

B

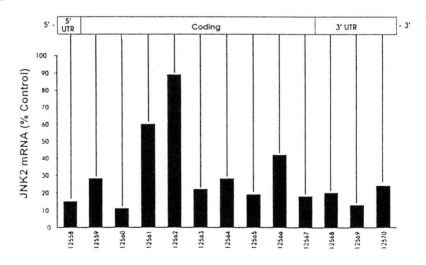

JNK2 Oligonucleotide n°

2A and B, JNK1AS[ISIS12539] eliminates JNK1 mRNA and protein but does not affect JNK2 mRNA and protein and, conversely, JNK2AS[ISIS12560] has no effect on JNK1 mRNA and protein but abolishes JNK2 mRNA and protein. In the Northern analysis we describe an example in which two other antisense oligonucleotides from the initial gene walk screening are less efficient in reducing mRNA steady state levels of their respective target gene (Fig. 2A). The Western analysis shows an example in which lipofection with the antisense oligonucleotides leads to complete elimination of their respective target steady state protein level whereas the scrambled sequence versions of the candidate oligonucleotide, JNK1Scr or JNK2Scr, have no effect. This test establishes that effective and specific reagents are available.

Determination of Window of Action of Antisense Oligonucleotides to Perform Phenotypic Studies

To assess phenotype or to employ additional lipofection treatment it is necessary to know how long the inhibition of expression lasts after a single treatment. The maximum time period available for observation of the effects of suppression of a protein after a single treatment depends on the stability of JNK1 and JNK2 proteins after elimination of mRNA and the duration of phosphorothioate oligonucleotides in the nucleus. Therefore, it is important to determine the half-life of JNK1 and JNK2 proteins and the duration of suppression of JNK1 and JNK2 proteins after a single treatment. Half-life may be determined by the following abbreviated procedure.

Abbreviated Half-Life Determination Procedure

A549 cells are seeded in six-well plates at 250,000 cells/well. One day later, cells are rinsed twice with PBS and incubated in the presence of 300 μl of "fresh" (no more than 3 weeks from supplier, stored at $-80°$)

FIG. 1. Example of "gene walk" survey for the elimination of JNK1 and JNK2 mRNA levels after treatment with phosphorothioate antisense oligonucleotides targeted to JNK1 or JNK2 mRNA. (A) JNK1 mRNA steady state levels in A549 cells after treatment with 26 different phosphorothioate antisense oligonucleotides targeted to JNK1 mRNA. JNK1AS[ISIS12539] gave the most consistent results for the elimination of JNK1 mRNA steady state levels. (B) is similar to (A), but was performed with 13 phosphorothioate antisense oligonucleotides complementary to the indicated regions of the JNK2 mRNA. JNK2AS[ISIS12560]. All oligonucleotides are arrayed relative to their complementary sequence along the JNK transcript. The probe was prepared by a random primer method using an expression vector for JNK1 and JNK2 and following standard protocols.

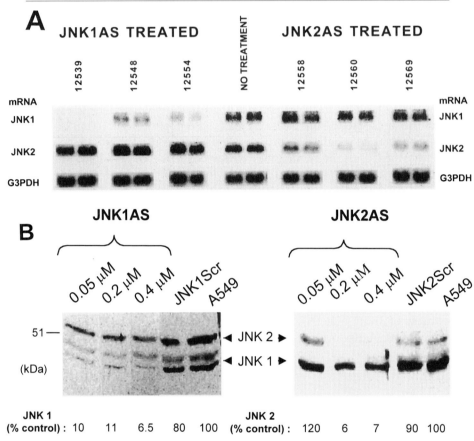

FIG. 2. "Cross-inhibition" experiment using JNK1 and JNK2 antisense oligonucleotides. (A) Northern analysis: A549 cells were treated with three different antisense oligonucleotides complementary to JNK1 mRNA (including the active antisense oligonucleotide JNK1AS[ISIS12539]) and three different antisense oligonucleotides complementary to JNK2 mRNA (including the active antisense oligonucleotide JNK2AS[ISIS12560]). Twenty-four hours after a 4-hr transfection with 0.4 μM antisense oligonucleotide, mRNA was prepared and examined by Northern analysis; the same membrane was hybridized successively with JNK1 probe, JNK2 probe, and the G3PDH probe. (B) Western analysis: A549 cells were treated with 0.05, 0.2, and 0.4 μM JNK1AS[ISIS15346] or JNK2AS[ISIS15353] and 0.4 μM concentration of their respective control oligonucleotides JNK1Scr[ISIS18076] and JNK2Scr[ISIS18078]. Cell extracts were prepared 36 hr after the transfection and examined by Western analysis, using "pan-JNK" antibody [anti-JNK1 antibodies (SC-571)].

methionine–lysine-free DMEM containing [^{35}S]methionine [Tran^{35}S-Label, 300 μCi/ml (ICN, Costa Mesa, CA) is recommended, which also contains a small amount of [^{35}S]cysteine] at 37° in 5% CO_2 in a Plexiglas radio-safe box containing a sheet of SO_2-absorbing carbon paper in order to avoid emission of radiolabeled gases. At various times, cells are washed once with PBS and then lysed by scraping the plates with radioimmunoprecipitation assay (RIPA) buffer [1% Nonidet P-40 (NP-40), 0.5% sodium deoxycholate, 0.1% sodium dodecyl sulfate (SDS), phenylmethylsulfonyl fluoride (PMSF, 100 μg/ml), 1 mM aprotinin, 1 mM sodium orthovanadate]. Cell extracts are immunoprecipitated first with JNK1-specific antibodies (e.g., Sc-474; Santa Cruz Biotechnology) to isolate JNK1 protein and second with pan-JNK antibodies recognizing JNK1 and JNK2 proteins (e.g., SC-571). After precipitation of the antibody–protein complex with 30 μl of washed *Staphylococcus* protein A-coated beads (Boehringer Manneheim, Indianapolis, IN) and three washes with cold RIPA buffer, 25 μl of Laemmli buffer is added to the beads. ^{35}S-Labeled proteins are resolved by 10% SDS–PAGE. The quantification of ^{35}S is carried out by digitalization of autoradiography of the dried gel, using a visible image capture system such as the Electrophoresis Documentation and Analysis System 120 (Kodak Scientific Imaging Systems, New Haven, CT) for autoradiographs. Two-dimensional array detection systems such as Ambis (San Diego, CA) are also excellent.

An example using this procedure is shown in Fig. 3. JNK1 has a much longer half-life than JNK2, with respective half-lives of 14.5 and 3.4 hr. Similar values were found for another human tumor cell line, T98G glioblastoma (O. Potapova, unpublished results, 1998). Thus, the long-lived nature of JNK1 means that the consequences of elimination of JNK1 mRNA may be considerably retarded compared with JNK2 and illustrates the importance of determining the half-life of target proteins even for closely related gene family members.

Determination of Duration of Suppression of JNK1 and JNK2 Proteins

It is essential to know how long the effect of a single treatment of cells with antisense lasts in order to plan additional treatment or to know when to terminate experiments. The duration of antisense effect may be determined by carrying out a time-course experiment by harvesting cells at different times after lipofection. After various times (e.g., 24, 48, and 72 hr after lipofection in six-well plates) cell lysates are prepared and analyzed by Western or Northern analysis.

As shown in Fig. 2, the kinetics of expression of JNK1 and JNK2 protein correlates well with the steady state mRNA level. Thus, when the JNK2

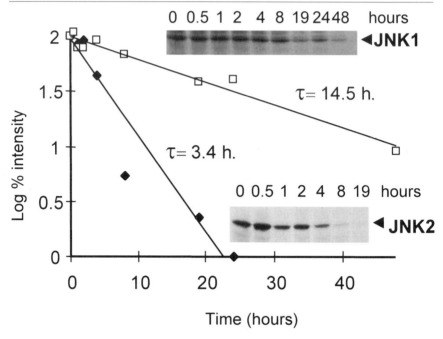

FIG. 3. Example of half-life determination JNK1 and JNK2 in A549 cells. A549 cells were incubated for 4 hr with [^{35}S]methionine. Cells extracts were prepared at the indicated time and immunoprecipitated first with JNK1-specific antibodies (SC-474) and second with "pan-JNK" antibody (SC-571). The half-life was determined by measuring the intensity of the radioactive band corresponding to JNK1 or JNK2, using the Electrophoresis Documentation and Analysis System 120 (Kodak Scientific Imaging Systems). Background values were subtracted in all cases.

mRNA level is back to normal at 72 hr, JNK2 protein is visible. For JNK1, the mRNA level is still strongly inhibited (40% of normal level only) at 72 hr and similarly no protein is detectable. The combined results define a "window" of 2 days (from 24 to 72 hr) when no JNK1 protein is present, even though JNK1 has a long half-life. For JNK2 the window is 12 to 48 hr. This long-term elimination of JNK1 and JNK2 proteins corresponds to approximately two times the cell cycle. It is, therefore, predictable that for cell-counting experiments and for cells that are growth regulated by the JNK pathway, the consequences of the 48-hr absence of JNK1 and JNK2 protein will be a permanent deficit in the expanding population and be detectable even after 5 days (120 hr) or more in a growth measurement experiment.[1,23] Second-generation oligonucelotides with increased stability and target affinity may persist longer within cells after a single lipofection and, therefore, provide extended windows.[33]

Determination of Total Jun N-Terminal Kinase Activity

Because it is possible that significant enzyme activity may be present in cells with little or no detectable protein as judged by conventional methods, it is recommended that selected time points during the window of antisense effect be checked by a direct enzyme activity assay.

Some important precautions must be considered in order to perform a kinase assay after the usual lipofection procedure. Because JNK is "stress activated," it is necessary to take several steps in order to prevent any nonspecific activation of the JNK pathway.

1. It is important to avoid any dryness of the wells when cells are washed at any time after the lipofection.
2. Allow the cells to "recover" in prewarmed 10% FBS medium for ≥6 hr after the transfection.
3. It is necessary to incubate the cells in prewarmed 0.5% FBS medium 15–18 hr prior to the kinase assay to promote a quasiquiescent state and prior to the application of UV or any addition of a test factor used to activate the JNK pathway.
4. Wash the cells twice with prewarmed PBS before adding the 0.5% FBS medium.
5. Cells must be covered with ≥55 μl/cm^2 of prewarmed PBS, which is transparent to UV-C wavelength. UV irradiation is performed using a UV-C source that provides a precise dose of UV-C (μJ/cm^2), for example, a Stratalinker UV cross-linker (Stratagene, La Jolla, CA). Furthermore, it is necessary when working with multiwell plates to leave the nonirradiated wells with the media and the plastic cover of the plate in order to prevent any false activation.

Assay

Cell lysates are prepared using whole cell extract buffer as described above (see Transfection Procedure) and 50 μg of total protein is used to perform the JNK assay. The kinase assay is performed exactly as described by Hibi *et al.*[5] Briefly, 50 μg of WCE is mixed with 10 μg of glutathione S-transferase–c-Jun (1–223) [GST–c-Jun(1–223)] for 3 hr at 4°. GST–c-Jun fusion proteins are previously expressed and purified from *Escherichia coli* and bind to glutathione–Sepharose 4B beads (Pharmacia Biotech, Piscataway, NJ). After four washes, beads are incubated with 30 μl of kinase reaction buffer [20 mM HEPES (pH 7.7), 20 mM MgCl$_2$, 20 mM β-glycerophosphate, 20 mM p-nitrophenyl phosphate, 0.1 mM Na$_3$VO$_4$, 2 mM dithiothreitol (DTT), 20 μM ATP, and 5 μCi of [γ-^{32}P]ATP] for 20 min at 30°. The reaction is stopped by addition of 20 μl of Laemmli sample buffer. The phosphorylated GST–c-Jun protein is eluted by boiling the sample for 5 min and resolved by 10% SDS–PAGE.

Phenotypic Studies of Antisense-Transfected Cells

Transformation

Transformation is typically characterized by assessing increased cell proliferation, loss of density-dependent growth, acquisition of anchorage-independent growth, and the ability to form tumors in animals. The addition of a growth factor (e.g., epidermal growth factor, EGF) or overexpression of an oncogene (e.g., Ras) triggers these important phenotypic changes. The JNK pathway is a growth-promoting pathway in several cancer cells as shown by several experiments performed in our laboratory[1,2,20,21,23] and by other groups working with animal model cells.[35–39]

Direct Proliferation Assay Procedure Using Jun N-Terminal Kinase Antisense Oligonucleotides

The choice of methods is determined by knowledge of the window of the antisense effect. Indirect but rapid methods of assessing cell growth such as the incorporation of tritiated thymidine into acid-insoluble DNA polymer or the measurement of the rate of protein accumulation can be carried out only during the window of the antisense effect (see above), as after that period, when the target protein reappears, the pretreatment kinetics are likely to be reestablished and no effect of the antisense treatment will be apparent. On the other hand, if the effect of elimination of one or more isoforms on total cell numbers is directly determined by, say, Coulter counting, then the impact of elimination of one or more JNKs will be apparent at any time, including days after the window period. This is because, for a binary expanding population, i.e., a mitotically expanding cell population, the loss of cell cycles during the 48-hr window is the equivalent of the loss of two or three typical cell division times, which cannot be recovered even though DNA synthesis and the rate of cell division may

[35] G. A. Rodrigues, M. Park, and J. Schlessinger, *EMBO J.* **16,** 2634 (1997).

[36] S. Tanaka, T. Ouchi, and H. Hanafusa, *Proc. Natl. Acad. Sci. U.S.A.* **94,** 2356 (1997).

[37] A. B. Raitano, J. R. Halpern, T. M. Hambuch, and C. L. Sawyers, *Proc. Natl. Acad. Sci. U.S.A.* **92,** 11746 (1995).

[38] W. D. Jarvis, K. L. Auer, M. Spector, G. Kunos, S. Grant, P. Hylemon, R. Mikkelsen, and P. Dent, *FEBS Lett.* **412,** 9 (1997).

[39] M. A. Antonyak, D. K. Moscatello, and A. J. Wong, *J. Biol. Chem.* **273,** 2817 (1998).

completely recover. This effect is illustrated for T98G glioblastoma cells in Fig. 4. Treatment with JNK1AS^{ISIS12539} leads to transient inhibition of DNA synthesis for 48–72 hr, consistent with the duration of the antisense effect, whereas the direct proliferation measurement reflect any previous loss of cell number. A simplified method of direct observation of proliferation is described here.

Assay. To determine the effect of JNK1AS^{ISIS12539} or JNK2AS^{ISIS12560} on cell growth on addition of EGF A549 cells are transfected with either JNK1AS^{ISIS12539} or JNK2AS^{ISIS12560} or their respective scrambled control oligonucleotides as described above (see Transfection Procedure). This experiment can be performed in 24-well plates. In such a small area (~2 cm^2) it is important to seed the cells uniformly in the well. Therefore, the cells should be seeded in prewarmed medium (37°) to avoid convection caused on incubation of cold medium in a small well. Immediately after the addition of the cells to the well, the plate should be gently agitated (but without a swirling motion) and allowed to sit for 5 min on a vibration-free surface. Microscope observation is recommended before returning the cells to the incubator. If the cells are not uniformly spread, resuspend them with a pipette. Uniform spreading results in a better efficiency of lipofection and a better response to EGF. The day after the lipofection, 10% FBS–DMEM is replaced by 0.5% FBS medium supplemented with 0.1 μM recombinant human EGF (rhEGF; R&D Systems, Minneapolis, MN). Five days after the addition of rhEGF, cells are harvested by addition of trypsin (ATV, 50 μl/cm^2; Irvine Scientific) and counted by transfer of the suspension to vials containing 9.9 ml of isotonic buffer, followed by counting using a Coulter (Hialeah, FL) counter (for instruments set for counting 0.2-ml aliquots and gated for the appropriate cell size, the approximate number of cells per well is 100× Raw counts, uncorrected for coincidence and noncellular particulate matter, usually >20%.

Analysis of Anchorage-Independent Growth Using Antisense Oligonucleotides

An excellent way to study a transformed phenotype *in vitro* is to test the ability of cells to grow in soft agar in an anchorage-independent environment; this is thought to be the best *in vitro* assay by which to test the *in vivo* ability to produce tumors.[40,41] Whereas normal cells need to adhere to a solid substratum, cancer cells lose this requirement. In previous stud-

[40] H. R. Herschman and D. W. Brankow, *Carcinogenesis* **8,** 993 (1987).
[41] H. Rubin, P. Arnstein, and B. M. Chu, *Cancer Res.* **44,** 5242 (1984).

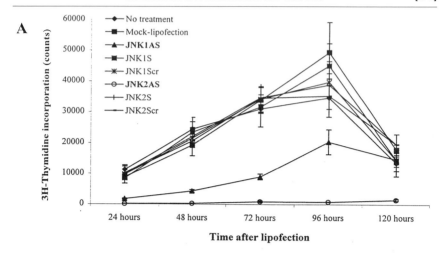

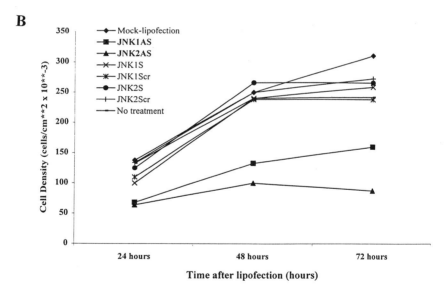

FIG. 4. DNA synthesis and growth inhibition of T98G glioblastoma cells treated with JNK1AS and JNK2AS. (A) Treatment with JNK11AS^{ISIS1253g} or JNK2AS^{ISIS12560} led to a specific inhibition of DNA synthesis in T98G glioblastoma cells during the 72-hr "window" of elimination of JNK1 and JNK2 proteins. For JNK2 DNA synthesis is resumed after 72 hr and follows the same slope as the controls. For JNK2 DNA synthesis is permanently altered. (B) Growth curve of T98G cells after antisense treatment is correlated with the inhibition of DNA synthesis.

ies,[23,42–44] it has been shown that oligonucleotides applied to cells grown in soft agar promote a dose-dependent and specific regulation of growth consistent with a specific effect of antisense oligonucleotides. We describe here a convenient method for the application of antisense oligonucleotides in soft agar, using our standard lipofection procedure.

Assay. Cells are transfected with the oligonucleotides according to the standard lipofection procedure (see above). Twenty-four hours after the transfection 2000 cells are seeded in soft agar in 12-well plates. First, a 0.6% (w/v) "bottom layer" of agarose is prepared from a sterile 1.2% agarose stock solution (low melting point agarose; BRL Life Technologies, Gaithersburg, MD) previously prewarmed to 45° by microwaving and mixed with an equal volume of 2× DMEM (prepared from powdered medium, filter sterilized; Irvine Scientific, at 37°). This mixture (260 μl/cm^2) is poured into the wells and allowed to solidify at room temperature. The top solution of agar at 0.3% (w/v) is prepared by adding 1 volume of 0.6% agarose stock solution prewarmed to 45° to 1 volume of the premixed and prewarmed solution containing 2× DMEM, 2% FBS, and 1× penicillin–streptomycin. Cells from the transfection are then added to this 0.3% agarose solution containing a 0.8 μM concentration of the phosphorothioate oligonucleotides, all at a temperature of 45 to 37°. When the combination of two oligonucleotides is used (JNK1 plus JNK2), each of the oligonucleotides is added at 0.4 μM. Next, 0.5 ml of DMEM, with or without 0.4 μM oligonucleotides, is layered over the cell-containing agar layer. Cells are then incubated for 3 weeks in a 37° incubator. Every 5 days a fresh solution of oligonucleotides and 0.1 μM fresh rhEGF, for the EGF-treated wells, is added to the medium. Live cells appear as full, round refractile balls when viewed in a dissecting microscope whereas damaged cells are smaller, wrinkled or speckled, dull, or lyse, leaving specks and a general view of decreased cell numbers.

After 21 days of incubation, the colonies are stained with *p*-iodotetrazolium violet [2-(*p*-iodophenyl)-3-(*p*-nitrophenyl)-5-phenyltetrazolium chloride]. For this, the top solution of medium is carefully removed and replaced by 131 μl/cm^2 of the 0.5-mg/ml aqueous stock solution of *p*-iodotetrazolium. The plate is incubated overnight at 37° and can be stored at 4° before analysis. Colonies are then scored and measured by determination of their area, using a dissecting microscope equipped with a telecamera connected

[42] J. R. Perez, K. A. Higgins-Sochaski, J. Y. Maltese, and R. Narayanan, *Mol. Cell. Biol.* **8,** 5326 (1994).
[43] G. Baldassarre, C. Bianco, G. Tortora, A. Ruggiero, M. Moasser, E. Dmitrovsky, A. R. Bianco, and F. Ciardiello, *Int. J. Cancer* **66,** 538 (1996).
[44] F. Ciardiello, R. Caputo, R. Bianco, V. Damiano, G. Pomatico, S. Pepe, A. R. Bianco, S. Agrawal, J. Mendelsohn, and G. Tortora, *J. Natl. Cancer Inst.* **90,** 1087 (1998).

to a display and a computer equipped with software allowing measurement of the size and number of colonies (e.g., IP Lab Spectrum software; Scanalytics, Billerica, MA).

We recommend using a stain such as *p*-iodotetrazolium violet, which will stain only living cells that are producing active mitochondrial lactate dehydrogenase; this provides information on the viability of the culture on day 21. Decreased colony formation and growth are readily apparent even at 21 days after the original lipofection with JNK2AS[ISIS12560].

Xenograft Formation Assay

Xenograft formation in athymic mice allows the assessment of tumorigenicity, i.e., the ability of inoculated cells to develop into a tumor, in an *in vivo* model. Several studies have employed antisense oligonucleotides in animal models in order to examine their effect on the regression of tumors.[26,27] Daily intraperitoneal injection of antisense oligonucleotides results in a prolonged increase in oligonucleotide concentration in serum (up to 6 hr).[45] Typically, oligonucleotides accumulate in certain organs such as kidney, liver, bone marrow, skeletal muscle, skin, and intestine.[30] It has also been shown that oligonucleotides accumulate into xenografts without the use of an additional delivery vehicle and specifically inhibit the expression of targeted proteins.[30,46] Furthermore, this inhibition occurs in a dose-dependent manner while no effect is detected using control oligonucleotides, consistent with a specific antisense effect. Antisense JNK has been used to demonstrate the role of JNK pathway in tumor formation and tumor growth.[47] The effect of inhibition of the JNK pathway was assessed by measuring the frequency of tumor appearance, the growth rate of subsequent tumors, and in some cases the frequency of regression of established tumors.

Assay. Using sterile techniques throughout, cells at 80% confluence are trypsinized, washed twice with cold PBS, and counted. The cells are resuspended at a concentration of 2×10^6–5×10^6 cells in 0.2 ml of cold PBS. The cells are then inoculated under the swabbed skin of 6- to 8-week-old athymic mice (e.g., Harlan Sprague Dawley, Indianapolis, IN), using a 26-gauge needle as soon as possible after harvesting. Any subcutaneous site on the back may be used, but the same site must be used on all mice. Once injected the remaining cells are returned to culture as a check on the

[45] P. L. Nicklin, D. Bayley, J. Giddings, S. J. Craig, L. L. Cummins, J. G. Hastewell, and J. A. Phillips, *Pharm. Res.* **15,** 581 (1998).

[46] P. Monia, J. F. Johnston, T. Geiger, M. Muller, and D. Fabbro, *Nature Med.* **2,** 668 (1996).

[47] F. Bost, O. Potapova, C. Liu, Y. M. Yang, W. Charbono, N. Dean, R. McKay, X. P. Lu, M. Pfahl, and D. Mercola, *Cancer Gene Ther.* **5,** S27 (1998).

sterility and viability of the cells: we examine the percentage of attached cells at 12 hr or the percentage of trypan blue-negative (living) cells [add a few drops of a 0.4% trypan blue solution prepared in 0.85% saline (GIBCO) directly to cultures of living cells and inspect for uptake of dye (dark blue) by cells (dead) in a dissecting microscope in 20 min]. This procedure assures that poor tumorigenicity is not due to the loss viability after harvesting. Tumor growth is monitored for tumor formation at least twice a week. Tumor sizes are measured by using a caliper. Oligonucleotide-treated animals are treated intraperitoneally 6 days/week with up to 25 mg of phosphorothiate backbone-based oligonucleotides per kilogram. At least four groups of animals should be considered: one injected with the active antisense compounds, one injected with a control oligonucleotides (scramble or sense), one injected with the solution used for dilution of the oligonucleotides (commonly PBS), and one group of untreated animals. Tumor volumes may be estimated by measurement of the greatest dimension, l, and the perpendicular dimension, w, and calculating the volume by using the formula $V \approx \pi/6(lw^2)$.[48] The number of animals per experimental group should be chosen to be large enough to achieve the desired statistical power.[49]

A tumor growth curve may be constructed by plots of estimated volume versus time. The early portions of such curves are most useful as large xenografts (>750–1000 mm^2) may have necrotic centers and, therefore, not representative solid tumor growth. A convenient means of assessment of such curves is to compare the mean tumor volume of one case with another at one selected time point or to compare groups at several time points representing a portion of the whole growth curve, using an analysis of variance (ANOVA) procedure. Many forms of convenient software are available, e.g., Systat of SPSS (www.spss.com[50]; Evanston, IL). After the definition of the growth curve, it is essential to sacrifice representative animals and immediately remove representative tumors for confirmatory analysis. Tumors should be removed by "blunt" dissection, which is usually rapid as subcutaneous xenografts of most cells grow as discrete masses readily separately from the overlying epithelium and underlying dermis. Tumors should be weight, analyzed to confirm the estimated volume and a portion of each tumor should be quickly frozen for subsequent Western analysis and/or enzyme activity assay. Another portion should be immersed in 10% formalin for subsequent paraffin embedding and histological exami-

[48] D. M. Klinman, A. K. Yi, S. L. Beaucage, J. Conover, and A. M. Krieg, *Proc. Natl. Acad. Sci. U.S.A.* **93,** 2879 (1996).
[49] S. A. Glantz, "Primer of Biostatistics," 3rd Ed. McGraw-Hill, New York, 1992.
[50] C. Smith, *Scientist* **12,** 20 (1998).

nation, using hematoxylin and eosin as the initial screening stain. Furthermore, it is recommended that one portion be kept for TUNEL [TdT (terminal deoxynucleotidyl transferase)-mediated dUTP nick end labeling] analysis if apoptosis is suspected. Frozen and fixed tissue provide material for most other common analyses including confirmation and estimation of the amount of antisense oligonucleotide in the tumor tissue.[51] The host mouse should be necropsied, which provides information on metatases, if any, and toxicity, if any. In the case of JNK antisense oligonucleotides tissue samples are kept to perform JNK assays. For the kinase assay, tissues are cut in small pieces, transferred to whole cell extract buffer (see above), and crushed with a tissue homogenizer. At all times the samples are kept on ice. After measuring the protein concentration of the samples a classic JNK assay is performed as described above.

Summary

    Methods for the selection and characterization of antisense oligonucleotides for specifically eliminating closely related gene family members are available. High-throughput semiautomated methods using 96-well plate formats and array technology and improved assays are under active development that will streamline many steps and will likely merge. Second-generation 20-mer antisense phosphorothioate oligonucleotides containing 2'-methoxyethyl groups at the first and last 6 nucleotides with improved nuclease resistance and RNA affinity are becoming available.

Acknowledgments

    We thank Myriam Bost and Eileen Adamson for help and support, and Leanne Shawler for excellent secretarial assistance. This work was supported by Grants CA56834 and CA76173 from the National Institutes of Health (to D.M.), la Ligue Nationale contre le Cancer (to D.M. and F.B.), and le Conseil Régional de Haute Normandie (to F.B.). This research was also supported in part by funds from the California Breast Cancer Research Program of the University of California (3CB-0246) to D.M., the American Cancer Society, Ray and Estelle Spehar Fellowship to F.B., and the fellowship program of the Sidney Kimmel Cancer Center.

[51] B. P. Monia, *Ciba Found. Symp.* **209,** 107 (1997).

[25] Evaluation of Biological Role of c-Jun N-Terminal Kinase Using An Antisense Approach

By CATHERINE L. CIOFFI and BRETT P. MONIA

Introduction

The mammalian mitogen-activated protein (MAP) kinase signaling pathway links extracellular stimuli to the regulation of cellular responses including cellular growth, differentiation, and apoptosis. MAP kinase family members are a conserved group of serine/threonine protein kinases that include c-Jun N-terminal kinase (JNK), extracellular regulated kinase (ERK), and p38 MAP kinases. While ERKs are activated primarily by phorbol esters and growth factors,[1,2] the JNKs and p38 MAP kinases, termed *stress kinases,* typically mediate responses to proinflammatory cytokines such as tumor necrosis factor α (TNF-α) and interleukin 1 and a variety of environmental stresses including ultraviolet light, heat shock, protein synthesis inhibitors, and ischemia/reperfusion.[3-5] Although JNK and p38 MAP kinases are activated by similar physical stresses, these stress kinases exhibit diverse substrate specificity and are differentially activated via upstream kinases.[6-8]

Activation of JNK by dual phosphorylation on tyrosine and threonine residues results in the expression and activation of transcription factors such as c-Jun, c-ATF2, and ELK-1.[3,8,9] Three major forms of JNK enzymes (JNK1, JNK2, and JNK3) are known to exist in mammalian cells and at least 10 JNK isoforms exist as a result of alternative splicing.[8] Whereas JNK1 and JNK2 are ubiquitously expressed, JNK3 is found predominantly

[1] R. J. Davis, *J. Biol. Chem.* **268,** 14553 (1993).

[2] M. J. Cobb and E. J. Goldsmith, *J. Biol. Chem.* **270,** 14873 (1995).

[3] J. M. Kyriakis, P. Banerjee, E. Nikolakaki, T. Dai, E. A. Rubie, M. F. Ahmad, J. Avruch, and J. R. Woodgett, *Nature (London)* **369,** 156 (1994).

[4] M. A. Bogoyevitch, J. Gillespie-Brown, A. J. Ketterman, S. J. Fuller, R. Ben-Levy, A. Ashworth, C. J. Marshall, and P. J. Sugden, *Circ. Res.* **79,** 162 (1996).

[5] B. W. Zanke, K. Boudreau, E. Rubie, E. Winnett, L. A. Tibbles, L. Zon, J. Kyriakis, F.-F. Liu, and J. R. Woodgett, *Curr. Biol.* **6,** 606 (1996).

[6] J. Rouse, P. Cohen, S. Trigon, M. Morang, A. Alonso-Llamazares, D. Zamanillo, T. Hunt, and A. R. Nebreda, *Cell* **78,** 1027 (1994).

[7] J. R. Woodgett, J. M. Kyriakis, J. Avruch, L. I. Zon, B. Zanke, and D. J. Templeton, *Phil. Trans. R. Soc. Lond.* **351,** 135 (1996).

[8] S. Gupta, T. Barrett, A. J. Whitmarsh, J. Cavanagh, H. K. Sluss, B. Derijard, and R. J. Davis, *EMBO J.* **15,** 2760 (1996).

[9] M. Hibi, A. Lin, T. Smeal, A. Minden, and M. Karin, *Genes Dev.* **7,** 2135 (1993).

METHODS IN ENZYMOLOGY, VOL. 314
0076-6879/99 $30.00

in brain.[8] Despite evidence linking the activation of the JNK signaling pathway to diverse cellular processes, little is known regarding the specific physiological role of the JNK isoforms.

One approach toward evaluating the function of a particular protein is to prevent the expression of that protein by the use of antisense oligonucleotides. The use of oligonucleotides to inhibit JNK gene expression specifically, and particularly the expression of specific JNK isoforms, in order to define the biological function of JNK would be extremely valuable. In this chapter, we describe oligonucleotides targeted against the human JNK sequence, which potently inhibit JNK mRNA and protein expression as well as JNK kinase activity. Furthermore, the biological consequences of oligonucleotide-mediated inhibition of JNK gene expression were evaluated in assays for reoxygenation-induced apoptosis in human kidney cells and TNF-α-induced E-selectin expression in human dermal microvascular cells.

General Considerations

Antisense oligonucleotides bind to complementary sequences of target mRNA or pre-mRNA by Watson–Crick base pairing and prevent the expression of the protein products encoded by that mRNA. Oligonucleotide-mediated regulation of gene expression can occur by several mechanisms including cleavage of targeted RNA by RNase H, an endonuclease that cleaves the RNA strand of RNA–DNA heteroduplexes.[10,11] A critical component of an antisense study should be the demonstration that the oligonucleotide is producing a pharmacological effect by an antisense mechanism of action. Initially, oligonucleotides should be carefully screened in order to identify the sequences that demonstrate the most potent affinity to hybridize to RNA. Second, the dose-dependent effects of oligonucleotides on target mRNA degradation (in the case of an RNase H-mediated mechanism) and inhibition of target protein should be measured. It should be noted that the optimal concentration of oligonucleotide will vary depending on cell type, cell culture conditions, half-life of the protein of interest, and oligonucleotide sequence, and therefore should be empirically determined. In addition, the inclusion of appropriate scrambled or mismatch control oligonucleotides in the studies is essential and can also serve to indicate a sequence-specific mode of action. Scrambled controls are oligonucleotides consisting of the scrambled sequence of each antisense oligonucleotide while maintaining the percent base composition of adenine,

---

[10] C. Helene and J.-J. Toulme, *Biochim. Biophys. Acta* **1049,** 99 (1990).
[11] S. T. Crooke, *Annu. Rev. Pharmacol. Toxicol.* **32,** 329 (1993).

guanine, thymine, and cytosine. A mismatch control oligonucleotide is generated by incorporating nonhybridizing base mismatches (typically one to six) into the oligonucleotide sequence. As demonstrated previously,[12,13] the incorporation of sequential base mismatches into the sequence should result in a significant loss of oligonucleotide potency. With properly conducted experiments, which address the aforementioned criteria, antisense oligonucleotides can serve as important research tools to identify biological function and address concept validation.

Procedures

Oligonucleotide Synthesis

Numerous *in vitro* and *in vivo* studies have demonstrated the utility of oligonucleotides to inhibit a diverse group of gene products in a target-selective manner. While oligonucleotides offer the advantage of target selectivity, they also offer the potential to inhibit the expression of one particular isoform of the same family while having no effect on the expression of related isoforms. Examples of this oligonucleotide-mediated isotype selectivity have been demonstrated previously for protein kinase C-α, A-Raf, C-Raf, and Ha-Ras.[13–16] For the present study, in order to generate isoform-selective inhibition of JNK gene expression, oligonucleotides were designed to hybridize to either human JNK1 or human JNK2 mRNA. JNK1 (ISIS 15347) and JNK2 (ISIS 15354) antisense oligonucleotides are both designed to hybridize within the coding region of their respective mRNAs. ISIS 15347 targets bases 219–238 of the human JNK1 mRNA and is homologous to JNK2 mRNA in only 12 of 20 bases. ISIS 15354 targets bases 563–582 of the human JNK2 mRNA and is homologous to JNK1 mRNA in 17 of 20 bases.

Oligonucleotides are synthesized on an Applied Biosystems (Foster City, CA) 380B automated DNA synthesizer using modified phosphoramidate chemistry. For the stepwise thiation of phosphite linkages, the oxidation step is in 0.2 M 1,2- [^{3}H] benzodithiol-3-one, 1, 1-dioxide in acetonitrile. After cleavage from the controlled pore glass and deblocking in concen-

[12] B. P. Monia, H. Sasmor, J. F. Johnston, S. M. Freier, E. A. Lesnik, M. Muller, T. Geiger, K.-H. Altmann, H. Moser, and D. Fabbro, *Proc. Natl. Acad. Sci. U.S.A.* **93,** 15481 (1996).

[13] C. L. Cioffi, M. Garay, J. F. Johnston, K. McGraw, R. T. Boggs, D. Hreniuk, and B. P. Monia, *Mol. Pharmacol.* **51,** 383 (1997).

[14] N. M. Dean and R. McKay, *Proc. Natl. Acad. Sci. U.S.A.* **91,** 11761 (1994).

[15] G. Chen, S. Oh, B. P. Monia, and D. W. Stacey, *J. Biol. Chem.* **271,** 28259 (1996).

[16] B. P. Monia, *in* "G Proteins, Cytoskeleton and Cancer" (H. Maruta and K. Kohama, eds.), p. 355. R. G. Landes, Austin, Texas, 1998.

trated ammonium hydroxide at 55° for 18 hr, the oligonucleotides are purified by reversed-phase high-performance liquid chromatography in methanol–water–sodium acetate. Only oligonucleotides judged to be >90% full-length material by capillary gel electrophoresis are used. Oligonucleotides incorporating 2'-*O*-methoxyethyl ribose substitutions exhibit improved RNA-binding affinity and increased nuclease resistance.[17] For this study, 2'-methoxyethyl mixed backbone oligonucleotides were prepared, which contain a central phosphorothioate oligodeoxynucleotide region that supports RNase H flanked by 2'-methoxyethyl-modified phosphodiester wings.[17] The sequence of the oligonucleotides used are as follows, with the areas in the sequence containing 2'-methoxyethyl modifications indicated by underlines:

ISIS 15347 (JNK1 AS): <u>CTCTCT</u>GTAGGCCC<u>GCTTGG</u>

ISIS 18077 (mismatch control for ISIS 15347): <u>CTTTCC</u>GTTG-GACC<u>CCTGGG</u>

ISIS 15354 (JNK2 AS): <u>GTCCGG</u>GCCAGGCC<u>AAAGTC</u>

ISIS 18079 (mismatch control for ISIS 15354): <u>GTGCGC</u>GCGAGCC-CGAAATC

Cell Culture

Primary human microvascular endothelial cells (HMVECs), primary human umbilical vein endothelial cells (HUVECs), and primary human renal proximal tubule epithelial cells (RPTECs) are purchased from Clonetics (San Diego, CA). HMVECs and HUVECs are cultivated in Clonetics endothelial basal medium supplemented with 10% fetal bovine serum (FBS) (HyClone, Logan, UT) and RPTECs are cultivated in Clonetics renal epithelial cell growth medium supplemented with a Clonetics Bulletkit containing 0.5% FBS, human recombinant epidermal growth factor (10 ng/ml), insulin (5 μg/ml), hydrocortisone (0.5 μg/ml), epinephrine (0.5 μg/ml), triiodothyronine (6.5 μg/ml), transferrin (10 μg/ml), gentamicin (50 μg/ml), and amphotericin B (50 ng/ml). All cells are grown at 37° in a 95% air/5% CO_2 humidified atmosphere and are subcultured by aspiration of the growth medium followed by a 30-sec rinse with a solution of 0.01% EDTA–0.025% trypsin.

Treatment with Oligonucleotides

Cells are grown in either 60- or 100-mm culture dishes and are 70–80% confluent at the time of oligonucleotide treatment. Cells are washed once

---

[17] K.-H. Altmann, D. Fabbro, N. M. Dean, T. Geiger, B. P. Monia, M. Muller, and P. Nicklin, *Biochem. Soc. Trans.* **24,** 630 (1996).

with phosphate-buffered saline (PBS) prewarmed to 37° and either Opti-MEM (Life Technologies, Gaithersburg, MD) or serum-free cell growth medium containing N-[1-(2,3-dioleoyloxy)propyl]-N,N,N-trimethyl ammonium/dioleoylphosphatidylethanolamine (DOTMA/DOPE, 20 μg/ml; Life Technologies) is added back to the cells. The oligonucleotides are then added at the desired concentration from a 10 mM stock solution and the two solutions are mixed by swirling of the culture dish. Alternatively, oligonucleotides from a 10 mM stock solution are added to polystyrene tubes containing serum-free growth medium and DOTMA/DOPE at a concentration of 0.25 μg/10 nmol of oligonucleotide. The mixture is vortexed and added immediately to cells which had been washed with PBS. Exposure of cells to oligonucleotides in the presence of lipofectin has been previously shown to facilitate the cellular uptake of oligonucleotide while decreasing the concentration of oligonucleotide necessary to elicit the desired effect[18]; this serves to minimize the potential toxic effects often observed at higher oligonucleotide concentrations. The cells are incubated for 3–4 hr at 37° and then the medium is replaced with prewarmed medium containing the appropriate amount of serum and growth factors (see Cell Culture, above). After an additional 24- to 48-hr incubation, the cells are harvested for assay.

Northern Blot Analysis

Total RNA is prepared from cells by the guanidinium isothiocyanate procedure[19] or by the Qiagen (Santa Clarita, CA) RNeasy method according to the manufacturer directions. RNA samples are quantified spectrophotometrically and electrophoresed through 1.2% agarose–formaldehyde gels and transferred to Hybond-N+ nucleic acid transfer membranes (Amersham, Arlington Heights, IL) by capillary diffusion for 12–14 hr. Immobilized RNA is cross-linked to the membrane by exposure to UV light using a Stratalinker (Stratagene, La Jolla, CA) and hybridized with ^{32}P-labeled JNK1-specific or ^{32}P-labeled JNK2-specific cDNA probes, which are prepared by asymmetric polymerase chain reaction (PCR) using specific cDNA templates. Hybridization is performed using Quick-Hyb (Stratagene) at 68°. After hybridization, membranes are washed twice for 15 min with the first wash solution [2× saline sodium citrate (SSC) containing 0.1% sodium dodecyl sulfate (SDS)] followed by a 30-min incubation with the second wash solution (0.1× SSC containing 0.1% SDS). Probes

[18] C. F. Bennett, M.-Y. Chiang, H. Chan, J. E. E. Shoemaker, and C. K. Mirabelli, *Mol. Pharmacol.* **41,** 1023 (1992).

[19] J. M. Chirgwin, A. E. Przybla, R. J. McDonald, and W. J. Rutter, *Biochemistry* **18,** 5294 (1979).

hybridized to mRNA transcripts are visualized and quantified using a Molecular Dynamics (Sunnyvale, CA) PhosphorImager. Blots are routinely stripped of radioactivity by boiling at 100° for 10 min in 0.1× SSC containing 0.1% SDS and reprobed with a ^{32}P-radiolabeled probe to the housekeeping gene, glyceraldehyde-3-phosphate dehydrogenase (G3PDH), to confirm equal loading.

Western Blot Analysis

For Western blot analysis, cells are lysed in radioimmunoprecipitation (RIPA) buffer [20 mM Tris-HCl (pH 7.5), 100 mM NaCl$_2$, 2.5 mM EDTA, 1 mM dithiothreitol (DTT), 1% (v/v) Triton X-100, aprotinin (100 units/ml), leupeptin (1 μg/ml), 1 mM phenylmethylsulfonyl fluoride (PHSF), 1 mM NaVO$_3$ 10 μM Na$_2$MoO$_4$, and 10 mM NaF; 300 μl, of RIPA extraction buffer is used per 100-mm dish]. Total protein concentration is quantified by Bradford assay, using the Bio-Rad (Hercules, CA) kit. Extracts are boiled in sodium dodecyl sulfate-polyacrylamide gel electrophoresis (SDS–PAGE) sample buffer and proteins (20–40 μg/lane) are separated on a 10% SDS–polyacrylamide minigel (Bio-Rad). The separated proteins are transferred to nitrocellulose membranes (Bio-Rad) and treated with blocking buffer [PBS containing 10% (w/v) dry milk (Carnation) and 0.2% Tween 20]. Membranes are probed with anti-JNK/SAPK antibody, which recognizes total JNK protein (1 : 1000 dilution; New England BioLabs, Beverly, MA) in blocking buffer (overnight at 4°). Anti-rabbit IgG conjugated with horseradish peroxidase is used as the second antibody (1 : 1000 dilution, 1 hr at room temperature) and immune complexes are visualized using an enhanced chemiluminescence kit (such as LumiGLO reagent; New England BioLabs) according to the manufacturer instructions. Blots are quantified by laser scanning densitometry.

Jun N-Terminal Kinase Activity Assay

Cells are grown in six-well tissue culture plates until ~80% confluent. After treatment of cells with TNF-α (5 ng/ml; R&D Systems, Minneapolis, MN) for 15 min, cells are washed twice with ice-cold PBS and suspended in lysis buffer [25 mM HEPES (pH 7.7), 0.3 M NaCl, 1.5 mM MgCl$_2$, 0.1% Triton X-100, PMSF (100 μg/ml), L-1-tosylamido-2-phenylethyl chloromethyl ketone (100 μg/ml), 1 mM EDTA, leupeptin (2 μg/ml) aprotinin (2 μg/ml), 20 mM β-glycerophosphate, and 0.1 mM Na$_3$VO$_4$]. Total protein concentration is quantified by Bradford assay, using the Bio-Rad kit. Typically, all preparations yield similar protein concentrations and are adjusted to provide equivalent amounts of cellular protein in all samples prior to analysis. For analysis of JNK1 and JNK2 isoforms, lysates, containing equal amounts of protein are incubated with JNK1-specific or JNK2-specific antibodies (Upstate Biotechnology, Lake Placid, NY) overnight at 4°. Anti-rabbit IgG conjugated with agarose beads is then added to cell extracts after JNK antibody treatment and wash steps and incubated for 2 hr at 4°. Complexes are pelleted by centrifugation and used for kinase assays. For the kinase assay, 50 μg of whole cell extract is mixed with 10 μg of glutathione S-transferase–c-Jun-(1–223) [GST–c-Jun-(1–223)] for 3 hr at 4°. The GST–c-Jun-(1–223) fusion protein is previously expressed and purified from *Escherichia coli* and bound to agarose beads (Sigma, St. Louis, MO). After four washes, the beads are incubated with 30 μl of kinase reaction buffer [20 mM HEPES (pH 7.7), 20 mM MgCl$_2$, 20 mM β-glycerophosphate, 20 mM p-nitrophenyl phosphate, 0.1 mM Na$_3$VO$_4$, 2 mM dithiothreitol, 20 μM ATP, and 5 μCi of [γ-^{32}P]ATP] for 20 min at 30°. The reaction is stopped by the addition of 20 μl of Laemmli sample buffer. The phosphorylated GST–c-Jun- (1–223) protein is eluted by boiling the sample for 5 min and separated by 10% SDS-PAGE. The quantification of ^{32}P-phosphorylated GST–c-Jun-(1–223) is carried out by digitalization using an Ambis scanner (Ambis, Billenca, MA) for dried SDS–polyacrylamide gels or using an Ultroscan XL (LKB, Gaithersburg, MD) for autoradiographs. Background values are subtracted in all cases.

Flow Cytometry Analysis

HMVECs or HUVECs are treated with oligonucleotides and allowed to recover for 48 hr, and then TNF-α (5 ng/ml; R&D Systems) is added for an additional 5 hr. Cells are then detached from the plates by brief exposure to 0.25% trypsin. Cells removed from the plate by this method express levels of E-selectin similar to those of cells removed from the plate by treatment with saline–EDTA solution. After a wash with DMEM supplemented with 10% FBS, cells are stained with a fluorescein isothiocyanate (FITC)-conjugated antibody that recognizes E-selectin (R&D Systems) at a concentration of 2 μg/ml diluted in PBS containing 2% bovine serum albumin and 0.2% sodium azide. Each step is performed at 4°. Cells are analyzed by flow cytometry using a Becton Dickinson (San Jose, CA) FACScan. E-selectin cell surface expression is calculated on the basis of the mean value of fluorescence intensity, using 3000–5000 cells stained with the E-selectin antibody. Results are expressed as percentage of control (cell surface expression induced by TNF-α in cells that are not treated with oligonucleotide) based on mean fluorescence intensity. Basal expression of E-selectin is undetectable in endothelial cells in the absence of TNF-α.

Quantification of Apoptosis by Fluorescence Microscopy Using DNA-Binding Dyes

Apoptosis is a regulated process of cell death characterized by distinct physiological events including membrane blebbing, nuclear and cytoplasmic shrinkage, DNA fragmentation, and chromatin condensation.[20] Apoptosis should be distinguished from necrosis, which is a mode of cell death accompanied by a rapid loss of cell integrity, membrane lysis, and an inflammatory response. One method of monitoring the extent of apoptosis or necrosis is by utilizing the differential uptake of the DNA-binding dyes acridine orange and ethidium bromide by viable and nonviable cells. Acridine orange enters viable cells and intercalates into the DNA, culminating in a green fluorescence in the nucleus. Conversely, ethidium bromide enters nonviable cells (cells whose membrane is compromised) and overwhelms the fluorescence of acridine orange to generate orange fluorescence. We have used the procedure described by McGahon et al.[21] with some modifications as described below.

Human RPTECs are plated into LabTek one-well Permanox chamber slides (Nalge, Naperville, IL) in 2 ml of growth medium. Hypoxia is induced with the use of a cell culture incubator perfused with 95% N_2/ 5% CO_2. The oxygen level is <1% and is monitored with a Fyrite gas analyzer (Bacharach, Pittsburg, PA). After 4 hr of hypoxia, cells are removed from the hypoxic incubator and reoxygenated by immediate replacement of hypoxic medium with normoxic renal epithelial cell growth medium (Clonetics) supplemented with 10% FBS. Stock solutions of ethidium bromide (Sigma) and acridine orange (Sigma) are prepared in PBS at a concentration of 100 μg/ml solution. Medium is aspirated from the cells and 50 μl of a 1:1 stock solution of ethidium bromide and acridine orange is added to 1 ml of PBS and immediately added to the cells on the chamber slide. Because dye uptake is instantaneous, a coverslip is placed onto the slide and the morphological features of apoptosis are monitored by fluorescence microscopy, using a microscope equipped with an FITC filter at ×600. Treated cells are quantified according to the following descriptions: normal nuclei (bright green chromatin with orga-

---

[20] A. J. Hale, C. A. Smith, L. C. Sutherland, V. E. A. Stoneman, V. L. Longthorne, A. C. Culhane, and G. T. Williams, *Eur. J. Biochem.* **236**, 1 (1996).

[21] A. J. McGahon, S. J. Martin, R. P. Bissonnette, A. Mahboubi, Y. Shi, R. J. Mogil, W. K. Nishioka, and D. R. Green, *in* "Cell Death" (L. M. Schwartz and B. A. Osborne, eds.), p. 153. Academic Press, San Diego, California, 1995.

nized structure), early apoptotic (bright green chromatin that is highly condensed or fragmented), late apoptotic (bright orange chromatin that is highly condensed or fragmented), or necrotic (bright orange chromatin with organized structure). At least 200 cells from randomly selected fields are counted and quantified for each data point. The percentage of apoptotic or necrotic cells is calculated as the number of apoptotic (or necrotic) cells per total cells counted $\times 100$.

Application of Procedures

Role of Jun N-Terminal Kinase in Apoptosis Induction

Cessation of blood flow (ischemia) followed by restoration of blood flow (reperfusion) is a major pathological event that is particularly common in the heart, kidney, and brain. Ischemia/reperfusion *in vivo* and hypoxia/reoxygenation *in vitro* have been shown to result in apoptotic cell death.[22,23] Apoptosis is a physiological mode of cell death characterized by a distinct sequence of morphological changes including condensation and fragmentation of nuclear chromatin, membrane blebbing, reduction in cell volume, and the eventual cellular disintegration into apoptotic bodies.[20] While apoptosis is essential for tissue homeostasis, deregulated apoptosis contributes to the pathogenesis of several diseases including atherosclerosis and myocardial infarction.[24] At present, the signaling pathways resulting in ischemia/reperfusion-induced apoptosis remain unknown. However, studies have suggested that the JNK signaling pathway contributes to reperfusion-induced apoptotic cell death. JNK is activated after reperfusion of ischemic tissue, but not during ischemia alone[4,22] and a good correlation exists between JNK activation and the onset of apoptosis after reperfusion injury in kidney and heart.[22,25] We applied an antisense approach to investigate definitively whether the JNK signaling pathway plays a critical role in reoxygenation-induced apoptosis.

Effect of JNK1 Antisense on JNK mRNA and Protein Levels in Human Kidney Cells. An oligonucleotide directed against human JNK1 (JNK1 AS) was utilized as a tool in order to assess the role of JNK in apoptosis, using

[22] T. Yin, G. Sandhu, C. D. Wolfgang, A. Burrier, R. L. Webb, D. F. Rigel, T. Hai, and J. Whelan, *J. Biol. Chem.* **272,** 19943 (1997).

[23] R. A. Gottlieb, K. O. Burleson, R. A. Kloner, B. M. Babior, and R. L. Engler, *J. Clin. Invest.* **94,** 1621 (1994).

[24] A. Haunstetter and S. Izumo, *Circ. Res.* **82,** 1111 (1998).

[25] T.-L. Yue, X.-L. Ma, X. Wang, A. M. Romanic, G.-L. Liu, C. Louden, J.-L. Gu, S. Kumar, G. Poste, R. R. Ruffolo, and G. Z. Feuerstein, *Circ. Res.* **82,** 166 (1998).

a cell culture model of ischemia/reperfusion. Initial experiments performed using Northern blot analysis revealed that both JNK1 and JNK2 mRNA were expressed in human renal proximal tubule epithelial cells (RPTECs). Exposure of RPTEC to increasing concentrations of JNK1 AS resulted in a potent, dose-dependent suppression of JNK1 mRNA levels (IC_{50} value <50 nM), whereas a slight non-dose-dependent inhibition of JNK2 mRNA was observed only after treatment with higher concentrations of JNK1 AS (Fig. 1A). In contrast, no inhibition of either JNK1 or JNK2 mRNA expression was observed after exposure of cells to a six-base mismatch control (Fig. 1A).

Both JNK1 and JNK2 genes produce alternatively spliced transcripts that encode proteins of ~46 and 54 kDa.[9] Thus, the proteins that migrate at 46 and 54 kDa represent a mixture of JNK isoforms.[8] In these experiments, lysates, prepared from RPTECs exposed to JNK1 AS, were analyzed by Western blot analysis using an antibody that does not discriminate between the individual JNK isoforms. Quantitative analysis of these Western blots revealed that treatment with increasing concentrations of JNK1 AS for 48 hr resulted in a marked, dose-dependent inhibition of p46-JNK protein levels whereas the expression of p54-JNK was reduced only to a slight degree at higher concentrations (Fig. 1B). Protein expression of p46-JNK or p54-JNK was not inhibited after exposure to the mismatch control oligonucleotide (Fig. 1B).

Effect of JNK1 Antisense on Apoptosis Induction by Hypoxia/Reoxygenation in Human Kidney Cells. Normoxic (control) cultures of RPTECs exhibited a small percentage (~5%) of apoptotic cells and exposure of the cells to 4 hr of hypoxia did not alter this number. However, when these hypoxic cells were reoxygenated for 15 hr, a significant increase in the number of apoptotic cells was observed, with the majority of apoptotic cells (>95%) classified as early apoptotic (Fig. 2). The percentage of necrotic cells typically observed in normoxic cultures (~2%) was not altered after exposure of cells to hypoxia or hypoxia/reoxygenation. However, when RPTECs were pretreated for 48 hr with 250 or 350 nM JNK1 AS, concentrations of oligonucleotides that resulted in a significant inhibition of JNK mRNA and protein expression, a profound suppression of reoxygenation-induced apoptosis, 61 and 72%, respectively, was noted (Fig. 2). In contrast, treatment with the JNK1 AS control oligonucleotide induced no change in the number of apoptotic cells occurring after hypoxia/reoxygenation, indicating that the JNK1 AS is working via a sequence-specific mode of action. These data demonstrate that JNK is required for reoxygenation-induced apoptosis. While inhibition of JNK1 mRNA expression may be sufficient to suppress reoxygenation-induced apoptosis, the requirement for both JNK1 and JNK2 in this response cannot be ruled out at present.

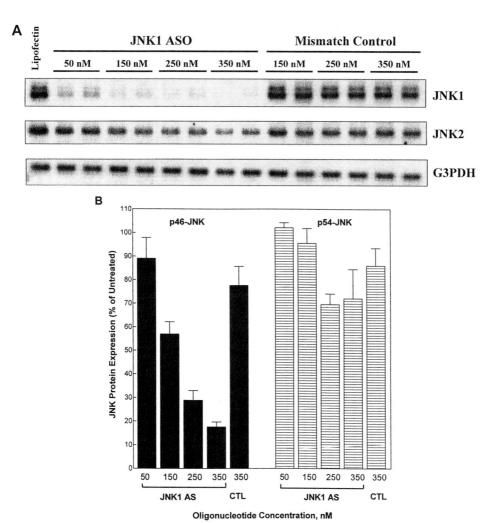

FIG. 1. Inhibition of JNK mRNA and protein expression by JNK1 antisense oligonucleotide and mismatched control in human kidney cells. (A) Cells were treated with increasing concentrations of JNK1 AS or a six-base mismatch control oligonucleotide for 24 hr. Northern analysis was performed on replicates from the same preparation of total cellular RNA (10 μg/lane). Each filter was hybridized with a JNK1 or JNK2 probe. The detection of multiple products on the Northern gel for JNK1 most likely is due to an alternatively spliced JNK1 transcript. To confirm equal loading, the same blot was stripped and blotted a second time with a probe for glyceraldehyde-3-phosphate dehydrogenase (G3PDH). "Lipofectin" indicates cells treated with cationic lipid only. (B) Quantitation of Western blot analysis of JNK protein levels in RPTE cells treated with JNK1 AS or a mismatch control oligonucleotide (CTL), which was obtained by densitometric scanning of lumigrams and expressed as a percentage of cells not treated with oligonucleotide. Protein extracts were prepared from cells exposed to the indicated oligonucleotide for 48 hr.

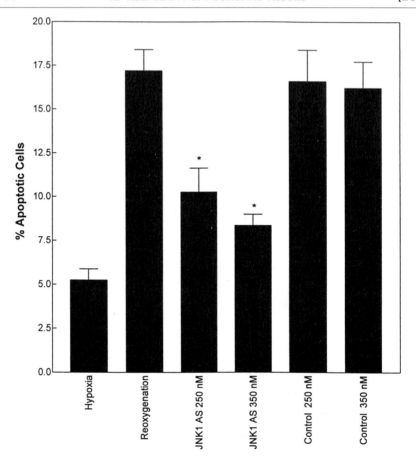

FIG. 2. Inhibition of reoxygenation-induced apoptosis by an antisense oligonucleotide directed against JNK1 in human kidney cells. Cells were incubated with 350 nM JNK1 AS or a six-base mismatch control oligonucleotide for 48 hr prior to exposure to hypoxia (4 hr)/reoxygenation (15 hr). The percentage of apoptotic cells was determined by fluorescent microscopy. Values shown are the mean obtained from three to six experiments. *Significantly different from untreated cells, $p < 0.05$.

This is because a slight suppression of JNK2 mRNA levels was also noted after treatment of RPTECs with JNK1 AS.

Role of Jun N-Terminal Kinase in Tumor Necrosis Factor α-Induced Signal Transduction

Inflammatory cytokines such as tumor necrosis factor alpha (TNF-α) and interleukin 1 (IL-1) play central roles in the regulation of immune and inflammatory responses. One of the important biological functions of these cytokines is to induce profound responses of the vascular endothelium in order to facilitate the adherence and recruitment of circulating leukocytes to local inflammatory tissues.[26] On binding to inflammatory cytokines, cytokine receptors on endothelial cells activate a variety of intracellular signaling pathways. Stimulation of these pathways results in the activation of NF-κB and AP-1 transcription factor families, which in turn upregulate the expression of specific genes including cell adhesion molecules such as E-selectin.[26-28] E-selectin is important for the initial attachment and rolling of leukocytes on the vascular endothelium during inflammation.

Two distinct membrane receptors for TNF-α, TNF receptor 1 (TNFR1) and TNF receptor 2 (TNFR2), have been identified. The fact that no intrinsic enzymatic activity has been associated with TNFR1 or TNFR2 has suggested that receptor-associated proteins participate in the downstream transduction of TNF-α signaling. Indeed, many such proteins have been identified that play critical roles in various aspects of TNF-α signaling.[29,30] Studies on the function of the downstream effectors of TNFR1 and TNFR2 are beginning to establish which signaling proteins are important for activation of NF-κB transcription and which are necessary for AP-1 transcription. Here, we applied an antisense approach to investigate which JNK isoform (JNK1 or JNK2) was critical for TNF-α-mediated induction of AP-1-mediated transcription of E-selectin.

Effect of JNK1 and JNK2 Antisense on JNK mRNA Levels in Human Endothelial Cells. Treatment of endothelial cells in culture with TNF-α results in the induction of E-selectin transcription and E-selectin cell surface protein expression.[31] In other studies, we have shown that the signaling mechanisms underlying this response to TNF-α involves all three groups of MAP kinases: the ERKs (extracellular regulated kinases), the JNKs, and p38 MAP kinases.[32] Utilizing both antisense and low molecular weight inhibitors, we have demonstrated that the JNK MAP kinases are the major MAP kinases involved in TNF-α stimulation of E-selectin.[32] These results prompted us to determine which JNK isoform was involved in this response.

To determine the expression of JNK isoforms in endothelial cells, JNK1,

[26] M. Introna and A. Mantovani, *Art. Thromb. Vasc. Biol.* **17,** 423 (1997).

[27] T. M. McIntyre, V. Modur, S. M. Prescott, and G. A. Zimmerman, *Thromb. Haemost.* **78,** 302 (1997).

[28] A. Mantovani, S. Sozzani, A. Vecchi, M. Introna, and P. Allavena, *Thromb. Haemost.* **78,** 406 (1997).

[29] H. Y. Song, C. H. Regnier, C. J. Kirschning, D. V. Goeddel, and M. Rothe, *Proc. Natl. Acad. Sci. U.S.A.* **94,** 9792 (1997).

[30] H. Hsu, H. Shu, M. Pan, and D. V. Goeddel, *Cell* **84,** 299 (1996).

[31] C. F. Bennett, T. P. Condon, S. Grimm, H. Chan, and M. Y. Chiang, *J. Immunol.* **152,** 35250 (1994).

[32] X. S. Xu, C. Vanderziel, C. F. Bennett, and B. P. Monia, *J. Biol. Chem.* **273,** 33230 (1998).

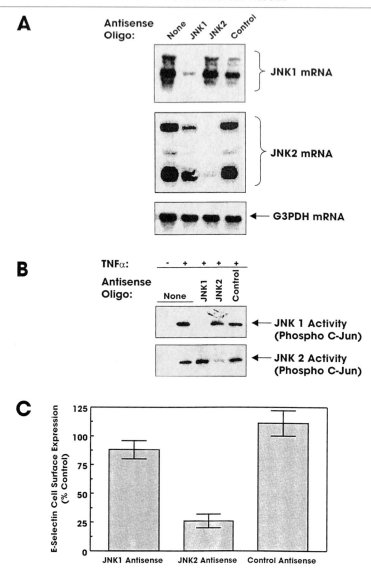

FIG. 3. Inhibition of JNK mRNA, TNF-α-mediated JNK activity, and TNF-α-induced E-selectin expression by JNK antisense oligonucleotides in human dermal endothelial cells. (A) Effect of oligonucleotides on JNK mRNA expression. Cells were exposed to JNK1 AS, JNK2 AS, or mismatch control oligonucleotide at 100 nM (JNK1 and control antisense) or 50 nM (JNK2 antisense) and allowed to recover for 24 hr. JNK1, JNK2, and G3PDH mRNA levels were determined by Northern blot analysis. (B) Effect of oligonucleotides on JNK activity. Cells were treated with JNK1 AS, JNK2 AS, or mismatch control oligonucleotide (Control) at the concentrations described in (A) and allowed to recover for 48 hr. Both JNK

JNK2, and JNK3 mRNA levels were measured by Northern blot analysis using isoform-specific probes after TNF-α treatment in human microvascular endothelial cells (HMVECs) and human umbilical vein endothelial cells (HUVECs). JNK3 expression was undetectable whereas JNK1 and JNK2 expression was detected and found to be expressed to comparable levels (Fig. 3A). The probes used in these studies detect on Northern gels multiple products that likely reflect alternatively spliced JNK1 and JNK2 transcripts. JNK1 AS and JNK2 AS were applied to HMVECs separately prior to TNF-α stimulation and the effects on JNK1 and JNK2 mRNA levels were determined. JNK1 and JNK2 AS treatment resulted in nearly complete inhibition of JNK1 and JNK2 mRNA expression, respectively (Fig. 3A). Furthermore, both antisense oligonucleotides were isoform specific at the employed concentrations. The JNK2 antisense molecule will inhibit JNK1 expression at higher oligonucleotide concentrations in certain cell types because it is complementary to JNK1 mRNA in 17 of its 20 bases.[32] However, under the conditions employed here, the JNK2 antisense specifically inhibits JNK2 expression without affecting JNK1 levels. Furthermore, a six-base mismatch control oligonucleotide had no effect on either JNK1 or JNK2 mRNA expression.

Effect of JNK1 and JNK2 Antisense on TNF-α-Mediated Induction of JNK Activity and E-Selectin Expression in Human Endothelial Cells. To determine whether treatment with JNK1 AS or JNK2 AS could inhibit JNK activity in HMVECs, isoform-specific antibodies that specifically recognize and immunoprecipitate JNK1 or JNK2 were employed to measure kinase activity of the isoforms against a c-Jun substrate. In these experiments, stimulation of HMVECs with TNF-α (5 ng/ml) resulted in a strong induction in JNK activity. Treatment of cells with either JNK1 AS or JNK2 AS for 48 hr prior to the kinase assay effectively reduced TNF-α-mediated induction of JNK activity in an isoform-specific manner (Fig. 3B).

E-selectin is rapidly and transiently induced by cytokines, with peak levels of expression appearing between 4 and 6 hr after cytokine exposure and a return to baseline levels occurring approximately 24 hr postexpo-

FIG. 3. (*continued*) oligonucleotides effectively and specifically inhibit expression of their intended targets at this concentration. TNF-α (5 ng/ml) was added for 15 min before cells were harvested. *In vitro* kinase assays for JNK using isoform-specific JNK antibodies were performed as described in Procedures. (C) Effect of oligonucleotides on TNF-α-mediated E-selectin induction. Cells were exposed to JNK1 AS, JNK2 AS, or mismatch control at the indicated concentrations in (A) and allowed to recover for 48 hr. TNF-α (5 ng/ml) was added for 15 hr before cells were harvested and analyzed by flow cytometry for E-selectin cell surface expression. Results shown are representative of two independent experiments. [Reprinted from Ref. 32 with permission of The American Society for Biochemistry and Molecular Biology.]

sure.[26] Therefore, in these experiments, HMVECs were exposed to either JNK1 AS or JNK2 AS, at concentrations of 20, 50, or 100 nM, prior to stimulation with TNF-α (5 ng/m) for 5 hr to induce E-selection expression. Under these experimental conditions, JNK2 AS treatment resulted in a substantially greater inhibition of TNF-α-mediated induction of E-selectin cell surface expression relative to JNK1 AS treatment whereas the mismatch control oligonucleotide did not significantly suppress E-selectin cell surface expression (Fig. 3C). Similar results were obtained in HUVECs under identical experimental conditions (data not shown). These results indicate that JNK2 function is critical for TNF-α-mediated induction of E-selectin in endothelial cells.

Conclusion

In this chapter, we demonstrate that an antisense approach is a useful strategy to potently and specifically inhibit gene expression in order to delineate the precise functional role of a target protein and identify potential pharmacological targets. Treatment of cultured cells with oligonucleotides targeted to either human JNK1 or human JNK2 resulted in a potent inhibition of JNK mRNA and protein expression as well as JNK kinase activity. Inhibition of JNK gene expression by antisense oligonucleotides revealed a significant suppression of hypoxia/reoxygenation-induced apoptosis in human kidney cells and TNF-α-induced E-selection expression in human endothelial cells. These results suggest that activation of JNK is an important, if not crucial, signal transduction protein for apoptosis induced by the cellular stress of hypoxia/reoxygenation. Furthermore, the results demonstrate that JNK2 is the JNK isoform responsible for TNF-α-mediated induction of E-selectin in endothelial cells.

[26] Role of Antisense in Kidney Cells

By MICHAEL B. GANZ

Introduction

Antisense oligodeoxynucleotides (AS-ODNs) have emerged as potential agents for blocking gene expression (reviewed in detail in Refs. 1 and

---

[1] S. T. Crooke and B. Lebleu, "Antisense Research and Applications," 1st Ed. CRC Press, London, 1993.

METHODS IN ENZYMOLOGY, VOL. 314   0076-6879/99 $30.00

2). Specifically, deoxynucleotides are composed of a nucleic acid sequence, which is complementary to the RNA target with which it is designed to interact. AS-ODNs used in the kidney cells are typically 15 to 30 nucleotides long and approximately 5 to 9 kDa in size. The high affinities of these deoxynucleotides for their RNA targets allow one to prevent the production of a single protein.[1,3] Once protein production has been altered one can then readily measure the effect of this maneuver on cell behavior. AS-ODNs modulate the information transfer from gene to the protein: in essence they alter the intermediary metabolism of RNA. AS-ODNs can alter transcription (by preventing binding of transcription factors), RNA processing (by binding to sequences necessary for RNA splicing), or translation (by binding to the translational initiation codon). There can, however, be nonspecific binding to RNA or DNA and there may be unintended effects of these agents or metabolites on transcription factors.

In the study of kidney function and kidney cell behavior, antisense technology has only recently been studied.[4–6] It has been applied to help define the regulatory mechanisms of ion transport (i.e., sodium channels) and function in addition to the pathologic processes (i.e., matrix deposition) that induce glomerular cell proliferation in proliferative glomerulonephritides and matrix formation in diseases such as diabetes mellitus. The techniques involving the use of liposomes, electroporation, and micropuncture have become the predominant strategies applied to gain entry of the deoxynucleotides into renal cells.[1,7] The approach to be utilized depends on the specific kidney cell type and the precise biological question to be asked. Moreover, the success of any given technique depends on both the deoxynucleotide and the cell type that is targeted. A number of investigators have begun using AS-ODNs both *in vitro* and *in vivo* to ascertain how tubules regulate volume homeostasis and which specific subcellular and/or transmembrane events are responsible for glomerular cell activation in progressive glomerular diseases. It has also become apparent that the use of AS-ODNs may be an effective strategy for the treatment of glomerular diseases resistant to conventional therapy. This chapter describes newly published and unpublished applications of these techniques in the study of kidney tubular function and glomerular cell function. Finally, strategies

[2] S. Agrawal, J. Temsamani, W. Galbraith, and J. Tang, *Clin. Pharmcokinet.* **28,** 7 (1995).

[3] R. W. Wagner, *Nature* (*London*) **372,** 333 (1994).

[4] H. Haller, C. Maasch, D. Dragun, M. Wellner, M. Von Janta-Lipinski, and F. C. Luft, *Kidney Int.* **53,** 1550 (1998).

[5] J. R. Gnarra and G. R. Dressler, *Cancer Res.* **55,** 4092 (1995).

[6] E. Imai, Y. Isaka, Y. Akagi, M. Akagi, T. Moriyama, M. Tkenaka, T. Kanaeko, M. Horio, A. Ando, Y. Okita, Y. Kanaeda, N. Ueda, and T. Kamata, *Contrib. Nephrol.* **118,** 86 (1996).

[7] E. L. R. Barry, F. A. Gesek, and P. A. Friedman, *BioTechniques* **76,** 721 (1993).

using AS-ODNs in altering glomerular mesangial cell behavior are presented and the advantages and disadvantages of each are delineated.

Role of Antisense in Renal Tubular Epithelial Cells

The kidney is composed of two critically diverse segments: renal tubules and the glomerulus,[8] which comprise the nephron. Normal function of these structures is essential for maintaining blood pressure, electrolyte homeostasis, and various other metabolic functions. Proximal renal tubules are epithelial cells that are important in the approximately 80% of all sodium, chloride, and water reabsorption occurs at this level of the nephron. These cells also readily transport other ions, such as calcium and phosphorus.[9-11] It has been well defined that electrolyte homeostasis along with volume regulation are maintained by specific transmembrane transport processes in these epithelial cells. In addition the regulatory transport processes that maintain normal serum potassium are found in the cells of the distal tubule. Alterations in the transport processes that modulate both water and electrolyte activity may be responsible, in part for the disordered Na^+ and Cl^- homeostasis in certain patients with hypertension and diseases such as cystic fibrosis.[10] Understanding the precise expression and molecular processes that regulate these transporters is important for understanding the disease state. Advances in antisense technology have allowed us to selectively study channels and ion transporters.

Most of the work to date on antisense in the kidney has been directed at mechanisms that regulate ion transport in renal tubular cells. In particular, the ability of the AS-ODNs to selectively block channel activity without affecting other transport phenomena may provide useful insights into the physiological regulation of that specific transport process (see Table I for a review).

A6 cells (*Xenopus* cell line) or mTAL cells have been utilized as model systems to examine the potential mechanisms by which epithelial Na^+ channels and basolateral Cl^- channels function.[12-14] By using the AS-ODNs to block the expression of the cystic fibrosis transmembrane conductance regulator (CFTR) (Cl^- channels) in A6 cells and then studying the short

[8] E. A. Van der Zee, J. J. Bolhuis, R. O. Solomonia, G. Horn, and P. G. M. Luiten, *Brain Res.* **676**, 41 (1995).

[9] S. Ohno, T. Baba, N. Terada *et al., Int. Rev. Cytol.* **166**, 181 (1996).

[10] T. A. Pressley, *Am. J. Kidney Dis.* **32**, 1084 (1998).

[11] W. H. Wang, S. C. Hebert, and G. Giebisch, *Annu. Rev. Physiol.* **59**, 413 (1997).

[12] L. Zimniak, C. J. Winters, R. B. Reeves, and T. E. Andreoli, *J. Am. Soc. Nephrol.* **6**, 358 (1995). [Abstract]

[13] T. R. Kleyman, C. Lin, K. A. McNulty, L. M. Gomez, R. T. Worrell, and D. C. Eaton, *J. Am. Soc. Nephrol.* **6**, 342 (1995). [Abstract]

[14] D. L. DeCoy, J. R. Snapper, and M. D. Breyer, *J. Clin. Invest.* **95**, 2749 (1995).

TABLE I
RENAL TUBULAR CELLS

Method	Mechanisms	Renal cells
Liposomes	Cationic lipids	COS-7
	Interact with RNA to form complexes that adsorb to cell surfaces, thereby fusing with plasma membrane	Metanephric epithelial
		Mouse renal tubular epithelial
		A6
		LLCPK1
	ODN enter cell after fusion	Cortical collecting
	Limitations: high variability of entry in cell	MTAL
Electroporation	Electrical permeability	A6
	Cells are permeabilized, allowing introduction of ODN	LLCPK1
		MTAL
	Limitations: cells are leaky and allow loss of cell contents	
Patch clamping	Path to cell surface	A6
	Pipette seal to cells allows introduction of ODN	Cortical collecting
	Limitations: difficulty of making seal with certain cells, and only one cell can be done at a time	

circuit current, investigators have been able to ascertain that the forskolin-activated amiloride-insensitive short circuit current is responsible for an increase in Na^+ channel open probability. As shown in Fig. 1, the introduction of AS-ODNs modulates channel activity and thereby one is able to determine open probability. The dark-colored shape reflects a change in the "channel" after the introduction of AS-ODNs into the epithelial cell. The difference is reflected in channel activity.

A similar approach has allowed investigators to examine the potential mechanism(s) by which Na^+ channel transport is regulated and its role in cystic fibrosis. These investigators generated the antisense to the channel and thereby measured the differences in activity in transport of ions in transfected versus nontransfected cells. This technique allows the investigators to focus selectively on one channel and define its role in transport and disease. In addition, others have utilized antisense strategy to determine the mere existence of ion channels in a specific cell type. This approach will allow the investigator to define new channels and thereby develop approaches to blocking the activity of those channels.[15,16] For example, by

[15] M. Kuwahara, S. Sasaki, S. Uchida, E. J. Cragoe, Jr., and F. Marumo, *Biochim. Biophys. Acta Mol. Cell Res.* **1220,** 132 (1994).

[16] P. Middleton, R. T. Worrell, and D. C. Eaton, *Fed. Am. Sci. Exp. Biol. J.* **9,** A388 (1995). [Abstract]

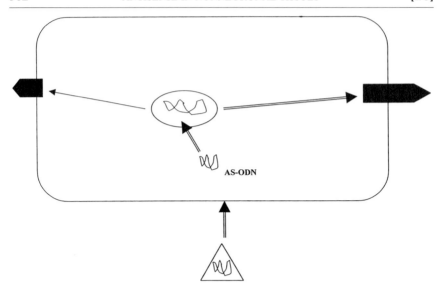

FIG. 1. Role of AS-ODN in channel activity for transfection of tubular cells. The double-line arrows represent changes in channel activity when new genetic material is inserted into the cell. This allows the investigator to examine selectively a specific transport function.

generating the RNA antisense sequence to specific Cl^- channels, investigators have been able to determine that the Clck3a and Clck3b (different Cl^- channel isotypes) encode basolateral mTAL Cl^- channels.[12,17] Therefore by inserting into a cell a characterized channel one can readily conclude what channel is responsible for the specific transport. Others have utilized this approach to identify the specific dopamine receptors in LLCPK1 cells. Finally, this technique has been used to define the family of proteins that is responsible for Na^+ channel activity in epithelial cells and the specific Na^+/Ca^+ exchanger (NACA 2 and 3) that is responsible for Na^+/Ca^{2+} transport in the distal convoluted tubule.[16,18,19]

Antisense has also been applied to the inflammatory disease state that is seen with tubular damage. Investigators have applied AS-ODNs in order to determine how intercellular adhesion molecules (ICAMs) may regulate tubular response to inflammatory mediators.[20] One such mediator, interleukin 1β (IL-1β), may be important for the initiation of acute interstitial disease as is seen in interstitial nephritis. Using liposomes, these investiga-

[17] J.-C. Martel, C. Cerruti, A. Fleckenstein, F. I. Carroll, M. J. Kuhar, and A. P. Patel, *Soc. Neurosci.* **21,** 376 (1995).

[18] K. E. White, F. A. Gesek, and P. A. Friedman, *J. Am. Soc. Nephrol.* **6,** 356 (1995). [Abstract]

[19] K. E. White, F. A. Gesek, R. F. Reilly, and P. A. Friedman, *Kidney Int.* **54,** 897 (1998).

[20] Y. Z. Ye, X. M. Chen, and B. Fu, *J. Am. Soc. Nephrol.* **9,** 546 (1998).

tors showed that AS-ODN to ICAM1 blocked interleukin 1β-induced ICAM expression.

Finally, AS-ODNs have been used for *in vivo* analysis of transport function. AS-ODN was infused and NaPi-2 cotransporter expression and activity were studied.[21] The investigators sought to demonstrate whether AS-ODN was cleared from the circulation of the kidney and whether they could exert an effect on proximal tubular transport function. In these *in vivo* studies rats were given an intravenous dose of AS-ODN to the NaPi-2 transporter and the effect on message for this transporter was assessed. A single injection led to a reduction of NaPi-2 expression and uptake of phosphate into the renal brush border membrane vesicles was greatly reduced. Although the mechanism by which AS-ODNs can enter the proximal tubule is ill defined, these studies suggested that systemically infused AS-ODNs could exert antisense effects in the renal proximal tubule. Therefore the potential clinical applicability becomes obvious as one develops ways to alter transport expression and activity in patients with disordered tubular function.

Role of Antisense in Glomerular Cells

In vitro and *in situ* data demonstrate that intrinsic glomerular cells are stimulated by inflammatory injury and participate in the progression of glomerular damage.[22] Changes in cell function characteristics of activation include proliferation, altered extracellular matrix synthesis, and *de novo* synthesis of proinflammatory mediators. The mechanisms mediating plasticity in differentiated glomerular cell phenotype are unclear, but it has been hypothesized that persisting cellular activation results in proliferation and altered extracellular matrix synthesis leading to progressive glomerulosclerosis and eventually in renal failure. A number of investigators have begun using these techniques; in particular, specific sequences directed against protooncogenes, growth factors, and protein kinase C (PKC) isoforms have been studied. In those studies inhibition of mediators of disease (platelet-derived growth factor, PDGF), production of matrix (collagen IV), and protein kinase C isoforms (α and β) was achieved and the biological behavior of the cell was then assessed.[23–25]

[21] R. Oberbauer, G. F. Schreiner, J. Biber, H. Murer, and T. W. Meyer, *Proc. Natl. Acad. Sci. U.S.A.* **93,** 4903 (1996).

[22] W. G. Couser and R. J. Johnson, *Am. J. Kidney Dis.* **23,** 193 (1994).

[23] N. Kashihara, Y. Maeshima, H. Sugiyama, T. Sekikawa, K. Okamoto, Y. Morita, K. Kanao, Y. Yamasaki, H. Makino, and Z. Ota, *J. Am. Soc. Nephrol.* **5,** 753 (1994).

[24] C. R. Albrightson, P. Nambi, and M. B. Ganz, *J. Am. Soc. Nephrol.* **6,** 782 (1995). [Abstract]

[25] Y. Maeshima, N. Kashihara, H. Sugiyama, T. Sekikawa, K. Okamoto, K. Kanao, Y. Morita, Y. Yamasaki, H. Makino, and Z. Ota, *J. Am. Soc. Nephrol.* **5,** 835 (1994).

We have studied the role of PKC isoforms in glomerular disease and have applied antisense technology to altering the expression of these enzymes. Previously we have been able to demonstrate that PKC β_{II} is upregulated by cytokines implicated in glomerular disease and, found in glomerular mesangial cells in proliferative human glomerular disease. Its expression becomes even more pronounced in these same cells as glomerulosclerosis ensues.[26–30] Moreover, a specific inhibitor of PKC β_{II} has been shown to modulate growth and matrix deposition by vascular smooth muscle cells in diabetic retinopathy.[31]

We therefore sought to knockout the activity of PKC β_{II} and study cell behavior. Two major technical difficulties encountered when examining PKC and mesangial cells have been the direct entry of oligonucleotides into the cell and the minimization of the anti-PKC effect of the entry vehicle, i.e., liposomes. We therefore used a phosphorothioate-modified antisense oligodeoxynucleotides (ODN). Kanwar and co-workers have successfully used this approach to ascertain the role of insulin-like growth factor I (IGF-I) receptor in cell and organ culture. These modified ODNs resist nuclease degradation, can be added (modified/unmodified) to media, and can be taken up by the cells via endocytosis. Antisense ODN complementary sequences overlapping the PKC β_{II} initiation codon were designed as detailed below.

A 32-mer ODN probe with the sequence derived from the full-length cDNA clone in antisense orientation was generated. Specificity of the antisense ODNs for the target nucleotide sequences has been ascertained by an S_1 nuclease protection assay. We have used computer analysis to avoid self-complementary sequences or sequences complementary to other proteins. T_m values are well over 37°. Depending on the effectiveness of these antisense ODNs, sites other than the initiation codon (e.g., 5' cap site, internal splice sites) may produce more effective antisense effects. Controls have included phosphorothioate ODNs with identical base pairs in the sense orientation and tRNA. We established the half-life of PKC β_{II}. We then harvested treated mesangial cells at various times after the addition of the AS-ODN.

To evaluate the time course of antisense inhibitions we add ODNs at an initial concentration of 0.1 μM to the medium and increase concentra-

[26] M. B. Ganz, R. Saxena and J. Grond, *J. Am. Soc. Nephrol.* **5,** 830 (1994). [Abstract]

[27] M. B. Ganz and P. Q. Barrett, *J. Am. Soc. Nephrol.* **4,** 439 (1993). [Abstract]

[28] M. B. Ganz, B. Saksa, R. Saxena, and J. R. Sedor, *Am. J. Physiol. Renal Fluid Electrolyte Physiol.* **271,** F108 (1996).

[29] R. Saxena, B. A. Saksa, and M. B. Ganz, *Am. J. Physiol.* **265,** F53 (1993).

[30] R. Saxena, B. A. Saksa, K. S. Hawkins, and M. B. Ganz, *FASEB J.* **8,** 646 (1994).

[31] H. Ishii, M. R. Jirousek, D. Koya, C. Takagi, P. Xia, A. Clermont, S.-E. Bursell, T. S. Kern, L. M. Ballas, W. F. Heath, L. E. Stramm, E. P. Feener, and G. L. King, *Science* **272,** 728 (1996).

TABLE II
UPTAKE OF ODN IN CELLS IN CULTURE

MCs in culture	No. of experiments	ODN-antisense (μM)	Percentage of cells with labeled ODNs
Day 3	2	0.5	85 ± 9
Day 5	3	0.5	89 ± 7.6
		ODN-scrambled sense (μM)	
Day 3	2	0.5	81 ± 7
Day 5	2	0.5	77 ± 12

tions until we reach a maximum concentration of 1 μM. The final concentration rests between 0.1 and 1.0 μM. To ascertain if and at what concentration the antisense ODN best inhibits PKC β_{II}, we perform reverse transcriptase-polymerase chain reaction (RT-PCR) on cells 12, 18, 24, 30, 36, and 48 hr after its addition. These experiments allowed us to ascertain the best concentration/time course for inhibiting PKC β_{II} synthesis and the conditions that maximize cellular uptake of ODNs. We have ascertained that the use of LipofectAMINE increases the uptake of antisense ODN (Table II).

We exposed stably transfected mesangial cells to antisense and sense (control) to PKC β_{II} and then measured [^{3}H]thymidine incorporation at 24, 28, and 72 hr as performed previously. These cells readily express the isoform PKC β_{II} and exhibit rapid proliferation and matrix deposition far greater than that normally seen in nontransfected mesangial cells. Figure 2 demonstrates that fetal bovine serum-induced cell proliferation in PKC β_{II}-transfected cells is readily blunted at all time points measured. These data demonstrate (1) the use of antisense to PKC β_{II}, and (2) the ability to modulate cell behavior and thereby study the biological role of PKC β_{II}. In addition, we sought to see if we can blunt collagen IV deposition with the AS-ODN to PKC β_{II} in a similar fashion to the antiproliferative effect of this antisense. As shown in Fig. 3, collagen IV production is also inhibited by blocking the expression of PKC β_{II}. The antisense to PKC β_{II} is specific: collagen I and III, fibronectin, and other matrix components are unaffected by this process along with ion transporter activity (data not shown). This clearly demonstrates the specificity of using such approaches to define potentially pathological processes in the cell and thereby allows us to study the behavior of the cell in disease.

In all antisense experimentation it is essential that the necessary controls be applied. We are aware of the potential difficulty of interpreting these experiments without proper oligonucleotide controls.[32,33] It has been re-

[32] Y. S. Kanwar, Z. Z. Liu, J. Wada, *Contrib. Nephrol.* **107,** 168 (1994).

[33] A. Kumar, K. Ota, J. Wada, E. I. Wallner, A. S. Charonis, F. A. Carone, and Y. S. Kanwar, *Kidney Int.* **52,** 620 (1997).

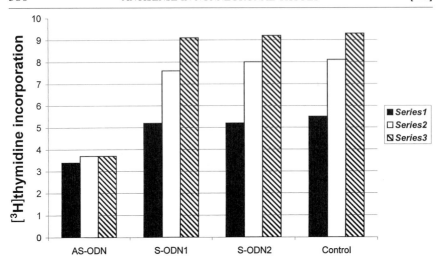

FIG. 2. The effects of AS-ODN on cell proliferation; a specific antisense to PKC β_{II} inhibits cell proliferation in mesangial cells. S-ODN-1 and S-ODN-2 are sense controls. Control was the addition of 0.1% bovine serum albumin. All cells were exposed to the mitogen–10% FBS. Series 1, 24 hr; series 2, 48 hr; series 3, 72 hr.

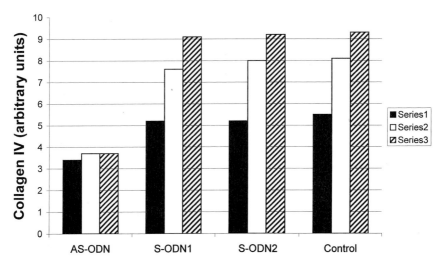

FIG. 3. The effects of AS-ODN on collagen IV production; specific antisense to PKC β_{II} inhibits collagen IV deposition as measured by ELISA in mesangial cells. S-ODN-1 and S-ODN-2 are sense controls. Control was the addition of 0.1% bovine serum albumin. All cells were exposed to the mitogen–10% FBS. Series 1, 24 hr; series 2, 48 hr; series 3, 72 hr.

ported that phosphorothioate oligonucleotides cause nonspecific effects. These toxic effects may limit the utility of the experiments. In addition to using a large number of unrelated oligonucleotides as controls, we measured nontargeted mRNAs to ascertain if they also degraded, thereby implying that the fluorescein-labeled ODN used led to a nonspecific response. We have also performed controls using cells that are not exposed to antisense and/or sense in addition to three scrambled sense oligonucleotides:

Sense: 5'-ATAATATAGATAATGATTAGTAATAATGG-3'
5'-GATGCCCATGCATATCGATCGGATTGCGAT-3'
5'-GATGATTTGAATGACCGGATTGCGAT-3'
5'-ATCGATCGATCGATCCATGCCG-3'

Additional controls of LipofectAMINE are studied by applying sense ODNs and LipofectAMINE to normal mesangial cells to ensure that this does not alter normal cell behavior itself; decreasing the concentration of LipofectAMINE (0.1 μM) has allowed us to minimize its effects.

In Vivo Use of Antisense

The preceding data demonstrate the utility of AS-ODNs *in vitro*. There are also examples of their efficacy as therapeutic agents *in vivo*. Investigators studying the role of transforming growth factor β (TGF-β), platelet-derived growth factor (PDGF), and nitric oxide (NO) have published *in vivo* applications of antisense technology.[16,34,35] Specifically, this approach has been used to deliver AS-ODNs via an injection into the renal artery by employing liposomal technology. Using the rat model of mesangial cell proliferation (Thy 1.1 model), there is a reduction in mesangial cell proliferation and matrix expansion when antisense was directed against TGF-β and/or PDGF. Intraarterial introduction of AS-ODNs directed against the inducible form of NO synthase (NOS) attenuated acute renal failure in rats subjected to renal ischemia.[35] Both of these studies demonstrate a novel method by which glomerular and renal tubular injury may be halted.

AS-ODN has also been generated against collagen-binding stress protein (HSP47). This specific heat shock protein is a collagen-specific molecular chaperone that appears to be important for the synthesis and secretion of the procollagen molecule. HSP47 expression is increased in glomerular disease and blocking its expression may halt matrix deposition and hence glomerulosclerosis.[36]

This approach has been applied as a therapeutic agent for peritoneal

[34] G. Wolf, G. Zahner, R. Schroeder, and A. K. Stahl, *Nephrol. Dial. Transplant* **11**, 263 (1996).
[35] E. Noiri, T. Peresleni, F. Miller, and M. S. Goligorsky, *J. Clin. Invest.* **97**, 2377 (1996).
[36] M. Sunamoto, K. Kuze, H. Tsuji, N. Ohishi, K. Yagi, K. Nagata, T. Kita, and T. Doi, *Lab. Invest.* **78**, 967 (1998).

mesothelial cell injury, which may occur with repeated infections in peritoneal dialysis.[37] These investigators generated a fluorescein isothiocyanate
(FITC)-labeled phosphorothioate ODN and introduced it into rat mesothelial cells using cationic liposomes. They targeted the PCNA (proliferating
cell nuclear antigen; a nuclear protein associated with the cell cycle) with
the antisense and cell proliferation *in vitro* was halted. When the AS-ODN
was introduced into the peritoneum of rats, fluorescence was noted only
in those infiltrating and mesothelial cells of rats with injury and not those
of normal rats. Inhibition of the total cell count and PCNA-positive cell
count was readily observed. These findings confirmed that this therapy may
be applicable to patients with peritoneal injury.

AS-ODNs have also been applied to experimental models of glomerular
disease. Proposed mediators of glomerular disease have been manipulated
by the use of AS-ODNs. Akagi *et al.* introduced AS-ODN to TGF-β_1.[6,38,39]
A multifunctional dimeric peptide, TGF-β_1 regulates biological processes
such as cell proliferation and differentiation. These investigators used a
variant of the anti-Thy 1.1 model of mesangioproliferative disease, as it has
been demonstrated that TGF-β_1 is upregulated. Using an HVJ–liposome
complex containing antisense, the investigators cross-clamped the aorta
and injected this complex into the kidney of nephritic animals. They found
that the AS-ODN accumulated predominantly in the mesangial cells and
that there was a marked decrease in TGF-β expression. However, only
30% of all glomeruli expressed AS-ODN and therefore enhanced delivery
will be needed to alter glomerular disease more effectively. Moreover,
better selective delivery systems (i.e., ones that enhance the selectivity of
the AS-ODN) need to be developed along with enhanced duration of the
gene expression (to minimize the need for repeated injections) before this
can be applied to human disease.

Approaches wherein one attempts to protect the kidney from injury
(ischemia) have been used with AS-ODNs.[35] Ischemic kidney disease is
responsible for a significant number of cases of morbidity and mortality in
hospitals. NO synthases are a presumed mediator of ischemic injury in dye-
induced or antibiotic-induced injury. AS-ODN was generated to NOS by
Noiri *et al.*[35] as a strategy to block this effect. By targeting NOS, a presumed
mediator of ischemic renal injury, Noiri *et al.* were able to provide not only
direct evidence of cytotoxic effects of NO (produced by inducible NOS)
in the course of acute renal injury, but also protective effects of the AS-

[37] T. Saksaki, . Yoshiyuki, H. Hatta, S. Nomura, and G. Osawa, *J. Am. Soc. Nephrol.* **9,**
523 (1999).
[38] Y. Akagi, Y. Isaka, M. Arai *et al., Kidney Int.* **50,** 148 (1996).
[39] N. Kashihara, Y. Maeshima, and H. Makino, *Exp. Nephrol.* **5,** 126 (1997).

ODN when used prior to ischemia. Obviously, the timing prior to the injury is important for preventing injury.

In conclusion, the ability of antisense technology to prevent the expression of specific proteins provides an effective tool not only for the study of normal transport processes in the normal kidney, but also for the study of the pathogenic mechanisms by which cell behavior is modified in disease. There are, however, many biological effects of AS-ODN that cannot be attributed to the antisense mechanism alone and therefore such effects may limit the conclusions that can be drawn and may also compromise the usefulness of deoxynucleotides as therapeutic agents. However, studies have shown that these limitations may be overcome and that *in vivo* targeting of certain proteins holds great promise in the treatment of kidney disease.

[27] Use of Antisense Techniques in Rat Renal Medulla

By DAVID L. MATTSON

Introduction

Experimental evidence indicates that the renal medulla plays an important role in the control of fluid and electrolyte homeostasis.[1,2] Alterations in renal medullary blood flow or tubular transport have been demonstrated to have a profound impact on renal excretory function, extracellular fluid volume, and the long-term regulation of arterial blood pressure. The paracrine, autocrine, and hormonal factors that regulate tubular and vascular function in this region of the kidney can therefore have a major influence on the regulation of arterial blood pressure. Of particular interest in our laboratory has been the physiological role(s) played by nitric oxide synthase (NOS) in the renal medulla. Unfortunately, functional studies that have attempted to examine the influence of NOS on renal medullary function have been complicated by the presence of three NOS isoforms, neuronal (nNOS), inducible (iNOS), and endothelial NOS (eNOS), in the tubular and/or vascular structures of this portion of the kidney.[3–7] One of the

[1] A. W. Cowley, Jr., *Am. J. Physiol.* **273,** R1 (1997).
[2] G. Bergstrom, G. Gothberg, G. Karlstrom, and J. Rudenstam, *Clin. Exp. Hypertension* **20,** 1 (1998).
[3] D. L. Mattson and D. J. Higgins, *Hypertension* **27,** 688 (1996).
[4] Y. Terada, K. Tomita, H. Nonoguchi, and F. Marumo, *J. Clin. Invest.* **90,** 659 (1992).
[5] K. Y. Ahn, M. G. Mohaupt, K. M. Madsen, and B. C. Kone, *Am. J. Physiol.* **267,** F748 (1994).

challenges presented when trying to determine the role of each NOS isoform in the control of renal function was to determine a means by which these similar enzymes could be individually manipulated. A second complication of these types of studies is the anatomy of the kidney, which makes the direct delivery of substances to the renal medulla extremely difficult. It therefore became necessary to adopt a strategy that could be used to selectively deliver to the renal medulla compounds that specifically target the individual NOS isoforms.

One approach to selectively block the individual NOS isoforms is to use enzyme inhibitors. Although compounds of various specificity are available for such studies, it is difficult to administer these compounds *in vivo* and ensure selective inhibition of the targeted isoform. As a means of avoiding some of the potential pitfalls associated with the use of enzyme inhibitors, we prepared an antisense oligonucleotide complementary to the mRNA for nNOS and determined the influence of antisense inhibition of nNOS in the renal medulla on renal medullary function and blood pressure.[8] To deliver the antisense oligonucleotides selectively to the renal medulla, a specially designed catheter was used that permitted the continuous infusion into this region of the kidney for periods of days to weeks. The functional results obtained with infusion of the antisense oligonucleotides were then compared with experimental results obtained after administration of 7-nitroindazole (7-NI), a specific enzymatic inhibitor of nNOS.

Materials and Methods

Experiments were performed on male Sprague-Dawley rats (300–350 g) obtained from Sasco (Madison, WI). The rats were housed in the Animal Resource Center at the Medical College of Wisconsin (Milwaukee, WI) with normal rat chow and tap water provided *ad libitum*. During the experimental protocol, unless otherwise noted, the tap water was replaced with 1% NaCl water in order to place the animals on a high sodium diet. All animal procedures were approved by the Medical College of Wisconsin Animal Care Committee, and the rats were closely monitored to ensure that none experienced undue stress or discomfort throughout the protocol.

[6] K. Ujiie, J. Yuen, L. Hogarth, R. Danziger, and R. A. Star, *Am. J. Physiol.* **267**, F296 (1994).
[7] M. G. Mohaupt, J. L. Elzie, K. Y. Ahn, W. L. Clapp, C. S. Wilcox, and B. C. Kone, *Kidney Int.* **46**, 653 (1994).
[8] D. L. Mattson and T. G. Bellehumeur, *Hypertension* **28**, 297 (1996).

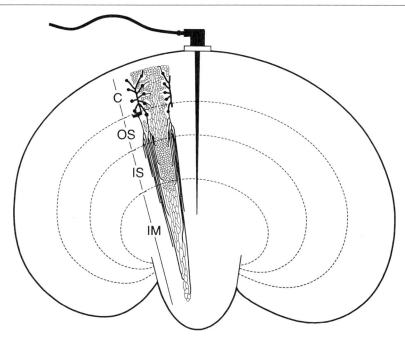

FIG. 1. A polyethylene catheter of tip diameter ~100 μm is inserted directly into the renal tissue to place the tip into the inner medulla. Gross anatomical regions of the kidney are designated as cortex (C), outer stripe (OS) and inner stripe (IS) of the outer medulla, and inner medulla (IM). [Adapted with permission from A. W. Cowley, Jr., *Am. J. Physiol.* **273**, R1 (1997).]

Renal Medullary Interstitial Infusion

To deliver antisense oligonucleotides and other compounds to the tubular and vascular structures of the renal medulla, a simple technique was developed that allowed selective infusion directly into the renal medullary interstitial space (Fig. 1).[9,10] A specially designed polyethylene catheter was constructed that consisted of a piece of polyethylene tubing (0.58-mm i.d., 0.965-mm o.d., PE-50; Becton Dickinson, Sparks, MD) heat fused to a second, smaller piece of polyethylene tubing (0.28-mm i.d., 0.61-mm o.d., PE-10; Becton Dickinson). The smaller piece of polyethylene tubing was stretched over hot air to taper the end to a tip diameter of ~100 μm and

---

[9] S.-H. Lu, R. J. Roman, D. L. Mattson, and A. W. Cowley, Jr., *Am. J. Physiol.* **263**, R1064 (1992).

[10] D. L. Mattson, R.J. Roman, and A.W. Cowley, Jr., *Hypertension* **19**, 766 (1992).

was inserted directly into the renal medullary interstitial space from the outer surface of the kidney. As depicted in Fig. 1, the end of the larger piece of polyethylene tubing was heat flared to provide a flange that rested on the surface of the kidney; a small drop of cyanoacrylate adhesive was used to secure and seal the catheter in place on the surface of the kidney after insertion into the renal tissue. Although the depth of insertion varied depending on kidney size, placement of the tip approximately 7 mm beneath the surface of the kidney was sufficient to position the tip in the inner medulla at a point just past the border of the outer medulla. Correct catheter placement was confirmed on conclusion of the experiment. After implantation the catheter was continuously infused with isotonic saline (with or without drug) at a rate of approximately 8.3 μl/min.[11,12]

Both autoradiographic and functional studies demonstrated that this technique can be used to localize infused compounds in the renal medullary interstitial space. Initial validation studies examined the distribution of [^{14}C]clentiazem, an analog of the calcium channel blocker diltiazem, after renal medullary interstitial infusion in anesthetized rats.[9] Comparison of the radioactivity distributed in the infused and contralateral kidney in this study demonstrated that more than 98% of the total counts were localized in the infused kidney. Furthermore, of the radioactivity in the infused kidney, greater than 92% was localized in the inner and outer medulla. This technique therefore permits the selective distribution of infused compounds into the medulla of the infused kidney. Further proof of the utility of this technique is a number of functional studies that have demonstrated selective modulation of renal medullary function during infusion of various pharmacological agents with this catheter.[8-14]

A second set of concerns with the implantation of this catheter into the renal tissue involves the possible deleterious effects on kidney function. Further experiments were then performed to examine the influence of both acute and long-term implantation of this catheter on renal function. In anesthetized animals, it was observed that there were no differences in urinary flow rate, sodium excretion, or glomerular filtration rate between the infused and contralateral kidney after acute implantation.[11] This experiment demonstrated a lack of deleterious effects on renal function in the hours immediately after implantation of this catheter. Additional experi-

[11] D. L. Mattson, and A. W. Cowley, Jr., *Hypertension* **21,** 961 (1993).
[12] A. W. Cowley, Jr., D. L. Mattson, S.-H. Lu, and R. J. Roman, *Hypertension* **25,** 663 (1995).
[13] D. L. Mattson, S.-H. Lu, K. Nakanishi, P. E. Papanek, and A. W. Cowley, Jr., *Am. J. Physiol.* **266,** H1918 (1994).
[14] S.-H. Lu, D. L. Mattson, and A. W. Cowley, Jr., *Hypertension* **23,** 337 (1994).

ments were performed to examine renal function in rats in which the medullary interstitial catheters were chronically implanted for an extended period.[13] A comparison of renal function between kidneys in rats in which one kidney was infused with saline into the renal medullary interstitial space for 2 weeks and the contralateral kidney was untouched demonstrated no effects on urinary flow rate, urinary sodium excretion, glomerular filtration rate, renal blood flow, or urine osmolality between the infused and contralateral kidney.[13] Of even greater interest, it was demonstrated that uninephrectomized rats with an interstitial infusion of saline produced a urine that was concentrated to the same extent (2730 ± 402 mOsm/kg H_2O, $n = 5$) as in control rats without the infusion (2671 ± 385 mOsm/kg H_2O, $n = 6$) after 48 hr of water restriction. These validation experiments demonstrated that neither short- nor long-term implantation and infusion into this medullary interstitial catheter alters the baseline level of renal function or the ability of the kidney to form a maximally concentrated urine.

The renal medullary interstitial infusion technique can be used to localize infused substances in the renal medulla. Moreover, the implantation of the catheter accompanied by continuous infusion does not adversely affect renal function. Despite the utility of this method, it should be recognized that this technique does have some limitations. The implantation of the catheter inevitably leads to the destruction of vascular and tubular structures in the path of the catheter. Although this damage appears to be minimal, the potential influence of a small amount of renal damage must be considered when designing these types of experiments. In addition, unilateral nephrectomy is necessary to carry out the experiments in conscious animals, so any potential consequences of nephrectomy may also influence the experimental results. This technique is also moderately labor intensive; the continuous infusion requires chronically instrumented animals with the equipment and facility capable of housing and maintaining them. Finally, it is important when performing experiments of this type to account for the possibility that compounds infused into the medullary interstitium can be absorbed by the renal capillaries and recirculate to the systemic circulation. The potential effects of recirculated compounds on other parts of the body should therefore be considered when designing these experiments and interpreting the experimental data. Despite these potential limitations, this is a technique we have used successfully in a number of experimental studies and is a convenient method to selectively deliver infused substances to this portion of the body.[8,14]

Surgical Preparation. Rats are anesthetized for chronic surgery with an intramuscular injection of ketamine (100 mg/kg) and acepromazine (2 mg/

kg), and the right kidney is removed. The unilateral nephrectomy is carried out so that the remaining kidney is the sole determinant of renal function.[8,13,14] A second surgery is performed 7 to 10 days after the nephrectomy to instrument the rats with chronic indwelling catheters. Catheters are placed in the abdominal aorta below the left renal artery via the femoral artery to measure blood pressure, in the vena cava via the femoral vein for intravenous infusion, and in the renal medullary interstitial space. The catheters are tunneled subcutaneously and exteriorized at the back of the neck in a piece of stainless steel spring. The spring is attached to a swivel device that allows each animal to move freely in the cage while being continuously infused. The animals receive a postoperative injection of penicillin (40,000 U, intramuscular) to prevent infection.

Western Blotting Protocols

Protein samples are electrophoretically size separated, using a discontinuous system[15] consisting of a 7.5% polyacrylamide resolving gel and a 5% polyacrylamide stacking gel. Molecular weight markers (range, ~40,000–200,000) are loaded into one lane as a size standard. Equivalent amounts of total protein from antisense-treated or control rats are added to adjacent lanes and the samples are electrophoresed at 200 V for 45–60 min on an 8 × 10 cm electrophoresis cell (Bio-Rad, Hercules, CA).

After separation, the proteins are electrophoretically transferred to a nitrocellulose membrane at 100 V for 1 hr. The membranes are washed in Tris-buffered saline (TBS), blocked with 5% nonfat dried milk in TBS (NFM–TBS) for 2 hr, and incubated with a 1 : 1000 dilution of monoclonal mouse anti-nNOS (Transduction Laboratories, Lexington, KY) in 2% NFM–TBS overnight at 4°. We previously demonstrated that this antibody binds to both rat nNOS and iNOS.[13] The membranes are then incubated with a horseradish peroxidase-labeled goat anti-mouse immunoglobulin G (IgG, 1 : 1000) in 2% NFM–TBS for 2 hr. The bound antibody is detected by chemiluminescence (Amersham, Arlington Heights, IL) on X-ray film. Binding of a monoclonal antibody against the structural protein β-actin (Sigma, St. Louis, MO) is used as a loading control. Membranes are stripped in a Tris-buffered solution containing 2% sodium dodecyl sulfate (SDS) and 100 mM 2-mercaptoethanol at 50° between incubations with different antibodies. Densitometry is performed using a Phospho-Imager personal densitometer (Molecular Dynamics, Sunnyvale, CA).

[15] U. K. Laemmli, *Nature* (*London*) **227,** 680 (1970).

Tissue Nitric Oxide Synthase Activity

The NOS enzyme assay is based on previously described methods.[16–18] The total tissue homogenate is incubated with 2 mM $CaCl_2$, 1 mM NADPH, 25 μM FAD, calmodulin (1.25 μg/ml), 10 μM BH_4, and [^3H] arginine (approximately 300,000 cpm, specific activity 68 Ci/mmol) in 20 mM HEPES buffer, pH 7.2, at 37° for 5 min. The arginine and converted citrulline are separated by isocratic reversed-phase high-performance liquid chromatography (HPLC) with a Supelco (Bellefonte, PA) LC_{18}-DB column (mobile phase, 11.5% methanol, 11.5% acetonitrile, 1% tetrahydrofuran, 0.1 M KH_2PO_4, pH 5.9). The amount of converted citrulline and the total counts are quantitated by radiochemical detection (Packard, Tampa, FL).

Antisense Oligonucleotides and Enzymatic Inhibitors

In the experiments described here, an antisense oligonucleotide complementary to the initiation region (bases 335–351, 5'-CAT GGT ATC TGT GTC CT-3') of the mRNA for nNOS[19] is infused directly into the renal medullary interstitial space in sterile saline (7.5 nmol/hr). A scrambled probe (5'-CGT CTC TGT TTG ATA CG-3') composed of the same constituent bases but in random order is used to control for nonspecific effects of oligonucleotide infusion. Oligonucleotides are synthesized in the Protein–Nucleic Acid Facility of the Medical College of Wisconsin with phosphorothioate bonds on the 5' and 3' ends (five on each end) and phosphodiester bonds in the remainder in an attempt to increase the oligonucleotide stability by decreasing degradation.[20] To inhibit renal medullary nNOS using a mechanistically different approach, 7-nitroindazole (7-NI; Biomol, Plymouth Meeting, PA), an inhibitor specific to nNOS when administered *in vivo*,[21–25] is infused into the medullary interstitial space of a

[16] D. L. Mattson, C. Maeda, T. D. Bachmann, and A. W. Cowley, Jr., *Hypertension* **31,** 15 (1998).

[17] D. S. Bredt and S. H. Snyder, *Proc. Natl. Acad. Sci. U.S.A.* **86,** 9030 (1989).

[18] M. Carlberg, *J. Neurosci. Methods* **52,** 165 (1994).

[19] D. S. Bredt, P. M. Hwang, P. E. Glatt, C. E. Lowenstein, R. R. Reed, and S. H. Snyder, *Nature (London)* **351,** 714 (1991).

[20] M. I. Phillips and R. Garcia, *Regul. Peptides* **59,** 131 (1995).

[21] R. C. Babbedge, P. A. Bland-Ward, S. L. Hart, and P. K. Moore, *Br. J. Pharmacol.* **110,** 225 (1993).

[22] A. G. B. Kovach, Z. Lohinai, J. Marczis, I. Balla, T. M. Dawson, and S. H. Snyder, *Ann. N.Y. Acad. Sci.* **738,** 348 (1994).

[23] B. Mayer, P. Klatt, E. R. Werner, and K. Schmidt, *Neuropharmacology* **33,** 1253 (1994).

[24] P. K. Moore, P. Wallace, Z. Gaffen, S. L. Hart, and R. C. Babbedge, *Br. J. Pharmacol.* **110,** 219 (1993).

[25] D. J. Wolff and B. J. Gribin, *Arch. Biochem. Biophys.* **311,** 300 (1994).

separate group of rats. The effects of 7-NI administration are compared with antisense inhibition to confirm the functional effects of the antisense oligonucleotide.

Results and Discussion

Functional Studies

The change in mean arterial pressure (MAP) after a 4-day renal medullary interstitial infusion of saline ($n = 6$), the scrambled oligonucleotide ($n = 8$, 7.5 nmol/hr), or the antisense oligonucleotide ($n = 9$, 7.5 nmol/hr) to rats on a high sodium intake is illustrated in Fig. 2. Infusion of saline vehicle or the scrambled oligonucleotide failed to alter MAP from the control values of 101 ± 3 and 97 ± 1 mmHg, respectively. In additional experiments administration of the antisense oligonucleotide did not significantly alter blood pressure in rats maintained on a low-sodium diet ($n =$

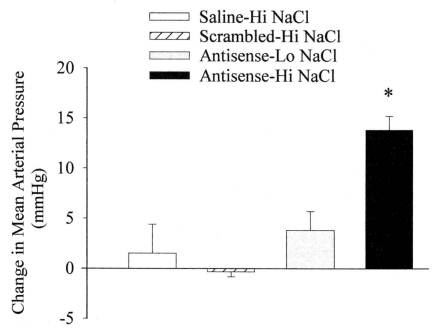

FIG. 2. Absolute change in mean arterial pressure from control in uninephrectomized Sprague-Dawley rats after 4 days of renal medullary interstitial infusion of saline (high-salt diet), the antisense oligonucleotide for nNOS (normal and high salt diets), and the scrambled oligonucleotide probe (high salt diet). *Significant difference ($p < 0.05$) from the control value. [Reprinted with permission from D. L. Mattson and T. G. Bellehumeur, *Hypertension* **28**, 297 (1996).]

5). In contrast to the results obtained with rats on a low-sodium diet, infusion of the antisense oligonucleotide to the Sprague-Dawley rats maintained on a high sodium intake led to a significant increase in MAP of approximately 14 ± 1 mmHg from a control value of 104 ± 4 mmHg.

Further experiments were performed to document the reversibility of the functional effects of antisense oligonucleotide administration. Renal medullary interstitial infusion of the antisense oligonucleotide for nNOS significantly increased blood pressure from a mean control value of 97 ± 3 to 112 ± 2 mmHg by day 5 of infusion (data not shown).[8] The average body weight of these animals progressively and significantly increased from a control level of 337 ± 5 to 392 ± 6 g on the final day of antisense infusion. After the termination of the antisense oligonucleotide infusion, mean arterial pressure decreased to a level not significantly different from control by the fifth postcontrol day. During the postcontrol period body weight did not return to control but also did not significantly increase from the final experimental period value, averaging 397 ± 5 g on the final postcontrol day.

These functional experiments demonstrated that the administration of an antisense oligonucleotide complementary to the initiation region of nNOS led to an increase in arterial blood pressure. The increase in arterial pressure was not mimicked by the scrambled oligonucleotide, produced a sodium intake-dependent change in arterial pressure, and was reversible when the antisense infusion was stopped. The increase in body weight after antisense oligonucleotide administration and the sodium dependence of the hypertension indicate that the elevation in arterial pressure was due to an alteration in renal function that led to a retention of sodium and water and a subsequent expansion of extracellular fluid volume. These observations are consistent with results we previously described during renal medullary interstitial infusion of the nonspecific NOS inhibitor N^{G}-nitro-L-arginine methyl ester.[13]

Documentation of Antisense Inhibition of Neuronal Nitric Oxide Synthase

To document the degree of nNOS inhibition in the renal medulla during antisense oligonucleotide administration, protein-blotting and enzymatic assays were performed on protein homogenates extracted from the renal medulla of antisense- and vehicle-treated rats.[8] A representative blot of protein homogenates from the renal medulla of control and antisense oligonucleotide-infused rats is illustrated in Fig. 3. The first four lanes (Fig. 3) were loaded with 100 μg of total protein from the renal medulla of control rats, the middle lane is a size marker (199-kDa marker shown), and the final four lanes were loaded with 100 μg of total protein from the renal medulla of antisense oligonucleotide-infused rats. The membrane was probed with an anti-nNOS antibody (top, Fig. 3) which binds to both nNOS and iNOS, and an antibody against the structural protein β-actin (bottom, Fig. 3)[3,8]. The intensity of the nNOS bands was reduced in the antisense-treated rats while there was no alteration in the level of iNOS or β-actin immunoreactive protein. Densitometric analysis (Fig. 4) of total medullary protein homogenates from seven antisense-treated and seven control rats demonstrated that nNOS was significantly reduced by 53% in the antisense-treated rats while there was no significant alteration in the intensity of iNOS, eNOS (blots not shown), or β-actin. Supporting the Western blot results are the NOS enzymatic assay data, which demonstrated that total renal medullary NOS activity was reduced by 30% in the renal medulla of the antisense-infused rats when compared with control rats. Furthermore, as a control for systemic spillover of the antisense oligonucleotide, no difference was detected in cerebellar NOS activity between control rats and those infused with the antisense oligonucleotide.

These biochemical data demonstrate that the antisense oligonucleotide had specific effects on nNOS. The oligonucleotide selectively decreased nNOS protein without altering three other gene products normally present in the rat renal medulla. The antisense oligonucleotide also significantly decreased total tissue NOS activity in the renal medulla without changing cerebellar NOS activity (which is primarily nNOS[3]) in these same animals. The lack of effects on cerebellar NOS activity suggests minimal systemic

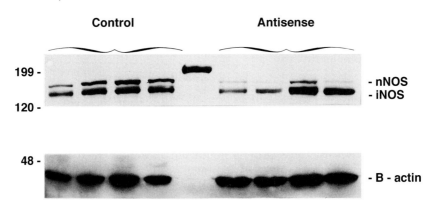

FIG. 3. Representative protein blot of 100 μg of total medullary protein from control rats (lanes 1–4) and rats infused directly into the renal medullary interstitial space with the antisense oligonucleotide for nNOS (lanes 6–9). The band in the middle (lane 5) is the 199-kDa molecular mass marker. The membrane was incubated with antibodies for nNOS and iNOS (top) and β-actin (bottom). [Reprinted with permission from D. L. Mattson and T. G. Bellehumeur, *Hypertension* **28,** 297 (1996).]

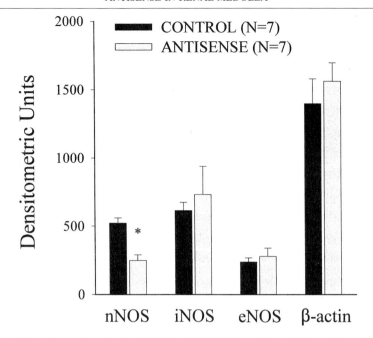

FIG. 4. Densitometric analysis of nNOS, eNOS, iNOS, and β-actin in medullary tissue of control rats and rats infused directly into the medullary interstitial space with an antisense oligonucleotide for nNOS. *Significant difference ($p < 0.05$) from control. [Reprinted with permission from D. L. Mattson and T. G. Bellehumeur, *Hypertension* **28,** 297 (1996).]

spillover of the interstitially infused oligonucleotide, although the influence of systemic infusion of the oligonucleotide on cerebellar and renal medullary NOS activity was not examined in this study. Together, these biochemical and physiology experiments support a selective influence of the antisense oligonucleotide to decrease the level of renal medullary nNOS and increase blood pressure in chronically treated rats.

Influence of 7-Nitroindazole on Blood Pressure and Nitric Oxide Synthase Enzymatic Activity

As a means of confirming the influence of nNOS in the renal medulla on blood pressure, a mechanistically different inhibitor of nNOS, the enzyme inhibitor 7-nitroindazole, was infused into the renal medullary interstitial space of an additional group of rats.[8] It was observed that the mean arterial pressure was increased by 15 ± 6 mmHg during the infusion of 7-NI ($400 \mu g/hr$, $n = 7$) while blood pressure was unaltered in rats infused with vehicle ($0.5\% NaCO_3$ in saline, $n = 5$) of the same period of time. The NOS enzymatic activity in the renal medulla of 7-NI-infused rats was decreased by

37% in comparison with vehicle-treated rats while total NOS activity was unaltered in the cerebellum of these animals. These functional and biochemical data from the rats chronically treated with the nNOS enzyme inhibitor 7-NI demonstrate that selective enzymatic inhibition of renal medullary nNOS with 7-NI also leads to increased blood pressure. The 7-NI data therefore support the functional effects of the antisense oligonucleotide because these two compounds with different mechanisms of action to reduce nNOS enzymatic activity caused the same functional effects in chronically treated rats maintained on a high-sodium diet.

Summary and Conclusion

The experiments outlined in this chapter utilized a novel infusion technique to deliver an antisense oligonucleotide (and an enzyme inhibitor) directly into the renal medullary interstitial space of conscious rats. Antisense treatment led to a selective decrease in nNOS protein and reduced total NOS enzymatic activity in the renal medulla of the infused rats while three other gene products found in the renal medulla (iNOS, eNOS, and β-actin) were unaltered. Physiological studies in rats demonstrated that infusion of the antisense oligonucleotide into the renal medullary interstitial space increased mean arterial pressure. The increase in blood pressure was dependent on the sodium intake of the rats, was not mimicked when a scrambled oligonucleotide was infused, and was reversible when the antisense infusion was stopped. To confirm the functional data obtained with the antisense oligonucleotide, renal medullary interstitial infusion of the nNOS enzyme inhibitor 7-NI was also shown to lead to a similar increase in arterial pressure and decrease in total NOS activity in the renal medulla. Together, the antisense oligonucleotide, the enzyme inhibitor, and the interstitial infusion technique were used to demonstrate that nNOS found in the renal medulla is important in the chronic regulation of arterial pressure.

The experiments summarized in this chapter outline a strategy that can potentially be used to examine the functional effects of many different proteins in this region of the body. Through the use of antisense oligonucleotides and other pharmacological agents, we can hope to gain a more comprehensive understanding of the factors that control renal medullary tubular and vascular function and consequently fluid and electrolyte homeostasis and blood pressure.

Acknowledgments

This work was partially supported by NIH grants HL-29587 and DK-50739 and American Heart Association Wisconsin Affiliate Grant 95-GS-76.

[28] Antisense Approaches to *in Vitro* Organ Culture

By TAKAHIRO OCHIYA and MASAAKI TERADA

Antisense technology has been greatly facilitated by the development of cancer therapy utilizing antisense oligodeoxynucleotides (AS-ODNs) representing a promising new line of pharmaceuticals.[1] However, prior to extensive use of natural ODNs *in vivo,* a number of problems must be addressed: (1) rapid degradation in the presence of serum, (2) poor membrane transport, (3) nonspecific inhibition of other genes, and (4) cytotoxicity at high doses. Several attempts have been made to overcome these limitations including a chemical modification of the backbone of ODNs, which resulted in an enhanced resistance to nucleases and longer survival times *in vivo.* An especially critical area lies in the development of a novel delivery system for an efficient transfer of ODNs into tissue or organs; this can be accomplished in a number of ways. Of the procedures examined to date, microinjection, electroporation, and chemical reagents, such as cationic liposomes and poly(L-lysine), and receptor-mediated endocytosis are believed to be effective methods.[2] As the main focus of this chapter, we present procedures for cationic liposome-mediated AS-ODN transfer into an *in vitro* organ culture system for mouse limb development.

Introduction

The mechanisms of pattern formation involve the integration of many developmental processes, including tissue-specific differentiation, control of cell growth, and cell-to-cell interaction. Studies of the developing limb represent a classic model for vertebrate pattern formation. Progress made in our understanding of the molecules involved in limb development has made the identification of many key molecules possible. In 1992, our group, among others, reported the expression of gene *Hst-1,* also known as *Fgf-4* (encoding fibroblast growth factor) in the apical ectodermal ridge (AER) of the developing mouse limb bud,[3] and several works have suggested its possible role in limb development.[4,5]

[1] R. W. Wagner, *Nature* (*London*) **327,** 333 (1994).

[2] J. P. Clarenc, G. Degols, J. P. Leonetti, P. Milhaud, and B. Lebleu, *Anti-Cancer Drug Design* **8,** 81 (1993).

[3] H. R. Suzuki, H. Sakamoto, T. Yoshida, T. Sugimura, M. Treada, and M. Solursh, *Dev. Biol.* **150,** 219 (1992).

[4] L. Niswander, S. Jeffrey, G. R. Martin, and C. Tickle, *Nature* (*London*) **371,** 609 (1994).

[5] R. L. Jhonson and C. J. Tabin, *Cell* **90,** 979 (1997).

Within this context, we established (1) a novel culture system for mouse limb development, (2) use of ODNs with a form of cationic liposome complex that allows enhanced incorporation of ODNs into a specific site compared with conventional methods, and (3) antisense ODNs complementary to *Hst-1/Fgf-4* to block limb development when using these new systems.[6] It should be noted that marked inhibition of the gene expression was attained when antisense ODNs were applied together with liposome.

Principle of Culture System for Mouse Limb Development Suitable for Antisense Strategy

To assess directly the role of Hst-1/Fgf-4 protein in limb development *in vivo,* we initially established a novel organ culture system to study mouse limb development *in vitro.* This system allows mouse limb bud formation by 9.5–10 days postcoitus (p.c.) embryos, parts of which are placed on a sheet of extracellular matrix in a defined medium and allowed to differentiate into a limb by 12.5 days p.c., i.e., within 4.5 days. Most significantly, this culturing system not only allows limb outgrowth but also provides us with a suitable model system to explore the key factors for limb development by controlling a variety of culture conditions, with or without additives, including a number of growth factors, inhibitors, synthetic peptides, and ODNs.

Limb Culture Method

Materials and Reagents

Limb Culture Plate. A limb culture (LC) plate coated with extracellular matrix is prepared (a mixture of laminin, 25 μg/ml; fibronectin, 50 μg/ml; vitronectin, 2 μg/ml; type I and type IV collagen, 3 mg/ml; heparan sulfate proteoglycans, 5 μg/ml). In brief, 1.0 ml of the extracellular matrix solution is poured into each well of a six-well plate and incubated overnight at 4°. After removal of the mixture, the wells are washed twice with phosphate-buffered saline [PBS(–)] and kept below −30°. Instead of this preparation, a Biocoat Matrigel plate (six-well plate; Becton Dickinson, San Jose, CA), or a growth factor-reduced Matrigel plate is also recommended. [Make certain that the reduced growth factor preparation still contains such things as basic fibroblast growth factor (bFGF) at 0–0.1 pg/ml.] Plates may also be prepared by adding 1 ml of Matrigel solution to each well of a six-well culture plate, which is then stored overnight at 4°. After removal of the

[6] T. Ochiya, H. Sakamoto, M. Tsukamoto, T. Sugimura, and M. Terada, *J. Cell. Biol.* **130,** 997 (1995).

Matrigel solution, the plate is washed twice with PBS(−), air dried, and stored below −30° until use.

Limb Differentiation Medium. Limb differentiation medium (LDM) contains Dulbecco's modified Eagle's medium (DMEM) with high glucose (GIBCO-BRL, Gaithersburg, MD); 2-mercaptoethanol, $10^{-4} M$; monothioglycerol, 200 μM; selenium, $10^{-7} M$; linoleic acid, 5 μg/ml; hydrocortisone, 4 μg/ml; and transferrin, 5 μg/ml. Methylcellulose can also be added at a final concentration of 1% to provide a firmer bed for the limb sections. All chemicals are purchased from Sigma (St. Louis, MO).

To determine the skeletal development of the limb culture, the samples are fixed with 10% formalin, stained with 0.1% alcian blue in 70% methanol at 37° for 72 hr, dehydrated in ethanol, and washed with methyl salicylate.[5]

Procedures for Limb Culture

1. Prior to beginning the procedure: (a) Place either an LC plate or Matrigel plate in an incubator and maintain for 30 min at 37°; (b) sterilize surgical tools; (c) prepare LDM and store it initially at 4°. Then warm the LDM to room temperature and then briefly to 37° in a CO_2 incubator prior to beginning the experiment; and (d) fill three bacterial petri dishes with PBS(−) containing antibiotics, and four dishes with serum-free DMEM containing antibiotics.

2. Sacrifice one ether-anesthetized pregnant mouse at a time (follow the same procedure for all of the embryos) by cervical dislocation, ethanol sterilize the abdomen, and quickly remove and place the embryonic sac in PBS plates while washing out any extraneous blood or tissue. Transfer the sac to the first DMEM plate, open the sac, and gently transfer the embryos to the second DMEM plate; carefully remove the extraembryonic membrane, and place each embryo in the third plate. Embryos are developmentally staged by the morphology of the hind limb bud, and day 9.5–12 p.c. embryos are used for this experiment.

3. Place the DMEM plate with a single embryo laying on its side (if a heart continues to beat for 1 week on the culture plate, limb development has been successfully completed) under a dissecting microscope and excise the head portion first, and then the tail segment. The microsurgery should be performed with a disposable scalpel (No. 21; Feather, Japan) and a single and decisive stroke should be used to make the excision rather than a swing motion (a sharp scalpel must be used). The cuts should be made just above the heart (upper) and through the lower third of the liver (lower). The planes of the upper and lower cuts should converge toward the abdomen of the embryo, and diverge toward the back. This will result in a trunk section with a cranial surface slightly depressed toward the abdomen and a flat caudal surface. The two surfaces can be distinguished visually without

the need of a microscope either by the liver mass conferring a more reddish color than the caudal portion, or by the orientation of the forelimbs, which at this stage (days 9–12 p.c.) should point toward the tail and away from the head. Next, separate the trunk segment from the rest of the embryo and place in the fourth DMEM plate.

4. Place the plates under a hood and transfer each embryo (two sections each) onto a separate plate and leave them in the hood for 10 min at room temperature. (It usually takes an approximate resting phase of 5 to 10 min before each trunk embryo can be placed on a culture plate. During this resting phase, the trunk embryos should be kept in serum-free DMEM at room temperature. This appears to allow the cut surfaces to become adhesive and helps the trunk segment to attach to the extracellular matrix (ECM). Do not subject the embryo to temperatures below 4° at any stage of this procedure.) Take each embryo and place it so that the cranial surface is facing up (forelimb trunk) or down (hindlimb trunk) in a prewarmed, dry plate and leave the plate inside the incubator (100% humidity) at 37° for 10 min (Fig. 1a). (This step is also needed to allow the trunk embryos to attach to the dishes.)

5. Finally, slowly add prewarmed (37°) LDM so that it just covers the sections and then leave the plate in a CO_2 incubator (5% CO_2–95% air, 100% humidity) and do not move them for at least 48 hr. (Add the prewarmed limb culture medium to the plate. Do not cover the explant completely with the medium.) Once this has been done, the sections can be viewed under the microscope and the medium renewed by replacing half of the volume from each well with fresh medium. (Replace the culture medium every 3–4 days, as needed. Frequent changes of the culture medium may cause detachment of the explant and subsequent loss of embryo development.)

6. If supplementing the LDM with serum for long-term culture of the limbs becomes necessary, serum should be added to a final concentration of 20% [10% fetal bovine serum (FBS) and 10% horse serum]. (When using serum in the culture, the quality of horse serum is important. Be sure to use a sufficient quantity that allows for good growth of the PC12 cells. Use an FBS that is suited for embryonic stem cell culture.) Likewise, methylcellulose can also be added at a final concentration of 1% to provide a firmer bed for the limb sections.

Comments. One of the critical problems in studying the signals that direct limb outgrowth is the lack of a suitable *in vitro* system in all experimental models excluding chick limb outgrowth *in ovo*. Here we have established a novel *in vitro* culture system in which a mouse limb bud is observed at 9.5–10 days p.c. and then placed on a sheet of extracellular matrix in a

(a)

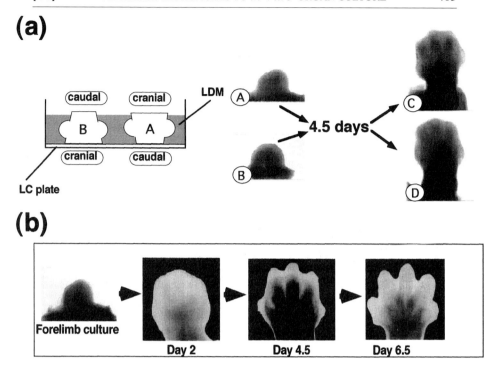

(b)

Fig. 1. Schematic representation of a novel culture system for limb development. (a) The body of a 9.5- to 10-day p.c. mouse embryo was cut up and placed on an LC plate at the indicated orientation shown at left (A, forelimb; B, hindlimb). The sections were cultured in serum-free LDM at 37°, 5% CO_2 for 4.5 days. The morphological pattern of cultured forelimb (C) and hindlimb (D) (right) resembled that of limbs from a 12.5-day p.c. embryo that developed *in vivo*. (b) Video recording image of the forelimb development based on limb culture system supplemented with 20% serum and 1.0% methylcellulose.

defined medium and allowed to differentiate into a limb at 12.5 days p.c., i.e., within 4.5 days (Fig. 1a). Alcian blue staining showed that normal skeletal development also occurred in culture. In the presence of serum, limb formation was much enhanced, and allows for observation of digit formation (Fig. 1b). Thus, this culturing system allows for normal limb development, as can be observed *in vivo*.

This culture system may contribute to a better understanding of the molecular basis of limb development and shed light on such processes as tissue-specific differentiation, control of cell growth, induction of apoptosis, and cell-to-cell communication.

Principle for Antisense Administration to *in Vitro* Organ Culture

In the next step, attempts were made to establish a procedure for targeted administration of AS-ODNs in our limb culture system. For this purpose, methods for the transduction of an effective dose of ODNs to the target site are essential. While attempting to enhance the cellular uptake and targeting of ODNs, we observed that some cationic lipids greatly enhanced the biological activity of AS-ODNs. It is also generally acknowledged that antisense experiments are difficult to interpret, and that numerous and rigorous controls are required before the conclusion can be drawn that inhibition of the expression of the target gene is indeed responsible for the observed antiproliferative effect. One of the most common problems associated with antisense experiments lies in the possible toxic nature of a given amount of ODN, and thus controls that ensure that nonspecific toxicity is not the cause of the observed inhibition are essential. To overcome these problems, we recommend the following steps be taken when using our experimental design.

1. The modification of ODNs should be done as phosphorothioate.

2. Purification of the ODNs should be done by HPLC column chromatography, followed by ethanol precipitation twice, prior to use.

3. When the cells are difficult to grow under serumless conditions, establish a minimum concentration of serum. The serum should be heat inactivated.

4. In the ODN–cell culture, change the medium containing ODNs every day. This ensures a continuous fresh supply of ODNs into the cells.

5. Use of cationic liposome is recommended to enhance the ODN incorporation into the cells. It is also well known that cationic lipid may protect ODNs from nuclease activity[7] and enhance cellular uptake of ODNs.[8]

6. Establishing an optimum ratio between ODNs and liposome on individual cell lines is needed. In our case involving use of LipofectAMINE (GIBCO-BRL), the concentration of liposome varied from 0, 2, 4, 6, 8, to 10 μg/ml.

7. A rescue of the cell growth inhibition through the addition of back-targeted gene products to the culture should be attempted. In addition, a "competition" experiment in which the effective antisense ODN is premixed with sense ODNs can be useful. Be that as it may, rigid controls that demonstrate the absence of nonspecific toxicity must be performed.

---

[7] A. M. Tari, M. Andreeff, H. D. Kleine, and G. Lopez-Berestein, *J. Mol. Med.* **74,** 623 (1996).
[8] C. F. Bennett, M.-Y. Chiang, H, Chan, J. E. E. Shoemaker, and C. K. Mirabelli, *Mol. Pharmacol.* **41,** 1023 (1992).

8. Another important point in this type of experiment concerns the concentration-dependent effect of antisense ODNs. ODNs should be effective at a concentration less than 10 μM. A linear effect can often be observed between 0.1 and 10 μM.

9. It may also be necessary to demonstrate the reversibility of the inhibitory effect on cell growth when antisense ODNs are removed from the culture medium or when there is no continuous supply of fresh ODNs.

Careful Choice of Target Sequence and Liposomes

Target Sequence. Many investigators have chosen to target the translation initiation site of mRNA. In the present experiment, phosphorothioate ODNs were synthesized with 18-mers against the ATG translation start site (5′-dCGGCCCGCGTTTCGCCAT-3′) and donor (5′-dAGAGCTC-CAGAAGACCTG-3′) and acceptor (5′-dCCCGCTGGCACTCACCAC-3′) sites for exon 2 of mouse *Hst-1/Fgf-4.* As a control, sense and scrambled ODNs were used. Choice of target sequences was examined to determine whether treating cultured NIH 3T3 cells transformed via the *Hst-1/Fgf-4* gene with antisense ODNs would cause them to revert to a normal morphology. Among the three tested synthetic types of phosphorothioate antisense ODNs, the most striking effects were obtained with antisense ODNs against the ATG site (20 to 30% inhibition in repeated experiments with ODNs alone); no significant effect was observed with the other two antisense ODNs. In contrast, the same sense and scrambled ODNs against the ATG site used as control revealed no effect. A sense sequence of 5′-dGCCGGGCGCAAAGCGGTA-3′, and a scrambled sequence of 5′-dAGCTGCTGCTGCCCC-3′, were used as a negative control. In the presence of cationic liposomes, the effect of *Hst-1/Fgf-4* AS-ODNs on transformed NIH 3T3 cells was three fold higher than that of ODNs alone.

Cationic Liposome. Several studies have shown that the potency of antisense ODNs can be greatly enhanced through the use of cationic liposomes that deliver the ODNs into the target site. To elucidate what sort of liposomes are best for ODN transfer to the AER of limbs, the efficacy of several cationic liposomes is evaluated by limb AER cell culture. Limb buds are removed from the embryo on day 12 p.c., and the AER is further dissected under a microscope. The tissues from the AER are trypsinized and cultured at a high cell density with LDM supplemented with 10% FBS, penicillin and (100 IU/ml), and streptomycin (100 μg/ml) (GIBCO-BRL) on gelatin-coated plates (Corning, Acton, MA). Three days after the cultivation, the cells are washed three times with PBS(−) followed by the addition of serum-free LDM. Uptake of fluorescein-labeled phosphorothioate ODNs by cells from the AER, using several kinds of liposomes, including

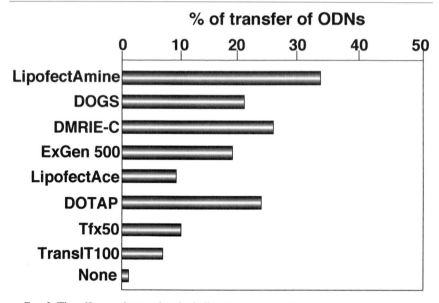

FIG. 2. The efficacy of several cationic liposomes was evaluated through the use of limb AER cell culture. Uptake of fluorescein-labeled phosphorothioate ODNs on cells from the AER, using several kinds of liposomes, was estimated. The results are represented as the percentage of fluorescein-positive cells among 200 cells counted, averaged over triplicate experiments.

LipofectAMINE (GIBCO-BRL), DOGS (Bios Sepra, Marlborough, MA), DMRIE-C (GIBCO-BRL), ExGen 500 (Euromedex), LipofectACE (GIBCO-BRL), DOTAP (Boehringer Mannheim, Indianapolis, IN), Tfx50 (Promega, Madison, WI), and Trans IT 100 (PanVera, Madison, WI), is estimated. Liposome–ODN complexes at a molar ratio of 5:1 are formed by incubating this mixture at room temperature for 10 to 15 min. Among the tested liposomes, lipofectAMINE reagent gives the highest yield for limb AER cells (Fig. 2). In general, in order to avoid biasing the efficacy of the ODNs transfer by the choice of liposome, it is necessary to establish the most suitable liposomes from experiment to experiment or from cell to cell.

Targeted Exposure of Oligodeoxynucleotides

For the limb organ culture, the same ODN–liposome mixture (5 μl/site) is applied directly to the surface of the limb bud (AER region) under a stereoscopic microscope [Nikon (Garden City, NY) SMZ-U] and incubated for 20 min in an incubator without culture medium. After several

A   B

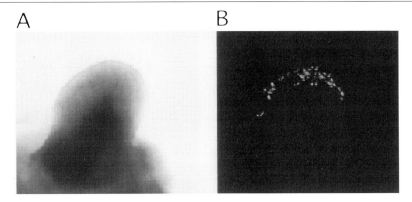

Fig. 3. Cationic lipids enhanced ODN transfer into the AER of an embryonic limb. Fluorescein-labeled ODNs were used to assess the ODN transfer efficiency into the AER region by LipofectAMINE cationic liposome *in vitro*. Limbs were observed by phase contrast (A) and fluorescence (B).

washings with phosphate-buffered saline to remove the excess ODN–liposome mixture, the limb differentiation medium is added to the explants and then cultured as previously described. By our procedure, it may be confirmed that fluorescein-labeled ODNs tend to accumulate mainly in the targeted AER region (Fig. 3). For the rescue experiment, antisense ODN-bearing limbs are cultured in the LDM supplemented with recombinant FGF-4 proteins (R&D Systems, Minneapolis, IN) at a final concentration of 800 ng/ml.

By using this *in vitro* organ culture system, we have treated 9.5–10 day p.c. embryos with antisense ODNs complementary to mouse *Hst-1* for 4.5 days. This treatment resulted in a blockage of forelimb and hindlimb outgrowth (Fig. 4b). In contrast, sense ODNs showed no inhibitory effects. Measurements of the proximal–distal (*P–D*) length of antisense ODN-treated limbs showed that the growth stage of forelimbs and hindlimbs at 9.5–10 days p.c. was slightly advanced, most closely resembling 10.5-day p.c. embryonal limbs prior to being stopped by *Hst-1/Fgf-4* antisense ODNs. In addition, we tested the amount of transcripts from a gene closely related to *Hst-1/Fgf-4* such as *Fgf-2* and *Fgf-8,* which are also expressed in the limbs of early embryonic stage. As shown in Fig. 4a, no apparent inhibitory effect on *Fgf-2* and *Fgf-8* transcripts was detected, indicating that our AS-ODN experiments have been done in a sequence-specific manner.

Comments. Our results provide direct evidence that *Hst-1/Fgf-4* plays a major role as a positive signal for mesenchyme proliferation of the limb, and is involved in proximodistal specification. Our ODN transfer method is different from widely used methods that simply add ODNs into the

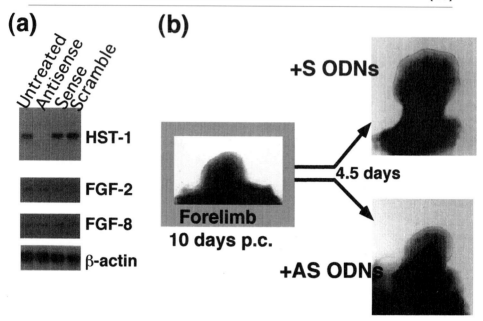

Fig. 4. *Hst-1/Fgf-4* AS-ODNs block limb development *in vitro*. (a) Expression of FGF mRNA of the limbs after exposure to AS-ODNs. ODN–liposome complexes were administered to the limb bud as described at a final concentration of 500 n*M* for 2.5 days in culture. At the end of the incubation period, 20 pairs of limb buds were pooled and then total mRNAs were extracted. About 25 μg of each RNA sample was subjected to RNA blot analyses using full-size mouse *Hst-1/Fgf-4* cDNA, *Fgf-2* cDNA, and *Fgf-8* cDNA as a probe. Treatment with ODNs did not affect the expression of β-actin transcripts. (b) Antisense (AS) or sense (S) ODN-treated limbs were cultured for 4.5 days *in vitro*. Although data are not shown, control scrambled ODNs result in no inhibitory effects on limb outgrowth.

culture medium in that we administered ODNs in the form of cationic liposome complexes. These complexes are well absorbed into the cultured explants without any great diffusion when placed on the desired small area, and then are highly incorporated into the cells. Thus, it is possible to introduce ODNs in a site-specific manner. Furthermore, this liposome transfer method makes it possible to use nanomolar amounts of ODNs, at concentrations 10- to 100-fold less than is commonly used compared with ODNs not complexed with liposome. In essence, we can see the effects of antisense ODNs with less cytotoxicity.

Acknowledgments

We could like to thank Dr. Francesco Ramirez from the Mount Sinai School of Medicine for valuable advice regarding the limb culture procedure. This work was supported in part by a Grant-in-Aid for a Comprehensive 10-Year Strategy for Cancer Control from the Ministry of Health and Welfare; by Grants-in Aid for Cancer Research from the Ministry of Health and Welfare; by the Senri Life Science Foundation; and by the Ministry of Education, Science, and Culture.

[29] In Vivo and in Vitro Antiproliferative Effects of Antisense Interleukin 10 Oligonucleotides

By GEORGE A. PARKER, BIHAI PENG, MAI HE, SUSAN GOULD-FOGERITE, CHUAN-CHU CHOU, and ELIZABETH S. RAVECHÉ

Introduction

Overview

Our laboratory has had a long-term interest in the autoimmune and proliferative disease processes of New Zealand Black (NZB) and related strains of mice, particularly the genetic basis of those processes.[1-6] During the course of those studies we isolated an immortal cell line that has the phenotype of a malignant CD5+ B-1 cell,[7] and found that interleukin 10 (IL-10) is a required growth factor for those malignant cells.[8] The biologic features of the murine B-1 cell are similar to the proliferative B-1 cells in human chronic lymphocytic leukemia (CLL), particularly the subset of CLL known as Richter's syndrome.[7,9] The biologic similarities between human CLL and the B-1 cell malignancy in NZB mice, and their shared dependence on IL-10 as a growth factor, suggested the NZB mouse neoplasm may be a model for human CLL and that antisense IL-10 may have some clinical utility in the treatment of CLL. Salient features of the major components of our studies are summarized below.

[1] E. Raveche, A. Steinberg, L. Klassen, and J. Tjio, *J. Exp. Med.* **147,** 1487 (1978).

[2] E. Raveche, J. Tjio, and A. Steinberg, *Cytogenet. Cell Genet.* **23,** 182 (1979).

[3] E. Raveche, O. Alabaster, J. Taurog, J. Tjio, and A. Steinberg, *J. Immunol.* **126,** 154 (1981).

[4] E. Raveche, E. Novotny, C. Hansen, J. Tjio, and A. Steinberg, *J. Exp. Med.* **153,** 1187 (1981).

[5] E. S. Raveche, P. Lalor, A. Stall, and J. Conroy, *J. Immunol.* **141,** 4133 (1988).

[6] M. Seldin, J. Conroy, A. Steinberg, L. D'Hoosteleare, and E. Raveche, *J. Exp. Med.* **166,** 1585 (1987).

[7] B. Peng, D. H. Sherr, F. Mahboudi, J. Hardin, Y. Wu, L. Sharer, and E. S. Raveche, *J. Immunol.* **153,** 1869 (1994).

[8] S. Ramachandra, R. A. Metcalf, T. Fredrickson, G. E. Marti, and E. Raveche, *J. Clin. Invest.* **98,** 1788 (1996).

[9] J. A. Phillips, K. Mehta, C. Fernandez, and E. S. Raveche, *Cancer Res.* **52,** 437 (1992).

CD5+ B-1 Cells

CD5$^+$ B-1 cells are lymphocytes that express a 67-kDa pan-T lymphocyte surface glycoprotein, designated CD5, plus surface antigens restricted to the B lymphocyte lineage.[10-13] B-1 cells typically have a CD5$^+$IgM$^+$IgD^{lo} phenotype, as opposed to the CD5$^-$IgM$^+$IgD^{hi} phenotype of adult (B-2) B lymphocytes. Murine and human CD5$^+$ B cells express CD5 at approximately 20% of the level expressed by T cells.[12-16] Expression of CD5 is defined by the reactivity of cells with monoclonal antibody (MAb) Ly-1 in mice,[17] or with Leu-1 or its equivalent in humans.[18] B-1 cells lack expression of other T cell-associated differentiation antigens such as CD4, CD8, and CD3, and, in mice, Thy-1.[19] In addition to expression of surface immunoglobulin (Ig), B-1 cells express other B cell surface antigens such as CD19, CD20, and CD21 at levels comparable to CD5$^-$ B cells.[10,13,16,20] Murine B-1 cells express other B cell surface antigens such as ThB,[21] the B cell isomer of Ly-5,[22] and a receptor for Ig Fc$_\varepsilon$.[23]

The CD5 antigen originally was thought to be a surface marker for the helper–inducer T lymphocyte subpopulation,[24] but was subsequently found to be expressed on all T lymphocytes.[25] CD5 antigen on certain B cell tumors of mice[26,27] and humans[28-31] was first thought to represent "lineage

[10] K. Hayakawa, R. Hardy, D. Parks, and L. Herzenberg, *J. Exp. Med.* **157,** 202 (1983).

[11] L. Herzenberg, A. Stall, P. Lalor, C. Sidma, W. Moore, D. Parks, and L. Herzenberg, *Immunol. Rev.* **93,** 81 (1986).

[12] R. Hardy and K. Hawakawa, *Immunol. Rev.* **93,** 53 (1986).

[13] N. Gadol and K. Ault, *Immunol. Rev.* **93,** 23 (1986).

[14] V. Manohar, E. Brown, W. Leiserson, and T. Chused, *J. Immunol.* **129,** 532 (1982).

[15] K. Hayakawa, R. Hardy, and L. Herzenberg, *J. Exp. Med.* **161,** 1554 (1985).

[16] T. Kipps and J. Vaughan, *J. Immunol.* **139,** 1060 (1987).

[17] J. Ledbetter and L. Herzenberg, *Immunol. Rev.* **47,** 63 (1979).

[18] P. Martin, J. Hansen, A. Siadak, and R. Nowinski, *J. Immunol.* **127,** 1920 (1981).

[19] T. J. Kipps, *Adv. Immunol.* **47,** 117 (1989).

[20] R. Hardy, K. Hayakawa, L. Herzenberg, H. Morse, W. Davidson, and L. Herzenberg, *Curr. Top. Microbiol. Immunol.* **113,** 231 (1984).

[21] L. Eckhardt and L. Herzenberg, *Immunogenetics* **11,** 275 (1980).

[22] R. Coffman, *Immunol. Rev.* **69,** 5 (1983).

[23] C. Dexter and R. Corley, *Eur. J. Immunol.* **17,** 867 (1987).

[24] H. Cantor and E. Boyse, *J. Exp. Med.* **141,** 1376 (1975).

[25] J. Ledbetter, R. Rouse, H. Micklem, and L. Herzenberg, *J. Exp. Med.* **152,** 280 (1980).

[26] L. Lanier, N. Warner, J. Ledbetter, and L. Herzenberg, *J. Immunol.* **127,** 1691 (1981).

[27] L. Lanier, E. Richie, A. Howell, and J. Allison, *Immunogenetics* **17,** 655 (1983).

[28] N. Gough, D. Kemp, B. Tyler, J. Adams, and S. Cory, *Proc. Natl. Acad. Sci. U.S.A.* **77,** 554 (1980).

[29] P. Martin, J. Hansen, R. Nowinski, and M. Brown, *Immunogenetics* **11,** 429 (1980).

[30] I. Royston, J. Majda, S. Baird, B. Meserve, and J. Griffiths, *J. Immunol.* **125,** 725 (1980).

[31] C. Wang, R. Good, P. Ammirati, G. Dymbort, and R. Evans, *J. Exp. Med.* **151,** 1539 (1980).

infidelity," but soon thereafter it was learned that CD5 may be expressed by normal B lymphocytes.[10,14,32]

Even when the best of immunofluorescent reagents and techniques are used, it is difficult to separate CD5$^+$ and CD5$^-$ cells into distinct populations via flow cytometry.[19] As a result, CD5$^+$ cell populations within a mixed cell population often are identified by subtracting an isotype control from the CD5$^+$ cell population.[19] CD5$^+$ B-1 cells tend to be larger than CD5$^-$ B cells, and thus exhibit greater forward- and side-angle light scatter on flow cytometric analysis.[19] Murine CD5$^+$ B-1 cells express higher levels of surface IgM and lower levels of surface IgD than other B lymphocytes.[33,34] Human and mouse CD5$^+$ B-1 cells commonly express CD11b, the receptor for C3bi that is a marker for cells of the monomyelocytic lineage.[16,35–37] CD5$^+$ B-1 cells survive after prolonged culture of mixed cell populations,[11,12] a feature that facilitates their isolation but complicates interpretation of experiments in which survival of cell populations is an end point.

Murine CD5$^+$ B-1 cells are essentially nonexistent in lymph nodes, blood, or bone marrow, and are rare in the spleen, but constitute the major lymphoid cell population of the peritoneal cavity.[10,38] CD5$^+$ B-1 cells may[39] or may not[10] be found in the thymus of mice.

Anatomic distribution of CD5$^+$ B-1 cells appears to be less defined in humans than in mice. CD5$^+$ B-1 cells may constitute 30% of the B cells in inflamed tonsils or lymph nodes,[19] but constitute less than 5% of the splenic B cells of human adults.[40] Immunohistochemical studies have shown CD5$^+$ B-1 cells to be distributed in a zone around the periphery of germinal centers in lymph nodes.[32,41] CD5$^+$ B-1 cells comprise 1–30% of circulating B cells.[12,13,16,42–47]

[32] F. Caligaris-Cappio, M. Gobbi, M. Bofill, and G. Janossy, *J. Exp. Med.* **155,** 623 (1982).

[33] K. Hayakawa, R. Hardy, M. Honda, L. Herzenberg, A. Steinberg, and L. Herzenberg, *Proc. Natl. Acad. Sci. U.S.A.* **81,** 2494 (1984).

[34] R. Hardy, K. Hayakawa, D. Parks, and L. Herzenberg, *Nature (London)* **306,** 270 (1983).

[35] D. Beller, T. Springer, and R. Schreiber, *J. Exp. Med.* **156,** 1000 (1982).

[36] N. Hogg and M. Horton, *in* "Leukocyte Typing III" (A. McMichael, P. Beverley, S. Cobbold, *et al,* eds.), p. 576. Oxford University Press, London, 1987.

[37] L. Herzenberg, A. Stall, J. Braun, D. Weaver, D. Baltimore, L. Herzenberg, and R. Grosschedl, *Nature (London)* **329,** 71 (1987).

[38] K. Hayakawa, R. Hardy, and L. Herzenberg, *Eur. J. Immunol.* **16,** 450 (1986).

[39] M. Miyama-Inaba, S. Kuma, K. Inaba, H. Ogata, H. Iwai, R. Yasumizu, S. Muramatsu, R. Steinman, and S. Ikehara, *J. Exp. Med.* **168,** 811 (1988).

[40] A. Freedman, A. Boyd, F. Bieber, J. Daley, K. Rosen, J. Horowitz, D. Levy, and L. Nadler, *Blood* **70,** 418 (1987).

[41] M. Gobbi, F. Caligaris-Cappio, and G. Janossy, *Br. J. Haematol.* **54,** 393 (1983).

[42] P. Casali, S. Burastero, M. Nakamura, G. Inghirami, and A. Notkins, *Science* **236,** 77 (1987).

[43] M. Dauphinee, Z. Tovar, and N. Talal, *Arthritis Rheum.* **31,** 642 (1988).

CD5[+] B-1 cells appear early in fetal development, and a nearly adult absolute number of cells is reached at an early age.[38] Subsequent development of other B lymphocyte populations results in a relative reduction in the CD5[+] B-1 cell population in the spleen of adult mice.[19] There is considerable evidence that CD5[+] B-1 cells represent a fetal B cell lineage.[48,49] In some strains of mice there is an age-associated increase in the absolute numbers and/or relative proportion of CD5[+] B-1 cells, with the propensity to develop CD5[+] B-1 cells being a relatively constant strain characteristic. In normal BALB/c mice the number of peritoneal CD5[+] B-1 cells increases from approximately 3×10^6 in 6- to 8-week-old mice to greater than 3×10^7 in mice more than 1 year of age, apparently as a result of oligoclonal expansions of CD5[+] B-1 cells.[50] Similar but accelerated clonal expansions are seen in NZB or (NZB $\times$ NZW) F_1 mice.[3,6,50,51]

Murine CD5[+] B-1 Cell Neoplasia

A number of B cell malignancies in aged mice have been noted to express CD5.[20,52–56] NZB mice have been proposed as a model for human chronic lymphocytic leukemia (CLL) because of the age-related clonal expansion of malignant CD5[+] B-1 cells in that strain of mice.[9,57] The malignant CD5[+] B-1 cell clones are slow-growing, long-lived, hyperdiploid cells that are present primarily in the spleen and peritoneal cavity.[9] Malignant

[44] R. Hardy, K. Hayakawa, M. Shimizu, K. Yamasaki, and T. Kishimoto, *Science* **236,** 81 (1987).

[45] R. Maini, C. Plater-Zyberk, and E. Andrew, *Rheum. Dis. Clin. North Am.* **13,** 319 (1987).

[46] C. Plater-Zyberk, R. Maini, K. Lam, T. Kennedy, and G. Janossy, *Arthritis Rheum.* **28,** 971 (1985).

[47] O. Taniguchi, H. Miyajima, T. Hirano, M. Noguchi, A. Ueda, H. Hashimoto, S. Hirose, and K. Okumura, *J. Clin. Immunol.* **7,** 441 (1987).

[48] R. R. Hardy and K. Hayakawa, *Adv. Immunol.* **55,** 297 (1994).

[49] K. Hayakawa, Y.-S. Li, R. Wasserman, S. Sauder, S. Shinton, and R. R. Hardy, *Ann. N.Y. Acad. Sci.* **815,** 15 (1997).

[50] A. Stall, M. Farinas, D. Tarlinton, P. Lalor, L. Herzenberg, S. Strober, and L. Herzenberg, *Proc. Natl. Acad. Sci. U.S.A.* **85,** 7312 (1988).

[51] D. Wofsy and N. Chiang, *Eur. J. Immunol.* **17,** 809 (1987).

[52] S. Slavin and S. Strober, *Nature (London)* **272,** 624 (1978).

[53] L. Lanier, M. Lynes, G. Haughton, and P. Wettstein, *Nature (London)* **271,** 554 (1978).

[54] L. Lanier, L. Arnold, R. Raybourne, S. Russell, M. Lynes, N. Warner, and G. Haughton, *Immunogenetics* **16,** 367 (1982).

[55] C. Pennell, L. Arnold, P. Lutz, N. LoCascio, P. Willoughby, and G. Haughton, *Proc. Natl. Acad. Sci. U.S.A.* **82,** 3799 (1985).

[56] W. Davidson, T. Fredrickson, E. Rudikoff, R. Coffman, J. Hartley, and H. Morse, *J. Immunol.* **133,** 744 (1984).

[57] H. Okamoto, H. Nishimura, A. Shinozaki, D. Zhang, S. Hirose, and T. Shirai, *Jpn. J. Cancer Res.* **84,** 1273 (1993).

B-1 cells derived from NZB mice produce significantly higher levels of murine IL-10 (mIL-10) mRNA than normal B-1 or B cells.[8] In addition to CD5⁺ B-1 cellular proliferations, NZB and related strains of mice have high levels of autoantibodies that react with autologous erythrocytes, thymocytes, and single-stranded DNA (ssDNA),[58–63] and thus are considered a model for lupus erythematosus. There is substantial evidence that the autoantibodies are produced by the expanded CD5⁺ B-1 cell population.[10,33,38]

Human CD5⁺ B-1 Cell Neoplasia

Chronic lymphocytic leukemia in humans, the most common form of leukemia in adults in the western world, involves an increase in the number of circulating CD5⁺ B-1 lymphocytes.[29,30,31,64] CLL cells produce IL-10 and there is evidence that IL-10 has a regulatory role in CLL.[65] The level of IL-10 production is correlated with the clinical course of CLL.[66]

As opposed to forms of leukemia that involve an uncontrolled proliferation of leukocytes, there is evidence to suggest that CLL is due to a defect in apoptosis.[67] CLL cells from human patients and malignant CD5⁺ B-1 cells from NZB mice readily undergo apoptosis when subjected to appropriate stimuli,[65,68] and thus appear to have intact apoptotic signaling pathways and effector mechanisms. Evidence suggests the malignant CD5⁺ B-1 cells may have a developmental arrest at some point prior to the stage at which apoptosis normally occurs,[12,15,48,49] which may explain the failure of apoptosis in cells that have intact apoptosis machinery. Regardless of the precise mechanism, failure of cells to undergo "normal" apoptosis results in an increased population of circulating lymphocytes, and the leukemic cells lack the overt proliferative phenotype associated with many forms of leukemia. Absence of the proliferative phenotype renders the CLL cells

[58] B. Andrews, R. Eisenberg, A. Theofilopoulos, S. Isui, C. Wilson, P. McConahey, E. Murphy, J. Roths, and F. Dixon, *J. Exp. Med.* **148,** 1198 (1978).
[59] D. DeHeer, E. Linder, and T. Edgington, *J. Immunol.* **120,** 825 (1978).
[60] S. Izui, P. McConahey, and F. Dixon, *J. Immunol.* **121,** 2213 (1978).
[61] T. Shirai and R. Mellors, *Proc. Natl. Acad. Sci. U.S.A.* **68,** 1412 (1971).
[62] H. Smith and A. Steinberg, *Annu. Rev. Immunol.* **1,** 175 (1983).
[63] A. Theofilopoulos and F. Dixon, *Adv. Immunol.* **37,** 269 (1985).
[64] C. Rozman and E. Montserrat, *N. Engl. J. Med.* **333,** 1052 (1995).
[65] H. Fernandes, W. Barchuk, C. Chou, C. Fernandes, and E. Raveche, *Oncol. Rep.* **2,** 985 (1995).
[66] W. Knauf, B. Ehlers, S. Bisson, and E. Thiel, *Blood* **86,** 4382 (1995).
[67] J. C. Reed, *Semin. Oncol.* **25,** 11 (1998).
[68] M. Zhang and E. Raveche, *Oncol. Rep.* **5,** 23 (1998).

less vulnerable to traditional chemotherapeutic agents, the majority of which are directed at various steps in cellular DNA synthesis or mitosis.

In addition to the common form of CLL, which involves an increase in morphologically mature but biologically immature CD5[+] B-1 cells,[69] patients with CLL occasionally develop an overtly proliferative large cell lymphoma, termed Richter's syndrome, which is thought to represent either continued evolution and genetic alterations in CLL clones,[70-75] or development of a second, unrelated malignancy.[76-79] If Richter's syndrome represents a second malignancy, it can be argued that CLL contributes to its development by presently an increased population of CD5[+] B-1 cells that represents a "fertile field" for the second oncogenic event.

Interleukin 10

Murine IL-10 (mIL-10), formerly known as cytokine synthesis inhibitory factor (CSIF), is a cytokine that is produced by T lymphocytes, macrophages, monocytes, mast cells, and B cells.[8,80,81] Murine IL-10 is produced by the CD4[+] helper T cell type 2 (Th2) subset, and has a downregulatory effect on cytokines produced by the Th1 subset.[82,83] Murine IL-10 has pleiotropic effects on many cell types, including thymocytes,[80] cytotoxic T

[69] K. A. Foon, K. R. Rai, and R. P. Gale, *Ann. Intern. Med.* **113,** 525 (1990).

[70] J. Brouet, J. Preud'Homme, G. Flandrin, N. Chelloul, and M. Seligmann, *J. Natl. Cancer Inst.* **56,** 631 (1976).

[71] G. Delsol, G. Laurent, E. Kuhlein, J. Familiades, F. Rigal, and J. Pris, *Am. J. Clin. Pathol.* **76,** 308 (1981).

[72] W. Chan and R. Dekmezian, *Cancer* **57,** 1971 (1986).

[73] L. Bertoli, H. Kubagawa, G. Borzillo, M. Mayumi, J. Prchal, J. Kearney, J. Durant, and M. Cooper, *Blood* **70,** 45 (1987).

[74] K. Miyamura, H. Osada, T. Yamauchi, M. Itoh, Y. Kodera, T. Suchi, T. Takahashi, and R. Ueda, *Cancer* **66,** 140 (1990).

[75] R. Schots, M.-F. Dehou, K. Jochmans, C. Heirman, M. de Waele, B. van Camp, and K. Thielemans, *Am. J. Clin. Pathol.* **95,** 571 (1991).

[76] T. Splinter, A. Bom-van Noorloos, and P. van Heerde, *Scand. J. Haematol.* **20,** 29 (1978).

[77] J. van Dongen, H. Hooijkaas, J. Michiels, G. Grosveld, A. de Klein, T. van der Kwast, M. Prins, J. Abels, and A. Hagemeijer, *Blood* **64,** 571 (1984).

[78] J. McDonnell, W. Beschorner, S. Staal, J. Spivak, and R. Mann, *Cancer* **58,** 2031 (1986).

[79] L. Trumper, D. Matthaie-Maurer, W. Knauf, and P. Moller, *Klin. Wochenschr.* **66,** 736 (1988).

[80] I. MacNeil, T. Suda, K. Moore, T. Mosmann, and A. Zlotnik, *J. Immunol.* **145,** 4167 (1990).

[81] K. Moore, A. O'Garra, R. de Waal Malefyt, P. Vieira, and T. Mosmann, *Annu. Rev. Immunol.* **11,** 165 (1993).

[82] D. Fiorentino, M. Bond, and T. Mosmann, *J. Exp. Med.* **170,** 2081 (1989).

[83] K. Moore, P. Vieira, D. Fiorentino, M. Trounstine, T. Khan, and T. Mosmann, *Science* **248,** 1230 (1990).

cells,[84] mast cells,[85] B cells,[86] and macrophages.[87] Interferon γ (IFN-γ) has a reciprocal inhibitory effect on IL-10 production by monocytes.[88]

The inhibitory effects of human IL-10 (hIL-10) on immunologic and inflammatory process are generally similar to those of mIL-10.[81,89–91] Human IL-10 inhibits, at the transcriptional level, the production of IFN-γ and granulocyte-macrophage colony-stimulating factor (GM-CSF) by peripheral blood mononuclear cells (PBMCs) activated by phytohemagglutinin (PHA) or anti-CD3 MAb.[92] Human IL-10 inhibits antigen-specific, macrophage-related T lymphocyte responses by downregulation of class II MHC molecules on macrophages.[93] Human IL-10 inhibits the production of the proinflammatory cytokines IL-1α, IL-1β, IL-6, IL-8, and tumor necrosis factor α (TNF-α), and the hematopoietic growth factors GM-CSF and granulocyte colony-stimulating factor (G-CSF) by monocytes stimulated with lipopolysaccharide (LPS), IFN-γ, or LPS plus IFN-γ.[94] In addition, hIL-10 has a downregulatory effect on its own production by stimulated monocytes.[94]

Human and murine IL-10 have extensive sequence homology to an open reading frame in the Epstein–Barr virus (EBV) genome, BCRF-1.[83,92] Expression of the open reading frame results in a protein, termed viral IL-10 (vIL-10), that shares many of the inhibitory effects of mIL-10 and hIL-10.[95] B cell transformation and proliferation in EBV-related Burkitts's lymphoma are related to the viral-associated elevation in IL-10 production. EBV-transformed B lymphocytes constitutively secrete IL-10 and there is

[84] W. Chen and A. Zlotnik, *J. Immunol.* **147,** 528 (1991).

[85] L. Thompson-Snipes, V. Dhar, M. Bond, T. Mosmann, K. Moore, and D. Rennick, *J. Exp. Med.* **173,** 507 (1991).

[86] N. Go, B. Castle, R. Barrett, R. Kastelein, W. Dang, T. Mosmann, K. Moore, and M. Howard, *J. Exp. Med.* **172,** 1625 (1990).

[87] D. Fiorentino, A. Zlotnik, T. Mosmann, M. Howard, and A. O'Garra, *J. Immunol.* **147,** 3815 (1991).

[88] P. Chomarat, M.-C. Rissoan, J. Banchereau, and P. Miossec, *J. Exp. Med.* **177,** 523 (1993).

[89] H. Spits and R. De Waal Malefyt, *Arch. Allergy Immunol.* **99,** 8 (1992).

[90] F. Rousset, E. Garcia, T. Defrance, C. Peronne, N. Vezzio, D. Hsu, R. Kastelein, K. Moore, and J. Banchereau, *Proc. Natl. Acad. Sci. U.S.A.* **89,** 1890 (1992).

[91] M. Bejarano, R. De Waal Malefyt, J. Abrams, M. Bigler, R. Bacchetta, J. De Vries, and M. Roncarolo, *Int. J. Immunol.* **4,** 1389 (1992).

[92] P. Vieira, R. de Waal Malefyt, M. Dang, K. Johnson, R. Kastelein, D. Fiorentino, J. de Vries, M. Roncarolo, T. Mosmann, and K. Moore, *Proc. Natl. Acad. Sci. U.S.A.* **88,** 1172 (1991).

[93] R. de Waal Malefyt, J. Haanen, H. Spits, A. Roncarolo, A. Velde, C. Figdor, K. Johnson, R. Kastelein, H. Yssel, and J. De Vries, *J. Exp. Med.* **174,** 915 (1991).

[94] R. I. de Waal Malefyt, J. Abrams, B. Bennet, C. G. Figdor, and J. E. de Vries, *J. Exp. Med.* **174,** 1209 (1991).

[95] D. Hsu, R. de Waal Malefyt, D. Fiorentino, M. Dang, P. Vieira, J. de Vries, H. Spits, T. Mosmann, and K. Moore, *Science* **250,** 830 (1990).

a relationship between IL-10 production by human malignant B cell lines and EBV expression.[96,97] Addition of vIL-10 antisense mRNA prevented EBV-induced B cell transformation *in vitro*.[98] There is evidence that IL-10 has a role in B cell transformation seen in human immunodeficiency virus (HIV)-associated B cell lymphomas,[97] and IL-10 antisense mRNA inhibits the growth of cultured non-Hodgkin's lymphoma cells from acquired immunodeficiency syndrome (AIDS) patients.[99]

Cochleates

Liposomes are self-assembling particles in which a lipid bilayer encapsulates a small amount of the surrounding aqueous medium. This propensity of phospholipids to form closed vesicles in aqueous media has been known for decades, and has been used for delivery of medically significant molecules as well as for the study of membrane physics and chemistry. The pharmacologic utility of conventional liposomes (CLs) is somewhat limited by their rapid uptake by phagocytic cells, a property that has been exploited to some degree in the treatment of phagocyte-based diseases such as leishmaniasis. Liposomes were initially regarded as optimal drug delivery systems, but early investigations yielded disappointing results. There is a resurgence of interest in liposomes as drug delivery systems, owing in part to advanced liposome technology that allows formulation of liposomes for specific purposes and targets. Nonreactive, sterically stabilized liposomes (SLs) and polymorphic liposomes are subclasses of a new generation of liposomes. Unlike liposomes, cochleates are anhydrous, multilayered crystals that resemble the morphology of the spirals of a cochlea.

Cochleates are essentially dehydrated fused liposomes that result from the action of calcium or other divalent cations on lipids such as phosphatidylserine. The resultant structure can be considered a stabilized intermediate of a calcium-mediated fusion event. Liposomes can condense DNA and increase transfection yields *in vitro* by several orders of magnitude.[100] Cochleates are considered to be particularly promising as a drug delivery system for gene therapy.[101] In the DNA–cochleate complex, the antisense oligonucleo-

[96] N. Burdin, C. Peronne, J. Banchereau, and F. Rousset, *J. Exp. Med.* **177,** 295 (1993).

[97] D. Benjamin, T. Knobloch, and M. Dayton, *Blood* **80,** 1289 (1992).

[98] I. Miyazaki, R. Cheung, and H. Dosch, *J. Exp. Med.* **178,** 439 (1993).

[99] R. Masood, Y. Zhang, M. W. Bond, D. T. Scadden, T. Moudgil, R. E. Law, M. H. Kaplan, B. Jung, B. M. Espina, Y. Lunardi-Iskandar, A. M. Levine, and P. S. Gill, *Blood* **85,** 3423 (1995).

[100] N. Zhu, D. Liggitt, Y. Liu, and R. Debs, *Science* **261,** 209 (1993).

[101] D. Lasic and D. Papahadjopoulos, *Science* **267,** 1275 (1995).

tides are not encapsulated but are simply complexed with the small vesicles by electrostatic interactions.[101] The exact mechanism by which cochleates deliver drugs or DNA to cells is not known, but is suspected to involve repeated fusion of the exterior layers of the cochleate with the plasma membrane, with subsequent intracellular release of a portion of the material entrapped within the cochleate.[102]

Materials and Methods

Malignant Murine CD5+ B-1 Cells

The cultured malignant B-1 cell line used in these experiments represents an aggressive clone that developed in a B-1 cell line previously maintained by *in vivo* passage.[7] The cell line is maintained in Iscove's modified Dulbecco's medium (IMDM) (GIBCO, Grand Island, NY) supplemented with 10% heat-inactivated fetal bovine serum (HyClone, Logan, UT) 2 mM L-glutamine (GIBCO), and Pen/Strep/Fungizone mix (BioWhittaker, Walkersville, MD).

Oligonucleotides

IL-10 antisense (ASI) and sense (SSI) mRNA, consisting of the same sequences as previously reported,[103] span regions near the initiation site of IL-10 translation. ASI consists of 5'-CAT TTC CGA TAA GGC TTG G and SSI consists of 5'-CCA AGC CTT ATC GGA AAT G (region 315–333). A scrambled oligonucleotide (SO) with the same base content (5'-TTT ATT GCG CAC GAC GGA T) as ASI is used as an additional control. All oligonucleotides are synthesized by the phosphoramadite procedure, including the thiosulfonate (Beaucage) procedure in the case of phosphoro-thioate-derivitized oligonucleotides, by Operon Technologies (Alameda, CA) or are a generous gift from C.-C. Chou (Schering-Plough Research Institute, Kenilworth, NJ).

Animals and Experimental Establishment of B-1 Cell Malignancies

(NZB × DBA/2) F_1 mice (Jackson Laboratories, Bar Harbor, ME), 2–10 months of age, are given 20×10^6 malignant B-1 cells via intravenous injection into the retroorbital venous plexus. Previous studies in our laboratory[7] have shown that mice with transplanted neoplastic B-1 cells consis-

---

[102] S. Fogerite and R. Mannino, *J. Liposome Res.* **6,** 357 (1996).

[103] B. Peng, N. H. Mehta, H. Fernandes, C.-C. Chou, and E. Raveche, *Leukemia Res.* **19,** 159 (1995).

tently develop hindlimb paralysis no later than 40 days posttransfer, which serves as an overt clinical indication of progression of the neoplasm. Additional studies have shown that localization of the neoplasm in the vertebral column and spinal cord is due to local proliferation of neoplastic cells at the site of the retroorbital inoculation, with penetration of tumor cells into the cerebrospinal space surrounding the olfactory lobes of the brain. Mice are housed in solid-bottom plastic cages with wood shavings and given a standard pelleted rodent ration (Laboratory Rodent Diet 5001; Purina Mills, St. Louis, MO) and reverse osmosis-purified water *ad libitum.* All aspects of animal care and experimental manipulation comply with applicable regulations[104] and the protocol is approved by the Institutional Care and Use Committee (IACUC).

Cochleate Preparation

Cochleate preparation is as previously described.[105] Briefly, phosphotidylserine and cholesterol (9:1) are dried to a thin film on the inner surface of a test tube by passing a stream of nitrogen over the chloroform-solubilized lipids. The drying step is conducted at low-level light intensity to prevent oxidation of lipids. The lipid film is separated from the surface of the test tube by agitation with extraction buffer (phosphate-buffered saline, pH 7.4), then sonicated to form small lipid particles. After addition of oligonucleotides to the lipid–extraction buffer mixture (1:10, oligonucleotide:buffer–lipid), gradual addition of calcium to a final level of 6 mM results in the formation of the "jelly-roll"-like cochleates with complexed oligonucleotides. The final oligonucleotide content of the cochleate preparation is 2500 μg/ml.

Treatment Regimens, Infusion Pump Studies

In the first series of experiments, (NZB × DBA/2) F_1 mice with experimentally transplanted B-1 neoplastic cells are given oligonucleotides via subcutaneously implanted osmotic pumps (model 2002; Alza, Palo Alto, CA). In experiment 1, a dose range-finding study, three groups of three mice each are given ASI, SSI, or SO oligonucleotides at dosage levels of 30, 100, or 300 μg/day and survival time is determined. Owing to the apparent dose-related increase in survival seen in experiment 1 (NZB × DBA/2) F_1 mice in experiment 2 are given ASI, SSI, or SO oligonucleotides

---

[104] National Research Council, "Guide for the Care and Use of Laboratory Animals." National Academy Press, Washington, D.C., 1996.

[105] S. Gould-Fogerite and R. Mannino, *in* "Liposome Technology" (G. Gregoriadis, ed.), Vol. III, pp. 262–275. CRC Press, Boca Raton, Florida, 1992.

at the highest dosage level of 300 μg/day for 28 days, for a total dose of 8.4 mg/mouse. Experiment 2 includes an additional group of three mice with implanted pumps that deliver only Tris–EDTA buffer. Mice from experiment 2 are either sacrificed on day 40, which is the approximate maximum survival time of untreated mice with transplanted tumor cells, or are monitored for survival. The 2-ml maximum volume limitation of the infusion pumps necessitates replacement of the pumps on day 14. Phosphorothioate-derivatized oligonucleotides are used in the osmotic pump infusion studies to retard oligonucleotide degradation by circulating nucleases.[106]

Treatment Regimens, Cochleate Studies

In experiment 3, three (NZB × DBA2) F_1 mice are given 500 μg of ASI oligonucleotides contained in cochleates on days 0 and 5, 250 μg on day 8, and 85 μg on day 14, for a total oligonucleotide dose of 1.335 mg/mouse. Controls for experiment 3 consist of a group of three untreated mice, and a group of three mice that receives cochleates containing no oligonucleotides. In experiment 4, two groups [each consisting of three (NZB × DBA/2) F_1 mice with transplanted B-1 neoplastic cells] are injected either subcutaneously or intraperitoneally with 500 μg of ASI or SSI oligonucleotides in cochleates at 3-day intervals for approximately 28 days, for a total oligonucleotide dose of 4 mg/mouse. End-point parameters in experiment 4 include survival time; gross necropsy examination; organ weight analysis of spleen, liver, and brain; and histopathologic examination of numerous organs and tissues. We presume that incorporation of oligonucleotides into cochleates would afford some level of protection to oligonucleotides, and therefore nonderivatized oligonucleotides are used for the cochleate studies. Injections of cochleates are given in the abdominal subcutis and the most superficial aspect of the peritoneal cavity, to avoid penetration of abdominal organs.

Flow Cytometric Analysis

Approximately 1 × 10^6 cells from spleen and liver are stained with fluorescein isothiocyanate (FITC)-conjugated goat anti-mouse IgM, anti-CD5bio/PE, or anti-B220 conjugated with Tricolor (Caltag, San Francisco, CA). Isotype antibodies are used as controls and for gating. Cells are analyzed on a FACScan flow cytometer using Lysis II software (Becton Dickinson, Sunnyvale, CA). For DNA content, cell suspensions are fixed in cold 70% ethanol and then stained with propidium iodide (PI, 50 μg/

[106] J.-P. Shaw, K. Kent, J. Bird, J. Fishback, and B. Froehler, *Nucleic Acids Res.* **19,** 747 (1991).

ml; Calbiochem, La Jolla, CA) for 30 min at 37°. The possibility of false-positive staining due to PI intercalation into double-stranded RNA species is minimized by addition of ribonuclease (1 mg/ml; Perkin-Elmer, Foster City, CA). At least 20,000 events are analyzed in each flow cytometric analysis.

Histologic Processing and Examination

Tissues selected for histologic examination are fixed in 10% phosphate-buffered formalin, processed through graded alcohols and a clearing agent, embedded in paraffin, sectioned at 6 μm, stained with hematoxylin and eosin, and examined by light microscopy. The spinal column with spinal cord *in situ* is decalcified with dilute nitric acid prior to histologic processing.

Results

On the basis of the *in vitro* studies employing various antisense IL-10 oligonucleotides, the sequence that gave optimal growth inhibition and apoptosis of malignant B-1 cells was chosen for use in the *in vivo* studies.[103] Initially, phosphorothioate-derivatized oligonucleotides were employed with constant *in vivo* infusion. Experiment 1 showed a dose-related increase in survival time in mice that received ASI oligonucleotide via infusion pump (Fig. 1). Mice given ASI at the highest dosage level of 300 μg/day survived to the scheduled termination of the study at 60 days, which was 20 days beyond the mean survival time of 40 days that was seen in preliminary studies.

On the basis of these results, a second experiment employing constant infusion of phosphorothioate-derivatized oligonucleotides at a dosage level of 300 μg/day was performed. Flow cytometric analysis of spleen and liver cells from experiment 2 revealed typical hyperdiploid malignant cells in mice given buffer alone, SSI, or SO oligonucleotide. Mice given ASI at 300 μg/day via infusion pump had no detectable increase in hyperdiploid cells in the spleen or liver (Fig. 2).

Histologic examination of liver, spleen, and spinal column/cord of mice from experiment 2 revealed normal histologic features in mice given ASI at 300 μg/day. In contrast, the liver of mice given SSI oligonucleotide or buffer alone had diffuse sinusoidal infiltrations of individualized mononuclear neoplastic cells, plus nodules composed of confluent masses of similar neoplastic cells (Fig. 3). The spleen of buffer- and SSI-treated mice consisted of a solid sheet of neoplastic cells that effaced the normal microarchitecture (Fig. 3). The lumbosacral vertebral column and surrounding musculature had a massive infiltration of neoplastic cells (Fig. 3). Associated degenera-

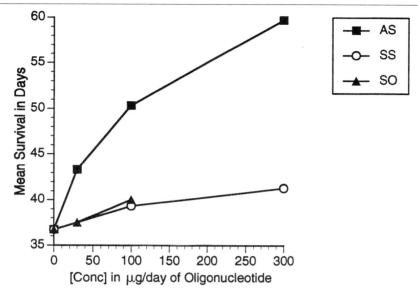

FIG. 1. Mean survival time of (NZB × DBA/2) F_1 mice after transfer of malignant B-1 cells and treatment with phosphorothioate-derivatized antisense IL-10 via subcutaneous miniosmotic pumps. Treatment with antisense IL-10 (AS), sense IL-10 (SS), or scrambled oligonucleotide (SO) was started at the time of tumor cell transfer.

tive changes in the spinal cord were considered to be the cause of the clinically apparent posterior paralysis.

In an effort to decrease the total amount of antisense IL-10 necessary to prevent the expansion of the malignant B-1 cells, lipid–matrix cochleates containing the antisense IL-10 oligonucleotides were employed in the next two *in vivo* experiments. In experiment 3, in which nonderivatized antisense IL-10 oligonucleotide were employed at a total dose of 1.335 mg/mouse, the antisense IL-10-treated mice had increased survival relative to mice given empty cochleates alone.

In experiment 4 all four control mice, which were given 500 μg of SSI oligonucleotide in cochleates every third day, had died by day 24 after transplantation of tumor cells. In contrast, only one of the mice given 500 μg of ASI oligonucleotide in cochleates every third day had died by day 24. The single ASI-treated mouse that died in experiment 4 had no gross evidence of neoplasia, and histologic evidence of neoplastic cell infiltration in various organs was only mild to moderate. For these reasons we suspect the single death in the ASI-treated group was due to causes unrelated to, or indirectly related to, progression of the neoplasm. Gross necropsy and histopathologic changes in SSI-treated mice from experiment 4 were similar

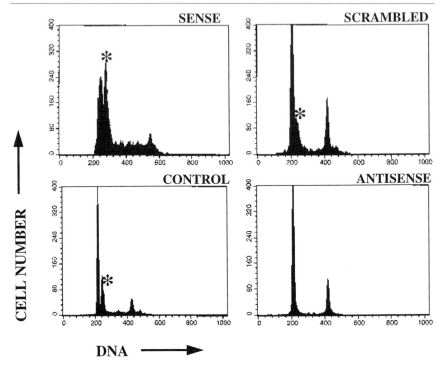

FIG. 2. Flow cytometric demonstration of hyperdiploid neoplastic cells in the spleen of antisense IL-10-treated and control mice with experimentally transplanted malignant B-1 cells. Spleens of recipient mice were removed on day 40 after transfer of malignant cells. Single-cell suspensions were stained with propidium iodide (PI) and analyzed by flow cytometric techniques for DNA content. Hyperdiploidy (identified by an asterisk, *) is a characteristic feature of malignant B-1 cells in this model. Mice were treated via surgically implanted miniosmotic pumps delivering antisense IL-10 oligonucleotide, sense IL-10 oligonucleotide, scrambled oligonucleotide, or buffer alone, as described in Materials and Methods.

to those in experiment 2. Livers of SSI-treated mice had diffuse sinusoidal and multifocal nodular infiltrations of mononuclear neoplastic cells, and spleens consisted of sheets of similar neoplastic cells. Pronounced neoplastic cell infiltrations were noted in and around the lumbosacral spinal column, including the bone marrow. Lymphosarcomatous infiltrations were commonly seen in the lung, kidney, brain, and various lymph nodes. The only histologic change associated with administration of the cochleates was multifocal histiocytic infiltrations on serosal surfaces of abdominal organs. The histiocytic infiltrates often were associated with accumulations of lipid-like material that was considered to be the remnants of partially degraded cochleates. In experiment 4, treatment with cochleates containing ASI

Spinal column   **Liver**   **Spleen**

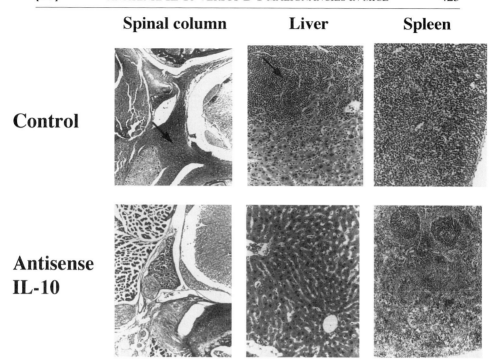

Control

**Antisense
IL-10**

FIG. 3. Photomicrographs of liver, spleen, and spinal column/cord of antisense IL-10-treated versus control mice with experimentally transferred malignant B-1 cells. Animals were sacrificed 34–40 days after transfer of tumor cells. Tissue specimens were fixed in buffered formalin and processed by standard histologic techniques, and paraffin-based sections were stained with hematoxylin and eosin. *Top:* Tissues from untreated mice. *Bottom:* Tissues from mice given antisense IL-10 via miniosmotic pump, as described in Materials and Methods. Note the prominent neoplastic cell infiltrations in extradural areas of the spinal column, nodular neoplastic cell proliferations and sinusoidal infiltrations in the liver, and diffuse neoplastic cell infiltration with microarchitectural effacement in the spleen. Original magnifications: spinal column, ×4, liver ×20, spleen, ×20.

oligonucleotide at a dosage level of 500 μg/day, administered every third day, was associated with lower spleen weights than were observed in mice given SSI in cochleates via the same regimen (Fig. 4). The altered organ weights in ASI-treated mice were attributable to the reduced neoplastic cell infiltration in the spleen of ASI-treated animals. The lack of splenic enlargement in the antisense IL-10 group was apparent when the data were expressed as direct organ weights, as a percentage of body weight, or as a percentage of brain weight (Fig. 4).

Overall survival of ASI- versus SSI-treated mice with transplanted B-1 malignancies is shown in Fig. 5. These data are combined results from the

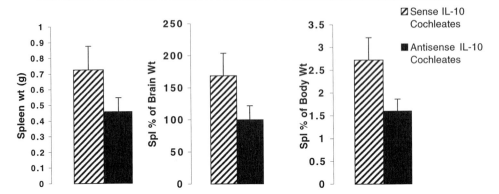

FIG. 4. Spleen weight analysis of (NZB × DBA/2) F_1 mice after experimental transfer of malignant B-1 cells and treatment with antisense (ASI) or sense (SSI) IL-10 oligonucleotides, as described in Materials and Methods. Each group consisted of four mice of similar age and weight.

constant infusion delivery as well as the cochleate delivery. Only 1 of 11 ASI-treated mice died spontaneously, and the overall gross necropsy and histopathologic findings in that mouse raise some question as to the precise cause of death. In contrast, the overwhelming majority of mice treated with sense oligonucleotide, scrambled oligonucleotide, or buffer had pronounced neoplastic proliferations that resulted in the demise of the mice.

Discussion

The preceding studies indicate that IL-10 mRNA antisense oligonucleotides can negatively influence the progression of experimentally transplanted B-1 cell malignancies in (NZB × DBA/2) F_1 mice. Antisense IL-10 treatment resulted in increased survival time, reduced tumor burden in liver and spleen as determined by organ weight analysis, and less severe organ involvement as assessed by histopathology. Antisense IL-10 was effective when administered by continuous subcutaneous infusion or by periodic subcutaneous/intraperitoneal injection of antisense-laden cochleates.

Demonstration of systemic effects of locally administered antisense IL-10 oligonucleotides in these studies has profound clinical implications owing to the implied possibility of administering antisense oligonucleotides via subcutaneous injection rather than intravenous infusion. Systemic effects of locally produced IL-10 have also been demonstrated after IL-10

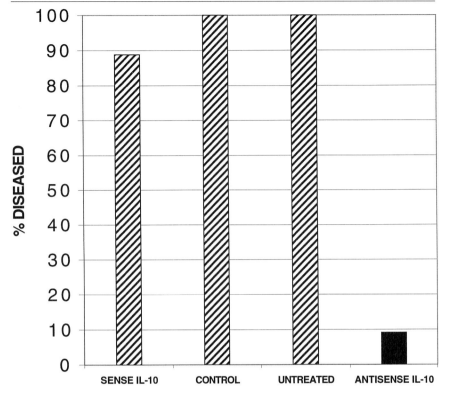

FIG. 5. Summary of survival of (NZB $\times$ DBA/2) F_1 mice with experimentally transferred malignant B-1 cells followed by treatment with antisense IL-10 via infusion pump or cochleate administration. Diseased mice are defined as mice that died before day 60 with hindleg paralysis, flow cytometric evidence of malignant B-1 cell clones, or histopathologic evidence of B-1 cell neoplasms. The percentages presented represent diseased mice per total number of mice studied. The groups are antisense IL-10-treated, sense IL-10-treated, control (receiving buffer via osmotic pump or injection of cochleates containing no oligonucleotides), and untreated mice that had no treatment after transfer of malignant B-1 cells.

gene transfection into keratinocytes.[107] While the negative effects of circulating nucleases on oligonucleotides are well known, results of these and other studies suggest the nuclease activity may be overcome by continuous administration of oligonucleotides. In this regard, it would be interesting to know the effects of a "decoy" oligonucleotide administered concurrently with our antisense IL-10 oligonucleotide.

A common, and valid, criticism of proposed clinical application of anti-

[107] X. Meng, D. Sawamura, K. Tamai, K. Hanada, H. Ishida, and I. Hashimoto, *J. Clin. Invest.* **101,** 1462 (1998).

sense therapy is the cost of administering the massive quantities of oligonucleotides suggested by most experimental protocols. For example, if antisense IL-10 were administered to a 77-kg patient at a dose level equivalent
to the 500-μg/dose used in mice, the drug cost alone would be approximately
$22,000 per dose. If antisense therapy is to have widespread clinical utility,
it is imperative that we find methods for reducing the overall treatment
cost. Phosphorothioate derivatization is known to prolong serum half-life
of oligonucleotides, and our studies suggest that incorporating oligonucleotides into cochleates affords some protection to the oligonucleotides. It is
possible that incorporation of phosphorothioate-derivitized oligonucleotides into cochleates could extend the half-life of the oligonucleotides even
more than sulfur addition alone. We are investigating the effects of lower
oligonucleotide dosage levels, derivitization of oligonucleotides, and effects
of coadministration of other molecules on the malignant B-1 cells in our
model system.

Although not specifically addressed in our experiments, the possibility
of toxicity or other untoward effects of oligonucleotide administration must
also be addressed. Oligonucleotides may be the ultimate of "natural" molecules, and thus direct chemical toxicity is not expected and has not been
shown in a number of studies. However, the desirable pharmacologic effects
of antisense oligonucleotides may have undesirable effects on overall
patient welfare, particularly if unbalanced, uncontrolled, or carried to ex-
tremes. For example, we presume the administration of antisense IL-10 in
our model system has a negative effect on normal B-1 cell populations,
and the overall clinical impact of those alterations is largely unknown at
this time. Repression of many biologic molecules, by whatever means,
commonly activates feedback mechanisms that can promote the production
of the molecule in question, increase or decrease production of its receptor
or accessory molecules, etc. The difficulties in dealing with this type of
interactive toxicity are particularly apparent when one considers the level
of redundancy, synergism, and pleiotropism commonly manifested by cytokines. Depending on the type and level of adaptive processes, sudden
withdrawal of the repressive effect of antisense oligonucleotides could also
result in adverse clinical effects.

A major advantage of antisense therapy versus any form of therapy
involving genomic manipulation is that antisense therapy is nonreplicative.
Untoward effects observed during the course of antisense therapy could
presumably be interrupted by withdrawal of treatment, administration of
sense oligonucleotides, or administration of the relevant protein.

Despite the cost and possible adverse effects, including the possibility
of a generalized increase in inflammatory responses in the case of IL-10,
the possibility of using antisense oligonucleotides directed at cytokines

has great therapeutic potential. In addition to the potential for systemic treatments to alter cytokine balance and inflammatory responses, there is great potential for treatment of solid tumors by local injection of antisense oligonucleotides and periodic treatments to reduce tumor burden in patients with CLL or other diffuse neoplasms that have targetable growth factors.

[30] Inhibition of c-ABL Expression in Hematopoietic Progenitor Cells Using Antisense Oligodeoxynucleotides

By VITTORIO ROSTI, CLAUDIA LUCOTTI, MARIO CAZZOLA, and GAETANO BERGAMASCHI

Introduction

Most oncogenic retroviruses originate from the recombination of naturally occurring retroviruses with cellular genes. When the cellular gene is involved in the regulation of cell proliferation or survival and recombination leads to deregulation of its expression or enhancement of the activity of its product, the resultant virus can provide infected cells with a growth or survival advantage leading to neoplasia. The first protooncogenes were discovered as the cellular homologes of transforming genes associated with retroviruses. The c-ABL protooncogene, for example, is a homolog to the transforming gene of Abelson murine leukemia virus (A-MuLV), originated from the recombination between the Moloney murine leukemia virus and the c-ABL gene of mouse.[1] Interest in the function of c-ABL mainly stems from the observation that the human gene is involved in the chromosomal translocation t(9;22)(q34;q11), which gives rise to the Philadelphia (Ph') chromosome,[2] the hallmark of chronic myeloid leukemia (CML). The translocation activates the transforming potential of c-ABL. In fact, human c-ABL is on the long arm of chromosome 9; after the Ph' translocation all c-ABL exons, except exon 1, move to chromosome 22. The breakpoint on chromosome 22 occurs within a 5.8-kb region called the major breakpoint cluster region (M-bcr), which is part of the BCR gene. Consequently, translocated c-ABL sequences are found downstream of the BCR gene exons 5' of the breakpoint on chromosome 22, and produce the BCR-ABL fusion gene. BCR-ABL is a functional gene that in CML cells is actively transcribed into an 8.5-kb chimeric mRNA coding for a new 210-kd

---

[1] H. T. Abelson and L. S. Rabstein, *Cancer Res.* **30**, 2213 (1970).
[2] J. D. Rowley, *Nature (London)* **243**, 290 (1973).

0076-6879/99 $30.00

← centromere

FIG. 1. Structure of the human c-ABL protooncogene. Two alternative first exons exist: exon 1a is 19 kb proximal to exon 2; exon 1b is ~200 kb proximal to exon 2 (see text for details). The number above each box denotes the exon number.

protein (p210) that contributes to the pathogenesis of the disease.[3] Although the exact mechanisms involved in cell transformation by BCR-ABL are not clearly defined, the gene and its product have been widely investigated. p210 is a tyrosine kinase[4] present in the cytoplasm of CML cells; *in vitro,* cells expressing the gene are resistant to apoptosis induced by growth factor deprivation,[5] cytotoxic drugs, and irradiation. These effects occur through activation of signal transduction pathways partially overlapping those activated by interleukin 3 and other growth factors.[6]

The c-ABL protooncogene encodes a p145 nonreceptor tyrosine kinase[7] ubiquitously expressed in mammalian cells. Two c-ABL mRNAs, 6.0 and 7.0 kb in size, exist in mammals. The main difference between the two mRNAs resides in the use of alternative 5' exons, 1a and 1b, respectively, which are linked to a common set of 3' exons (Fig. 1). Alternative splicing causing differential usage of exons 1a and 1b produces proteins with variable N-terminal domains. p145 molecules derived from exon 1a-containing mRNA (type I isoform) are mainly found in the cytoplasm, whereas molecules containing amino acids encoded by exon 1b (type IV isoform) are found in the nucleus. The N terminus of type IV c-ABL contains a myristoylation signal that is absent in type I c-ABL, but its role in nuclear localization is unknown, although deletion of this region causes a redistribution of the protein from the nucleus to the cytoplasm and cell transformation. However, other nuclear localization signals exist in the common C terminus of p145. The C terminus contains several functional domains

[3] A. de Klein, A. G. van Kessel, G. Grosveld, C. R. Bartram, A. Hagemeijer, D. Bootsma, N. K. Spurr, N. Heisterkamp, J. Groffen, and J. R. Stephenson, *Nature (London)* **300,** 765 (1982).

[4] Y. Ben-Neriah, G. Q. Daley, A.-M. Mes-Masson, A. N. Witte, and D. Baltimore, *Science* **233,** 212 (1986).

[5] G. Q. Daley and D. Baltimore, *Proc. Natl. Acad. Sci. U.S.A.* **85,** 9312 (1988).

[6] U. Matulonis, R. Salgia, K. Okuda, B. Druker, and J. D. Griffin, *Exp. Hematol.* **21,** 1460 (1993).

[7] J. B. Konopka and O. N. Witte, *Mol. Cell. Biol.* **5,** 3116 (1985).

essential for c-ABL activity such as an actin-binding domain and a DNA-binding domain. Between the N and the C termini of the protein there is a kinase-regulatory region, containing the src homology domains (SH2 and SH3) that are necessary for interaction of p145 with other proteins to form macromolecular complexes, and the tyrosine kinase domain. Functionally, the DNA-binding domain has been thoroughly investigated.[8] This domain has a high affinity for the EP DNA element, a conserved 15-base pair (bp) sequence that is found in the enhancer and promoter regions of several cellular and viral genes, as in the hepatitis B virus, and is essential for their function. Regulation of c-ABL-binding affinity for DNA is finely tuned during the cell cycle; in fact, the affinity is reduced after phosphorylation of c-ABL on critical serine and threonine residues by the cdc2 kinase during the M phase of the cell cycle and by formation of a complex with the retinoblastoma (RB) gene product. On the other hand, during the S phase of the cell cycle c-ABL becomes phosphorylated on tyrosine residues, shows increased DNA-binding affinity, and is activated as a tyrosine kinase. These S-phase events are also associated with the release of c-ABL from the RB complex. The reversible binding of c-ABL to DNA suggests that it may participate in a cell cycle-dependent regulation of gene transcription. Potential substrates and regulatory mechanisms for the cytoplasmic isoform of p145 are essentially unknown, but available knowledge and the presence of the protein in both cell cytoplasm and nucleus suggest that c-ABL may couple signaling pathways in the cytoplasm with nuclear events such as transcription.

Although several structural and biochemical aspects of p145 have been characterized, its function in the cell is less clear. c-ABL is expressed ubiquitously and throughout development. Disruption of the gene in *Drosophila* and mice affects development, although the animals survive embryogenesis. At birth mice homozygous for deletions of c-*abl* are runted and show B and T cell lymphopenia with thymic and splenic atrophy. Neonatal death invariably occurs within 1 to 2 weeks after birth.[9] However, it appears that in most cell types, with the possible exception of lymphocytes, c-*abl* function is redundant and other tyrosine kinases compensate for its absence. The influence of c-ABL on cellular behavior has been investigated using different experimental models, often providing contradictory results. According to some authors c-ABL has growth-inhibitory[10] and proapoptotic effects that are activated in response to DNA damage induced by irradiation

[8] P. J. Welch and J. Y. J. Wang, *Cell* **75,** 779 (1993).

[9] V. L. Tybulewicz, C. E. Crawford, P. K. Jackson, R. T. Bronson, and R. C. Mulligan, *Cell* **65,** 1153 (1991).

[10] C. L. Sawyers, J. McLaughlin, A. Goga, M. Havlik, and O. Witte, *Cell* **77,** 121 (1994).

or other genotoxic agents such as the antimetabolite 1-β-D-arabinofurano-sylcytosine (Ara-C).[11] c-ABL induced growth arrest occurs in the G_1 phase of the cell cycle and is dependent on the RB and p53 tumor suppressor gene products.[12] More recently, interactions have been shown between c-ABL and the ATM gene product (ATM is the gene mutated in the congenital disorder ataxia teleangectasia); accordingly, an involvement of ATM in c-ABL activation by DNA damage has been suggested, and the loss of such activation in subjects with the disease might be responsible for a defective G_1/S checkpoint after radiation damage and hypersensitivity of AT cells to ionizing radiation.[13] Other investigators, however, described c-ABL as an antiapoptotic gene,[14] necessary in lymphoid differentiation[9] and for the normal G_1/S progression of the cell cycle.[8,15]

In the following section we describe the experimental procedures we have used for the investigation of the role of c-ABL in normal human bone marrow-derived hematopoietic progenitor cells, using antisense oligodeoxy-nucleotides directed against the two types of c-ABL mRNA. First, we have assessed the consequences of the suppression of the expression of c-ABL on the growth of clonogenic progenitor cells (colony-forming units, CFC), Second, we have used flow cytometry and the Ara-C suicide approach to investigate the effects on the cell cycle of CD34+ cells (which so far are considered representative of the entire hematopoietic progenitor cell population), and single classes of CFC, respectively.

Materials and Methods

Cell Separation Procedures

Bone marrow aspirates (3–5 ml, from posterior iliac crest), obtained from bone marrow donors after informed consent, are drawn into heparinized syringes. Samples are diluted 1:4 with Iscove's modified Dulbecco's medium (IMDM) and layered onto an equal volume of Ficoll-Paque (Pharmacia Biotech, Uppsala, Sweden) gradient (1077 g/ml) in 50-ml Falcon

[11] Y. Huang, Z.-M. Yuan, T. Ishiko, S. Nakada, T. Utsugisawa, T. Kato, S. Kharbanda, and D. W. Kufe, *Oncogene* **15,** 1947 (1997).

[12] S.-T. Wen, P. K. Jackson, and R. A. Van Etten, *EMBO J.* **15,** 1583 (1996).

[13] T. Shafman, K. K. Khanna, P. Kedar, K. Spring, S. Kozlov, T. Yen, K. Hobson, M. Gatei, N. Zhang, D. Watters, M. Egerton, Y. Shiloh, S. Kharbanda, D. Kufe, and M. F. Lavin, *Nature (London)* **387,** 520 (1997).

[14] R. Daniel, P. M. C. Wong, and S.-W. Chung, *Cell Growth. Differ.* **7,** 1141 (1996).

[15] V. Rosti, G. Bergamaschi, C. Lucotti, M. Danova, C. Carlo-Stella, F. Locatelli, L. Tonon, G. Mazzini, and M. Cazzola, *Blood* **86,** 3387 (1995).

tubes. Cells are centrifuged at 400*g* for 30 min, at room temperature and interface cells are collected through a fine-bore Pasteur pipette. Interface cells are washed three times in IMDM–2% Fetal calf serum (FCS) and then either depleted of adherent cells or directly selected for their CD34+ fraction. For the former purpose mononuclear cells which we refer to from now on as light-density bone marrow cells, LDBMCs) are resuspended in IMDM–20% FCS (HyClone, Logan, UT) and incubated in tissue culture-treated flasks for two additional 1-hr periods at 37° and 5% CO_2. Non-adherent LDBMCs (AC⁻LDBMCs) are then collected, washed once in IMDM–2% FCS, resuspended in the same medium, and used for the experiments.

The selection of CD34+ cells is carried out using the MiniMACS magnetic cell-sorting device (Miltenyi Biotec GmbH, Bergisch Gladbach, Germany), according to the manufacturer specifications. For this procedure we work with a high number of LDBMCs ($>5 \times 10^7$): in fact, CD34+ cells represent on average 0.5–3% of mononuclear bone marrow cells. Starting with such a number of LDBMCs at the end of the selection the recovery of CD34+ cells is sufficient to perform a reasonable number of experiments. The following procedure is described for 10^8 LDBMCs; if the procedure is carried out with a smaller or larger number of cells the volumes must be reduced or increased proportionally. Briefly, 10^8 LDBMCs are pelleted and resuspended in 300 μl of phosphate-buffered saline (PBS)–0.5% FCS–5 m*M* EDTA, pH 7.2 (washing buffer). EDTA helps to reduce aggregation. One hundred microliters of reagent A1 (human IgG, which is necessary to prevent aspecific binding of the anti-CD34 antibody) and 100 μl of reagent A2 (a mouse IgG_1 anti-CD34 antibody, clone QBEND/10) are added to the cells and gently but thoroughly mixed; the suspension is then incubated for 15 min at 4°. Cells are then washed once with 5 ml of washing buffer. After washing the buffer is removed and cells are resuspended in 400 μl of fresh washing buffer. One hundred microliters of reagent B (colloidal superparamagnetic MACS Microbeads, recognizing the anti-CD34 antibody) is then added and the cell suspension incubated for 15 min at 6° to 12°. After washing, resuspend the pellet in 400 μl of fresh buffer and add the suspension to a prefilled (with buffer) MiniMACS column placed on the magnet. Let the cells go through the column, then wash the column four times with 500 μl of buffer. Remove the column from the magnet, place it on a tube, and wash it with 1 ml of buffer in order to elute the CD34+ cells (repeat the last step twice to recover all of the positive cells). The collected CD34+ cells are then washed once, counted, and resuspended in the appropriate medium (see below). Using this procedure the purity of the selected CD34+ cells is on average >95% and the recovery is on average ~60%.

TABLE I
OLIGODEOXYNUCLEOTIDES AND PRIMERS

Nucleotide/Primer	Sequence
Oligodeoxynucleotides	
1a AS	5' CTTCAGGCAGATCTCCAA 3'
1b AS	5' TTTTCCAGGCTGCTGCCC 3'
1a S	5' TTGGAGATCTGCCTGAAG 3'
1b S	5' GCCTTCAGCGCGCCAGTA 3'
Primers	
A3	5' TGTGATTATAGCCTAAGACCCGGAG 3'
A2	5' GGTACCGAATTCAGCGGCCAGTAGCATCTGACTT 3'
β-Actin 3'	5' GGTGTACTTGCGCTCAGGAGGAGC 3'
β-Actin 5'	5' GCACTCTTCCAGCCTTCCTTCCTG 3'

Oligonucleotides

Antisense experiments are performed using commercial, unmodified oligodeoxynucleotides, purified by gel filtration (Med Probe, Oslo, Norway). Because no absolute rules exist for the choice of effective antisense oligonucleotides, as far as oligonucleotide length and sequence are concerned, we chose to use 18-nucleotide sequences corresponding to codons 2 to 7 of c-ABL mRNA.[16] Sequences corresponding to the first codons of a mRNA have been shown to be usually effective in antisense experiments, although mRNA-associated proteins, tertiary structure, and local folding potentially influence the access of an oligonucleotide to its target sequence. Because c-ABL has two alternative first exons,[17] two antisense (AS) oligonucleotides complementary to sequences from exons 1a and 1b are used at the same time.[16] The corresponding sense (S) oligonucleotides are used as controls. Oligonucleotide sequences are reported in Table I.

Incubation of Hematopoietic Progenitor Cells with Oligonucleotides

AC⁻LDBMCs (10^6/ml) or CD34⁺ (2×10^5/ml) cells are incubated with sense or antisense anti-c-ABL oligodeoxynucleotides for 18 to 48 hr in IMDM containing 10% heat-inactivated FBS at 37°, 5% CO_2 in air. Several batches of FBS are usually screened in a preliminary experiment to identify those batches with low nuclease activity. Once a suitable batch has been identified, it is used throughout all of the experiments. In some experiments incubation with anti-c-ABL oligodeoxynucleotides is performed in serum-

[16] V. Rosti, G. Bergamaschi, L. Ponchio, and M. Cazzola, *Leukemia* **6**, 1 (1992).
[17] E. Shtivelman, B. Lifshitz, R. P. Gale, B. A. Roe, and E. Canaani, *Cell* **47**, 277 (1986).

free medium made of IMDM containing $5 \times 10^{-5} M$ 2-mercaptoethanol, human insulin (10 μg/ml; Sigma Chemicals, St. Louis, MO), iron-saturated human transferrin (200 μg/ml; ICN Pharmaceuticals, Costa Mesa, CA), deionized bovine serum albumin (20 mg/ml; Stem Cell Technologies, Vancouver, BC, Canada), and low-density lipoproteins (40 μg/ml; Sigma) in the presence of stem cell factor (SCF, 100 ng/ml), interleukin 3 (IL-3, 20 ng/ml) (both from Genzyme, Cambridge, MA) and granulocyte colony-stimulating factor (G-CSF, 20 ng/ml; Amgen, Thousands Oak, CA). Sense and antisense oligodeoxynucleotides are used at a final concentration of 14 μM, which proved to be optimal in preliminary experiments. When the effects of c-ABL suppression on the cycling status of CFCs or CD34$^+$ cells are investigated, either the Ara-C suicide approach (for CFCs) or the cytofluorimetric analysis of DNA content and bromodeoxyuridine (BrdU) incorporation (for the whole CD34$^+$ cell population) is performed (see below).

Assessment of Suppression of c-ABL Expression by Reverse Transcriptase-Polymerase Chain Reaction

Effects of oligonucleotide treatment on c-ABL expression are evaluated by means of a competitive reverse transcriptase-polymerase chain reaction (RT-PCR) in which sequences from c-ABL and from the β-actin housekeeping gene are simultaneously reverse transcribed and amplified in the same reaction tube.[15] Briefly, after a 24-hr incubation of cells with oligonucleotides, total RNA is isolated from 2×10^4 to 4×10^4 CD34$^+$ cells using the Trizol reagent from GIBCO-BRL (Gaithersburg, MD). Reverse transcription is performed in a final volume of 20 μl using 250 ng of each 3' primer corresponding to the third exon of c-ABL (primer A3) and to nucleotides 1032–1055 of the β-actin c-DNA (see Table I for sequences), 100 U of Moloney murine leukemia virus reverse transcriptase, and 20 U of RNasin (both from GIBCO-BRL), bovine serum albumin (0.2 mg/ml; Promega Biotec, Madison, WI), 0.01 M dithiothreitol, 50 mM Tris-HCl (pH 8.3), 75 mM KCl, 3.0 mM MgCl$_2$, and a 0.4 mM concentration of each deoxynucleotide triphosphate (dNTP) (Boehringer GmbH, Mannheim, Germany). RNA and primers are initially incubated for 3 min at 80°, followed by 30 min at 69°, and then transferred to ice; on addition of the other reagents, previously combined to form a master mix, reverse transcription is allowed to proceed for 60 min at 42°, followed by inactivation of the reverse transcriptase by incubation at 95° for 10 min. Ten microliters of first-strand cDNA are subsequently amplified in a final volume of 50 μl containing 125 ng of the reverse transcription primers, 250 ng of 5' primers corresponding to the second exon of c-ABL (A2) and to nucleotides 819–842 of β-actin c-DNA (see Table I for sequences), a 0.1 mM concentration of each dNTP, 10 mM Tris-HCl (pH 8.3), 50 mM KCl, 1.5 mM MgCl$_2$, 1 U of Dynazyme thermostable DNA polymerase (Finnzymes OY, Espoo, Finland). After an initial denaturation step of 2 min at 94°, 40 PCR cycles are performed, each involving 45 sec at 94°, 25 sec at 61°, and 2 min at 72°. In the final cycle elongation at 72° is prolonged for 7 min. PCR products are run on a 2% agarose gel in 1× TBE buffer, containing ethidium bromide at 0.5 μg/ml. β-Actin bands are directly visualized on the ethidium bromide-stained gel. Because of its low level of expression compared with β-actin, c-ABL amplification product can be detected after Southern blotting of the gel onto a nylon filter (Hybond-N+; Amersham Pharmacia Biotech, Uppsala, Sweden) and hybridization with a ^{32}P-labeled probe spanning the second and the third c-ABL exons, which has been previously synthesized in our laboratory by an RT-PCR as described above, in which the β-actin primers have been omitted. Prehybridization and hybridization procedures are performed according to the filter manufacturer specifications; labeling is performed with a random primed DNA labeling kit (Boehringer GmbH), using 50 ng of DNA and 10 μCi of high specific activity [α-^{32}P] dCTP. Bands are visualized by exposure of the filter to Eastman Kodak (Rochester, NY) autoradiography film.

Clonogenic Assays

AC$^-$LDBMCs or CD34$^+$ cells are assayed for their content in committed hematopoietic progenitor cells (CFU-GEMM, BFU-E, CFU-GM, globally referred as CFCs, colony-forming cells) using a clonogenic assay according to Fauser and Messner[18] with modifications. Briefly, 2 × 10^4 AC$^-$LDBMCs or 5 × 10^3 CD34$^+$ cells are cultured in a 1-ml aliquot of IMDM containing 30% FBS, 5 × 10^{-5} M 2-mercaptoethanol (Sigma Chemicals), IL-3 (10 ng/ml), granulocyte-macrophage colony-stimulating factor (GM-CSF, 10 ng/ml), SCF (50 ng/ml) (all growth factors are from Genzyme), 3 IU of erythropoietin (Boehringer GmbH), and 0.9% (w/v) methylcellulose (Stem Cell Technologies). One-milliliter aliquots are seeded in 35-mm petri dishes (non-tissue-culture treated) and incubated for 14 days at 37°, 5% CO$_2$, in a fully humidified atmosphere. Assays are always performed in duplicate or triplicate. The number of the different types of colonies (erythroid or BFU-E, granulocyte–macrophage or CFU-GM, and mixed or CFU-GEMM) is scored by an inverted microscope at the end of incubation according to morphology: a BFU-E is defined as one colony (of

---

[18] A. A. Fauser and H. A. Messner, *Blood* **52**, 1243 (1978).

at least 500 cells) or a cluster of two or more subcolonies that appear red (owing to their hemoglobinization); a CFU-GM is defined as a colony of at least 40 cells that appears white or slightly brownish; a CFU-GEMM is a colony that is composed of a mixture of erythroid (hemoglobinized), granulocytic, and macrophage (white–brown) cells. Sometimes the identification of a colony is uncertain: in this case the colony is removed from the methylcellulose, a Cytospin is created and stained with May-Grünwald Giemsa (or Wright), and identified on the basis of the cells (myeloid or erythroid or both) that are recognized on the slide.

S-Phase Assessment by Cytosine Arabinoside Suicide

Assessment of the proportion of progenitor cells (CFCs) in the S phase of the cell cycle is done using the cytosine arabonoside (Ara-C) suicide approach. Ara-C has been used extensively *in vitro* for its S phase-specific cytotoxicity, in preference to the traditional [^{3}H]thymidine suicide approach. The results obtained using the two different approaches have been shown to be superimposable.[19] AC⁻LDBMCs or CD34$^+$ cells are incubated with anti-c-ABL oligodeoxynucleotides as described above for 48 hr and Ara-C (1×10^{-6} M) is added for the last 24 hr. This concentration of Ara-C was chosen according to preliminary experiments carried out in our laboratory and according to published data[19] showing that a plateau of specific killing (with a minimal aspecific toxic effect) was reached at 1×10^{-6} M. At the end of incubation cells are transferred to a tube, washed once with fresh medium, and resuspended in IMDM–2% FCS, after which appropriate quantities from each sample are assayed for their CFC content in a clonogenic assay (see above).

The proportion of CFCs killed by Ara-C (i.e., the proportion of progenitor cells in S phase) is calculated as follows: $[(X-Y) \times 100]/X$, where X is the number of CFCs grown in control culture and Y is the number of CFCs grown after exposure to Ara-C.

Flow Cytometric Analysis of Cell-Cycle Phase Distribution and
  Bromodeoxyuridine Uptake

After incubation with anti-c-ABL oligodeoxynucleotides, CD34$^+$ cells are fixed with ice-cold 70% ethanol, washed once, and stained for 30 min at room temperature with propidium iodide (50 μg/ml; Calbiochem, San Diego, CA) in PBS and in the presence of 0.1% Triton X-100 and 1% (w/v) RNase A (both from Sigma Chemical). Monoparametric conventional cell cycle analysis is then performed on nuclei from control and treated cells,

[19] H. D. Preisler and J. Epstein, *Br. J. Haematol.* **47,** 519 (1981).

using a flow cytometer, and the percentage frequencies of cells in the different phases of the cell cycle are calculated using a mathematical model.[20] By this approach it is also possible to evaluate the presence of apoptotic cells, according to the principle that the reduced stainability of apoptotic cells with DNA fluorochromes is a consequence of partial loss of low molecular weight DNA. For biparametric BrdU/DNA analysis, at the end of incubation with anti-c-ABL oligomers, CD34$^+$ cells are pulse labeled with 30 μM BrdU for 30 min, washed once, and fixed with 70% ethanol as described above. Cells are then washed with PBS and the DNA denatured with 2 N HCl for 20 min at room temperature. BrdU uptake is detected after a 30-min incubation with 100 μl of an anti-BrdU monoclonal antibody (Becton Dickinson, San Jose, CA) diluted 1 : 10 in PBS, followed by incubation with 100 μl of a FITC-labeled goat antimouse monoclonal antibody (Becton Dickinson, San Jose, CA) diluted 1 : 50 in PBS. Finally, cells are washed twice with PBS at room temperature and counterstained for 3 hr at room temperature with propidium iodide (5 μg/ml). Bivariate measurement of green fluorescence (BrdU-positive cells) versus red fluorescence (DNA content) is performed with a flow cytometer.

Results and Discussion

Suppression of the expression of c-ABL in human normal bone marrow-derived CD34$^+$ cells using oligodeoxynucleotides directed against the two alternative first exons resulted in the absence of c-ABL mRNA when evaluated by RT-PCR. Inhibition of c-ABL expression in normal human hematopoietic progenitor cells resulted in a significant inhibition of the growth of granulocyte–macrophage colony-forming units (CFU-GM) whereas erythroid (BFU-E) and mixed (CFU-GEMM) clonogenic progenitor cells were not affected.[15,16] Exposure to anti-c-ABL oligonucleotides also reduced the proportion of CD34$^+$ cells in the S phase of the cell cyle and BrdU labeling in a significant manner (from 19 to 7% and from 13 to 6%, respectively; $p < 0.05$ in either case). Flow cytometry showed that treated CD34$^+$ cells accumulated in the G_0/G_1 phase of the cell cycle with no evidence of arrest during the S phase: in fact, the bivariate DNA/BrdU analysis failed to show significant aliquots of cells with a DNA content typical of the S phase, but not incorporating BrdU. By the same cytofluorimetric approach (confirmed also by DNA agarose gel electrophoresis) we found no evidence of apoptosis, indicating that the growth arrest of CD34$^+$ cells due to the inhibition of c-ABL expression did not result in the induction of programmed cell death, but rather in an arrest and accumulation of

---

[20] J. Fried, *Comput. Biomed. Res.* **9,** 263 (1976).

these cells in the G_0/G_1 phase of the cell cycle.[15] On the basis of these findings we have also indicated whether c-ABL-related block of the entry into S phase of hematopoietic progenitor cells could spare these cells from the cytotoxic effects of an S phase-specific drug such as cytosine arabinoside (Ara-C). Twenty-four-hour exposure to Ara-C alone or in the presence of sense (control) oligodeoxynucleotides resulted in an expected >80% killing of clonogenic progenitor cells. However, addition of Ara-C to hematopoietic progenitor cells in the presence of antisense oligodeoxynucleotides reduced significantly (~40%) the killing of clonogenic progenitor cells, indicating that c-ABL inhibition protects CFCs from the cytotoxic effects of Ara-C, likely because fewer hematopoietic progenitor cells are found in the S phase of the cell cycle.[21] Taken together our results (1) confirm that c-ABL expression is necessary for the differentiation and proliferation of granulocyte-macrophage colony-forming units (CFU-GM); and (2) strongly suggest that the product of c-ABL is critical for entry of $CD34^+$ hematopoietic progenitor cells into the S phase of the cell cycle, possibly playing a specific role in the transition from the G_1 to the S phase. They also suggest that inhibition of c-ABL expression might represent a means to protect normal hematopoietic progenitor cells from S phase-specific cytotoxic agents. However, the possibility must be considered that cell protection by AS oligonucleotides is not related to abrogation of c-ABL effects on the cell cycle but to suppression of c-ABL proapoptotic effect.[11] In fact, the mechanisms responsible for the proapoptotic effects of c-ABL in response to Ara-C are not clear, and it cannot be ruled out that such effects are related to a permissive effect of c-ABL on cell cycle progression.

Our results are in agreement with other data indicating a permissive role for c-ABL in the progression of the cell through different phases of the cell cycle; however, several reports suggest that c-ABL can exert growth-suppressive functions (see Introduction). A synthesis of these experimental results on the biological function of c-ABL is not simple. Discrepancies might be due to the use of different models. We cannot rule out that, *in vitro,* c-ABL has both growth-promoting and growth-suppressive functions, depending on the cell type, the cellular environment, and the level of expression of the gene in different phases of the cell cycle. In fact, the gene can be necessary for normal cell proliferation, but its overexpression at critical points in the cell cycle may cause growth arrest[10] or apoptosis.[11] The same inhibitory effect on the cell growth might also be exerted by c-ABL when DNA damage would require a block of or a slowing in the cell cycle progression to allow for DNA

[21] C. Lucotti, G. Bergamaschi, A. Novella, S. Casula, V. Rosti, and M. Cazzola, *Blood* **90**(Suppl. 1), 50a (1997).

repair.[13] However, *in vivo,* mice homozygous for an inactivating mutation of the c-*abl* gene had thymus and spleen atrophy and lymphopenia[9]: these findings are more consistent with a role for c-*abl* as a positive rather than a negative regulator of the cell growth, at least in the mouse.

In conclusion, although significant progress has been made in elucidating the biological and biochemical functions of c-ABL, further studies are needed to finally clarify its role in the regulation of the cell growth. A better knowledge of c-ABL function in hematopoietic progenitor cells might also be helpful in understanding the pathogenesis of chronic myeloid leukemia.

Acknowledgments

We thank Sabina Casula for helping with clonogenic assays and Ara-C suicide experiments and Marco Danova for helping with flow cytometry experiments. This work was supported by the AIRC (Milan, Italy) and the IRCCS Policlinico San Matteo (Pavia, Italy).

[31] Cellular Pharmacology of Antisense Oligodeoxynucleotides

By Christoph Schumacher

Introduction

Antisense oligodeoxynucleotides (ODNs) offer the potential to block the expression of specific genes within cells and are therefore an essential strategy to study gene function in a cell-based system. Inhibition of gene expression by antisense ODNs relies on the ability of the ODNs to hybridize specifically to a complementary messenger ribonucleic acid (mRNA) sequence and to decrease translation of the mRNA.[1]

Messenger RNA is a short-lived intermediary in the transfer of genetic information from deoxyribonucleic acid (DNA) to protein. Transcriptional activity and RNA decay rates determine the steady state level of an mRNA sequence and therefore its translation into an amino acid sequence. The protein decay rate will subsequently determine the steady state protein level of the target gene and, ultimately, its effect on cell function. Therefore, inhibition of gene expression with antisense molecules reflects a serial

[1] J. F. Milligan, M. D. Matteucci, and J. C. Martin, *J. Med. Chem.* **36,** 1923 (1993).

time- and dose-dependent mechanism to achieve modulation of cellular functions.

Antisense molecules can be designed for any mRNA using only Watson–Crick base-pairing rules. However, because of the inaccessibility of some secondary or tertiary RNA structures to oligonucleotides as well as putative nonspecific ODN–protein interactions, rigorous biochemical evidence must demonstrate that biological effects arise from the anticipated reduction in target gene expression. New ODN sequences are best evaluated in cellular assays, where membrane permeability, intracellular stability, RNase H-mediated degradation of targeted mRNA, cell toxicity, and potency can be evaluated. Both intracellular target mRNA and protein levels are quantified in dose–response as well as in time–response studies for a direct indication of specific ODN activity. Application of mismatched ODN analogs should provide experimental evidence of sequence-specific biological ODN effects. Finally, functional cell assays are correlated with target protein expression and activity.

In this chapter, we outline the procedures to assess the biological responses elicited by an antisense molecule in a cell-based system. The application of antisense compounds to intact cells and the protocols for a dose–response study and a time–response study are described using CGP 69846A,[2] a phosphorothioate antisense ODN compound targeted to human Raf-1 kinase. The efficacy of the antisense ODN to downregulate protein expression is ultimately evaluated by a kinetic comparison with the endogenous protein turnover rate.

CGP 69846A was shown to reduce specifically, in a concentration- and time-dependent manner, the cellular levels of Raf-1 transcripts in cultured human coronary artery smooth muscle cells (SMCs). This resulted in a significant suppression of cell proliferation by reducing the rate of cell cycle progression in the absence of any detectable cytotoxic effects.[3] The reduction of cellular protein levels after antisense treatment correlated kinetically with the endogenous protein turnover half-life. The pharmacodynamic effects of an antisense molecule on a short-lived mRNA that encodes a long-lived protein are therefore predicted by its duration of intracellular activity. Pharmacologically, in order to achieve and maintain a decreased steady state target protein level, we suggest a repeated ODN application within the time range of the ODN decay rates.

[2] B. P. Monia, H. Sasmor, J. F. Johnston, S. M. Freier, E. A. Lesnik, M. Muller, T. Geiger, K.-H. Altmann, H. Moser, and D. Fabbro, *Proc. Natl. Acad. Sci. U.S.A.* **93,** 15481 (1996).
[3] C. Schumacher, C. L. Cioffi, H. Sharif, W. Haston, B. P. Monia, and L. Wennogle, *Mol. Pharmacol.* **53,** 97 (1998).

Application of Antisense Compounds to Intact Cells

The phosphorothioate ODN has become the standard backbone for antisense research because of its ease of synthesis and excellent nuclease stability relative to phosphodiesters. In this class of oligonucleotides, one of the oxygen atoms in the phosphate group is replaced with a sulfur atom. ODNs, available through several commercial institutions, are usually designed in lengths from 18 to 22 nucleotides on the basis of a statistical evaluation of the length necessary to avoid random matches in the human genome and nuclease susceptibility.[4–6] Common targets on the RNA transcript include the 5' CAP site, the 5' and 3' untranslated regions, the translation initiation site, internal intron/exon splice junctions, and the polyadenylation site.[7] However, antisense targets are not limited to these sites. To date, screening multiple ODN sequences complementary to various sites along the entire message in cell-based assays (message walk) has been the most reliable way to identify effective antisense sequences.[8,9]

In vitro cellular uptake of ODN is time dependent and influenced by cell type, cell culture conditions, and chemical properties of the applied ODNs. Although the mechanism of cellular ODN uptake is not clearly understood, it has been shown that cationic lipids significantly enhance uptake of negatively charged phosphorothioate oligonucleotides in cells.[10,11] The molar ratio of delivery vehicle to ODN is critical to achieve optimal cellular uptake of the liposome particles and cytoplasmic release of ODN molecules while avoiding potentially toxic cellular side effects of the cationic lipids. The subcellular distribution of ODNs is also influenced by cationic lipid vehicles, which is important because the antisense effect is believed to occur in the nucleus.[5,12] As shown by Bennett *et al.*,[13] cellular uptake of ODN liposomes occurs in a concentration-dependent manner and becomes saturated within 4 hr. At this point, approximately 8 pmol of ODN mole-

[4] D. E. Szymkowski, *Drug Discovery Today* **1**, 415 (1996).

[5] C. A. Stein, *Antisense Nucleic Acid Drug Dev.* **8**, 129 (1998).

[6] R. W. Wagner, *Nature (London)* **372**, 333 (1994).

[7] S. T. Crooke, *FASEB J.* **7**, 533 (1993).

[8] B. P. Monia, J. F. Johnston, T. Geiger, M. Muller, and D. Fabro, *Nature Med.* **2**, 668 (1996).

[9] S. P. Ho, Y. Bao, T. Lesher, R. Malhotra, L. Y. Ma, S. J. Fluharty, and R. R. Sakai, *Nature Biotechnol.* **16**, 59 (1998).

[10] P. L. Felgner, T. R. Gadek, M. Holm, R. Roman, H. W. Chan, M. Wenz, J. P. Northrup, G. M. Ringold, and M. Danielson, *Proc. Natl. Acad. Sci. U.S.A.* **84**, 7413 (1987).

[11] C. F. Bennet, *in* "Delivery Strategies for Antisense Oligonucleotide Therapeutics" (S. Akhtar, ed.), p. 223. CRC Press, Boca Raton, Florida, 1995.

[12] T. Woolf, *Nature Biotechnol.* **14**, 824 (1996).

[13] C. F. Bennett, M.-Y. Chiang, H. Chan, J. E. E. Shoemaker, and C. K. Mirabelli, *Mol. Pharmacol.* **41**, 1023 (1992).

cules is associated with 10^6 cells after incubation in the presence of 1 μM ODN.[13] Assuming a steady state concentration of 1 fmol of a particular mRNA in 10^6 cells, which corresponds to approximately 1000 transcript copies per cell, sufficient ODN molecules should be accessible in the nucleus for antisense effects in this time and concentration frame of ODN cell exposure.

Protocols for Antisense Application to Intact Cells

Materials

Custom-designed phosphorothioate ODNs targeted to an mRNA of interest: Can be obtained from Life Technologies (Gaithersburg, MD) or Genosys (The Woodlands, TX)

ODN vehicles such as Lipofectin: N-[1-(2,3-dioleoyloxy)propyl]-N,N,N-trimethylammonium chloride/dioleoylphosphatidylethanol-amine (DOTMA/DOPE) solution from Life Technologies

Cells of interest with appropriate cell culture equipment and media

Oligodeoxynucleotide Design

Complementary oligonucleotide sequences of 18 to 22 bases in length can be derived from the entire targeted mRNA sequence. Protein-binding G-quartet sequence motifs, however, should be avoided.[14] For example, in order to target Raf-1, 34 phosphorothioate ODNs of 20 bases in length were designed by Isis Pharmaceuticals (Carlsbad, CA).[8] The CGP 69846A/ISIS 5132 sequence TCCCGCCTGTGACATGCATT hybridizes to the 3' untranslated sequence of Raf-1 mRNA. The mismatched control analog of CGP 69846A, TCCCGC_GCACTTG_ATGCATT, contains seven base changes (underlined) within the CGP 69846A sequence.[2]

Cell Culture

Cells of choice can be cultured according to standard protocols. For the study described here, primary human coronary artery SMCs (passages 4–7) from Clonetics (San Diego, CA) are grown in Falcon Primaria tissue culture flasks (75 cm²/250 ml) from Becton Dickinson (Franklin Lakes, NJ) in Clonetics smooth muscle cell basal labeling medium (SmBM) supplemented with 5% fetal bovine serum (FBS), human recombinant epidermal growth factor (0.5 ng/ml), insulin (5 μg/ml), human recombinant fibroblast growth factor (2 ng/ml), gentamicin (50 μg/ml), and amphotericin B (50

---

[14] C. A. Stein, *Nature Med.* **11,** 1119 (1995).

ng/ml) at 37° in a 5% CO_2 humidified atmosphere. Aliquots of approximately 50,000 cells are subcultured into 60-mm dishes (Falcon) by aspiration of the growth medium followed by a 30-sec rinse with a solution containing 0.5 mM EDTA and trypsin (0.25 mg/ml). Cell expansion resumes for 48 hr prior to ODN treatment.

Treatment of Cells with Oligodeoxynucleotides

The number of cells subjected to ODN treatment is limited by the sensitivity of the subsequent mRNA/protein analysis and by the time in which cells reach confluency. ODNs can be applied either to quiescent or cycling cells as required for the functional assay. In our study, approximately 200,000 cells are incubated with ODNs at a concentration of 10–400 nM in SmBM containing DOTMA/DOPE solution at a concentration of 0.25 mg/10 nM ODN, which corresponds to 0.175 mM N-[1-(2,3-dioleoyloxy)-propyl]-N,N,N-trimethylammonium chloride and 0.175 mM dioleoylphosphatidylethanolamine/10 nM ODN, or a 35 : 1 molar ratio of Lipofectin to ODN. After 4 hr, the medium is removed and replaced with SmBM containing 5% FBS. Twenty-four hours after ODN treatment the cells can be harvested to assess antisense effects on mRNA levels by Northern blotting.

Dose–Response Study: Assessment of Necessary Oligodeoxynucleotide Concentration for Inhibition of Gene Expression

It is essential first to demonstrate positively the concentration-dependent cellular effects of the applied ODN molecules, which indicate a specific antisense mechanism. For ODNs that have RNase H-activating ability, Northern blot analysis is the best choice to detect a selective loss of the target RNA. This antisense mechanism offers the advantage that the ODN activity is directly detectable within 24 hr after application in a dose–response study. For translation-inhibiting ODNs, Western blot analysis will represent the first and most direct assessment of a specific antisense mechanism. However, the analysis of this mechanism is strongly time dependent owing to the wide range of endogenous protein decay rates. A dose–response study will have to be generated after completion of a time–response assessment. Therefore, the identification of a phosphorothioate ODN that induces RNase H-mediated mRNA cleavage on hybridization facilitates activity and specificity assessments.

A quantitative assessment of concentration-dependent inhibition of mRNA function can be obtained by analyzing Northern or Western blot autoradiographs, using a densitometer, and normalizing the antisense ODN effects to mismatched control ODN values. The normalized densitometric mRNA and protein values can then be fitted to a general sigmoidal dose–response equation [Eq. (1), below] by nonlinear regression analysis to calculate approximate 50% inhibition concentration (IC_{50}) values as shown in Fig. 1.

The IC_{50} value, as expressed in nM concentration units, may be used to indicate the ODN concentration needed in the cell culture medium to achieve efficient inhibition of gene expression in the following time course and, ultimately, in the functional studies. We recommend exposing cells to an ODN concentration which exceeds by at least twofold the measured IC_{50} concentration.

Protocols for Dose–Response Study

Materials

Total RNA isolation system such as the TRI reagent from Molecular Research Center (Cincinnati, OH), or ToTally RNA from Ambion (Austin, TX)

Spectrophotometer for RNA quantitation

Agarose gel electrophoresis equipment

Nylon nucleic acid transfer membranes such as Hybond-N+ from Amersham Life Science (Arlington Heights, IL)

Polymerase chain reaction (PCR) primer for amplification of a complementary DNA (cDNA) probe template from a cDNA library such as Marathon cDNA from Clontech (Palo Alto, CA); several companies such as Life Technologies or Genosys offer custom primer delivery services

Glycerol-3-phosphate dehydrogenase (G3PDH) control cDNA template from Clontech

Random prime cDNA labeling kit from Promega (Madison, WI) or Amersham Life Science, [α-^{32}P]dCTP from Amersham Life Science for radiolabeled probe synthesis

Hybridization solution such as Quickhyb from Stratagene (La Jolla, CA) or ExpressHyb from Clontech

Densitometric scanner from Molecular Dynamics (Sunnyvale, CA)

Software Prism from GraphPad Software (San Diego, CA) for kinetic calculations and representations

Bradford protein quantitation solution from Bio-Rad (Hercules, CA)

Sodium dodecyl sulfate–polyacrylamide gel electrophoresis (SDS–PAGE) and protein transfer equipment

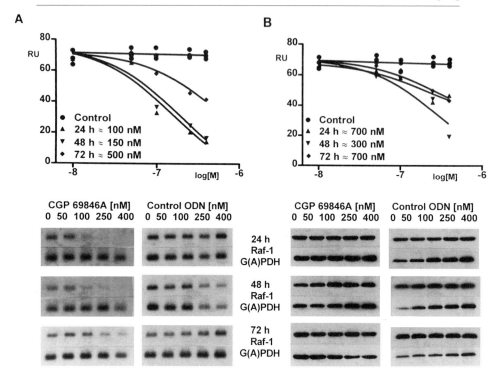

FIG. 1. Dose–response study for CGP 69846A. Dose–response analysis of Raf-1 mRNA (A) and protein (B) expression in human coronary artery SMCs after exposure to CGP 69846A and mismatched control ODNs. Cells were exposed to the indicated concentrations of ODN and analyzed for mRNA and protein expression by Northern and Western blotting, respectively. Cellular mRNA abundance of Raf-1 and G3PDH as well as Raf-1 and GAPDH protein levels were assessed 24, 48, and 72 hr after antisense treatment. The normalized quantitative Raf-1 mRNA and protein units, indicated as relative units (RU), were analyzed by nonlinear regression to calculate 50% inhibitory concentration (IC_{50}) values for CGP 69846A. IC_{50} values for the mismatch control ODN were not applicable. [Reproduced with data points from Schumacher *et al.*,[3] with permission of the publisher.] Dose-dependent catalysis of mRNA within 24 hr after antisense ODN exposure with an IC_{50} value of 100 nM in presented. Therefore, an ODN concentration of 400 nM was used in the following time-course and functional studies to downregulate mRNA levels. In comparison, the antisense ODN effect on protein expression was delayed in onset and of short duration, which indicates the targeting of a long-lived protein and transient mRNA suppression.

Immobilon-P, polyvinylidene difluoride protein transfer membranes from Millipore (Bedford, MA)

Enhanced chemiluminescence (ECL) detection system from Amersham Life Science or New England Nuclear (Boston, MA)

Specific antibodies to targeted gene product
Glyceraldehyde-3-phosphate dehydrogenase (GAPDH) monoclonal
    antibodies from Advanced ImmunoChemical (Long Beach, CA) as
    control reagent
Radioimmunoprecipitation assay (RIPA) buffer [contains 20 mM Tris-
    HCl (pH 7.5), 100 mM NaCl, 2.5 mM EDTA, 1 mM dithiothreitol
    (DTT), 1% (v/v) Triton X-100, aprotinin (100 kallikrein-inactivating
    units/ml), leupeptin (1.0 μg/ml), 1 mM phenylmethylsulfonyl fluo-
    ride (PMSF), 1 mM sodium orthovanadate, 10 μM sodium molyb-
    date, and 10 mM sodium fluoride]. All buffer chemicals can be
    obtained from Sigma (St. Louis, MO)
Binding buffer [contains Tris-buffered saline, 1 mM EDTA, 0.1%
    (v/v) Tween 20, 2% (w/v) bovine serum albumin (BSA), 1 mM
    dithiothreitol, and 0.02% (w/v) sodium azide (Sigma)]

Northern Blot Analysis

The isolation of total RNA used in our study is achieved using the TRI
system according to the manufacturer protocol. Purification of poly(A)$^+$
RNA is recommended only for low-copy messages. Briefly, monolayer cell
cultures are lysed directly in 60-mm culture dishes by addition of 0.8 ml of
TRI reagent supplemented with 8 μl of microcarrier gel. The lysate is mixed
with 0.1 ml of 1-bromo-3-chloropropane and centrifuged at 12,000g for 15
min at 4°. Total RNA is precipitated from the separated aqueous phase by
the addition of 2-propanol and the pellet dissolved in stabilized formamide.
RNA samples are quantitated spectrophotometrically by measuring the
optic density at 260 nm. The yield of this procedure ranges from 12 to 17
μg of total RNA isolated from approximately 10^6 primary human coronary
artery SMCs. The purity of these RNA preparations as assessed by the
OD$_{260/280}$ ratio ranges between 1.7 and 1.9. Five micrograms of total RNA
is fractionated by agarose–formaldehyde denaturing gel electrophoresis
and transferred to nylon membranes. Labeled Raf-1 cDNAs are synthesized
in vitro in the presence of [α-^{32}P]dCTP from a human Raf-1 cDNA template,
using random primers and Klenow enzyme, and are used to probe the
Northern blots. Hybridization analysis is carried out in Quickhyb solution
at 65° and visualized by autoradiography. The blots are then stripped of
radioactivity and reprobed with a ^{32}P random prime-labeled G3PDH cDNA
probe to confirm equal loading. Raf-1 mRNA of 3.6-kb molecular size is
quantified and normalized to G3PDH mRNA (1.4 kb) using densitometric
scanner analysis of autoradiograms exposed in the linear range of film
density.

Western Blot Analysis

The monolayer cell cultures expanded in 60-mm dishes are harvested in RIPA buffer. Lysates are cleared of particulate material by centrifugation at 10,000g for 10 min at 4°. Protein concentrations of cell lysates are determined by the Bradford method. The yield of extracted protein from approximately 10^6 primary human coronary artery SMCs ranges between 100 and 150 μg. A 20-μg aliquot of total cellular protein is fractionated by SDS–PAGE, transferred onto polyvinylidene difluoride membranes, and probed with anti-Raf-1 monoclonal antibodies from Transduction Laboratories (Lexington, KY) diluted in binding buffer. Bound primary antibodies are detected with peroxidase-labeled secondary antibodies, using the ECL method and autoradiography according to the protocol provided by the manufacturer. The membranes are stripped of Raf-1 immunodetectors and reprobed with anti-GAPDH monoclonal antibodies. Raf-1 protein levels of 74-kDa molecular size are quantified and normalized to GAPDH protein (36 kDa) levels by densitometric scanner analysis of autoradiograms.

Regression Analysis

The normalized densitometric scanner data from Northern and Western blot autoradiograms are fitted to a four-parameter first-order logistic equation by nonlinear regression, using the software Prism to calculate IC_{50} values.

$$RU = Min + [(Max - Min)/(1 + 10^{\log IC_{50} - Conc.})] \tag{1}$$

Min represents the normalized densitometric scanner value expressed in relative units (RU) at the top plateau whereas Max is the RU value at the bottom plateau. The logarithm of the IC_{50} concentration (log IC_{50}) is the ODN concentration (Conc.) that induces a response halfway between top and bottom plateau.

Time–Response Study: Assessment of Optimal Time Point to Perform
  Functional Study

The downregulation of a long-lived protein with antisense ODNs requires sustained inhibition of mRNA function, which depends ultimately on the duration of intracellular ODN activity. The three parameters of antisense efficacy, i.e., protein decay, mRNA inhibition, and ODN stability, can be assessed in a time-course study as shown in Fig. 2. After cell exposure to a single ODN treatment of 4 hr at the predetermined ODN concentration, mRNA and protein levels of the targeted gene are quantified at various time points. A rapid drop in target RNA levels indicates an efficient induction of

RNase H catalysis whereas a slow onset of protein level reduction reflects a long intrinsic protein turnover rate. The recovery of decreased mRNA levels back to endogenous levels results from decreasing intracellular ODN activity. The mRNA recovery with time is reflected by the weak and transient reduction of protein levels. The mRNA recovery rate can be calculated by subjecting the normalized density values to nonlinear regression analysis and can be interpreted as the half-life of active intracellular ODNs. To achieve sustained inhibition of mRNA and hence reduction of protein levels at the endogenous protein decay rate, a repeated application of antisense molecules is required as demonstrated in Fig. 2. Pharmacologically, we suggest a repeated ODN application within the time range of the ODN decay rates.

The biological consequences of downregulation of the targeted gene product are evaluated on the basis of the predetermined dose and time parameters of the ODN application. The functional assay should be conducted at the dose that produces the greatest reduction in mRNA (determined in dose–response study) at the time in which the greatest reduction in protein occurs (determined in the time-course study) as outlined in Fig. 2B.

To assess differential effects and side effects as well as direct and indirect effects of antisense ODN application, we recommend evaluation of the reduction of gene expression in complementary biological assays. The viability of a cell subjected to the antisense treatment must be assured in order to differentiate between toxic side effects and specific gene function. For instance, inhibition of cell proliferation can be validated by assays for apoptosis, necrosis, and cell cycle arrest.

The sensitivity of the chosen functional assay may also reflect the effects of gradual protein level reduction. Hence, a time-course correlation between functional response and protein expression may validate the significance of a targeted signaling protein.

Protocol for Time-Response Study

Repeated Oligodeoxynucleotide Application to Cell Culture

Approximately 200,000 human coronary artery SMCs are incubated for 4 hr with CGP 69846A or control ODN at a concentration of 400 nM in SmBM containing 10 mg of DOTMA/DOPE solution. Subsequently, the medium is replaced with SmBM containing 5% FBS and Raf-1 mRNA and protein values are assessed at 24-hr intervals by Northern and Western blotting, respectively. The decay rate of CGP 69846A intracellular activity over time is interpreted as equivalent to the mRNA doubling time of

A

Raf-1 mRNA levels in response to one (T) ODN application

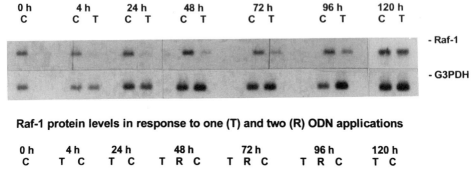

Raf-1 protein levels in response to one (T) and two (R) ODN applications

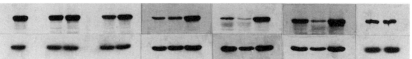

B

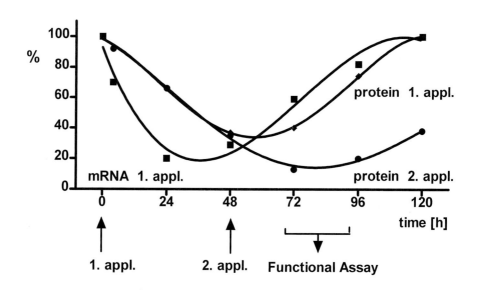

recovering Raf-1 mRNA density values (normalized to control ODN). Thus, the CGP 69846A decay rate is calculated by nonlinear regression analysis to be ~50 hr. Therefore, after 48 hr the growth medium is removed and the cells reexposed for 4 hr to 400 nM CGP 69846A or control ODN in SmBM containing 10 mg of DOTMA/DOPE solution. Subsequently, the cell culture resumes in growth medium for an additional 72 hr.

Recommended Cellular Viability Assays

1. Cell proliferation assays based on the quantification of the succinate–tetrazolium reductase system in the mitochondrial respiratory chain or assays based on the monitoring of DNA synthesis by the incorporation of bromodeoxyuridine or tritiated thymidine into cellular DNA.

2. Apoptosis detection systems such as the assessment of DNA fragmentation by the TUNEL (TdT-mediated dUTP nick-end labeling) assay or the identification of cell membrane changes by annexin V detection.

3. Cytotoxicity detection systems such as the measurement of cellular lactate dehydrogenase release or the quantification of differential uptake/exclusion of dyes such as trypan blue and propidium iodide by microscopic counting.

Kinetic evaluation of Antisense Oligodeoxynucleotide Activity

The pharmacodynamic activity of an antisense ODN on cellular gene expression and function is reflected by a serial dose- and time-dependent mechanism. The ODN activity is reversible and can be extended by repeated

---

FIG. 2. Time–course study for CGP 69846A. Time course of Raf-1 mRNA and protein expression in human coronary artery SMCs after treatment with CGP 69846A and mismatched control ODN. Cells were treated with ODN at a concentration of 400 nM, reexposed to growth stimulation, and analyzed for *Raf-1* gene expression at the indicated time points. (A) Raf-1 and G3PDH/GAPDH levels were measured by Northern and Western blotting and represented as normalized values in (B). Samples indicated by (T) were subjected to a single antisense treatment. Samples indicated by (R) were subjected to an additional treatment period after 48 hr. Control cell samples (C) were mock treated. [Reproduced with data points from Schumacher *et al.*,[3] with permission of the publisher.] (B) The normalized Raf-1 time-point values were subjected to a nonlinear regression analysis. In response to a single ODN application, mRNA level dropped within 30 hr before recovering to endogenous levels due to diminishing ODN activity. In comparison, inhibition of protein expression was delayed in onset and limited by *de novo* synthesis from recovering mRNA levels. A second ODN application within the time range of the ODN decay rate (48 hr) was necessary to achieve further and sustained suppression of protein expression. Functional assays for the targeted gene were ultimately performed within the time window of maximally reduced cellular protein levels.

A   **B**

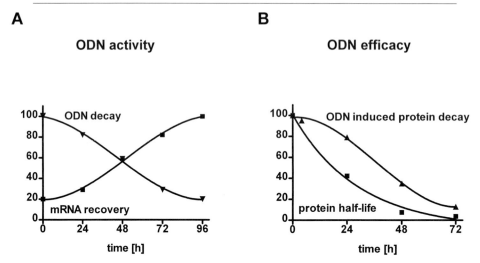

FIG. 3. Kinetic evaluations of antisense activity. (A) mRNA recovery points after a single antisense ODN treatment as measured in the time-course study in Fig. 2 were used to calculate an mRNA doubling time by nonlinear regression analysis. The mRNA doubling time of 50 hr was interpreted by inversion as the half-time of intracellular ODN activity. The ODN decay rate may indicate the time interval for repeated ODN applications. (B) Metabolic turnover rate of Raf-1 protein in correlation with antisense-induced protein level reduction. Cells were metabolically labeled with [^{35}S]methionine/[^{35}S]cysteine and subsequently Raf-1 was immunoprecipitated at the indicated time points. Immunoprecipitates were analyzed by Western blotting. The Raf-1 density values indicated as relative units (RU) were subjected to nonlinear regression analysis to calculate the decay half-time ($t_{1/2}$). The endogenous Raf-1 turnover half-life of ~30 hr correlated kinetically with the antisense-induced protein decay rate (50% decay in ~33 hr as derived from Fig. 2), indicating that the efficiency of CGP 69846A in decreasing Raf-1 protein levels was rate limited by the endogenous protein turnover rate. [Reproduced with data points from Schumacher et al.,[3] with permission of the publisher.]

applications to achieve downregulation of long-lived proteins as shown in Fig. 2. Nonlinear regression analysis of the recovering mRNA density values (normalized to control) may generate an mRNA doubling time. The mRNA doubling time can be interpreted as the half-life of active intracellular ODN molecules because the intrinsic mRNA turnover rate is presumed to occur faster, as shown in Fig. 3A. The chemical half-life of an average mRNA species in mammalian cells is estimated to last only several hours.[15] The intracellular activity of phosphorothioate ODNs may be time limited by their susceptibility to nucleolytic degradation or to cellular excretion.

To assess the efficacy of an antisense ODN compound in decreasing protein levels, the endogenous protein turnover rate in cycling cells can be

[15] M. F. Tuite, *Nature* (*London*) **382,** 577 (1996).

measured as shown in Fig. 3B. Assuming an antisense treatment suppresses cellular protein synthesis entirely, one would expect the antisense-induced decay rate to be equal to the endogenous protein turnover rate. The endogenous protein turnover rate can be assessed using metabolically pulse–chase labeled cells and subsequently monitoring the decomposition of immuno-purified protein over time. The 50% protein decay rate is calculated by nonlinear regression analysis of the densitometric protein values. An endogenous protein decay rate comparable to the antisense-induced protein decay rate may indicate that the efficiency of the antisense ODN in decreasing protein levels is rate limited by the endogenous protein turnover rate. The pharmacodynamic effects of the antisense ODN are therefore limited by the duration of its intracellular activity rather than its ability to transiently decrease mRNA levels.

Protocol for Endogenous Protein Turnover Determination

Materials

[^{35}S]Methionine/[^{35}S]cysteine labeling mix from New England Nuclear
Methionine/cysteine-free cell culture medium
Protein G beads from Pharmacia (Piscataway, NJ)
Western blot protocol
PhosphorImager with Storm system from Molecular Dynamics
Software Prism from GraphPad Software for kinetic calculations and
    representations

Metabolic Turnover Rate Determination of Raf-1 Protein

Human coronary artery SMCs are grown to 40% confluence and washed twice with methionine/cysteine-free medium. The cells are incubated with [^{35}S]methionine/[^{35}S]cysteine labeling mix (0.4 mCi/ml) in medium devoid of methionine and cysteine for 2 hr. After addition of an equal volume of 10% FBS in SmBM, the incubation is extended to 16 hr. After replacement with 5% FBS in SmBM, cell lysates are prepared after various time periods as described in Western blot analysis. Total cell lysates are incubated for 3 hr at 4° with 50 μl of protein G bead-conjugated Raf-1 antibodies or an unrelated control antiserum. The immunocomplexes are collected by centrifugation, washed twice with RIPA buffer, subjected to SDS-PAGE, and transferred to polyvinylidene difluoride membranes. Radioactivity is quantitated by phosphoimaging, using a Storm system. The phosphoimager-quantitated protein turnover data are subjected to a one-phase exponential decay equation by nonlinear regression analysis to calculate half-time values.

Acknowledgments

The author gratefully acknowledges Teresa E. Gerlock and Catherine Cioffi for critical reading of the manuscript and Lawrence Wennogle for technical advice. The author thanks ISIS Pharmaceuticals (Carlsbad, CA) for supplying CGP 69846A and the control ODN.

[32] Application of Antisense Oligodeoxynucleotides for Suppression of Na^+/Ca^{2+} Exchange

By BEAT SCHWALLER, MARCEL EGGER, PETER LIPP, and ERNST NIGGLI

Introduction and Principle of Method

Ca^{2+} homeostasis within many different types of cells depends on the activity of an Na^+/Ca^{2+} exchanger located in the plasmalemma.[1] Functional studies pertaining to the Na^+/Ca^{2+} exchanger are notoriously difficult to undertake, as well as to interpret, at both the cellular and molecular levels. In cells exhibiting complex Ca^{2+} signaling systems with several different pathways for Ca^{2+} entry and removal, it has been a great challenge to elucidate the physiological role of the Ca^{2+} fluxes via Na^+/Ca^{2+} exchange. Cardiac muscle cells represent one such case, where studies on the physiological role of this transporter have been complicated by the complexity of cardiac Ca^{2+} signaling.[2] Thus far no specific pharmacological inhibitor of the Na^+/Ca^{2+} exchanger has been discovered. All known inhibitors also block Ca^{2+} and/or K^+ channels, and thus cannot be used to distinguish different Ca^{2+} signal transduction pathways or to elucidate the cellular role of the Na^+/Ca^{2+} exchange in cardiac muscle cells.

The lack of a specific inhibitor has also impeded attempts to understand the molecular mode of operation of the Na^+/Ca^{2+} exchanger.[3,4] At the molecular level Na^+/Ca^{2+} exchange is known to comprise a cycle of several reaction steps that lead to the electrogenic translocation of ions across the cell membrane with a stoichiometry of $3\ Na^+ : 1\ Ca^{2+}$. But the sequence of

[1] J. P. Reeves, *J. Bioenerg. Biomembr.* **30,** 151 (1998).
[2] D. Schulze, P. Kofuji, R. Hadley, M. S. Kirby, R. S. Kieval, A. Doering, E. Niggli, and W. J. Lederer, *Cardiovasc. Res.* **27,** 1726 (1993).
[3] E. Niggli and W. J. Lederer, *Nature (London)* **349,** 621 (1991).
[4] D. W. Hilgemann, D. A. Nicoll, and K. D. Philipson, *Nature (London)* **352,** 715 (1991).

events actually involved remains to be clarified. Several computer models have been developed with this aim in view, and two fundamentally different transport mechanisms have been proposed: (1) a consecutive exchange cycle in which Na^+ and Ca^{2+} are moved across the membrane in two separate steps; and (2) a simultaneous mechanism in which both Na^+ and Ca^{2+} must bind to the transporter molecule before it can undergo a molecular rearrangement involving ion translocation.[5]

In this chapter, we describe an alternative approach for investigating the role of the Na^+/Ca^{2+} exchanger. Because no specific pharmacological tools are available, we have designed antisense oligodeoxynucleotides (AS-ODNs) directed against the Na^+/Ca^{2+} exchanger mRNA of rat myocytes that specifically downregulate this transport protein.[6,7] The antisense approach is evaluated here, with particular emphasis being given to uptake, specificity, and cytotoxicity of the AS-ODNs, and to the methodology implemented to check for potential pitfalls. Exposure of cultured neonatal rat myocytes to the AS-ODNs resulted in a rapid (within 24 hr) suppression of Na^+/Ca^{2+} exchange activity. This was revealed by measuring Ca^{2+} transport capacity using ratiometric confocal Ca^{2+} imaging and flash photolysis of caged Ca^{2+}, as well as by monitoring membrane currents generated by electrogenic Na^+/Ca^{2+} exchange using the voltage-clamp technique in the whole-cell mode. The antisense approach permits not only the gleaning of important information concerning the function of the Na^+/Ca^{2+} exchanger but also facilitates an investigation of the role of other Ca^{2+} extrusion systems in rat neonatal myocytes.

Strategies for Design of Antisense Oligodeoxynucleotides and for Antisense Experiments

Initially, the design of antisense-oligodeoxynucleotides (AS-ODNs) seemed to be a relatively simple undertaking. A short DNA molecule (15–20 nucleotides in length), which was complementary to a part of the mRNA of interest, was believed to bind to and inhibit the target RNA via Watson–Crick base pairing, thereby blocking expression of the selected protein. It was assumed that an AS-ODN of 17 or more nucleotides would be sufficiently specific to ensure, with high probability, a perfect match for the desired target gene alone (for review, see Refs. 8–10). Hence, a search in the gene databank (e.g., GenBank update and EMBL update) using the sequences of the designed AS-ODNs was deemed to be adequate for the

[5] P. Läuger, *J. Membr. Biol.* **99,** 1 (1987).

[6] P. Lipp, B. Schwaller, and E. Niggli, *FEBS Lett.* **364,** 198 (1995).

[7] E. Niggli, B. Schwaller, and P. Lipp, *Ann. N.Y. Acad. Sci.* **779,** 93 (1996).

experimental needs. However, this notion has since been overturned by the observation of many nonspecific effects exerted by AS-ODNs, sometimes of an entirely unexpected nature (reviewed in Refs. 8, 11, and 12). These include suppression or stimulation of cell proliferation, inhibition of viral entry into cells, and changes in cell adhesiveness. Furthermore, the effective blocking of the specific mRNA by AS-ODNs depends critically on the choice of region. On the basis of data presented in several reports, Milner et al.[13] have concluded that the structure of the native RNA dramatically restricts the binding of ODNs. Of the 1938 ODNs (with various lengths up to 17 nucleotides) tested against a stretch of 122 nucleotides at the 5' end of the β-globin gene, a small number became stably bound to the target RNA. This suggests that accessibility is one of the most important factors governing successful interaction between AS-ODNs and RNA.[13] Systematic testing of 9 AS-ODNs against the proliferation-associated nucleolar antigen p120 mRNA,[14] and of 34 AS-ODNs against c-raf kinase mRNA,[15] has revealed only a small fraction (less than 10%) of AS-ODNs to exert an antisense effect. In the latter study, only one AS-ODN manifested more than a fivefold reduction in c-raf kinase mRNA. Given the tremendous interest in AS-ODNs and the problems encountered with nonspecific effects, guidelines have been laid down by editors of journals devoted to antisense research (e.g., Antisense Research and Development[16]) and these are now widely complied with. If the function of the targeted protein is unknown, then the task of pinpointing the observed "antisense effect" to its inhibition is an awesome one, requiring the careful consideration of appropriate controls.[17] The same holds true for the inhibition of a protein whose function is known; but in this case, the availability of specific functional assays facilitates the design of control experiments. ODN variants to be tested would include (1) sense control [maintaining structural features (e.g., palindromes, stem loops, and class), but not composition], (2) inverse antisense control [reverse 5'-to-3' orientation compared with AS-ODN (maintaining structural features and composition, but no capacity to anneal

[8] A. D. Branch, Trends Biochem. Sci. 23, 45 (1998).
[9] C. Hélène and J.-J. Toulmé, Biochim. Biophys. Acta 1049, 99 (1990).
[10] J. F. Milligan, R. J. Jones, B. C. Froehler, and M. D. Matteucci, Ann. N.Y. Acad. Sci. 716, 228 (1994).
[11] C. A. Stein, Nature Med. 1, 1119 (1995).
[12] C. A. Stein, Trends Biotechnol. 14, 147 (1996).
[13] N. Milner, K. U. Mir, and E. M. Southern, Nature Biotechnol. 15, 537 (1997).
[14] L. Perlaky, Y. Saijo, R. K. Busch, C. F. Bennett, C. K. Mirabelli, S. T. Crooke, and H. Busch, Anti-Cancer Drug Design 8, 3 (1993).
[15] B. P. Monia, J. F. Johnston, T. Geiger, M. Muller, and D. Fabbro, Nature Med. 2, 668 (1996).
[16] C. A. Stein and A. M. Krieg, Antisense Res. Dev. 4, 67 (1994).
[17] C. A. Stein, Antisense Nucleic Acid Drug Dev. 8, 129 (1998).

to the target mRNA), (3) scrambled control (maintaining composition, but not structural features), (4) mismatched control (a few mismatches within the AS-ODN prevents perfect annealing to the target mRNA and demonstrates target hybridization selectivity), and (5) mismatched target control (addition of AS-ODNs to a cell line in which the gene is mutated or not expressed, to demonstrate the lack of nonsequence specificity). Additional experiments must be performed to confirm that the AS-ODNs are taken up by the control cells.

Experimental Procedures

Design of Phosphorothioate Oligodeoxynucleotides

In many studies, AS-ODNs have been targeted against the region around, and often preceding, the AUG start codon, but in several other studies potent AS-ODNs have been yielded by targeting against the 3' untranslated region (3' UTR) of the mRNA.[14,18] Many ODNs usually need to be tested before an efficacious and specific AS-ODN can be identified. The rat cardiac Na^+/Ca^{2+} exchanger mRNA consists of 3' 155 nucleotides and has a fairly short 3' UTR, only 177 nucleotides in length, before the poly(A) tail.[19] Comparison of the 3' UTR sequences for the cardiac Na^+/Ca^{2+} exchangers in different species has revealed a high degree of homology in certain sections (Fig. 1A), and Lipman[20] has postulated that these highly conserved regions may be associated with the regulation of mRNA stability. The same authors have also hypothesized that the conserved regions could form long, perfect duplexes with endogenous antisense transcripts, which would render them accessible targets for AS-ODNs. A hypothetical structure for the 3' UTR of the rat cardiac Na^+/Ca^{2+} exchanger has been determined using the RNAdraw program[21] (Fig. 1B). We have chosen a sequence in the 3' UTR of the cardiac Na^+/Ca^{2+} exchanger (positions 3065–3083, numbering according to the rat sequence[19]; see enlargement in Fig. 1B) that is 100% conserved in the rat, dog and human (Fig. 1A). The sequences of the antisense (AS), nonsense (NS, reverse 5'-to-3' orientation compared with antisense), and mismatched (MM, containing three mismatches when compared with the AS-ODN) ODNs are given in Fig. 1C. In the literature, NS-ODNs are often referred to by the more

[18] C. Cazenave, C. A. Stein, N. Loreau, N. T. Thuong, L. M. Neckers, C. Subasinghe, C. Helene, J. S. Cohen, and J. J. Toulme, *Nucleic Acids Res.* **17**, 4255 (1989).
[19] W. Low, J. Kasir, and H. Rahamimoff, *FEBS Lett.* **316**, 63 (1993).
[20] D. J. Lipman, *Nucleic Acids Res.* **25**, 3580 (1997).
[21] O. Matzura and A. Wennborg, *Comput. Appl. Biosci.* **12**, 247 (1996).

A

```
        2977
RAT   TAAAGGAACAATCAAGATATAATAAATTTATATATATATGTATACATATATATATACATAAAA ATTATGTATAATGAACAGAGGAAA
DOG   TAAAGGAACAATCA GATATAGTAAATTTATATATATATACGT      ATATATATACATAAAAATTATGTATAATGAACAGAGGAAA
HUM   TAAAGGAACTATCA GATATAGTAAATTTATATATATAC          ATATATATACATAAAA ATTATGTATAATGGACAGAGGAAA
BOV   TAAAGGAACAATCA GATGTAGTAAATTTATATATATATAC  ATATATATATATACATAAAA ATTATGTATAATGAACAGAGGAAA

                                                                                            3155
RAT   CTGACATTTGTCATGTTCACTTAACCTGCTGATGGAATCCAGCTTCAAGAACGTACTCTGTACTAGGCCGGAAGTCAGAAACCATCATCTCCAAA..
DOG   CTGACATTTGTCATGTTCACTTA CCTGCTGATGGAATCCAGCTTCAAGAGCATACTCTGTACTAGGGCTGAAGTGAGAAACCATCACCTCC...
HUM   CTGACATTTGTCATGTTCACTTA CCTGCTGATGGAATCCAGCTTCAAGAGCATACTCTGTACTAGGGCCGAAGTAAAAAACCATCACCTCC...
BOV   CTGGCATTTGTCATGTCCACCCA CCTGCTGATGGAATCCAGCTTCAAGAGCAGACTCTGTACTAGGGCCGGAGAGAGAAGGCATCACCTCC...

DOG   CATT CCCAGGGGCGT CATCACATTGAACAAGGCATGGAGGCAGGGGCATCTTTGCAGCTCAGCCTAGAAGGACTGTGTTCTGGAATTC..
HUM   CATT CCCAGGGGCAT CATCATGTTCAACAAGGCATGGAGGCAGGGCCATCTTTGCAGCTCAGTCTAGAAGGGCTGCACTCTCAAA...
BOV   CGTTTCCCAGGGGCGTTCGTCTTGTTGAACCAGGCATGGAGGCAGGGCCATCTTTACGTCAGCTCAGCCCAGAAGCGGTGTGTTCTCCC...
```

B

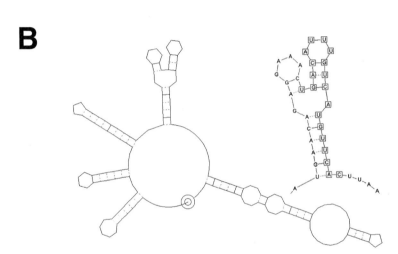

C

```
                                                            3065
Na-Ca exchanger cDNA                              5'  TGACATTTGTCATGTTCAC  3'

Antisense-oligodeoxynucleotide         AS    3'  ACTGTAAACAGTACAAGTG  5'
Non-sense-oligodeoxynucleotide         NS    5'  ACTGTAAACAGTACAAGTG  3'
Mismatched-oligodeoxynucleotide        MM    3'  AATGTACACAGTACAATTG  5'
Control oligodeoxynucleotide           CR    5'  GGCCAGCCATGGCGAGCCG  3'
```

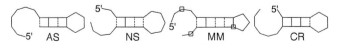

logical name *inverse-antisense* (IAS) ODNs; but because we introduced the term NS-ODN in a previous publication,[6] we will continue to use it here for the sake of consistency. In addition to the AS-, NS-, and MM-ODNs a completely unrelated ODN, which is complementary to the start-codon region of the EF-hand calcium-binding protein calretinin (CR),[22] was used as a control. Each of the ODNs used is 19 nucleotides in length, the G/C content of the Na$^+$/Ca^{2+} exchanger ODNs (AS, NS, and MM) being approximately 30%. One important aspect to be considered is the secondary structure of the ODNs[23]; all were thus tested for hairpin formation, using the RNAdraw program.[21] The results are presented in Fig. 1C.

Synthesis and Purification of Oligodeoxynucleotides

Phosphorothioate ODNs are synthesized on a Gene Assembler A 470 (Pharmacia, Uppsala, Sweden) using standard phosphoramidite chemistry and the specific thiolating reagent. They are either purified on a Mono Q HR5/5 anion-exchange column using a fast protein liquid chromatography (FPLC) system (Pharmacia) or simply desalted by passing them through NAP-5 columns (Pharmacia). For the FPLC purification, ODNs are dissolved in buffer A (10 mM NaOH; 0.1 M NaCl; pH 12), applied to the column, and then eluted with 30 ml of 35–70% buffer B (10 mM NaOH; 1 M NaCl; pH 12) at a flow rate of 1 ml/min. A typical chromatogram is

---

[22] J.-C. Gander, V. Gotzos, B. Fellay, and B. Schwaller, *Exp. Cell Res.* **225,** 399 (1996).
[23] M. Mitsuhashi, *J. Gastroenterol.* **32,** 282 (1997).

FIG. 1. (A) Sequence comparison of the 3′ untranslated regions (3′ UTR) of the cardiac Na$^+$/Ca^{2+} exchanger in the rat, dog, human (hum), and cow (bov). Numbering corresponds to the rat cDNA sequence published by Low *et al.*[19] The stop codon (boldface) and the A's of the poly(A) signal (italics) are depicted. One of the regions that is 100% conserved in the rat, dog, and human (underlined) was chosen for the AS-ODN. (B) Hypothetical structure of the 3′ UTR determined using the RNAdraw program.[21] The onset of the 3′ UTR is circled. G/C pairing is indicated by solid lines, A/U pairing by dashed lines. An enlargement of the AS-ODN-binding region (boxed nucleotides) is furnished. (C) Nucleotide sequence (top) and hypothetical structure (determined using the RNAdraw program,[21] bottom) of the four ODNs used in this study. The sequence of the antisense oligodeoxynucleotide (AS-ODN) is complementary to that of the 3′ UTR of the rat Na$^+$/Ca^{2+} exchanger from positions 3065–3083. The sequences of the nonsense (NS) (inverse 5′-to-3′ orientation compared with the AS-ODN) and mismatched (MM) [three mismatches (underlined) compared with the AS-ODN] ODNs are also represented. The sequence of the control ODN (AS-ODN against the calcium-binding protein calretinin) is given on the bottom line.

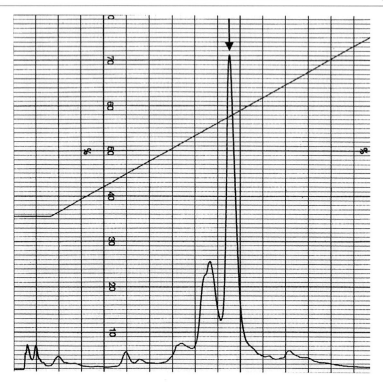

Fig. 2. Chromatogram of ODN purification on an anion-exchange column (Mono Q HR5/5), using an FPLC system from Pharmacia. The peak elution of the pure, full-length ODN is marked by an arrow.

depicted in Fig. 2; the peak fraction used for the experiments is indicated by an arrow. A good separation from shorter ODNs ($n - 1, n - 2$, etc.) is usually achieved. ODNs of interest eluted at 0.5–0.6 M NaCl. Fractions are concentrated in a SpeedVac (Savant; Holbrook, NY), dissolved in water, and then desalted on PD-10 columns (Pharmacia). The concentration of the ODNs is measured photometrically at 260 nm (1 OD = 30 μg/ml). Stock solutions of purified ODNs (100–300 μM) are stored at 4° (or at −70°, if not required immediately). One batch of control ODNs against calretinin is in addition labeled at the 5′ end with fluorescein, using the FluoPrime fluorescein amidite (Pharmacia), and used to monitor the uptake of ODNs in the different cell types. Phosphorothioate ODNs (either fully phosphothiolated or with thio linkages between only three and five nucleotides at the 5′ and 3′ ends) can be purchased from many commercial suppliers. It is important to obtain preferably high-performance liquid chromatography (HPLC)-purified ODNs, because even small amounts of impurities from the ODN synthesis can have nonspecific effects.

Cell Culture

Neonatal Rat Cardiac Myocytes

Primary cultures of neonatal rat cardiac myocytes are prepared by established methods.[24] Briefly, the hearts are rapidly removed from decapitated 2- to 3-day-old Wistar rats, cut into 1-mm^3 pieces, and gently stirred in Hanks' balanced salt solution (GIBCO, Basel, Switzerland) containing 4% trypsin (Roche Diagnostics, Rotkreuz, Switzerland) and pancreatin (120 mg/liter; Sigma, St. Louis, MO) at 35°. Cells are harvested from the supernatant every 15 min until the tissue has disintegrated. Isolated cells are sedimented by centrifugation (at 1000 rpm for 3 min) and resuspended in medium M199 (GIBCO) supplemented with penicillin (20 U/ml), streptomycin (20 μg/ml), and 10% fetal calf serum (FCS). A drop of cell suspension is placed on a round coverslip in a petri dish. The AS-ODNs are added at an initial concentration of 1 or 3 μM after 2 hr. Control cultures are established in parallel. Petri dishes are maintained in an incubator (1.5% CO$_2$) at 37° until required for experimental purposes.

WiDr (Colon Carcinoma) Cells

The human colon adenocarcinoma cell line WiDr [American Type Culture Collection (ATCC, Manassas, VA): (ATCC CCL 218) is cultivated as a monolayer at 37°, in a 5% CO$_2$/95% air atmosphere with a relative humidity of 98%. The culture medium (EF medium) is a 1:1 mixture of EMED (an enriched Dulbecco's modified Eagle's medium) and FMED (modified Ham's F-12 nutrient mixture), containing 10% fetal calf serum, penicillin (10^4 units/liter), and streptomycin (10 mg/liter). The cells are regularly subcultured by trypsination [0.25% (w/v) trypsin in Ca^{2+}- and Mg^{2+}-free phosphate-buffered saline].

Sf9 Insect Cells

Cell Growth. Sf9 cells from the insect *Spodoptera frugiperda* are grown to near confluence (logarithmic growth phase; 98% viability) in Grace's insect medium (GIBCO) supplemented with 10% fetal calf serum and

[24] S. Rohr, *Pflügers Arch.* **416,** 201 (1990).

gentamicin (50 μg/ml; Sigma) under ambient atmospheric conditions (22–27°; CO_2 is not essential). They are resuspended at an initial density of 10^6 cells/ml[25] in the same medium containing 0.1% pluronic F-68 (Sigma) to reduce the shear stress, and transferred to a 100-ml spinner flask, where they are maintained at ambient temperature (22–27°) with constant stirring at 60–70 rpm. The cells are counted every 7 days and an appropriate aliquot withdrawn to contain 2–2.5 × 10^6 cells/ml when required.

Expression of the Na$^+$/Ca^{2+} Exchanger and Antisense Oligodeoxynucleotide Uptake. The human cardiac Na$^+$/Ca^{2+} exchanger is cloned and expressed in HEK293 cells,[26] the cDNA being then subcloned into baculovirus for expression in Sf9 cells.[26,27] For expression of the human cardiac Na$^+$/Ca^{2+} exchanger and AS-ODN uptake, Sf9 cells at a density of 1–2 × 10^6 cells/ml of serum-free Grace's medium are transferred from the suspension culture to 25-cm^2 culture flasks. They are allowed to attach to the flasks for 40–60 min at 27°. The supernatant is then replaced with fresh complete Grace's medium and AS-ODNs are added at an initial concentration of 1–5 μM after 24 hr (70% confluent monolayer), in the absence or presence of 1,2-dioleoyl-3-trimethylammoniumpropane (DOTAP, 10 μg/ml; Roche Diagnostics). After incubation overnight, the Sf9 cells are infected at a multiplicity of infection (MOI) of 5–10 from a recombinant baculoviral stock (0.5–1 × 10^8 PFU/ml; viral titer determined by end-point dilution). The supernatant (containing AS-ODNs and the viral inoculum) is not replaced during the ensuing 48 hr. To facilitate infection, the volumes of virus and culture medium should be kept to a minimum, just sufficient to cover the cells. Control cultures are established in parallel. The culture flasks are maintained at 27° in an incubator. Tests for the functional expression of the human Na$^+$/Ca^{2+} exchanger in Sf9 cells are described in detail elsewhere.[28]

Cellular Uptake of Oligodeoxynucleotides

Endocytosis-mediated uptake of ODNs is assumed to be the major mechanism in cultured cells and also *in vivo*.[29] Many factors including the

[25] M. D. Summers and G. E. Smith, *in* "Texas Agriculture Experiment Station Bulletin," Vol. 1555. College Station, Texas, 1987.

[26] P. Kofuji, R. W. Hadley, R. Kieval, W. J. Lederer, and D. H. Schulze, *Am. J. Physiol.* **263**, 1241 (1992).

[27] Z. Li, C. D. Smith, J. R. Smolley, J. H. Bridge, J. S. Frank, and K. D. Philipson, *J. Biol. Chem.* **267**, 7828 (1992).

[28] M. Egger, A. Ruknudin, P. Lipp, P. Kofuji, W. J. Lederer, D. H. Schulze, and E. Niggli, *Cell Calcium* **25**, 9 (1999).

[29] C. A. Stein, *Antisense Nucleic Acid Drug Dev.* **7**, 207 (1997).

length and composition of the ODNs, cell type, and composition of the culture medium are known to influence this process.[30] Optimal conditions thus must be established for each cell system in hand. Using some cell types, the quantity of ODNs taken up from the bathing medium is sufficient to produce an antisense effect; with others uptake of ODNs must be facilitated by various means. These include the application of cationic lipids[31–33] such as those used for transfection experiments [e.g., DOTAP (Roche Diagnostics) and Lipofectin and LipofectAMINE (GIBCO-BRL, Gaithersburg, MD)], microinjection by pipette,[34] and brief permeabilization with streptolysin O.[35] To check for ODN uptake, an AS-ODN against the calcium-binding protein calretinin[22] labeled with fluorescein at the 5′ end is used. The uptake of this ODN into either neonatal rat cardiac myocytes or Sf9 insect cells is investigated. Results pertaining to the kinetics of uptake in these two cell types are presented in Fig. 3. Neonatal rat cardiac myocytes are incubated for 5–18 hr with the fluorescein-labeled ODNs (final concentration; 1 μM), which are added directly to the culture medium. The time course of cellular uptake is monitored by confocal microscopy. Fluorescence intensity is highest in the nucleus and much lower in the cytoplasm of the myocytes, indicating that the ODNs localize preferentially in the former (Fig. 3A6 and 3B7). Because the plasma membrane is virtually devoid of staining, nonspecific binding of ODNs to its outer or inner faces can be excluded. Under the same conditions, the uptake of the fluorescein-labeled ODN (final concentration, 1 μM) by Sf9 cells is poor (Fig. 3A1 and 3B2). An uptake comparable to that observed in myocytes occurs only in the presence of the cationic lipid DOTAP (10 μg/ml) or under conditions of alkaline culture medium (pH 8) (Fig. 3C4 and 3C5, respectively). In the absence of DOTAP, a sufficient quantity of ODNs is taken up only when their final concentration is raised by 10- to 20-fold within the bathing medium. Under those conditions, uptake is still inhomogeneous after 18 hr (Fig. 3C3).

[30] J. Temsamani, M. Kubert, J. Tang, A. Padmapriya, and S. Agrawal, *Antisense Res. Dev.* **4**, 35 (1994).

[31] J. H. Felgner, R. Kumar, C. N. Sridhar, C. J. Wheeler, Y. J. Tsai, R. Border, P. Ramsey, M. Martin, and P. L. Felgner, *J. Biol. Chem.* **269**, 2550 (1994).

[32] S. Capaccioli, G. DiPasquale, E. Mini, T. Mazzei, and A. Quattrone, *Biochem. Biophys. Res. Commun.* **197**, 818 (1993).

[33] M. J. Hope, B. Mui, S. Ansell, and Q. F. Ahkong, *Mol. Membr. Biol.* **15**, 1 (1998).

[34] J. P. Leonetti, N. Mechti, G. Degols, G. Gagnor, and B. Lebleu, *Proc. Natl. Acad. Sci. U.S.A.* **88**, 2702 (1991).

[35] K. E. White, F. A. Gesek, and P. A. Friedman, *Ann. N.Y. Acad. Sci.* **779**, 115 (1996).

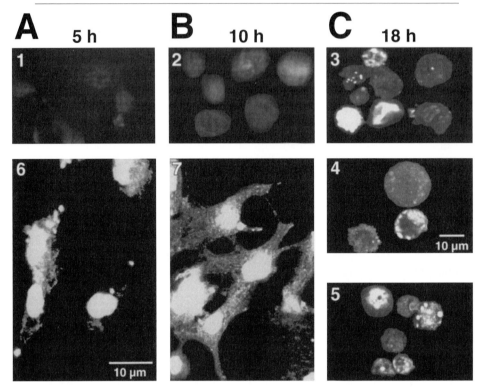

FIG. 3. Uptake of ODNs by Sf9 cells (A1–C5) and neonatal rat myocytes (A6 and B7) monitored using the fluorescein-labeled AS-ODNs against the Ca^{2+}-binding protein calretinin. Confocal images were taken after 5 hr (A), 10 hr (B), and 18 hr (C). Sf9 cells incubated with the fluorescent ODN at a concentration of 1 μM have taken up little of the ODN (A1 and B2) compared with neonatal rat myocytes (A6 and B7). At 10 μM, uptake of ODNs by Sf9 cells was enhanced (C3), as it was also in the presence of DOTAP (10 $\mu g/ml$; C4) or when the pH of the culture medium was rendered alkaline (pH 8; C5).

Oligodeoxynucleotide Cytotoxicity in WiDr Cells

For cell growth assays, WiDr cells are cultured for 24 hr in 96-well plates (7000 cells/well). The supernatant is then removed and replaced by fresh medium containing ODNs (at a final concentration of 3 μM). The number of viable cells is determined after 24, 48, and 96 hr, using the MTT assay.[36] With this colorimetric method, the dye 3-(4,5-dimethyldiazol-2-yl)-2,5-diphenyltetrazolium bromide (MTT) is reduced to a blue crystalline

[36] M. C. Alley, D. A. Scudiero, A. Monks, M. L. Hursey, M. J. Czerwinski, D. L. Fine, B. J. Abbott, J. G. Mayo, R. H. Shoemaker, and M. R. Boyd, *Cancer Res.* **48,** 589 (1988).

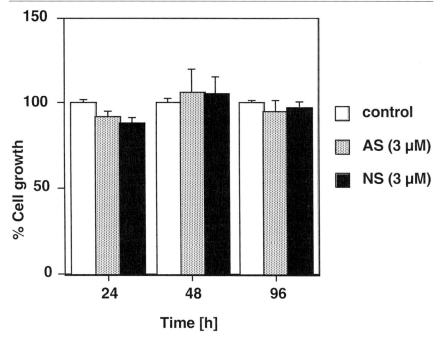

FIG. 4. Cytotoxicity assay. WiDr cells (7000 cell/well) were grown in 96-well microtiter plates for 24–96 hr and the cell number was determined by MTT assay. The value for untreated control cells was set at 100% for each time point.

product (formazan) by the intracellular dehydrogenases of viable cells. The crystals are then dissolved in dimethyl sulfoxide (DMSO) and the solutions photometrically measured at 546 nm. The results from two experiments are represented in Fig. 4. While a small, yet not significant, decrease in cell growth in both AS- and NS-ODN-treated WiDr cells is observed 24 hr after the addition of the ODNs, cell growth of ODN-treated cells is not different from control cells at 48 hr. At that time the antisense effect on the Na$^+$/Ca^{2+} exchange is most pronounced in rat cardiac cells. Furthermore, no long-term cytotoxic effects are detectable, as determined 96 hr after the application of ODNs to the culture medium. In addition, the viability of ODN-treated cells (trypan blue exclusion test) is in no way compromised (data not presented). These results indicate that the ODNs targeted against the Na$^+$/Ca^{2+} exchanger have no growth-related side effects on this cell type. AS-ODNs directed against calretinin, on the other hand, inhibit cell growth by approximately 60% at 48 hr (3 μM final concentration) as previously reported[22] (data not shown). See *below* for experiments excluding cytotoxicity in cardiac myocytes.

Na^+/Ca^{2+} Exchanger Function

Solutions for Na^+/Ca^{2+} Exchange Measurements in Neonatal Rat Myocytes

The superfusion solution contains (in mM): NaCl 140, KCl 5, CaCl$_2$ 2, MgCl$_2$ 1, HEPES 10, glucose 10. The pH is adjusted to pH 7.4 with NaOH. Extracellular solutions are changed with a half-time ($t_{1/2}$) of less than 0.5 sec, using a rapid switching system. Li$^+$ is used as a substitute for Na$^+$ in Na^+_o-free solutions. Ryanodine (10 μM) and thapsigargin (0.2 μM) are added 30 min prior to an experiment to inhibit the sarcoplasmic reticulum Ca^{2+} release and Ca^{2+} uptake function and to prevent interference from this Ca^{2+} store with the Ca^{2+} extrusion by the Na$^+$/Ca^{2+} exchange. The pipette-filling solution contains (in mM): cesium aspartate 120, NaCl 10, K-ATP 4, tetraethylammonium-Cl 20, HEPES 10, K$_5$-fluo-3 0.033, (NH$_4$)$_4$-fura-red 0.066 (Molecular Probes, Eugene, OR). For some experiments (in mM), Na$_4$-DM-nitrophen 2 (Calbiochem, Lucerne, Switzerland), reduced glutathione (GSH) 2, and CaCl$_2$ 0.5 are added to the pipette-filling solutions. The pH is adjusted to pH 7.2 with CsOH. All experiments are carried out at ambient temperature (20–22°).

Calcium Measurements

Intracellular Ca^{2+} is measured ratiometrically with a confocal microscope, using a mixture of two fluorescent Ca^{2+} indicators. Cells are dialyzed with the salt form of fluo-3 and fura-red in the whole-cell recording mode of the patch-clamp technique. In some experiments cells are loaded with fluo-3-AM (Teflabs, Austin, TX). This is achieved by exposing the myocytes to the AM-ester form of this indicator at a final concentration of 3–5 μM for 20–30 min. Ca^{2+} is calculated from the fluo-3/fura-red ratio or according to the self-ratio method assuming a resting [Ca^{2+}]$_i$ of 100 nM and a K_d of fluo-3 for Ca^{2+} of 400 nM. In addition to image sequences, rapid line-scans are performed with the laser-scanning confocal microscope (MRC600; Bio-Rad, Glattbrugg, Switzerland) to monitor the Ca^{2+} concentration with a high temporal resolution (up to 500 Hz). Absolute [Ca^{2+}]$_i$ concentrations are calculated from the fluorescence signals using an *in vivo* calibration curve. The set-up, as well as the ratiometric techniques, have been described in detail elsewhere.[37] Measurements of normalized Ca^{2+} concentration changes are represented as means ± standard errors in the figures.[4,8,9]

---

[37] P. Lipp, C. Lüscher, and E. Niggli, *Cell Calcium* **19,** 255 (1996).

Effect of Antisense Oligonucleotides on Na$^+$/Ca^{2+} Exchange Function

Measurements of Na$^+$/Ca^{2+} Exchanger Activity via the Reverse Mode.
For functional studies, cultures of control cells and of those exposed to
ODNs (1 or 3 μM added at time zero) are set up in parallel. For an initial
examination of Na$^+$/Ca^{2+} exchange activity, the Na$^+$ concentration gradient
across the sarcolemma of voltage-clamped cells is reversed by completely
replacing extracellular Na$^+$ with Li$^+$. In the presence of extracellular Ca^{2+}
this maneuver induces Ca^{2+} entry into cells via the "reverse mode" of
Na$^+$/Ca^{2+} exchange. The activity of the exchanger in the reverse mode is
estimated by determining the initial rate of increase in [Ca^{2+}]$_i$ (V_{up}), measured by ratiometric confocal microscopy in the line-scan mode. All experiments are performed after preincubating the cultures with ryanodine and
thapsigargin to inhibit the sarcoplasmic reticulum. In the presence of these
blockers, Ca^{2+} entry and removal from the cytosol are almost exclusively
due to Na$^+$/Ca^{2+} exchange as suggested by the absence of the [Ca^{2+}]$_i$ decay
when the Na$^+$/Ca^{2+} exchange is completely suppressed (see below).

Figure 5 depicts representative Ca^{2+} signals recorded in a control cell
(A) and in a myocyte that had been exposed to AS-ODNs (B). During the
Na^+_o-free period (indicated by the horizontal line in Fig. 5), [Ca^{2+}]$_i$ increases
by about 3 μM. Reapplication of Na^+_o leads to a monotonic decay of
[Ca^{2+}]$_i$ to resting levels. In the cell exposed to the AS-ODNs at 3 μM for
24 hr [Ca^{2+}]$_i$ rises more gradually and to a significantly lower final level,
indicating that its Na$^+$/Ca^{2+} exchange activity has been almost completely suppressed.

Flash Photolysis of Caged Ca^{2+}. As a consequence of the almost complete inhibition of the Na$^+$/Ca^{2+} exchange activity after 48 hr, many cells
could not be loaded with Ca^{2+} by Na^+_o removal and thus the Ca^{2+} efflux
mode of the exchanger could not be analyzed by this approach. We therefore opted to elevate [Ca^{2+}]$_i$ by applying an alternative technique. Intracellular Ca^{2+} concentration jumps are induced with UV flash photolysis of
caged Ca^{2+}. For this purpose, single cells are dialyzed with a pipette-filling
solution containing 2 mM DM-nitrophen and 0.5 mM Ca^{2+} (in addition to
the Ca^{2+} indicators fluo-3 and fura-red). During these experiments, a confocal image is taken every 2 sec, and membrane currents recorded simultaneously. Pertinent data are presented in Fig. 6 (control cell) and Fig. 7 [cell
exposed to AS-ODNs (3 μM) for 48 hr]. Superfusion of control cells with
medium lacking Na^+_o elicits a rise in [Ca^{2+}]$_i$ to about 1.8 μM, but the same
procedure has virtually no effect on [Ca^{2+}]$_i$ in cells exposed to AS-ODNs,
indicating that Na$^+$/Ca^{2+} exchange activity has been almost completely
suppressed. On application of a UV flash, [Ca^{2+}]$_i$ jumps to 1.5 μM in the
control cell (Fig. 6A) and decays to basal levels within 6 sec. In the cell

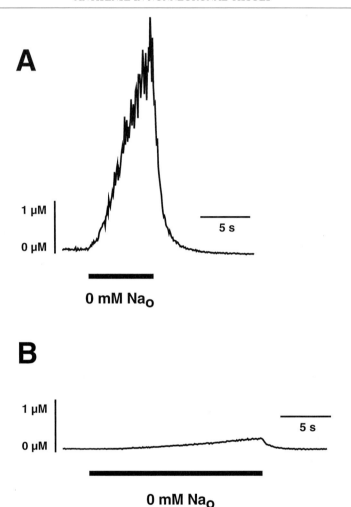

FIG. 5. Suppression of reverse mode of Na^+/Ca^{2+} exchange after exposure to AS-ODNs. Removal of extracellular Na^+_o induces a reversal of the Na^+/Ca^{2+} exchange. The resulting Ca^{2+} entry leads to an increase in the intracellular Ca^{2+} concentration in a control cell (A). After exposure to 3 μM AS-ODN for 24 hr, the velocity of increase is dramatically reduced, indicating that Na^+/Ca^{2+}-exchange activity has been suppressed (B). $[Ca^{2+}]_i$ was recorded ratiometrically in the line-scan mode of the confocal microscope, in the presence of ryanodine and thapsigargin. [Modified from P. Lipp, B. Schwaller, and E. Niggli, *FEBS Lett.* **364,** 198 (1995).]

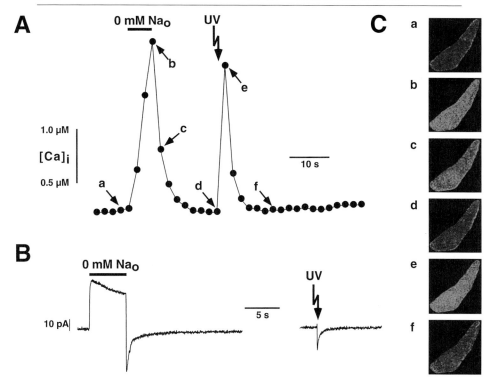

FIG. 6. Ca^{2+} signals and Na^+/Ca^{2+}-exchange currents (I_{NaCa}) induced in a control myocyte by removing Na^+_o, and by flash photolysis of caged Ca^{2+}. Both procedures induced transient Ca^{2+} signals in a control cell (A). During Na^+_o removal, an outward membrane current was observed. On readdition of Na^+_o and after the photolytic flash, transient inward currents corresponding to I_{NaCa} were elicited, as expected (B). $[Ca^{2+}]_i$ was recorded in the confocal imaging mode. Frames were taken every 0.5 sec [filled circles in (A)]; representative images are shown in (C); [(A) and (B) are modified from P. Lipp, B. Schwaller, and E. Niggli, *FEBS Lett.* **364**, 198 (1995).]

exposed to AS-ODNs (Fig. 7A), a similar flash results in a comparable jump in $[Ca^{2+}]_i$, but no appreciable decay occurs during the time of monitoring (approximately 1 min). These findings suggest that (1) Na^+/Ca^{2+} exchange is almost completely absent and that (2) all other Ca^{2+} transport systems are not contributing significantly to the removal of Ca^{2+} under these conditions (i.e., in the presence of ryanodine and thapsigargin).

The Na^+/Ca^{2+} exchanger is an electrogenic transporter with a stoichiometry of $3\,Na^+ : 1\,Ca^{2+}$. An outward membrane current is therefore generated in the Ca^{2+} influx mode, and an inward current is produced when Ca^{2+} is removed from the cytosol. The changes in the membrane current during

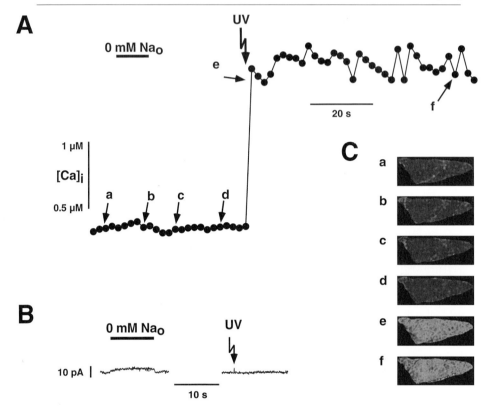

FIG. 7. Ca^{2+} signals and Na^+/Ca^{2+}-exchange currents (I_{NaCa}) in a myocyte exposed to 3 μM AS-ODN for 48 hr. In this cell Na^+_o removal failed to increase $[Ca^{2+}]_i$, indicating that Na^+/Ca^{2+} exchange activity had been almost completely suppressed during the Ca^{2+} influx mode (A). The photolytic jump in $[Ca^{2+}]_i$ decayed only slowly. This finding suggests that Ca^{2+} removal via the Na^+/Ca^{2+} exchanger had been abolished. Na^+_o removal and flash photolytic changes in $[Ca^{2+}]_i$ had virtually no effect on the membrane current (B). Several frames of the imaging sequence are represented in (C); [(A) and (B) are modified from P Lipp, B. Schwaller, and E. Niggli, *FEBS Lett.* **364,** 198 (1995).]

and after Na^+_o removal and after photorelease of Ca^{2+} are also shown in Figs. 6 and 7. During Na^+_o removal, a slowly decaying outward current is recorded from the control cell (Fig. 6B), but not in the cell exposed to AS-ODNs (Fig. 7B). After readdition of Na^+_o and after the photolytic jump of $[Ca^{2+}]_i$, a significant inward current is activated in the control cell (Fig. 6B), which reflects the removal of this cation via the Na^+/Ca^{2+} exchanger.[38] The inward current is transient because $[Ca^{2+}]_i$ is lowered by means of the

[38] P. Lipp and L. Pott, *J. Physiol. (London)* **397,** 601 (1988).

exchanger itself. Neither the reapplication of Na^+_o nor flash photolysis of caged Ca^{2+} generates a detectable transient inward current in the cell exposed to AS-ODNs (Fig. 7B).

Time Course and Concentration Dependence of Antisense Effect

The onset of inhibition by AS-ODNs depends on several regulated processes. These include the uptake of ODNs (see above), the kinetics of binding to the target RNA, degradation of RNA/DNA heteroduplexes by RNase H, and the stability of ODNs in the extra- and intracellular milieus. Because AS-ODNs should inhibit only the *de novo* synthesis of the Na^+/Ca^{2+} exchanger, and should therefore be without effect on the protein already present in the plasma membrane, the onset of the antisense effect will also be influenced by the functional half-life of the exchanger protein; this is distinct from the biological half-life, which is a measure only of the physical presence of a protein (and can be estimated by biochemical methods, such as binding assays, Western blot analysis).

We perform the functional studies 24 and 48 hr after adding AS-ODNs at 1 or 3 μM to the bathing medium of neonatal cardiac myocytes, by measuring the velocity of the increase in $[Ca^{2+}]_i$ (V_{up}) under conditions when external Na^+ was removed. In control cells, the mean V_{up} is 350 ± 35 nM/sec as determined during the first 2 sec of zero $[Na^+]_o$. However, in cells exposed to 3 μM AS-ODNs, V_{up} is only 22.0% of the control value after 24 hr and 3.4% after 48 hr, respectively (see Fig. 8B for a summary of normalized data). After 48 hr five of the seven cells monitored exhibit no detectable increase in $[Ca^{2+}]_i$ whereas the other two cells show low Na^+/Ca^{2+} exchange activities. Furthermore, after removal of AS-ODNs from the culture medium V_{up} recovers to 71.3% of the control value after 24 hr and to 103.4% after 48 hr (Fig. 8C).

When cells are incubated with the lower concentration of AS-ODNs (1 μM), V_{up} decreases to 21.8% of the control value after 24 hr, but rises to 47.3% after 48 hr (Fig. 8A). The lower concentration of AS-ODNs (1 μM) is thus insufficient to sustain the blockage of the Na^+/Ca^{2+} exchanger expression for 48 hr. It is interesting to note that the initial inhibition at 24 hr is almost identical to that elicited using the higher concentration of AS-ODNs (3 μM). We assume that the intra- or extracellular degradation of AS-ODNs is fairly rapid and that only when they are present at higher concentrations are sufficient intact AS-ODN molecules available within the cells to exert an antisense effect.

Experiments with Control Oligodeoxynucleotides

A series of control experiments is performed to exclude direct or indirect nonspecific effects of the nucleotide on the Na^+/Ca^{2+} exchanger as outlined

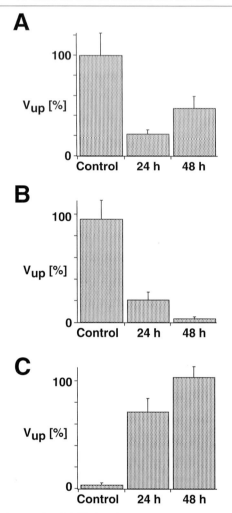

FIG. 8. Summary of normalized Ca^{2+} signals recorded in cultured cardiac myocytes. The initial rate of rise in $[Ca^{2+}]_i$ was used as a measure of Na^+/Ca^{2+} exchange activity (V_{up}) and was compared with that in control cultures (100%) and in those exposed to AS-ODNs. When cultures were exposed to the AS-ODNs at 1 μM, partial recovery of the Na^+/Ca^{2+} exchange activity was observed within 48 hr (A). Using a concentration of 3 μM AS-ODN, the activity was more depressed at 48 hr than after 24 hr (B). When the AS-ODNs were removed after 48 hr, Na^+/Ca^{2+}-exchange activity recovered to control levels within 48 hr (C). [Results from (B) and (C): modified from P. Lipp, B. Schwaller, and E. Niggli, *FEBS Lett.* **364**, 198 (1995).]

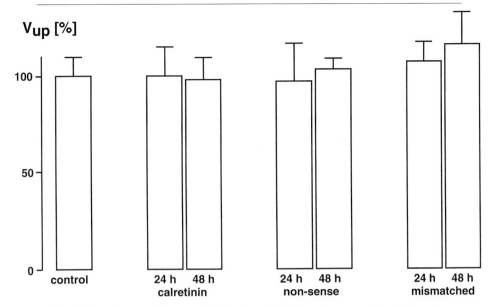

FIG. 9. Experiments with control ODNs. Several ODNs not corresponding to the antisense sequence of the Na$^+$/Ca^{2+} exchanger were tested at a concentration of 3 μM. None of these ODNs had an inhibitory effect after 24 or 48 hr, indicating that the inhibition observed with the AS-ODN was sequence specific. See text for details. [Modified from P. Lipp, B. Schwaller, and E. Niggli, *FEBS Lett.* **364**, 198 (1995).]

in Strategies (above). Three control ODNs are used: (1) a nonsense (NS) ODN, with a reversed 5'-to-3' orientation compared with the antisense ODN but maintaining the same base composition; (2) a mismatched (MM) ODN containing three mismatches compared with the antisense ODN (see Fig. 1C), and that should not anneal efficiently to the target sequence, because the annealing temperature (T_m) is too low; and (3) an ODN directed against the calcium-binding protein calretinin (CR),[22] which is not expressed in cardiac cells. None of these control ODNs should significantly affect Na$^+$/Ca^{2+} exchange function. Neonatal rat cardiac myocytes are exposed to each of these control ODNs (3 μM) for 24 and 48 hr. As anticipated, none of these significantly suppresses Na$^+$/Ca^{2+} exchange activity (Fig. 9), indicating that the inhibition observed in the presence of AS-ODNs results from a sequence-specific interaction with the Na$^+$/Ca^{2+} exchanger mRNA. In addition, the Na$^+$/Ca^{2+} exchanger ODNs (3 μM), including the AS-ODNs, have no effect on the viability and cell growth of the colon adenocarcinoma cell line WiDr (Fig. 4). That this concentration of ODNs is adequate for efficient inhibition of calretinin expression has been demonstrated previously[22] using the same calretinin-specific AS-ODNs.

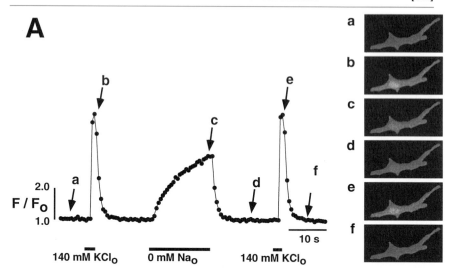

A

F / F$_O$

2.0

1.0

a

b

c

d

e

f

10 s

140 mM KCl$_O$ 0 mM Na$_O$ 140 mM KCl$_O$

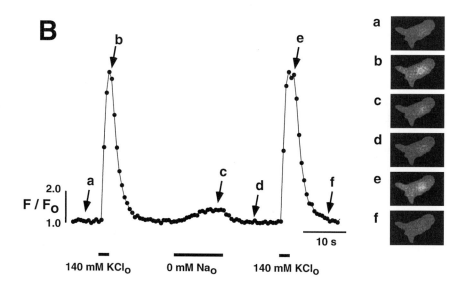

B

F / F$_O$

2.0

1.0

a

b

c

d

e

f

10 s

140 mM KCl$_O$ 0 mM Na$_O$ 140 mM KCl$_O$

Cytotoxicity Assays in Cardiac Myocytes

Several control experiments are carried out to verify that the applied ODNs do not exert nonspecific cytotoxic effects. In experiments with voltage-clamped cells, sodium currents (I_{Na}) are routinely activated using appropriate voltage-clamp protocols in order to confirm the viability and cardiac muscular origin of each cell tested. Cells that are loaded with fluorescent Ca^{2+} indicators in the AM-ester form are briefly superfused with a solution containing 140 mM KCl to depolarize the cells and to activate Ca^{2+} influx via L-type Ca^{2+} channels (Fig. 10A). While the rate of increase of [Ca^{2+}]$_i$ during Na^+_o withdrawal is considerably lowered in cells exposed to AS-ODNs (Fig. 10B), the Ca^{2+} transient induced by KCl depolarization remains intact; only its rate of decay is slowed down (because of the lower Na$^+$/Ca^{2+} exchange activity). These findings confirm that cultured cardiac myocytes exposed to 3 μM AS-ODN for 48 hr are viable. Indirect evidence of the absence of cytotoxic effects can also be derived from two other sets of experiments: (1) control experiments with NS- and MM-ODNs (see Fig. 9) do not affect the Na$^+$/Ca^{2+} exchange activity, which suggests that the inhibitory effect of the AS-ODNs is sequence specific; (2) in recovery experiments the reappearance of Na$^+$/Ca^{2+} exchange activity follows a time course comparable to that of its disappearance, indicating that the cultured cells remain viable after exposure to the ODNs (Fig. 8C).

Conclusions

In this chapter, we have presented a method for investigating the role of the Na$^+$/Ca^{2+} exchanger. The antisense technique—like all other methods—has its disadvantages as well as its advantages. One extremely important aspect to be considered in antisense studies is the specificity of the effect induced. This begins with the design of AS-ODNs, an undertaking that is still poorly conceptualized, even though several tools have been

---

FIG. 10. Viability test in cultured cardiac myocytes. Cells were briefly superfused with a depolarizing KCl solution that led to Ca^{2+} entry via L-type Ca^{2+} channels (A). After Na$^+$ removal, [Ca^{2+}]$_i$ rose gradually, as in Fig. 6. In cells exposed to 3 μM AS-ODN for 48 hr KCl also induced a Ca^{2+} transient, which decayed more slowly than in control cells (B). The increase in [Ca^{2+}]$_i$ observed during Na^+_o removal was significantly depressed, which indicates that Na$^+$/Ca^{2+}-exchange activity had been inhibited. Cells thus exhibited normal voltage-dependent Ca^{2+} influx via Ca^{2+} channels, even though their Na$^+$/Ca^{2+}-exchange activity had been specifically suppressed. For these experiments, cells were loaded with fluo-3 by exposure to fluo3-AM and preincubated with ryanodine and thapsigargin to ablate sarcoplasmic reticulum function.

developed to approach it from a more "rational" point of view.[21,23,39–41] Not only the sequence specificity, but also the length and chemical nature of the ODNs (e.g., phosphorothioate linkages, side-chain modifications, and backbone chemistry), must be taken into account. The next step is the planning of experiments, which should include an assessment of the uptake and cytotoxicity (concentration dependence) of the AS-ODNs, as well as monitoring of the time course and reversibility (recovery) of the effect induced. Careful consideration of appropriate control experiments, including the testing of NS- and MM-ODNs, is also an absolute requirement. Only when all of these aspects have been addressed can meaningful information be obtained.

In our study, we have demonstrated that downregulation of the Na^+/Ca^{2+} exchanger in neonatal rat cardiac myocytes led to a significant suppression of its activity, which was virtually complete within 48 hr. When the Na^+/Ca^{2+} exchanger was run in the "backward" mode, no Ca^{2+} influx was observed in AS-ODN-treated cells, which indicated that the functional proteins had been almost completely knocked out.

This notion was supported in experiments in which $[Ca^{2+}]_i$ was artificially increased by flash photolysis of caged Ca^{2+}. In untreated cells or myocytes incubated with various control ODNs the Ca^{2+} was rapidly removed, whereas in AS-ODN-treated myocytes the photoreleased Ca^{2+} remained in the cytosol. In all cells the sarcoplasmic reticulum Ca^{2+} release and Ca^{2+} uptake were inhibited by ryanodine and thapsigargin. These results demonstrate that the Na^+/Ca^{2+} exchanger represents the major pathway for Ca^{2+} extrusion from this cell type.

In conclusion, carefully designed antisense experiments represent a novel approach to basic research and may furnish us with an invaluable tool in some instances. Looking to the future, this approach could pave the way to the development of new therapeutic strategies.

Acknowledgments

This work was supported by a grant from the Swiss National Science Foundation (Grant 31-50564.97 to E.N.). We gratefully acknowledge the technical assistance of A. Wyss (Hoffmann-LaRoche, Basel), B. Herrmann (Institute of Histology and General Embryology, Fribourg), and J. Gygax and M. Herrenschwand (Institute of Physiology, Bern). We would like to thank Dr. H. Porzig from the Institute of Pharmacology, University of Bern, for helpful discussions.

[39] S. Agrawal and Q. Zhao, *Antisense Nucleic Acid Drug Dev.* **8**, 135 (1998).
[40] S. Akhtar, *J. Drug Targeting* **5**, 225 (1998).
[41] S. T. Crooke, *Anticancer Drug Des.* **12**, 311 (1997).

[33] Optimizing Efficacy of Antisense Oligodeoxynucleotides Targeting Inhibitors of Apoptosis

By ANNEMARIE ZIEGLER, A. PAULA SIMÕES-WUEST,
and UWE ZANGEMEISTER-WITTKE

It is now widely accepted that defects in the regulation of a genetically controlled suicide process, also known as programmed cell death or apoptosis, can contribute to multiple pathologies including cancer. The search for gene products involved in the regulation of apoptosis has identified numerous proteins, and the list continues to grow. In particular, much interest has been devoted to members of the *bcl-2* gene family of apoptosis regulators, because of their contribution to oncogenesis and drug resistance. This family of structurally related proteins includes members that exert opposing functions in apoptosis control. The Bcl-2 protein functions in normal and neoplastic cells to inhibit or delay apoptosis induced by a variety of endogenous and exogenous stimuli.[1,2] The related Bcl-xL protein also acts as a negative regulator of apoptosis, and protects from cell death where Bcl-2 is ineffective.[3] In contrast to normal tissues, Bcl-2 and Bcl-xL are frequently coexpressed in tumors, where they contribute to resistance to a broad range of anticancer agents. Pro- and antiapoptotic Bcl-2 family members can heterodimerize and seemingly titrate one another's function, suggesting that their relative concentration may act as a rheostat for the death program.[2] This suggests that cell survival or death after an apoptotic stimulus is determined by competitive interactions between distinct pairs of death agonists (e.g., Bax and Bad) and antagonists (e.g., Bcl-2 and Bcl-xL). Therefore, any approach to turn the rheostat in favor of cell death, e.g., by repression of Bcl-2 or Bcl-xL expression, could be therapeutically useful. In addition to Bcl-2 and Bcl-xL, inhibitors of apoptosis proteins (IAPs) have been described that are directly engaged in the caspase activation pathway.[4,5] Further investigations are necessary to determine whether XIAP, the c-IAPs, and survivin represent additional targets for the antisense approach to cancer therapy.

[1] J. C. Reed, *Nature* (*London*) **387,** 773 (1997).
[2] J. M. Adams and S. Cory, *Science* **281,** 1322 (1998).
[3] P. L. Simonian, D. A. Grillot, and G. Nunez, *Blood* **90,** 1208 (1997).
[4] Q. L. Deveraux, N. Roy, H. R. Stennicke, A. T. Van, Q. Zhou, S. M. Srinivasula, E. S. Alnemri, G. S. Salvesen, and J. C. Reed, *EMBO J.* **17,** 2215 (1998).
[5] G. Ambrosini, C. Adida, and D. C. Altieri, *Nature Med.* **3,** 917 (1997).

0076-6879/99 $30.00

Antisense Oligodeoxynucleotide Approach to Repression
of Gene Expression

Many of the initial limitations imputed to antisense oligodeoxynucleotides (ODNs), such as nuclease sensitivity, limited cellular uptake and
distribution, non-sequence-specific effects, and low RNA-binding affinity,
have now either been solved by chemical modification of the ODNs or
have proved not to be functionally limiting.[6,7] In addition, improvements
in the manufacturing process have dramatically reduced the overall costs
of ODN synthesis.[8] In the case of gene products involved in the negative
regulation of apoptosis, the use of antisense ODNs appears particularly
attractive. First, a number of genes have been identified at the sequence
level, and their contribution to tumor development and drug resistance has
been assessed *in vitro* and in animal models. The expression pattern of
some antiapoptotic proteins has been analyzed in numerous cell lines and
tumor types. Second, owing to the mechanisms of action by which antisense
ODNs exert their effects, overlapping toxicity should be largely precluded
in combined treatment modalities. This gains particular weight in the light
of clinical trials, which have demonstrated good tolerance and few side
effects of ODNs in patients. Furthermore, we demonstrated a synergistic
effect between an antisense ODN targeting *bcl-2* and conventional anticancer agents on cultured lung cancer cells.[9] Finally, experiments with antisense
ODNs targeting *bcl-2*,[10,11] and more recently also *bcl-xL*,[12] have met the
expectancies with regard to specificity of action and therapeutic efficacy.
Because the success of the antisense approach depends not only on the
experimental setting, but also on the proper target site selection, we address
both issues in the following sections. A flowchart summarizing the different
steps leading from ODN design to their experimental application in tissue
culture is presented in Fig. 1.

Antisense Oligodeoxynucleotides: General Considerations

Before discussing the criteria for selection of an antisense target site,
two additional ODN features are considered: ODN length and backbone

[6] S. T. Crooke, *Antisense Res. Appl.* **131,** 1 (1998).

[7] S. T. Crooke, *Antisense Nucleic Acid Drug Dev.* **8,** 115 (1998).

[8] C. F. Bennett, *Biochem. Pharmacol.* **55,** 9 (1998).

[9] U. Zangemeister-Wittke, T. Schenker, G. H. Luedke, and R. A. Stahel, *Br. J. Cancer* **78,**
1035 (1998).

[10] A. Ziegler, G. H. Luedke, D. Fabbro, K. H. Altmann, and U. Zangemeister-Wittke, *J. Natl.
Cancer Inst.* **89,** 1027 (1997).

[11] A. Webb, D. Cunningham, F. Cotter, P. A. Clarke, F. Distefano, P. Ross, M. Corbo, and
Z. Dziewanowska, *Lancet* **349,** 1137 (1997).

[12] M. J. Pollman, J. L. Hall, M. J. Mann, L. Zhang, and G. H. Gibbons, *Nature Med.* **4,** 422 (1998).

**Design antisense and control sequences based on the secondary structure
of the target mRNA or by use of high density ODN arrays**

**Check sequences for homology to other human genes
by use of a searchable database (Advanced BLASTN)**

Synthesize and purify ODNs; store aliquots at −20°

**Prepare cell cultures for ODN transfection (check for optimal
growth conditions, avoid mycoplasma infections)**

**Optimize transfection conditions by analyzing cells in an easy and fast to perform
test system (western blotting, flow cytometry) to define the optimal:**

**Cell density
Cationic lipid
Lipid/ODN ratio
Serum concentration**

Transfect cells and examine antisense effect in secondary test systems

FIG. 1. Flowchart describing the most critical steps for selection and experimental application of antisense ODNs.

modifications. The specificity of the antisense ODN approach is based on the estimate that any sequence of at least 13 bases in RNA, and 17 bases in DNA, is represented only once within the human genome.[13] In 1978, Zamecnik and Stephenson[14] first demonstrated that a 13-mer antisense ODN could block replication of Rous sarcoma virus in cell cultures, and further studies since have confirmed the ability of antisense ODNs to inhibit gene expression in a specific manner.[10,15–17] Currently, most ODNs used to repress cancer-related gene expression are 15 to 25 bases long. Although

[13] C. Helene and J. J. Toulme, *Biochim. Biophys. Acta* **1049,** 99 (1990).
[14] P. C. Zamecnik and M. L. Stephenson, *Proc. Natl. Acad. Sci. U.S.A.* **75,** 280 (1978).
[15] B. P. Monia, H. Sasmor, J. F. Johnston, S. M. Freier, E. A. Lesnik, M. Muller, T. Geiger, K. H. Altmann, H. Moser, and D. Fabbro, *Proc. Natl. Acad. Sci. U.S.A.* **93,** 15481 (1996).

we did not pursue this issue in detail, comparison of two ODNs targeting the same sequence within the *bcl-2* transcript revealed that a 20-mer was more effective at reducing Bcl-2 protein levels than a shorter 18-mer.[10] Therefore, the cost saving brought by the synthesis of shorter ODNs should be evaluated carefully. It is advisable that minimal ODN lengths be kept according to the above-mentioned estimates.

In initial experiments, antisense action was limited by the susceptibility of ODNs harboring conventional phosphodiester bonds to degradation by nucleases. This deficiency was successfully overcome by the introduction of chemical modifications in the ODN backbone that prevent nuclease cleavage. In particular, the replacement of the oxygen atom of the phosphate group by a sulfur atom resulted in increased antisense stability, RNA hybridization efficiency, and RNase H-activating capacity.[6,18] Such phosphorothioate ODNs are now widely used to repress cancer-related gene expression. Further improvements have been achieved by introducing modifications in the ODN sugar moiety.[19,20] These ODNs are superior to 2-deoxyphosphorothioates in terms of RNA-binding affinity and possibly cellular uptake. Moreover, their reduced unspecific toxicity should allow one to better discern the true antisense effect on cells. Because the drawback of ribose-modified ODNs is their lower potency to activate RNase H, gapmers or hemimers with only part of the nucleosides being modified seem more promising. Some of the new analogs are currently under preclinical investigation, and preliminary data suggest that improved RNA-binding affinity indeed translates into better gene repression efficacy. However, at the current standing of knowledge, the choice of phosphorothioate ODNs appears adequate for experimental purposes.

Rational Design of Antisense Oligodeoxynucleotides

The development of approaches for the selection of antisense target sites should allow the design of ODNs with optimal potency. However, preferred target regions on mRNA have often been identified by empirical

[16] N. Dean, R. McKay, L. Miraglia, R. Howard, S. Cooper, J. Giddings, P. Nicklin, L. Meister, R. Ziel, T. Geiger, M. Muller, and D. Fabbro, *Cancer Res.* **56,** 3499 (1996).

[17] T. Skorski, M. Nieborowska Skorska, P. Wlodarski, D. Perrotti, G. Hoser, J. Kawiak, M. Krajewski, L. Christensen, R. V. Iozzo, and B. Calabretta, *J. Natl. Cancer Inst.* **89,** 124 (1997).

[18] C. A. Stein, K. Mori, S. L. Loke, C. Subasinghe, K. Shinozuka, J. S. Cohen, and L. M. Neckers, *Gene* **72,** 333 (1988).

[19] M. Matteucci, *CIBA Found. Symp.* **209,** 5 (1997).

[20] S. M. Freier and K. H. Altmann, *Nucleic Acids Res.* **25,** 4429 (1997).

means owing to the inherently limited number of ODNs that may be screened by gene walking. One rationale to design effective ODNs is based on the assumption that mRNA is single stranded at the AUG site to allow ribosomal entry, and thus should be accessible for ODN hybridization. ODNs targeting the AUG start codon have indeed been used with some success, but owing to the limited information available on mRNA secondary structure, other sites are likely to exist that could serve as more effective targets. For instance, antisense ODNs targeting sequences at the 5' and 3' untranslated regions of the c-*myc* and the c-*raf* mRNAs, respectively, proved most potent in screening experiments.[21,22] Developments in the preparation of high-density ODN arrays on solid supports[23] may overcome these limitations, and help identify more accessible target sites. However, this technology is not available in most academic institutions and is rather expensive when applied to large target molecules, such as the 5-kb *bcl-2* mRNA. A more feasible approach is the use of computer programs to predict the secondary structure of mRNA molecules, such as the RNAdraw program developed by Matzura and Wennborg.[24] In this program, the optimal structure/base pair probability matrix/heat curve calculation algorithms were imported from the Vienna RNA package V1.1.[25] The implemented dynamic programming algorithm is based on work by Zuker and Stiegler,[26] and the base pair probability algorithm on work by McCaskill.[27] Energy parameters are taken from Turner and Sugimoto,[28] Freier *et al.*,[29] Jaeger *et al.*,[30] and He *et. al.*[31] Using the predicted secondary structure, the most accessible hybridization sites can be expected at regions where no complementary base pairing occurs, and the RNA sequence presents in a single-strand conformation. As an example, the structures of part of the *bcl-2* mRNA (bases 1085–2319), and of the entire *bcl-xL* mRNA, are shown

[21] T. A. Bacon and E. Wickstrom, *Oncogene Res.* **6,** 13 (1991).

[22] B. P. Monia, J. F. Johnston, T. Geiger, M. Muller, and D. Fabbro, *Nature Med.* **2,** 668 (1996).

[23] E. M. Southern, G. S. Case, J. K. Elder, M. Johnson, K. U. Mir, L. Wang, and J. C. Williams, *Nucleic Acids Res.* **22,** 1368 (1994).

[24] O. Matzura and A. Wennborg, *Comput. Appl. Biosci.* **12,** 247 (1996).

[25] I. L. Hofacker, W. Fontana, P. F. Stadler, L. S. Bonhoeffer, M. Tacker, and P. Schuster, *Chem. Monthly* **167** (1994).

[26] M. Zuker and P. Stiegler, *Nucleic Acids Res.* **9,** 133 (1981).

[27] J. S. McCaskill, *Biopolymers* **29,** 1105 (1990).

[28] D. H. Turner and N. Sugimoto, *Annu. Rev. Biophys. Chem.* **167** (1988).

[29] S. M. Freier, R. Kierzek, J. A. Jaeger, N. Sugimoto, M. H. Caruthers, T. Neilson, and D. H. Turner, *Proc. Natl. Acad. Sci. U.S.A.* **83,** 9373 (1986).

[30] J. A. Jaeger, D. H. Turner, and M. Zuker, *Proc. Natl. Acad. Sci. U.S.A.* **86,** 7706 (1989).

[31] L. He, R. Kierzek, J. SantaLucia, Jr., A. E. Walter, and D. H. Turner, *Biochemistry* **30,** 11124 (1991).

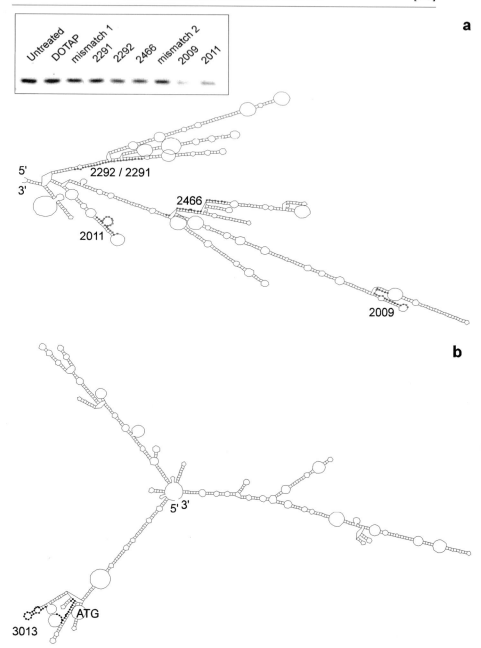

in Fig. 2a and b,[10,12,32,32a] respectively. This portion of the *bcl-2* mRNA reveals numerous unpaired single-stranded loops, all of which might represent potential target sites. Alignment along the *bcl-2* mRNA of different antisense ODNs that had been designed empirically and tested *in vitro* demonstrated that regions encompassing at least two small single-stranded loops linked by short double-stranded spacer sequences, represented effective hybridization sites for antisense ODNs. Such a structural motif was identified in the AUG region (2292), the midcoding region (2009), and the translation termination site (2011). On the basis of this observation, we also have designed antisense ODNs targeting similar structural motifs in the *bcl-xL* mRNA, which is composed of 926 bases (Fig. 2b). Although the design of antisense ODNs based on the predicted secondary structure of the mRNA does not pay regard to other parameters that contribute to antisense potency, such as the three-dimensional conformation *in vivo,* or RNase H-recruiting capacity, this approach is simple and feasible, and it helps to reduce the number of target sites that must be tested experimentally. The value of computer programs for the design of antisense ODNs is also suggested by the study of Robinson *et al.*[32] Using a program to assure negligible self-complementarity, the authors designed an antisense ODN (ODN 2466) that hybridized to a sequence located 80 bases downstream of the translation initiation site of the *bcl-2* mRNA (Fig. 2a). This ODN effectively inhibited the growth of non-small cell lung cancer cells.

Once antisense sequences have been selected on the basis of the secondary structure of an mRNA molecule, appropriate control sequences should be prepared and tested. In some cases, reported cytotoxic effects of antisense ODNs were related to nonantisense mechanisms, especially for phosphorothioates, the sulfur backbone of which binds to heparin and heparin-

---

[32] L. A. Robinson, L. J. Smith, M. P. Fontaine, H. D. Kay, C. P. Mountjoy, and S. J. Pirruccello, *Ann. Thorac. Surg.* **60,** 1583 (1995).
[32a] B. Dibbert, I. Daigle, D. Braun, C. Schranz, M. Weber, K. Blaser, U. ZangemeisterWittke, A. N. Akbar, and H. U. Simon, *Blood* **92,** 778 (1998).

FIG. 2. Secondary structure of part of the human *bcl-2* mRNA (bases 1085–2319; a) and of the entire *bcl-xL* mRNA (b) as predicted by the RNAdraw program. The hybridization sites for some effective antisense ODNs are shown. The inset in (a) shows a Western blot of Bcl-2 protein levels after treatment of SW2 small cell lung cancer cells with *bcl-2* antisense ODNs. Note that the effective antisense ODNs hybridize to a structural motif that comprises at least two short single-stranded loops separated by a short base-paired segment. The *bcl2* and *bcl-xL* antisense ODNs have been described in detail elsewhere (2291, 2292, 2009, and 2011[10]; 2466[32]; 3013[32a]; ATG[12]).

binding proteins on the cell surface.[33] To be accepted as a general approach for the treatment of human diseases, it is absolutely crucial to separate nonantisense from truly antisense effects. This is particularly compelling for antisense ODNs designed to repress the expression of survival factors, such as the apoptosis inhibitors Bcl-2 and Bcl-xL. Therefore, appropriate concept validation experiments, such as the proof of mRNA and protein downregulation, should always be included in the study. An ideal control system would be the use of cells lacking expression of the respective target gene. Should such a direct proof be difficult to obtain, e.g., we were unable to find any tumor cell line that completely lacks Bcl-2 expression, the use of adequate sequence control ODNs can serve as acceptable substitutes. Most studies have used sense, scrambled, or base-mismatched sequences for head-to-head comparison with the respective antisense ODNs. Because it cannot be excluded that the base sequence of an ODN also affects its uptake into cells, 5'–3' reversed sequence ODNs should be tested as an additional control.

Subsequently, antisense and control ODNs must be searched for homology to other human gene sequences to avoid repression of nontarget genes. The availability of sequences in a searchable database represents a convenient and easy-to-use tool for potential targets in the human genome. We have used a World Wide Web browser to access the NCBI BLAST page (http://www.ncbi.nlm.nih.gov/BLAST/). The homology search was done at low stringency by use of the Advanced BLASTN 1.4.9MP program applied to a nonredundant database compiling the sequences of GenBank, EMBL, DDBJ, and PDB (expect = 1000, filtering = none). ODNs (antisense and control) with more than 70% homology to other gene sequences should be discarded.

Purification of Oligodeoxynucleotides

Phosphorothioate ODNs produced by the phosphoramidite chemistry can now be purchased from commercial providers at reasonable costs. Because crude products obtained from a DNA synthesizer usually contain 70% impurity, further purification is mandatory. This can be achieved by high-pressure liquid chromatography [reversed-phase purification on a Waters (Milford, MA) machine equipped with a C_{18} column, and elution with a 5 to 25% gradient of acetonitrile], followed by anion-exchange chromatography [e.g., Millipore (Danvers, MA) Q-15 anion exchanger]. ODN preparations can further be ethanol precipitated, redissolved, and desalted by

---

[33] M. A. Guvakova, L. A. Yakubov, I. Vlodavsky, J. L. Tonkinson, and C. A. Stein, *J. Biol. Chem.* **270**, 2620 (1995).

ultrafiltration, which can translate into reduced unspecific toxicity. If possible, analyze purified ODNs by mass spectrometry. It is recommended to store small aliquots of the purified ODNs at $-20°$ in 10 mM Tris-HCl (pH 7.4)–1 mM EDTA. Owing to the rather complicated synthesis and purification procedure of phosphorothioate ODNs, different batches may vary with regard to their quality and, hence, their extent of antisense-specific and nonspecific effects on cells. To avoid any false interpretation of an antisense effect, two or three different preparations should be tested in addition to the control ODNs mentioned above.

Transfection of Cultured Cells

A major barrier for the action of antisense ODNs is the impermeability of cellular membranes to polyanions. To hybridize with the target mRNA, ODNs must distribute in the cytoplasm and enter the nucleus at sufficiently high concentrations.[34] Certain cell types seem to possess surface receptors for highly charged polyanions, which mediate their internalization by an endocytotic pathway.[35,36] However, current evidence suggests that most cultured cell types do not take up ODNs at all. This has been circumvented by the delivery of ODNs as complexes with cationic lipids, such as 1,2-dioleoyl-3-trimethylammoniumpropane (DOTAP) or N-[1-(2,3-dioleoyl-oxy)propyl]trimethylammoniumchloride (DOTMA), which improves ODN uptake and efficiency, and reduces the amounts required in experimental settings by a factor of 100.[37] This is illustrated in Fig. 3, which shows the intracellular localization of a fluorescein isothiocyanate (FITC)-labeled *bcl-xL* antisense ODN delivered either as naked DNA, or in the form of complexes with a cationic lipid. However, even in the presence of transfection reagents, the cellular uptake of ODNs is cell-type specific and the optimization for certain cell lines does not necessarily lead to improved uptake in others. Further complexity is added by the effects of cationic lipids themselves, which can be manifold and cell-type specific. For *in vitro* studies, it remains essential to establish well-defined conditions to assure the ultimate proof of the antisense concept. Surprisingly, ODN uptake does not seem to be a limiting step for antisense activity *in vivo*, and transfection reagents are dispensable in most animal models. The *in vitro* transfection

[34] P. Lorenz, B. F. Baker, C. F. Bennett, and D. L. Spector, *Mol. Biol. Cell* **9**, 1007 (1998).
[35] L. A. Yakubov, E. A. Deeva, V. F. Zarytova, E. M. Ivanova, A. S. Ryte, L. V. Yurchenko, and V. V. Vlassov, *Proc. Natl. Acad. Sci. U.S.A.* **86**, 6454 (1989).
[36] S. L. Loke, C. A. Stein, X. Zhang, K. Mori, M. Nakanishi, C. Subasinghe, J. S. Cohen, and L. M. Neckers, *Proc. Natl. Acad. Sci. U.S.A.* **86**, 3474 (1989).
[37] C. F. Bennett, M. Y. Chiang, H. Chan, J. E. Shoemaker, and C. K. Mirabelli, *Mol. Pharmacol.* **41**, 1023 (1992).

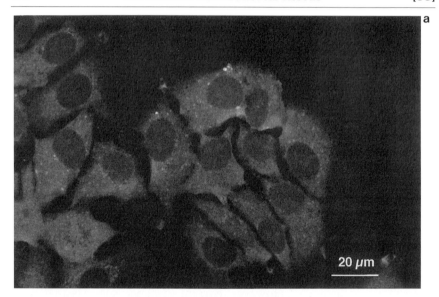

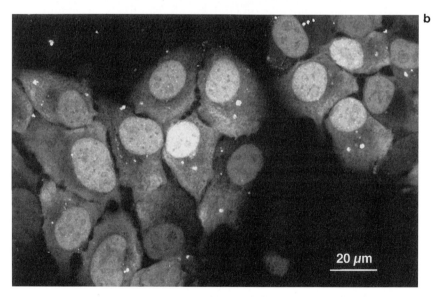

Fig. 3. Intracellular distribution of an FITC-labeled *bcl-xL* antisense ODN (600 n*M*) in ZR-75-1 breast carcinoma cells 20 hr after transfection, (a) delivered as naked DNA or (b) in the form of complexes with the cationic lipid Lipofectin. In the presence of Lipofectin, most of the ODN localized to the nucleus, whereas in the absence of the carrier lipid only small amounts of ODN were diffusely distributed in the cytoplasm. Photographs were taken on a conical laser microscope [Zeiss (Oberkochen, Germany) Axioplan equipped with an MRC 600 confocal system (Bio-Rad, Hercules, CA)].

problem remains a challenge for the successful general implementation of the antisense approach.

A variety of transfection reagents are now commercially available. The most frequently used are cationic lipids, which readily form complexes with polyanionic nucleic acid molecules. Because transfection conditions must be optimized for each cell type individually, it is not possible to give general recommendations for the use of cationic lipids on cultured cells, and this must be determined rather empirically. Thus, enhancing the transfection efficiency for antisense ODNs can be laborious. The following parameters are considered to be most critical.

1. Quality and purity of the ODN preparation (the use of low-quality batches may increase the unspecific toxicity of phosphorothioate ODNs)

2. Presence and quality of supplements in the culture medium, such as serum and antibiotics

3. Parameters related to cell growth, such as cell viability and doubling time

In our hands, DOTAP works well with cells growing in suspension, whereas Lipofectin, a 1 : 1 (w/w) liposome formulation of the cationic lipids DOTMA and DOPE (dioleoylphosphatidylethanolamine), yields more reproducible results with cells growing as monolayers. As an alternative, adherent cells can be trypsinized and harvested as single-cell suspensions for the treatment with DOTAP. Under optimized conditions, both cationic lipids should reduce the ODN concentrations necessary to attain measurable effects to the nanomolar range. In this regard, DOTAP is more effective than Lipofectin, but requires harsher transfection conditions. In the next sections, two transfection protocols are described that have been optimized for cells growing in suspension (SW2 small cell lung carcinoma) and as monolayers (A549 non-small cell lung carcinoma and ZR-75-1 ductal breast carcinoma), respectively.

Transfection of SW2 Cells Growing in Suspension by Use of Cationic
  Lipid DOTAP

Materials

DOTAP, 1 mg/ml (Boehringer GmbH, Mannheim, Germany)
Cell culture medium: RPMI 1640 (Euroclone, UK) or equivalent; 10% Fetalclone II, heat inactivated (HyClone, Logan, UT) or equivalent; 2 mM L-glutamine
ODNs: Working stocks at approximately 500 μM dilution in 10 mM Tris-HCl (pH 7.4), 1 mM
EDTA
HEPES-buffered saline (HBS)

Method

1. One to 2-hr prior to transfection, cells are plated into tissue culture plates at a density of 4×10^5 cells/ml.

2. To prepare the transfection reagent, carefully mix (do not vortex) equal volumes of ODNs (diluted to 6 μM in HBS) and DOTAP (diluted to 0.16 mg/ml in HBS), and allow to complex for 10 min at room temperature.

3. Dilute the DOTAP–ODN mixture into 9 volumes of cell culture medium without serum. Add to an equal volume of cells to achieve a final density of 2×10^5 cells/ml [the final concentrations are 150 nM (ODN) and 4 μg/ml (DOTAP)]. The effective ODN concentration range is between 30 and 300 nM. Transfection should be performed serum free during the first 4 hr. After this, Fetalclone II is added to achieve a final serum concentration of 5–10%. If cells do not tolerate serum-free conditions, serum can be added (not to exceed 5%) during the first 4 hr.

4. If higher ODN concentrations are used, the molar DOTAP : ODN ratio should be kept constant. In addition, if secondary test systems require varying numbers of cells for optimal analysis (see below), the ratio between the amount of ODN and the cell number should also be kept constant for the same ODN. As an example, if transfections are performed in six-well plates, the number of cells per well should be 10^6, and the volumes of the 6 μM ODN and 0.16-mg/ml DOTAP solutions mentioned above should be 125 μl each.

In our hands, optimal transfection times are 18–24 hr for extraction of RNA (Northern blot analysis), 24–48 hr for preparation of protein extracts (Western blot analysis or analysis by flow cytometry), and 24–96 hr for measurement of apoptosis and cell viability.

Transfection of Adherent A549 Non-Small Cell Lung Carcinoma
   and ZR-75-1 Breast Carcinoma Cell Lines by Use of Cationic
   Lipid Lipofectin

Materials

   Lipofectin (GIBCO-BRL, Life Technologies, Gaithersburg, MD)
   Cell culture medium: RPMI 1640 (Euroclone) or equivalent; 10% Fetalclone II, heat inactivated (HyClone) or equivalent; 2 mM L-glutamine
   ODNs: Working stocks at approximately 500 μM in 10 mM Tris-HCl (pH 7.4), 1 mM EDTA (store at $-20°$)

Method

1. One day before transfection, seed 2×10^5 cells in 1 ml of culture medium (avoid the presence of antibiotics) into each well of a six-well tissue culture plate [for ZR-75-1 cells, medium is further supplemented with glucose (4.5 g/liter), 10 mM HEPES, and 1.0 mM pyruvate].

2. Incubate cells overnight at 37°.

3. For each well, prepare the following solutions in 2-ml sterile plastic tubes: *Solution A*, Dilute the adequate amount of the ODN stock solution in serum-free medium to obtain a final concentration of 6.45 μM in a total volume of 186 μl; *Solution B*, add 18.6 μl of Lipofectin to 167 μl of pre-warmed (37°) serum-free culture medium and incubate for 30 min.

4. Combine solutions A and B, mix by pipetting up and down several times, and incubate for 10 min at room temperature.

5. Add 628 μl of serum-free medium, mix as before and transfer the total volume (1 ml) to the well. Note that this mixing proportions will yield a final ODN concentration of 600 nM. In case other concentrations are required, the ratio between ODN and Lipofectin must be kept constant. This also applies to the concentration of Lipofectin in solution B.

6. Incubate the cells with ODNs for 20 hr (avoid the presence of antibiotics in the culture medium during transfection).

7. Harvest the cells or carefully replace the transfection medium with 2 ml of normal cell culture medium (avoid the loss of cells while removing medium) and further incubate the cells for various time periods prior to use in secondary test systems (see description of DOTAP transfection, above).

Secondary Test Systems

After antisense treatment, cells must be analyzed in secondary test systems to prove the efficacy of the gene repression approach. It is beyond the scope of this chapter to provide the reader with a description of the different molecular and cellular test systems available. The most direct and reliable proof of an antisense effect can be provided only by determination of mRNA levels by Northern blotting or (semi-)quantitative polymerase chain reaction (PCR). In addition, expression of the target protein should be quantitated by Western blotting or flow cytometry. A number of mono- and polyclonal antibodies are commercially available for the detection of the apoptosis inhibitors Bcl-2 and Bcl-xL. The monoclonal mouse anti-human Bcl-2 clone 124 from Dako A/S (Glostrup, Denmark) works well for Western blotting, flow cytometry, and immunohistochemistry. For the detection of Bcl-xL, we use the polyclonal rabbit anti-Bcl-x antibody from Transduction Laboratories (Lexington, KY) for Western blotting, whereas the polyclonal rabbit anti-Bcl-x antibody L-19 from Santa Cruz Biotechnology (Santa Cruz, CA) is well suited for flow cytometry. If the ultimate goal of a study is to repress the expression of apoptosis inhibitors that might serve as survival factors for cells, the antisense effect should also be assessed in cell viability or apoptosis assays. Because Bcl-2 and Bcl-xL are engaged in the rheostat for the death program,[2] depending on the cell type their repression may either directly induce apoptosis or sensitize cells to unrelated cytotoxic agents, such as chemotherapeutic drugs. Unfortunately, the unspecific toxicity of phosphorothioate ODNs may also induce apoptosis in cells, making it difficult to determine the true antisense effect that results from repression of apoptosis inhibitors. Optimization of the transfection conditions as described above may help to reduce the unwanted effects of these compounds.

With the exception of follicular lymphoma cells,[38] in which Bcl-2 is strongly overexpressed owing to a chromosomal translocation, most other tumor types express apoptosis inhibitors at comparatively low levels. Depending on the test system, these levels might be under the detection limit, raising the false interpretation of negativity. Moreover, the downregulation of a target, the basal level of which is barely detectable, makes it difficult to allocate the antisense effect. Our experience is that there is hardly any tumor cell that is negative for Bcl-2 or Bcl-xL on the transcriptional and protein levels if highly sensitive detection methods, such as RT-PCR or overexposure of Western blots, are used. Because even extremely low levels of Bcl-2 and Bcl-xL might be sufficient to inhibit apoptosis in cells, this issue must be carefully addressed.

---

[38] Y. Tsujimoto and C. M. Croce, *Proc. Natl. Acad. Sci. U.S.A.* **83,** 5214 (1986).

Section IV

Antisense in Therapy

[34] *In Vitro* and *in Vivo* Modulation of Transforming Growth Factor β_1 Gene Expression by Antisense Oligomer

By Hun-Taeg Chung, Dong-Hwan Sohn, Byung-Min Choi, Ji-Chang Yoo, Hyun-Ock Pae, and Chang-Duk Jun

Introduction

Antisense technology is emerging as an effective means of lowering the levels of specific gene products.[1–3] It is based on the findings that these "antisense" sequences hybridize to specific RNA transcripts, disrupting normal RNA processing, stability, and translation, thereby preventing the expression of a targeted gene. Administration of antisense oligonucleotides or transfer of expression constructs capable of producing intracellular antisense sequences complementary to the mRNA of interest have been shown to block the translation of specific genes *in vitro* and *in vivo*. No other drug discovery method is better than the antisense principle in terms of duration to identify a specific lead compound with fewer side effects.

Transforming growth factor β_1 (TGF-β_1) is a 25,000 MW homodimeric protein secreted by a variety of transformed and nontransformed cells, including platelets, lymphocytes, and macrophages.[4–8] TGF-β_1 is known to be a powerful immunomodulatory agent. One report indicates that TGF-β_1 plays a role in ablating the respiratory burst of activated macrophages without impairing their phagocytic function.[9] Concurrent with this deactivation of macrophages, the growth factor also inhibits their ability to produce

[1] A. Harel-Bellan, D. R. Ferris, M. Vinocour, J. T. Holt, and W. L. Farrar, *J. Immunol.* **140,** 2431 (1988).

[2] D. A. Melton, *Proc. Natl. Acad. Sci. U.S.A.* **82,** 144 (1985).

[3] C. A. Stein and J. S. Cohen, *Cancer Res.* **48,** 2659 (1998).

[4] R. Derynck, J. A. Jarrett, E. A. Chen, D. H. Eaton, J. R. Bell, R. K. Associan, A. B. Roberts, M. B. Sporn, and D. V. Goeddel, *Nature* (*London*) **316,** 701 (1985).

[5] R. K. Assoian, A. Komoriya, C. A. Meyers, D. M. Miller, and M. B. Sporn, *J. Biol. Chem.* **258,** 7155 (1983).

[6] J. H. Kehrl, L. M. Wakefield, A. B. Roberts, S. Jakowlew, M. Alvarez-Mon, R. Derynck, M. B. Sporn, and A. S. Fauci, *J. Exp. Med.* **163,** 1037 (1986).

[7] J. H. Kehrl, A. B. Roberts, L. M. Wakefield, S. Jakowlew, M. B. Sporn, and A. S. Fauci, *J. Immunol.* **137,** 3855 (1986).

[8] S. M. Wahl, D. A. Hunt, H. L. Wong, S. Dougherty, N. McCartney-Francis, L. M. Wahl, L. Ellingsworth, J. A. Schmidt, G. Hall, A. B. Roberts, and M. B. Sporn, *J. Immunol.* **140,** 3026 (1988).

[9] S. Tsunawaki, M. Sporn, and C. F. Nathan, *Nature* (*London*) **334,** 260 (1988).

nitric oxide (NO).[10] Otherwise, it has been reported that TGF-β_1 enhances the wound healing process, but induces the overproduction of fibrous tissue and the formation of granulation tissue owing to its highly fibrogenic nature.[11,12]

Because antisense oligomer inhibits expression of specific genes,[1–3] this approach provides the opportunity to examine the role of the expression of TGF-β_1 during macrophage activation or wound healing processes. Thus, we will introduce *in vitro* and *in vivo* modulation of TGF-β_1 gene expression by antisense oligomer complementary to TGF-β_1 mRNA.

The materials with which the antisense approach can be applied are broadly divided into antisense oligodeoxynucleotides (ODNs), antisense RNAs, and catalytic RNAs or ribozymes.[13] Antisense sequences can be engineered either chemically (antisense ODNs)[14] or genetically (antisense RNAs).[15] This chapter focuses on antisense ODNs, which offer important stability and synthetic advantages over the use of antisense RNAs and is intended as an introduction to practical approaches in the use of antisense ODNs.

Using the following methods, we introduce *in vitro* and *in vivo* modulation of TGF-β_1 gene expression by antisense ODNs targeting the TGF-β_1 translation initiation region. First, we describe the evidence of the cellular uptake of antisense TGF-β_1 and its effect on TGF-β_1 expression in cultured macrophages. Second, we present the skin absorption of antisense and scar-removing effect by the inhibition of TGF-β_1 expression on the skin.

Synthetic Oligonucleotides

Antisense 25-mer phosphorothioate ODNs are synthesized on a Gene Assembler Special (Pharmacia LKB Biotechnology, Uppsala, Sweden) and purified by perfusion chromatography (PerSeptive Biosystems, Framingham, MA). The oligomers are lyophilized, resuspended in TE [10 mM Tris (pH 7.5)–1 mM EDTA (pH 8.0)], and quantified by spectrophotometry and gel electrophoresis. Phosphorothioate ODNs from the translation initiation region of the human TGF-β_1 gene are used in this study. The sequences are as follows: antisense oligomers (5'-CAG CCC GGA GGG CGG CAT

[10] Y. Vodovotz, C. Bogdan, J. Paik, Q. W. Xie, and C. F. Nathan, *J. Exp. Med.* **178,** 605 (1993).

[11] Q. Daniela, Jr., L. B. Nanney, J. A. Ditestein, and J. M. Davidson, *J. Invest. Dermatol.* **97,** 34 (1991).

[12] W. A. Border and E. Ruoslahti, *J. Clin. Invest.* **90,** 1 (1992).

[13] A. Colman, *J. Cell Sci.* **97,** 399 (1990).

[14] E. Uhlmann and A. Peyman, *Chem. Rev.* **90,** 543 (1990).

[15] D. A. Melton, P. A. Krieg, M. R. Rebagliati, T. Maniatis, K. Zinn, and M. R. Green, *Nucleic Acids Res.* **12,** 7035 (1984).

GGG GGA G-3') and sense oligomers (3'-GTC GGG CCT CCC GCC GTA CCC CCT C-5'). 5'-End fluorescense labeling of phosphorothioate ODN is done by DNA synthesizer with Fluoreprime (Pharmacia LKB Biotechnology).

In Vitro Modulation of Gene Expression

Cellular Uptake of Antisense Transforming Growth Factor β_1 Oligodeoxynucleotide

To determine cellular uptake of oligomers, murine peritoneal macrophages on coverslips are loaded with fluorescein isothiocyanate (FITC) end-labeled phosphorotioate ODNs (20 μg/ml) at room temperature and the changes of green fluorescence are measured at various time intervals using an ACAS Interactive Laser Cytometer (Meridian Instruments, Okomos, MI) equipped with a 200-mW argon-ion laser emitting light at 488 nm. For the pseudocolor images, black–blue represents low fluorescence while orange–red illustrates a high degree of fluorescence. Figure 1 (see color insert) represents the kinetics of cellular uptake of antisense TGF-β_1 ODNs in murine peritoneal macrophages. The antisense ODNs are detectable as early as 5 min after incubation and rapidly continue to increase for 30 min.

Effects of Antisense Transforming Growth Factor β_1 Oligodeoxynucleotides on Expression of Transforming Growth Factor β_1 in Activated Macrophages

Northern blot analysis is employed to determine whether the antisense ODNs targeted to TGF-β_1 mRNA can block the expression of TGF-β_1 gene that has been upregulated by the activation of macrophages. Thus, murine peritoneal macrophages are cultured for 6 hr with a biologically active phorbol ester (phorbol myristate acetate, PMA; 200 nM) in the presence or absence of antisense ODNs (20 μg/ml).

A cDNA (1 kb) probe of part of the rat TGF-β_1 gene is random prime labeled using the Megaprime labeling system. Total RNA is prepared by the LiCl–urea method, electrophoresed in 1.2% agarose–formaldehyde gels, and transferred to nylon membranes in 20× SSC (saline sodium citrate). After prehybridization, the filters are hybridized with random [α-^{32}P]dCTP-labeled probes in 50% formamide, 4× SSC, 1× Denhardt's solution, and salmon sperm DNA (10 μg/ml) for 16 hr at 42°. The filters are then washed, dried, and examined by autoradiography. On the basis of the Northern hybridization data, TGF-β_1 mRNA levels considerably decrease

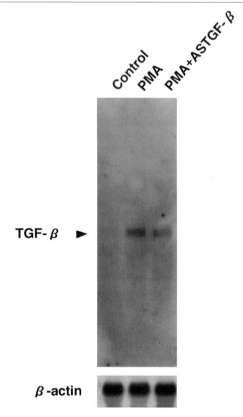

Fig. 2. Effects of antisense ODNs complementary to TGF-β_1 mRNA on the expression of TGF-β_1 gene induced by PMA. Thioglycolate-elicited peritoneal macrophages were plated at 2×10^7 cells/100-mm diameter petri dish and were cultured either in medium alone or in medium that contained PMA (200 nM) or PMA plus antisense ODNs (20 mg/ml). After 12 hr of incubation, total RNA was prepared and TGF-β_1 mRNA was analyzed by Northern hybridization. Blots were hybridized with the indicated [^{32}P]dCTP-radiolabeled cDNA and exposed to X-ray film for 24–48 hr.

in antisense ODNs-treated cells (Fig. 2). Simultaneous determination of β-actin mRNA in the control and antisense-treated cells shows no major differences.

In Vivo Modulation of Gene Expression

Distribution of Antisense Oligonucleotides into Skin

Damaged skin is prepared by removing the stratum corneum by repeated application of cellophane tape until the skin glistens. The diffusion

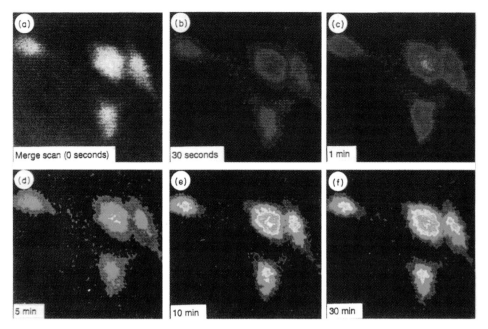

Fɪɢ. 1. The kinetics of cellular uptake of antisense ODNs in murine peritoneal macrophages. Thioglycolate-elicited macrophages (1×10^6) were cultured with 5′ end FITC-labeled phosphothioate antisense ODNs complementary to TGF-β_1 mRNA and the changes in green fluorescence were measured at various time intervals using an ACAS interactive laser cytometer. For the pseudocolor images, black–blue represents low fluorescence while orange–red illustrates a high degree of fluorescence.

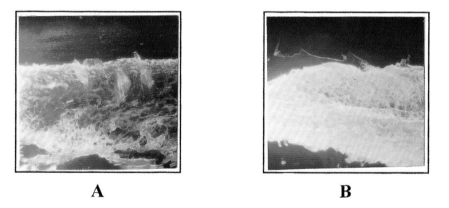

A　　　　　　　　　　　**B**

Fɪɢ. 3. Fluorescence photomicrograph of fluorescence-labeled antisense PS-ODN within skin 12 hr after permeation study (original magnification, ×40): (A) normal skin; (B) damaged skin.

TABLE I
CUMULATIVE AMOUNT (μg/cm^2) OF ANTISENSE
ODNs PENETRATED ACROSS RAT SKIN[a]

Time (hr)	Normal rat skin	Damaged rat skin
1	0.000	0.041 ± 0.040
4	0.028	0.279 ± 0.330
8	0.038	0.555 ± 0.560
12	0.077	0.772 ± 0.690

[a] Each value represents the mean ± SEM ($n = 7$).

study is performed by use of a Franz diffusion cell. The receptor compartment of the diffusion cell is filled with 10 ml of phosphate-buffered saline and kept at 37°. The available surface area for distribution is 1.77 cm^2. The contents of the receptor compartment are mixed with a Teflon magnetic stirring bar. Twelve hours after treatment, the skin is washed and then fixed with liquid paraffin. Fluorescence photomicrographs of skin are obtained using a reversed-contrast microscope with ×40 magnification. ODNs can be distributed across tape-striped damaged skin, while a small amount of ODNs penetrate into the normal skin (Fig. 3, see color insert). The removal of stratum corneum markedly increases the diffusion of ODNs.

Percutaneous Absorption of Antisense Oligonucleotides

ODN is 5'-end labeled with ^{35}S-labeled rATP and T4 polynucleotide kinase. The method used to study skin transport is similar to that of the distribution study, except that radiolabeled ODN is used. ^{35}S-Labeled ODN is applied to the donor side and 100-μl aliquots are withdrawn at the desired time. The aliquot is mixed with 6 ml of scintillation cocktail and determined using a liquid scintillation system [Beckman (Fullerton, CA) LS 7800]. The skin permeation parameters are calculated according to the following equations:

$$D = l^2/6T$$
$$J_s = DK_mC_s/l$$
$$K_p = DK_m/l$$

where J_s is the penetration rate, D denotes the diffusion constant within skin, K_m is the skin/vehicle partition coefficient of ODN, and T represents the lag period calculated from the intercept of the flux with time axis. K_p denotes the permeability coefficient through the stratum corneum and C_s is the ODN concentration. The cumulative amount of ODN penetrated in 12 hr is 0.772 ± 0.690 μg/cm^2 in damaged skin and 0.077 μg/cm^2 in normal skin (Table I). The penetrated amount is increased about sixfold by removal

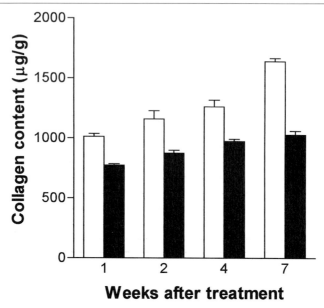

FIG. 4. Effect of antisense ODN on the collagen content (μg/g) of wounds at various times after wounding. Open columns are collagen contents in wound tissue treated with sense ODN and black columns are collagen contents in normal tissue treated with antisense ODN.

of stratum corneum, which is largely responsible for controlling the permeation.

Effects of Antisense Transforming Growth Factor β_1 Oligodeoxynucleotide on Scar Formation

For analysis of the area of both scar and fibrosis, macro- or microphotographs are taken with a video camera, and relative quantitative analysis is made using Image-Pro Plus software (Image-Pro Plus; Media Cybernetics, Silver Spring, MD). Statistical analyses are carried out automatically using Image-Pro Plus software. On the basis of a relative quantitative analysis of the area, less scar (approximately 20% of control at 7 weeks) and fibrosis (approximately 30% of control at 4 weeks) are found (data not shown).

The specimens of wound tissue are weighed and hydrolyzed in 6 M hydrochloric acid. An aliquot of the samples is derivatized using chloramine-T solution and Erhlich reagent and measured at 558 nm. A standard calibration curve is prepared using *trans*-4-hydroxy-L-proline.[16] Analysis of the

[16] I. S. Jamall, V. N. Fimelli, and S. S. Quehee, *Anal. Biochem.* **112,** 70 (1981).

wound tissue for hydroxyproline content reveals an increase in the collagen content in wounded skin compared with normal skin. However, the wound tissue treated with antisense TGF-β_1 ODNs contains much less collagen than does the wound tissue treated with sense TGF-β_1 ODNs in the same animal (Fig. 4).

Conclusions

Administration of antisense TGF-β_1 ODNs to cultured macrophages shows rapid kinetics of cellular uptake and inhibition of the expression of the TGF-β_1 gene. In addition, topical administration of antisense oligomers to wounded skin effectively reduces subsequent formation of excessive fibrous tissue. Taken together, the rapid cellular uptake with relative stability of phosphorothioate oligomers and reduction of excessive fibrous tissue after topical application on damaged skin suggest potential applicability of the antisense approach as a therapeutic modality *in vivo*.

[35] Analysis of Cancer Gene Functions through Gene Inhibition with Antisense Oligonucleotides

By CHERYL ROBINSON-BENION, ROY A. JENSEN, and JEFFREY T. HOLT

Introduction

Functional analysis of human cancer susceptibility genes has been greatly facilitated by the use of gene inhibition methods, including homologous recombination in mice and antisense methodologies in human cells. Although homologous recombination has clearcut advantages over antisense inhibition methods in mouse experimental models, antisense oligonucleotide approaches continue to serve as a rapid and important means of gene inhibition in human cell lines; and are clearly of great importance for the study of poorly conserved genes with different function between mouse and human.

The breast cancer susceptibility genes are examples of genes with different functions in mouse versus human model systems. Mouse BRCA1 is only 57% homologous to human BRCA1[1-3] and BRCA1 appears to function

---

[1] L. C. Gowen, B. L. Johnson, A. M. Latour, K. K. Sulik, and B. H. Koller, *Nature Genet.* **12,** 191 (1996).

[2] R. Hakem, J. L. de la Pomba, C. Sirard, R. Mo, M. Woo, A. Hakem, A. Wakeham, J. Potter, A. Reitmair, F. Billia, E. Firpo, C. Hui, J. Roberts, J. Rossant, and T. W. Mak, *Cell* **85,** 1009 (1996).

differently in the two systems. BRCA1 has been shown to be a powerful growth suppressor in both yeast and human systems,[4–8] but is required for cellular proliferation during mouse development. Although mice carrying homozygous BRCA1 mutations die early in gestation, there are reports of patients who are homozygous for BRCA1 mutations.[9] Further, humans heterozygous for BRCA1 mutations develop breast cancer early in life, but mice heterozygous for BRCA1 mutations do not develop cancers. Well-described differences in DNA repair between mouse and human cells clearly complicate the ability to extrapolate results from mouse models to patients.[10–12] Mouse and human cells show differences in the amount of damage sustained per given DNA-damaging dose, in the kinetics of DNA repair and in cellular survival at a given dose of a DNA-damaging agent. BRCA2 is also poorly conserved between mouse and human,[13] although some clear functional similarities have been described.[14–16]

Principle of Method

Oligonucleotide-based antisense methods use single-stranded oligonucleotides composed of either unmodified or modified bases as molecular tools to inhibit the expression of target genes. Because unmodified deoxyri-

[3] C. Y. Liu, A. Flesken-Nikitin, S. Li, Y. Zeng, an W.-H. Lee, *Genes Dev.* **10**, 1835 (1996).

[4] M. E. Thompson, R. A. Jensen, P. S. Obermiller, D. L. Page, and J. T. Holt, *Nature Genet.* **9**, 444 (1995).

[5] J. T. Holt, M. E. Thompson, C. Szabo, C. Robinson-Benion, C. L. Arteaga, M. C. King, and R. A. Jensen, *Nature Genet.* **12**, 298 (1996).

[6] J. S. Humphrey, A. Salim, M. R. Erdos, F. S. Collins, L. C. Brody, and R. D. Klausner, *Proc. Natl. Acad. Sci. U.S.A.* **94**, 5820 (1997).

[7] K. Somasundaram, H. Zhang, Y.-X. Zeng, Y. Houvras, Y. Peng, H. Zhang, G. S. Wu, J. D. Licht, B. L. Weber, and W. S. El-Deiry, *Nature (London)* **389**, 187 (1997).

[8] T. F. Burke, K. S. Cocke, S. J. Lemke, E. Angleton, G. W. Becker, and R. P. Beckmann, *Oncogene* **16**, 1031 (1998).

[9] M. Boyd, F. Harris, R. McFarlane, H. R. Davidson, and D. M. Black, *Nature (London)* **375**, 541 (1995).

[10] M. Namba, K. Nishitani, and T. Kimoto, *Jpn. J. Exp. Med.* **47**, 263 (1997).

[11] T. Yagi, *Mutat. Res.* **96**, 89 (1982).

[12] D. G. Walton, A. B. Acton, and H. F. Stich, *Mutat. Res.* **129**, 129 (1984).

[13] S. V. Tavtigian, J. Simard, J. Rommens, F. Couch, D. Shattuck-Eidens, S. Neuhausen, S. Merajver, S. Thorlacius, and K. Offit, *Nature Genet.* **12**, 333 (1996).

[14] S. K. Sharan, M. Morimatsu, U. Albrecht, D. S. Lim, E. Regel, C. Dinh, A. Sands, G. Eichele, P. Hasty, and A. Bradley, *Nature (London)* **386**, 804 (1997).

[15] F. Connor, D. Bertwistle, P. J. Mee, G. M. Ross, S. Swift, E. Grigorieva, V. L. Tybulewicz, and A. Ashworth, *Nature Genet.* **17**, 423 (1997).

[16] K. J. Patel, V. P. Yu, H. Lee, A. Corcoran, F. C. Thistlethwaite, M. J. Evans, W. H. Colledge, L. S. Friedman, B. A. Ponder, and A. R. Venkitaraman, *Mol. Cell* **1**, 347 (1998).

bonucleotides are often rapidly degraded in culture systems, a variety of modified oligonucleotides have been studied.[17] Phosphorothioate deoxyribonucleotides are a commonly employed type of modified oligonucleotide with reasonable *in vitro* stability, although investigators must be concerned about reported nonantisense effects of phosphorothioate oligonucleotides.[18]

Materials and Reagents

For oligonucleotide antisense experiments to be successful and free of experimental artifacts, it is important to use pure oligonucleotides preparations as well as nuclease-free culture conditions. Antisense oligonucleotides may be synthesized by investigators or obtained from commercial sources. However, purification by repeated lyophilization, chromatography, or gel purification is often necessary regardless of the source of oligonucleotides (one should not assume that oligonucleotides are pure and free of volatile contaminants simply because they are purchased commercially). After an appropriate purification step, we lyophilize oligonucleotides in sterile distilled water two times to remove volatile molecules produced in synthetic or purification procedures. Both modified and unmodified deoxyribonucleotides may be resuspended in 10 mM HEPES (N-2-hydroxyethylpiperazine-N'-2-ethanesulfonic acid)-buffered saline at pH 7.4. It is essential that the pH of oligonucleotide preparations be tested because the addition of even small quantities of acidic or basic solutions has profound effects on the growth and viability of tissue culture cells.

Modified phosphorothioate deoxyribonucleotides use in the experiments described in this chapter were synthesized by the Midland Certified Reagent Company (Midland, TX), and then visualized by UV shadowing after electrophoresis on denaturing 20% polyacrylamide gels by our prior method.[19] These oligonucleotide preparations were twice lyophilized after resuspension in distilled water and then resuspended in HEPES, pH 7.4, as described above.

Methods

Antisense Inhibition with Oligonucleotides

Criteria for design of antisense oligonucleotides include (1) site of target sequence within the gene, (2) length of antisense oligonucleotide, and (3) secondary structure of antisense oligonucleotide.

---

[17] C. Robinson-Benion and J. T. Holt, *Methods Enzymol.* **254,** 363 (1995).

[18] J. T. Holt, *Nature Med.* **1,** 407 (1995).

[19] J. T. Holt, R. L. Redner, and A. W. Nienhuis, *Mol. Cell. Biol.* **8,** 963 (1988).

The target sequence should be within a region of the mRNA that is unbound by protein and theoretically free of secondary structure. Sequences within the 5′ untranslated region or translation initiation region are often used as antisense target sequences. To obtain T_m optimal for inhibition of target genes in culture cells grown at 37° we generally employ phosphorothioate oligonucleotides between 17 and 22 bases in length.

Demonstration of Antisense Inhibition

To be confident that the observed results in an antisense experiment are due to a true antisense-mediated inhibition of gene expression, it is important to perform a quantitative assay to determine the degree of inhibition of gene expression. The antisense oligonucleotide is designed to hybridize to the translation initiation site of the human BRCA2 gene (nucleotides 229–245 from GenBank accession number HSU43746). The sequence of the antisense oligonucleotide synthesized by Midland as described above is 5′-TTGGATCCAATAGGCAT-3′. The sequence of the complementary sense oligonucleotide is 5′-ATGCCTATTGGATCCAA-3′. An oligonucleotide with a shuffled sequence containing four mismatched bases compared with the antisense oligonucleotide has sequence 5′-TTGGAATCATCAG-GCAT-3′. BxPC-3 carcinoma cells are grown in Dulbecco's modified Eagle's medium (DMEM) with 10% fetal bovine serum (FBS) and oligonucleotides are introduced by transfection with Lipofectin (Life Technologies, Gaithersburg, MD). Prior to transfection the cells are washed with serum-free growth medium, and the modified phosphorothioate oligonucleotides are diluted in 100 μl of serum-free medium (in sufficient quantity to obtain a final concentration of 1 μM oligonucleotide) and then mixed with Lipofectin solution B in 12 × 75 ml sterile tubes as described by the manufacturer.[20] The oligonucleotide–liposomal precipitate is incubated at room temperature for 30 min and then added to the cultured cells, which are incubated for an additional 48 hr to obtain antisense inhibition. RNA is isolated from culture cells by the guanadinium thiocyanate method and the total RNA from 2 million cells is hybridized with 3 million cpm of labeled probe as described previously.[21] Control oligonucleotides are employed to assure that the observed results correlate with antisense inhibition as determined by nuclease protection assay.

The inhibition of BRCA2 mRNA levels is determined by ribonuclease protection assay, employing a BRCA2 probe fragment of exons 3–8 cloned into pBluescript SK(+) (Stratagene, La Jolla, CA). This plasmid is digested with *Pst*I restriction enzyme and transcribed with T3 polymerase in the

[20] D. W. Abbott, M. L. Freeman, and J. T. Holt, *J. Natl. Cancer Inst.* **90**, 978 (1998).
[21] R. A. Jensen, D. L. Page, and J. T. Holt, *Proc. Natl. Acad. Sci. U.S.A.* **91**, 9257 (1994).

presence of ^{32}P-labeled guanosine triphosphate (^{32}P]GTP). A control human glyceraldehyde-3-phosphate dehydrogenase (GADPH) probe is a cloned SacI–XbaI fragment used as a control for RNA loading as previously described.[22] The GADPH probe is transcribed with T7 polymerase in the presence of ^{32}P]GTP. Two micrograms of RNA from each sample is hybridized with the mixed labeled probes for BRCA2 and GADPH in 50% formamide at 55° for 3 hr. The samples are treated with RNase A and T1 (Life Technologies) to degrade the unhybridized probes, and then the DNAs are ethanol precipitated, washed with 70% ethanol, and resolved on an 8% denaturing polyacrylamide gel containing 6 M urea.

The results of this experiment are presented in Fig. 1, and demonstrate substantially lower BRCA2 expression than in those cells transfected with either sense or shuffled oligonucleotide (Fig. 1, lanes 3–6). The purpose of this experiment was to determine whether antisense inhibition of BRCA2 results in increased sensitivity of cells to the drug mitoxantrone. For this reason, it was important to show whether mitoxantrone treatment affected either BRCA2 gene expression or affected antisense inhibition of the BRCA2 gene. It appears that mitoxantrone exposure has no effect on the degree of BRCA2 gene inhibition (Fig. 1, lanes 7–10).

Effects of BRCA2 Inhibition on Drug Sensitivity

Having demonstrated that antisense oligonucleotides directed against BRCA2 produced inhibition of BRCA2 mRNA steady state levels (but control oligonucleotides had no effect), we next assayed the effect of BRCA2 gene inhibition on drug sensitivity. For these studies BxPC-3 human pancreatic cancer cells are seeded as single-cell suspensions and then exposed to log dilutions of mitoxantrone for 1 hr at 37°. The cells are then washed and replated in their appropriate complete growth medium (described above) for 3 weeks, stained, and the colonies quantified. The 50% inhibitory concentration (IC $_{50}$) is then determined as the drug concentration that causes a 50% inhibition of cell proliferation. Relative survival is computed by standardization to the average survival of the other carcinoma cell lines, performed in triplicate. Drug sensitivities of BxPC-3 cells transfected with sense, antisense, or shuffled oligonucleotide sequence are determined as described above.

Figure 2 demonstrates that cells treated with antisense BRCA2 are significantly more sensitive to mitoxantrone than those cells treated with the control oligonucleotides for sense or shuffled BRCA2 (Fig. 2). These results are significant at the $p < 0.01$ level.

[22] M. E. Thompson, R. A. Jensen, P. S. Obermiller, D. L. Page, and J. T. Holt, *Nature Genet.* **9,** 444 (1995).

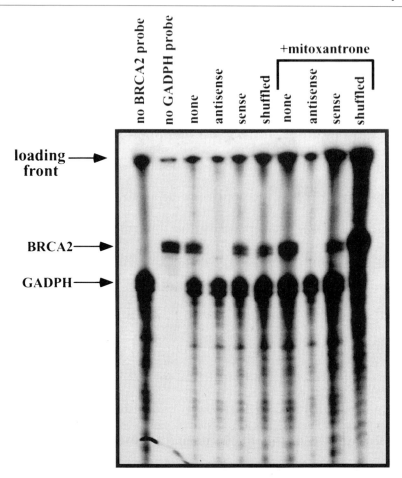

FIG. 1. Inhibition of BRCA2 messenger RNA levels. Nuclease protection assay analyzing extent of BRCA2 gene inhibition by antisense. BxPC-3 cells (expressing normal BRCA2) were transfected with antisense (lanes 4, 8), sense (lanes 5, 9), or shuffled (lanes 6, 10) BRCA2 antisense single-stranded DNA sequences. Cells that received no exogenous DNA are shown in lanes 3 and 7. Nuclease protection assays were then performed to determine BRCA2 messenger RNA levels after transfection with each single-stranded DNA. In addition, cells were also transfected in the presence of mitoxantrone to determine if mitoxantrone affected antisense oligonucleotide-mediated inhibition of BRCA2 messenger RNA levels (lanes 7–10). The upper band represents BRCA2 mRNA. To control for equivalent RNA loading, the levels of glyceraldehyde-3-phosphate dehydrogenase (GADPH) mRNA were also measured (lower band). [Adapted from Fig. 5A in Ref. 20.]

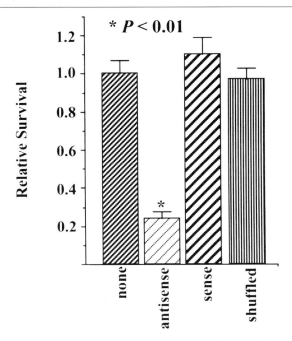

FIG. 2. Relative survival of cells treated with antisense BRCA2 DNAs plus mitoxantrone. BxPC-3 cells transfected with antisense, sense, or shuffled BRCA2 DNA sequences or un-transfected BxPC-3 cells were exposed to mitoxantrone, and cell survival was determined. Results are presented as relative survival (two-sided $p < 0.01$ by Student t test). [Adapted from Fig. 5B in Ref. 20.]

Concluding Remarks

Antisense oligonucleotides can serve as effective tools for gene inhibition. The use of control oligonucleotides directed against the opposite strand (sense) or against a shuffled sequence provides evidence that the observed effects are the result of antisense inhibition and not merely a consequence of nonspecific toxicity resulting from either oligonucleotides or from contaminants arising in oligonucleotide preparations. It is crucial to employ a number of control oligonucleotides in these types of experiments as well as to demonstrate gene inhibition by a quantitative method.

Nuclease protection assays provide a simple and quantitative method for assaying mRNA levels and are consequently an excellent method for analyzing the extent of gene inhibition in antisense oligonucleotide experiments. Northern blots are also quantitative studies but generally require reasonably large numbers of cells so that several micrograms of RNA can be loaded for each experimental point. Although polymerase chain reaction

(PCR)-based assays are useful in a wide variety of experimental studies that evaluate the presence or absence of a signal, the semiquantitative nature of PCR studies (even those containing internal controls) makes them less desirable than nuclease protection assays in antisense studies because the degree of inhibition in antisense experiments is rarely absolute.

The experiments presented here provide an example of how oligonucleotide-mediated antisense methods can be used to analyze gene function. Cancer predisposition genes are presumed to have important functions because mutations result in serious malignancies, but the cellular or biochemical functions of these genes are unfortunately rarely apparent at their initial discovery. Antisense oligonucleotide approaches such as those presented here can consequently serve as useful adjuncts to a variety of gene transfer studies and biochemical reconstitution approaches. In addition, they can be readily adapted for human cellular model systems so that one can rapidly test whether hypotheses developed from mouse knockout models are relevant and predictive for human cancers.

Acknowledgments

We thank Patrice Obermiller for expert technical assistance; this work was supported by NIH RO1CA51735 (J.T.H.) and the Frances Williams Preston Laboratory of the T. J. Martell Foundation (J.T.H.).

[36] Dosimetry and Optimization of *in Vivo* Targeting with Radiolabeled Antisense Oligodeoxynucleotides: Oligonucleotide Radiotherapy

By KALEVI J. A. KAIREMO, ANTTI P. JEKUNEN, and MIKKO TENHUNEN

Introduction

Antisense oligomers may serve as a vehicle for carrying cytotoxic or radioactive compounds into a particular location. Radiolabeled oligodeoxynucleotides (ODNs) have the potential of having both direct antisense inhibition and radiation effects. In this chapter, we extend conventional use of radioactive ODNs from limited biokinetic studies to therapeutic use. We present calculations from macroscopic biodistribution data to cell nuclear doses.

Previously we have shown theoretically that oligonucleotide therapy may be effective with internally labeled (^{32}P, ^{33}P, and ^{35}S) oligodeoxynucleotide phosphorothioates. In addition, Auger- and γ-emitting radionuclides ^{51}Cr, ^{67}Ga, ^{111}In, ^{114m}In, ^{123}I, ^{125}I, ^{131}I, and ^{201}Tl have been evaluated.

Subcellular biodistribution was used in evaluation of the best targeting inside the cell with one oligomer: ^{35}S gave the lowest variation of nuclear dose in the different cell dimensions we studied (nuclear diameter, 6–16 μm; cellular diameter, 12–20 μm). Therefore, cell nuclear targets should be treated with short-range β emitters, such as ^{35}S or ^{33}P, suitable for internal labeling of oligonucleotides in optimal oligonucleotide radiotherapy. In addition, dual labeling with ^{32}P and ^{35}S may provide therapeutic benefits when treating small and large tumors simultaneously. Further, *in vivo* development of labeled anticancer ODNs, optimally with ^{33}P and ^{35}S, is highly indicated.

Background and Significance

While most cytotoxic and biological drugs in the treatment of human cancer are lacking specificity, antisense ODNs are derived against the most specific target, a DNA/RNA sequence. Antisense ODNs constitute an attractive tool for specific therapeutic applications.[1] An antisense ODN binds to its complementary counterpart in mRNA, as reverse complementary strands of DNA or RNA hybridize with a specific target mRNA sequence. Hybridization forms a duplex that prevents mRNA utilization and also promotes its destruction by RNase H. This results in the inhibition of mRNA translation into protein and can block all kinds of cellular processes. Binding to DNA requires formation of a triplex helix with the target DNA duplex.[2] Specificity of triplex-forming processes is comparable with that of complementary strand pairing in DNA duplex. The primary target for ionizing radiation is nuclear DNA, and the radiotoxicity of Auger electron emitters is mainly due to low-energy electrons with short ranges (1–10 nm). If the decays occur in close proximity to DNA, the combined action of the Auger electrons results in molecular damage to DNA.[3] Antisense techniques also have limitations, e.g., the need for a viral vector to deliver long antisense sequences into cells and lack of stability. However, the use of ODNs without a viral vector is relatively simple, and synthesis of ODNs is almost routine.

Several nuclease-resistant ODNs have been developed, because cellular nucleases are highly efficient at digesting phosphodiester ODNs. In phosphorothioate ODNs, a nonbridging oxygen atom is replaced by a sulfur

[1] R. F. Wagner, *Nature (London)* **333,** 3372 (1994).
[2] M. D. Frank-Kamenetskii and S. M. Mirkin, *Annu. Rev. Biochem.* **64,** 65 (1995).
[3] K. S. R. Sastry, *Med. Phys.* **19,** 1361 (1992).

atom, whereas a methyl group replaces one of the nonbridging oxygen atoms in methyl phosphonate ODNs.[4] There are several other types of ODNs that are nuclease resistant: ODNs having a methyl group at the 2′-O position on the ribose moiety[5] or those where the phosphate backbones are replaced by polyamide structures to which the nitrogenous bases are linked.[6] Ribozymes have been used as antisense oncogenes, and they provide specificity and small size consisting of 40 to 50 bases.[7]

Antisense ODNs have been studied *in vivo* with regard to their pharmacokinetics and their pharmacological and toxicological properties. The pharmacokinetics of various ODNs in animal models have already been determined.[8,9] ODNs are rapidly absorbed from parenteral sites. They bind to serum albumin and other proteins with low affinity and distribute to all peripheral tissues, with the kidneys and liver accumulating most of the drug. They are cleared by slow metabolism with an elimination half-life up to 50 hr. The biokinetics of GEM91 phosphorothioate oligodeoxynucleotide have been evaluated in six patients with acquired immunodeficiency syndrome (AIDS), where the plasma mean residence time varied from 24.7 to 49.6 hr, the mean being 41.7 ± 3.6 hr.[10] Several antisense radiopharmaceuticals have been prepared.[11] They have been labeled with ^{111}In, ^{99m}Tc, and ^{67}Ga. Phosphorothioate oligodeoxynucleotides are relatively resistant to destruction by nucleases, they have good aqueous solubility, they hybridize efficiently with target RNA with relatively high specificity, they are relatively efficiently taken up by cells, and they are widely used in automated oligonucleotide synthesizers.[12] The selection of an appropriate nucleotide sequence is of importance in inducing translational arrest. The therapeutic possibilities of radiolabeled antisense oligodeoxynucleotides are still unknown, and one of the basic questions in the therapeutic use of labeled oligonucleotides is the optimal source of radiation. Phosphorothioate oligo-

---

[4] P. S. Miller and P. O. P. Ts'o, *Anticancer Drug Design* **2**, 117 (1987).

[5] B. Monia, J. Johanston, and D. Eckert, *J. Biol. Chem.* **267**, 19954 (1992).

[6] P. E. Nielsen, M. Egholm, and R. H. Berg, *Science* **252**, 1497 (1991).

[7] K. J. Scanlon, H. Ishida, and M. Kashani-Sabet, *Proc. Natl. Acad. Sci. U.S.A.* **91**, 11123 (1994).

[8] S. Agrawal, J. Temsamani, and J. Y. Tang, *Proc. Natl. Acad. Sci. U.S.A.* **88**, 7595 (1991).

[9] H. Sands, L. J. Gorey-Feret, A. J. Cocuzza, F. W. Hobbs, D. Chidester, and G. L. Trainor, *Mol. Pharmacol.* **45**, 932 (1994).

[10] R. Zhang R, J. Yan, H. Shahinian, G. Amin, Z. Lu, T. Liu, M. S. Saag, Z. Jiang, J. Temsamani, B. R. Martin, P. J. Schechter, S. Agrawal, and R. B. Diasio, *Clin. Pharmacol. Ther.* **58**, 44 (1995).

[11] M. K. Dewanjee, A. K. Ghafouripour, M. Kapadvanjwala, and A. T. Samy, *BioTechniques* **16**, 844 (1994)

[12] R. Zhang, R. B. Diasio, Z. Lu, T. Liu, Z. Jiang, W. M. Galbraith, and S. Agrawal, *Biochem. Pharmacol.* **49**, 929 (1995).

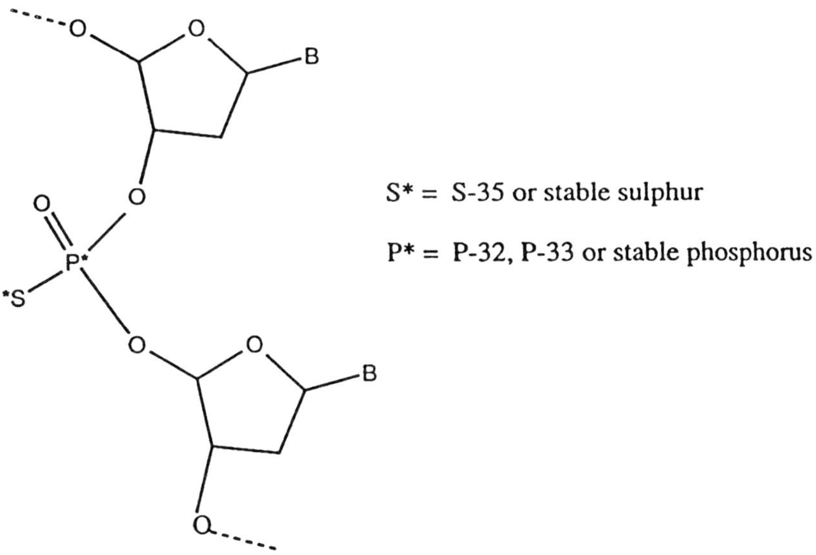

S* = S-35 or stable sulphur

P* = P-32, P-33 or stable phosphorus

FIG. 1. The structural formulas of oligonucleotide analogs using internally labeled oligonu-cleotide phosphorothioate either with ^{32}P, ^{33}P, or ^{35}S. B, Base. Reprinted with permission from Oxford University Press, *Anti-Cancer Drug Design* **11,** 439 (1996).]

deoxynucleotides labeled internally either with ^{35}S or $^{32/33}$P do not require any extra coupling techniques, as is the case with transition metals.

We have previously shown by using the biodistribution data of oligonu-cleotide phosphorothioates in a xenograft model that this type of therapy can theoretically be given with ^{32}P and ^{33}P.[13] Figure 1 demonstrates the structure of oligonucleotide phosphorothioates. Here we extend the analysis to several more isotopes, e.g., β- and Auger-emitting radionuclides. We calculated tumor and organ doses, and *in vivo* subcellular tissue distribution for oligodeoxynucleotide phosphorothioates.

We estimated those dosimetric properties of oligonucleotides that, at the cellular level, could be predicted from existing data on the characteriza-tion of phosphorolabeled ODNs. Predetermined situations differed from each other by nuclear and cell sizes. This made it possible to assess relative radiation exposures in these variable cellular dimensions. We used 10 radio-nuclides with different physical properties in our calculations. Four prede-termined cellular dimensions were used, even though, in reality, cell and

---

[13] K. J. A. Kairemo, M. Tenhunen, and A. P. Jekunen, *Antisense Nucleic Acid Drug Dev.* **6,** 215 (1996)

nuclear sizes may vary in a particular tumor type. We chose to keep mathematics simple and thus to obtain more understandable results. Calculations can give a recommendable source for the labeling of a ODN, and thus allow a proper selection of the optimal label. This requires estimation about benefits from different isotopes in several cell models with different cell dimensions.

The aim of this chapter is to discuss the internal radiation doses derived from cellular data, and discuss the suitability of several radionuclides for labeling ODNs. The role of oligonucleotide radiotherapy is also discussed by proposing a speculative treatment regimen.

Dose Calculation: From Macro to Micro

The accumulated dose from internally administrated radionuclides is usually estimated using the principles and formalism of the Medical Internal Radiation Dose (MIRD) scheme.[14] In the MIRD scheme the dose D [in grays (Gy)] at the target organ (T) is calculated as a sum of the dose component from the target organ itself (T ← T) and dose components from different source organs (N_k) to the target (T ← N_k):

$$D = \tilde{A}_T \cdot S(T \leftarrow T) + \sum_k \tilde{A}_{N,k} S(T \leftarrow N_k) \tag{1}$$

where $\tilde{A}$ represents the cumulated activity [unit, becquerel · seconds (Bq · sec)], i.e., number of nuclear disintegrations in an organ or target during the time interval of interest. S represents the fraction of dose per nuclear disintegration (unit, Gy/Bq · sec) that emits from the source organ or target and is absorbed in the target. The absorbed dose in the target organ can be macroscopically divided into two components: the dose from the penetrative radiation (such as X-rays and γ quanta) and the dose from the nonpenetrative radiation (such as β particles). For β-emitting radionuclides the dose rate $\dot{D}$ (unit, Gy/sec) can be calculated in a spherically symmetric situation as shown in Eq. (2):

$$\dot{D} = A \Delta \Phi_\beta(r) \tag{2}$$

where the dose rate in a selected point is calculated at a distance r from the source activity A. $\Delta = n_\beta E_{av}$ gives the emitted energy per nuclear disintegration with an average number of emitted particles n_β and average β energy E_{av}. $\Phi_\beta(r)$ gives the specific absorbed energy at the selected point.

[14] R. Loevinger and M. A. Berman, "A Revised Schema for Calculating the Absorbed Dose from Biologically Distributed Radionuclides." MIRD pamphlet no. 1, revised. Society of Nuclear Medicine, New York, 1976.

If the activity concentration C (activity per unit mass: Bq/kg or Bq/g) is constant inside a spherical volume, Eq. (2) can be written as

$$\dot{D} = C\Delta\phi_\beta(r) \tag{3}$$

where absorbed fraction $\phi_\beta(r)$ for spherical geometry can be calculated by the method of Leichner,[15] using a volume integral:

$$\phi_\beta(r) = G_0 \int_V \frac{e^{-\mu'r}}{4\pi r^2}\{1 + [d_1(\mu'r) + d_2(\mu'r)^2 + d_3(\mu'r)^3]e^{-(d_4-1)\mu'r}\}\,dV \tag{4}$$

with energy-specific constants μ', d_1, d_2, d_3, d_4, and G_0. This method was used to calculate the dose at the center of a water-equivalent sphere with a uniform ^{32}P, ^{33}P, or ^{35}S activity concentration inside and zero activity outside as a function of sphere mass. Owing to the continuous energy spectrum of β particles the central dose is expected to increase until a radius larger than the maximum range of the β particles is reached. With this method it is possible to compare theoretically the dosimetry of β-emitting nuclides in the treatment of tumors of different size. Figure 2 shows the effect of different β energies of ^{32}P, ^{33}P, and ^{35}S. The local absorption of ^{33}P is higher than that of ^{32}P if the tumor mass is smaller than 300 μg (Fig. 2). The local absorption of ^{35}S is higher than that of ^{32}P if the mass is less than 80 μg (Fig. 2).

In addition to the tumor dosimetry it is equally important to consider the dosimetry in the normal tissue for the optimization of the radiation therapy. For that reason we have made dose calculations using published biodistribution data on oligonucleotides in the mouse with 15-mer ^{111}In-labeled oligonucleotide sequence coupled with diethylenetriamine pentaacetate isothiocyanate[16] or in the rat with 25-mer oligodeoxynucleotide phosphorothioate GEM91[12] and 27-mer oligonucleotide phosphorothioate.[17] The pharmakokinetics of the compounds were not expected to change with phosphorus or sulfur labeling. The activity concentrations were calculated using the estimated relative organ weights of the whole animal for the liver, kidneys, bone marrow, and tumor of 9, 2, 3, and 5%, respectively.[17,18] The cumulative concentration, i.e., the area under the time–concentration curve, was estimated either from the published time–concentration data by supposing monoexponential elimination after the

[15] P. K. Leichner, *J. Nucl. Med.* **35**, 1721 (1994).

[16] M. K. Dewanjee, A. K. Ghafouripour, M. Kapadvanjwala, S. Dewanjee, A. N. Serafini, D. M. Lopez, and G. N. Sfakianakis, *J. Nucl. Med.* **35**, 1054 (1994).

[17] P. L. Iversen, J. Mata, W. G. Tracewell, and G. Zon, *Antisense Res. Dev.* **4**, 43 (1994).

[18] P. A. Cossum, H. Sasmor, D. Dellinger, L. Truong, L. Cummins, S. R. Owens, P. M. Markham, J. P. Shea, and S. Crooke, *J. Pharm. Exp. Ther.* **267**, 1181 (1993).

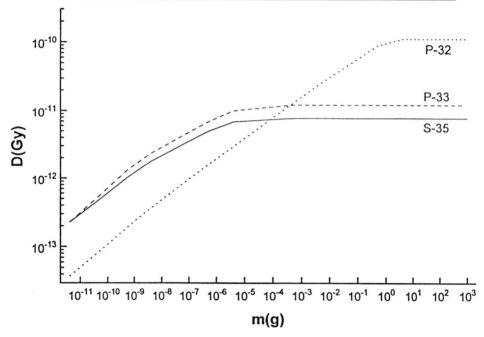

Fig. 2. The dose at the center of a water-equivalent sphere having a cumulated activity concentration of 1 Bq · sec/g for ^{32}P (dotted line), ^{33}P (dashed line), and ^{35}S (solid line) plotted as a function of the sphere mass. If the mass of the sphere is smaller than 300 μg, the central dose is higher for ^{33}P than for ^{32}P. [Reprinted with permission from Oxford University Press, *Anti-Cancer Drug Design* **11,** 439 (1996).]

maximum uptake, by using published biological half-lives, or directly, when available, by using the published areas under the time–concentration curve. The basic physical data on radionuclides ^{32}P, ^{33}P, and ^{35}S were taken from ICRP report 38[19] and β energy-specific constants of Leichner[14] were used in Eq. (4). The results for the estimated organ doses are presented in Table I. There is only a slight difference between the doses of ^{33}P and ^{35}S whereas the doses of ^{32}P are approximately 10-fold higher owing to the greater disintegration energy.

In addition, the theoretical calculations for 1-μg, 1-mg, and 1-g tumors were performed in the mouse. The tumor data are presented in Table II. Large differences can be seen in small tumors ($<$1 g); both ^{35}S and ^{33}P deliver much higher tumor doses than ^{32}P.

The previous analysis deals with the "macroscopic" absorbed dose and

[19] ICRP, "Radionuclide Transformations: Energy and Intensity of Emissions," Publication 38. Pergamon Press, New York, 1983.

TABLE I

ESTIMATED ORGAN DOSES FOR ^{32}P/^{33}P/^{35}S-LABELED OLIGONUCLEOTIDES WITH AN ADMINISTERED DOSE OF 1 MBq OF 100%-LABELED OLIGONUCLEOTIDE[a]

Radionuclide	Rat, anti-HPV, 21-mer		Rat, anti-HIV-1, 25-mer			Rat, anti-HIV-1, 27-mer		Mouse, c-myc oncogene in breast cancer, 15-mer		
	Liver	Bone marrow	Liver	Kidney	Bone marrow	Liver	Kidney	Liver	Kidney	Tumor
^{32}P	0.3	1.5	0.2	2.0	0.45	0.5	0.6	5.0	14	11
^{33}P	0.03	0.19	0.02	0.2	0.06	0.06	0.07	0.7	2.0	1.5
^{35}S	0.02	0.14	0.01	0.14	0.04	0.04	0.04	0.4	1.3	1.0

[a] Doses presented in grays. Data derived from Cossum et al. (1993),[18] Iversen et al (1994),[17] Dewanjee et al. (1994),[16] and Zhang et al. (1995).[10] Reprinted with permission from Oxford University Press, Anti-Cancer Drug Design 11, 439 (1996).

TABLE II
Radiation Doses[a] of the Mouse Tumor Xenograft Model[b] after Intravenous Injection of Radiooligonucleotide, When Kidney Dose Was 5 Gy[c,d]

	^{32}P	^{33}P	^{35}S
Liver	1.93	1.97	2.06
1 g	4.9	5.1	5.5
1 mg	0.73	5.1	5.5
1 μg	0.08	3.1	3.9

[a] Doses presented in grays.
[b] Model is that of Dewanjee et al. (1994).[16]
[c] The tumor doses have been calculated for 1-g, 1-mg, and 1-μg tumors using internally labeled oligonucleotide phosphorothioate either with ^{32}P, ^{33}P, or ^{35}S.
[d] Reprinted with permission from Oxford University Press, Anti-Cancer Drug Design **11**, 439 (1996).

does not give the complete description of radiobiologically relevant properties of the dose distribution. At the cellular level the primary target for the initiation of the radiation effect seems to be the chromosomal DNA. Being dependent on the particle energies, the subcellular dose distribution is modulated by the nonuniform biodistribution of the labeled oligonucleotide inside the cell as well. On a cellular scale the activity distribution inside the adjacent cells is also significant. Using MIRD principles we have calculated the total nuclear S factor, corresponding to the dose at the DNA, from the self-absorption process nucleus ← nucleus (N ← N), absorbed dose from cytoplasm to the nucleus (N ← Cy), and cell surface to nucleus (N ← Cs). The nuclear S factor can be written as a sum:

$$S(N \leftarrow N, Cy, CS) = p_N S(N \leftarrow N) + p_{Cy} S(N \leftarrow Cy) + p_{Cs} S(N \leftarrow Cs) \quad (5)$$

where the p values represent the relative activity content of the nucleus, cytoplasm, and cell surface.

To obtain a comprehensive picture of the nuclear level dose distribution of different, potentially useful clinical radionuclides we selected the following radionuclides for calculation: ^{32}P, ^{35}S, chromium -51 (^{51}Cr), gallium-67 (^{67}Ga), indium-111 and -114m (^{111}In and ^{114m}In), iodine-123, -125, and -131 (^{123}I, ^{125}I, and ^{131}I), and thallium-201 (^{201}Tl). The calculation was performed for spherical cell forms of different cellular and nuclear radii [r_C, r_N: 6 and 3 μm (I), 6 and 5 μm (II), 8 and 5 μm (III), as well as 10 and 8 μm (IV)], using the cellular level S factors of Goddu et al.[20] In these situations the (spherical) nuclear volume fractions of the whole (spherical) cell are 13% (I), 58% (II), 24% (III), and 51% (IV), respectively. Three different activity distributions were considered: published cellular level distribution of oligo-nucleotide phosphorothioates ISIS 2105 and ISIS 2922[21] and the uniform activity distribution. The activity outside the cell was not taken into account. This assumption shows the maximum possible differences between the selected radionuclides. The calculated data for all radionuclides are presented in Table III. These numbers demonstrate that Auger-emitting isotopes do not offer any benefit as compared with ^{35}S. On the other hand, ^{35}S has the smallest variation in tumor doses of different cellular dimensions.

The absorbed dose distribution does not merely describe the biological response of the cell or tissue to radiation. With an externally delivered radiation dose in radiotherapy of cancer it is known that total dose and dose rate (fraction size and interval) and total treatment time are the most important external factors affecting radiation reaction of healthy tissue and tumor response. With internally administered radionuclides the same variables are—no doubt—equally important and Howell et al.[22] have introduced the relative efficacy factor (RE) for radioimmunotherapy to take into account the dose–rate effect. Even if the biokinetics and activity distribution of two tracers labeled with different radionuclides were similar, the differences in physical dose distribution and decay time lead to differences in dose–rate distribution that can be biologically significant. For the instant uptake and monoexponential washout with effective half-life T_e the relative efficacy of a radionuclide can be expressed as

$$RE = 1 + \frac{r_0}{\ln 2}\left(\frac{\beta}{\alpha}\right)\frac{T_\mu T_e}{T_\mu + T_e} \qquad (6)$$

where r_0 is the initial dose–rate in an organ and T_μ is the time constant for the sublethal radiation damage repair. This model is based on the assumption of linear–quadratic (LQ) cell survival with parameters α and β. Furthermore, Howell et al.[22] defined a relative advantage factor (RAF) to calculate the relative benefit of longer lived radionuclides over shorter lived nuclides:

$$RAF = \frac{(r_{0,t}/r_{0,B})_S}{(r_{0,t}/r_{0,B})_l} \qquad (7)$$

[20] S. M. Goddu, R. W. Howell, and D. V. Rao, J. Nucl. Med. 35, 303 (1993).
[21] R. M. Crooke, M. J. Graham, M. E. Cooke, and S. T. Crooke, J. Pharm. Exp. Ther. 275, 462 (1995).
[22] R. W. Howell, S. M. Goddu, and D. V. Rao, J. Nucl. Med. 35, 303 (1994).

TABLE III

CALCULATED NUCLEAR DOSES RELATIVE TO THE UNIFORM ACTIVITY DISTRIBUTION γ, AUGER FOR DIFFERENT RADIONUCLIDES AS A FUNCTION OF CELLULAR AND NUCLEAR RADII AND OLIGONUCLEOTIDE PHOSPHOROTHIOATES ISIS 2105 AND ISIS 2922[a]

Radionuclide (major components of emission)	Decay model (half-life, days)	r_C/r_H (μm)	S factor (N ← N, Cs, Cy) Uniform distribution (Gy/Bq·sec)	Nuclear dose relative to the uniform concentration distribution	
				D (ISIS 2105) (%)	D (ISIS) 2922 (%)
^{32}P	β^-	6/3	2.19×10^{-4}	89	108
		6/5	1.92×10^{-4}	138	164
β^-	14.3	8/5	1.16×10^{-4}	76	87
		10/8	6.94×10^{-4}	61	72
^{35}S	β^-	6/3	9.11×10^{-4}	89	110
		6/5	7.92×10^{-4}	61	67
β^-	87.4	8/5	4.54×10^{-4}	75	87
		10/8	2.61×10^{-4}	62	68
^{51}Cr	EC	6/3	6.70×10^{-4}	84	140
		6/5	6.47×10^{-4}	27	38
γ, Auger	27.8	8/5	2.68×10^{-4}	48	79
		10/8	1.34×10^{-4}	25	40
^{67}Ga	EC	6/3	1.29×10^{-3}	85	128
		6/5	1.23×10^{-3}	39	49
γ, Auger	3.26	8/5	5.42×10^{-4}	55	80
		10/8	2.87×10^{-4}	37	48
^{111}In	EC	6/3	9.93×10^{-4}	87	133
		6/5	9.39×10^{-4}	36	46
γ, Auger	2.83	8/5	4.47×10^{-4}	59	83
		10/8	2.39×10^{-4}	40	51
^{114n}In	EC, IT	6/3	1.03×10^{-3}	87	124
		6/5	9.48×10^{-4}	45	54
γ, Auger	49.5	8/5	4.89×10^{-4}	65	84
		10/8	2.76×10^{-4}	51	59
^{123}I	EC	6/3	1.04×10^{-3}	86	135
		6/5	9.80×10^{-4}	33	43
γ, Auger	0.55	8/5	4.53×10^{-4}	57	82
		10/8	2.48×10^{-4}	38	50
^{125}I	EC	6/3	2.30×10^{-3}	86	135
		6/5	2.20×10^{-3}	32	42
γ, Auger	60.1	8/5	9.97×10^{-4}	55	82
		10/8	5.41×10^{-4}	37	49
^{131}I	β^-	6/3	5.52×10^{-4}	88	111
		6/5	4.87×10^{-4}	60	66

[a] Reprinted with permission from Stockton Press, *Cancer Gene Ther.* **5,** 410 (1998).

TABLE III (*continued*)

Radionuclide (major components of emission)	Decay model (half-life, days)	r_C/r_H (μm)	S factor (N ← N, Cs, Cy) Uniform distribution (Gy/Bq·sec)	Nuclear dose relative to the uniform concentration distribution	
				D (ISIS 2105) (%)	D (ISIS) 2922) (%)
$\beta^-\gamma$, Auger	8.09	8/5	2.81×10^{-4}	73	86
		10/8	1.64×10^{-4}	63	89
^{201}Tl	EC	6/3	2.99×10^{-3}	85	128
		6/5	2.77×10^{-3}	39	48
γ, Auger	3.04	8/5	1.26×10^{-3}	56	80
		10/8	6.64×10^{-4}	38	49

where the subscripts t and B refer to the initial dose rate in tumor and total body (healthy tissue) for a shorter-lived (s) and longer lived (l) radionuclide. RE and RAF were calculated using the radiobiologically adequate values of $\alpha/\beta = 10$ Gy and $T_\mu = 0.5$ hr for the tumor and $\alpha/\beta = 3$ Gy and $T_\mu = 1.5$ for the normal tissues. In RAF calculations for the relative benefit in the tumor/kidney and tumor/liver effects both ^{33}P and ^{35}S were compared with ^{32}P. Relative advantage factors are shown in Table IV for kidney and liver doses. As Table IV demonstrates, there are no major differences in the radiation dose to normal organs.

Discussion

The ideal dose in external radiation therapy has traditionally been defined as the dose that gives as many cures as possible before an exponen-

TABLE IV
RELATIVE ADVANTAGE FACTORS[a] FOR KIDNEY
AND LIVER DOSES USING INTERNALLY LABELED
OLIGONUCLEOTIDE PHOSPHOROTHIOATE EITHER
WITH ^{32}P, ^{33}P, OR ^{35}S[b]

Organ	^{32}P	^{33}P	^{35}S
Liver	1.000	1.001	0.999
Kidney	1.000	0.997	1.000

[a] Calculated according to Eq. (4).
[b] RAFs are calculated according to Eq. (7). Reprinted with permission form Oxford University Press, *Anti-Cancer Drug Design* **11**, 439 (1996).

tial increase in complications occurs. The dose is always dependent on the nature of the complications. Of course, the worst complication of radiotherapy is tumor recurrence. Late effects, e.g., necrosis, fibrosis, fistulas, ulceration, and damage to specific organs, are the dose-limiting factors, which depend on irradiated volume, the total dose delivered, and the fraction size. Enhanced local control is obtained when radiotherapy is followed by or administered simultaneously with adjuvant chemotherapy in locally advanced cancer. Combination treatment is based on attempting to increase the therapeutic index. External radiation may be replaceable by oligonucleotide radiotherapy, which is highly specific, minimizing the radiation effects in normal tissue and dramatically reducing complications (Fig. 3, see color insert). The therapeutic index is increased as dose-limiting, late side effects do not occur or occur only minimally. Two terms widely used in the context of radiotherapy also need to be kept in mind when radiotargeting on a nanometer scale, i.e., "radionanotargeting": *radiosensitivity,* which is initial sensitivity of the cells to radiation, and *radioresponsiveness,* which is the clinical appearance of tumor regression after radiation.

There is great pharmaceutical potential for the use of synthetic antisense ODNs in the treatment of AIDS and cancer because of their specificity. Of particular interest are studies showing that antisense phosphorothioate ODNs can inhibit the human immunodeficiency virus type 1 (HIV-1).[23,24] The effectiveness of ODNs as chemotherapeutic agents depends on their properties, such as cellular uptake and stability against nucleases. To improve antisense oligonucleotide efficiency, chemical modifications have been developed, and improvement of oligonucleotide uptake has been achieved with different systems of vectorization, including liposomes (neutral, cationic, immunoliposome), nanoparticles, and covalent attachment in a carrier.[25] Polyalkylcyanoacrylate nanoparticles have been introduced as polymeric carriers of oncogene-targeting antisense DNA.[26] Our aim was to optimize radiation exposure and to select the radiation source that provides the highest amount of radiation to the subcellular target (nucleus), and to diminish the radiation in the surrounding tissue.

The envisioned therapeutic use of radiolabeled antisense ODNs is based on the assumption that an appropriate amount of radiation is delivered to

[23] S. Agrawal, J. Goodchild, M. P. Civeira, A. H. Thornton, P. S. Sarin, and P. C. Zamecnik, *Proc. Natl. Acad. Sci. U.S.A.* **85,** 7079 (1988).

[24] M. Matsukura, G. Zon, K. Shinozuka, M. Robert-Guroff, T. Shimada, C. A. Stein, H. Mitsuya, F. Wong-Staal, J. S. Cohen, and S. Broder, *Proc. Natl. Acad. Sci. U.S.A.* **86,** 4244 (1989).

[25] C. Lefebvre-d'Hellencourt, L. Diaw, and M. Guenounou, *Eur. Cytokine Network* **6,** 7 (1995).

[26] G. Schwab, C. Chavany, I. Duroux, G. Goubin, J. Lebeau, C. Helene, and T. Saison-Behmoaras, *Proc. Natl. Acad. Sci. U.S.A.* **91,** 10460 (1994).

a targeted location of specific sequence of cellular DNA or RNA causing local damage (Fig. 4, see color insert). In addition, the radiotherapeutic effect may be enhanced by antisense-mediated inhibition of gene function. Tumor-specific activity will be obtained by hitting appropriate targets, such as anti-oncogenes or tumor-suppressor genes, e.g., *p53* mutations. The number of targets may be increased in many ways, e.g., by gene transfer, and thus the efficacy of nucleotide radiotherapy can be improved. Targets with different cellular locations have been described, such as mRNA translation sites, pre-mRNA splicing sites, on the DNA molecules themselves.[1] A fast transformation in the nuclei and a high concentration of antisense ODNs on nuclear structures were observed in spite of the type of ATP pool, temperature, or excess unlabeled oligomer. Accumulation of the ODN in the nuclei essentially uses a set of proteins. Use of antisense ODNs to inactivate genes still has several difficulties to be resolved and requires improvements: delivery of the ODN into cells and entry into an appropriate intracellular compartment, nonsequence specificity, optimizing pharmacokinetic properties, and designing new and better ODN backbones. Radionanotargeting has a limited range from the radiation source, resulting in a rapid dose fall-off effect and sparing of the surrounding tissues. That means restrictions in applications. Thus local applications in lung cancer may include hilar tumors adherent to major vasculature, attachment of tumors to mediastinal structures, extensive tumor involvement of the chest wall, spine, or paravertebral tissue, and recurrent or metastatic endobronchial lesions.

Here, we approach optimizing the label of ODNs with Auger-emitting radionuclides by calculating subcellular dose distribution. We show that for subcellular targeting, internal labels ^{35}S and ^{32}P give the lowest variation in estimated absorbed nuclear dose in our cell model with defined dimensions (nuclear diameter, 6–16 μm; cellular diameter, 12–20 μm). The doses vary considerably depending on cellular dimensions when using Auger-emitting isotopes; however, in small cells they may give a high dose (Table III). In tumors cell dimensions may vary and therefore these Auger-emitting isotopes should be applied only when nuclear target circumstances are well characterized. High-energy β-emitter ^{32}P gives the nuclear dose closest to uniform distribution in varying cell sizes, but this is due to its high energy. We have previously noted[13,27] that when using ^{32}P-labeled ODNs cells other than target cells will be destroyed because of the long-range effect. This is not the case when using β-emitters ^{33}P and ^{35}S, which are optimal when targets are less than 300 μm in diameter.[13]

[27] K. J. A. Kairemo, M. Tenhunen, and A. P. Jekunen, *Anti-Cancer Drug Design* **11,** 439 (1996).

It has been hypothesized[28] that the protein association of ODNs could determine the final antisense activity. If the association is weak, ODNs may be free to reassociate with target RNA or DNA, but if the association with nuclear protein is too strong, it may hinder the antisense ODNs from the effective formation of triplex complexes. It is evident that exogenous ODNs can enter the cytoplasm via binding to cell surface receptors, and after entering the cytoplasm ODNs are transported to the nucleus.[29] There is an interesting piece of information differentiating different kinds of oligonucleotides: the apparent endocytic uptake of labeled methyl phosphonate oligonucleotides could not be blocked by competition with unlabeled methyl phosphonate or phosphodiester oligonucleotides, or by ATP. This contrasts with radiolabeled phosphodiester oligonucleotides, whose uptake can be completely blocked with unlabeled competitor. In addition, uptake of phosphodiester oligonucleotides, but not of methyl phosphonate oligonucleotides, could be blocked by acidification of the cytosol.[28] These observations emphasize that there are distinct forms of uptake for phosphodiester oligonucleotides and methyl phosphonate oligonucleotides, implicating that calculations should be done again for other types of ODNs. ODNs appear not to use passive diffusion in entering into the cell interior, probably because of their electric charge.[28] The internalization can be slowed by metabolic inhibitors, and it is temperature dependent.[30] Although it is not fully understood how ODNs enter cells, two major pathways have been determined: adsorptive endocytosis and fluid-phase endocytosis. Adsorptive endocytosis accounts for a relatively small percentage of ODNs internalized, while fluid-phase endocytosis is the major mechanism, although it is relatively inefficient. Endocytic internalization can be mediated by nonspecific association with plasma membrane proteins, as has been demonstrated for poly-L-lysine-conjugated ODNs.[31] The strong anionic nature of unmodified phosphodiester ODNs makes their nonspecific association with the largely anionic plasma membrane unlikely. Various ODN-binding proteins have been identified on the cell membrane: e.g., 75 kDa[29] and 79 kDa.[30]

Preliminary studies using ^{125}I-labeled ODNs for mammalian cell lines suggest that oligonucleotides delivered with liposomes give a lower nuclear dose than DNA-incorporated ^{125}I-UdR.[32] However, our calculations indi-

---

[28] Y. Shoji Y. S. Akhtar, and A. Periasamy, *Nucleic Acids Res.* **19,** 5543 (1991).

[29] D. A. Geselowitz and L. M. Neckers, *Antisense Res. Dev.* **2,** 17 (1992).

[30] L. A. Yakubov, E. A. Deeva, V. F. Zarytova, E. M. Ivanova, A. S. Ryte, V. L. Yurchenko, and V. V. Vlassov, *Proc. Natl. Acad. Sci. U.S.A.* **86,** 6454 (1989).

[31] J. P. Leonetti, N. Mechti, G. Degols, C. Gagnor, and B. Lebleu, *Proc. Natl. Acad. Sci. U.S.A.* **88,** 2702 (1989).

[32] O. A. Sedelnikova, I. G. Panyutin, A. R. Thierry, and R. D. Neumann, *J. Nucl. Med.* **39,** 1412 (1998).

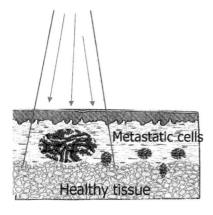

External radiation therapy

Radionucleotide therapy

A   **B**

FIG. 3. (A) Exposure area of external radiation. Metastases not located in the field are left outside, while healthy normal tissue is radiated. (B) Exposure area of successful oligoradiotherapy. Metastases are treated while irradiation is low in normal tissue, except that surrounding the tumor.

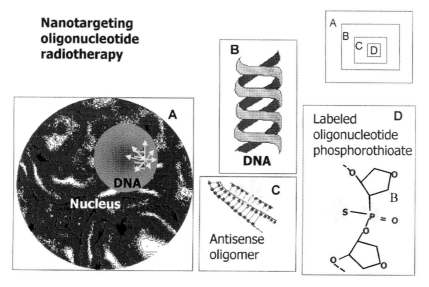

Nanotargeting oligonucleotide radiotherapy

FIG. 4. *Nanotargeting* describes a specific way of targeting small molecules (nanometer scale) by antibodies or antisense oligonucleotides. Isotopes in oligonucleotide phosphorothioates are substituted for sulfur or phosphorus atoms. Emitted radiation of isotopes affects structures close to the binding site of the oligonucleotide. Antisense oligonucleotides serve as vehicles for radioisotopes, enhancing targeted efficacy.

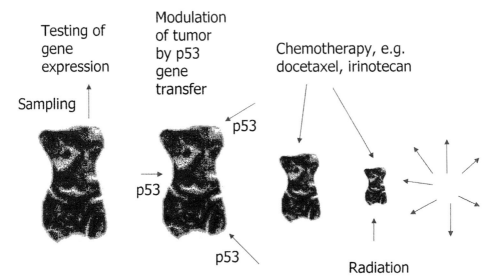

Fɪɢ. 5. A speculative treatment regimen consisting of nanotargeting performed by oligonu-cleotide radiotherapy. Testing of gene expression is done by microarray assays, and a transcript of *p53* was selected for transfection. Increasing suppressor gene *p53* expression in tumor cells improves sensitivity of tumor cells to ODNs established by retroviral and adenoviral vectors. Docetaxel and irinotecan are examples of new efficacious drugs in a variety of tumor types with a new mode of action: prevention of depolymerization of tubulin and specific DNA topoisomerase I inhibition, respectively. These drugs have a favorable interaction with radi-ation.

cate that there are several radionuclides more optimal than [125]I. For example, ^{35}S and ^{32}P doses seem to concentrate more efficiently around the nucleus than ^{125}I, which may be of practical importance in delivering the effective doses to the nuclear target. It should be emphasized that the behavior of the radiation at short distances is crucial. Therefore, the radionuclides ^{35}S and ^{33}P appear to be more suitable than ^{125}I for cell destruction at short distances. It would be important in oligonucleotide radiotherapy to achieve the highest possible uptake in the target cell and minimal radiation toxicity to surrounding cells. We conclude that ^{32}P and ^{35}S have more favorable radiation distribution than other radiolabels for oligonucleotides owing to the better dose uniformity.[125]I and other radiolabels should not be used unless better specificity is achieved.

Antisense technology has begun to provide an alternative approach for manipulating the expression of specific genes. Because many cell types spontaneously take up antisense oligonucleotides from the culture medium, this technology is particularly useful for altering gene expression in organisms for which the transformation pathway is unknown. Studies done by Thomas and Price[33] with ciliated protozoa illustrate clearly that dead cells have a significant affinity for ODNs. If this observation is transferable to *in vivo* models, tumors with necrosis should have increased uptake of ODNs, so that the radiation dose delivered by radiolabeled ODNs would increase in necrotic areas. This would also be useful in diagnostic imaging, but no easy procedure for imaging with ^{32}P, ^{33}P, or ^{35}S exists. However, other possibilities for imaging are available, but the biodistribution may be dependent on the coupled radiometal. ODNs have been labeled with the positron emitter ^{18}F, and thus early kinetics of the first hour can be calculated in patients.[34]

Phosphorothioate ODNs demonstrate rapid disappearance from plasma within 1 hr, and a biexponential elimination. However, $T_{1/2\beta}$ is apparently longer for the phosphorothioate ODNs. Although a phosphorothioate ODN leaves the plasma rapidly, it requires days to leave the whole body. There is also significant extravascular accumulation of greater than 50% of the injected dose over a period of 3 to 12 hr. Furthermore, the tissue uptake is not at saturation point, as some uptake occurs even at 28 days during continuous infusion.[35] The ODNs are extensively eliminated in the urine over the first 3 days after bolus injection. Distribution to, and accumulation

[33] L. L. Thomas and C. M. Price, *Antisense Res. Dev.* **2,** 251 (1992).
[34] B. Tavitian, S. Terrazino, B. Kühnast, S. Marzabal, O. Stettler, F. Dollé, J.-R. Deverre, A. Jobert, F. Hinnen, B. Bendriem, C. Crouzel, and L. Di Giamberardino, *Nature Med.* **4,** 467 (1998).
[35] P. L. Iversen, J. Mata, W. G. Tracewell, and G. Zon, *Antisense Res. Dev.* **4,** 43 (1994).

by, tissues reveals the tissue-specific rates and extent of ODN accumulation after a single injection.[17,35]

Because the half-lives of all of the investigated radionuclides are long, no remarkable differences in radiobiological effects can be seen between them (Table III). In practice, the relative advantage factors for all of the nuclides are similar and equal to 1.0. Because the biodistribution was not supposed to change with different radiolabels, the organ dose ratios kidney/liver, kidney/tumor, and liver/tumor did not change either. This assumption was adequate because we have used all of the possible radionuclides suitable for internal labeling of oligodeoxynucleotide phosphorothioates.

We are aware of several simplifications concerning our calculations of the energy deposited from a radioactive source in an organ. Owing to the limited biodistribution data we could not use multicompartmental modeling. Instead, monoexponential kinetic behavior was applied. Because the internal labeling procedure with ^{32}P, ^{33}P, or ^{35}S does not change the sequence specificity or other radiopharmacokinetic properties of oligodeoxynucleotides remarkably, their effects on the absorbed dose remain similar. In addition, the metabolic fate of β-emitting ODNs requires further experimental studies assessing several points such as chemical degradation. The number and amount of radioactive species should be determined by HPLC or other separation techniques. Furthermore, the metabolic rates are dependent on water solubility, nucleotide chain length, nature of radionucleotide backbone, and terminal modification necessary for radiolabeling with metallic radionucleotides. In our case, the biodistribution data on the animal tumor model were derived from indium-labeled compounds. However, sequence specificity determines most of the kinetic characteristics of ODNs.

It is possible to use a mixture of radioisotopes to ensure a complete coverage of targets in more than one location, e.g., targeting nuclear-related and cellular RNA at the same time. In addition, modern imaging technique allows visual control over kinetic events. Sometimes the target is dense, e.g., in the nucleus, or it can be diffusely spread around the cellular area. Dual labeling with ^{32}P and ^{35}S may provide therapeutic benefits when treating smaller and larger targets simultaneously (Fig. 2). Further *in vivo* development, especially with ^{33}P and ^{35}S labels for ODNs, is highly indicated. ^{33}P and ^{35}S have some benefits over ^{32}P, because the organ doses remain smaller and thus the therapeutic index may be wider. Critical organ exposure remains 10-fold lower with ^{33}P and ^{35}S than with ^{32}P. Moreover, ^{33}P and ^{35}S concentrate more efficiently around the target than ^{32}P, which could be of practical importance for the delivery of effective doses to the tumor. This can clearly be seen from the shape of the dose-versus-mass curves. It can be said that the behavior of the radiation at small distances is crucial. Therefore, the radionuclides ^{33}P and ^{35}S are more suitable than

TABLE V
SCHEMATIC PRESENTATION OF MAIN CHARACTERISTICS IN DIFFERENT CELLULAR TARGETS

Factor	Outer membrane receptor	RNA	Nuclear
Accessibility	Easy	Moderate	Difficult
Target	Stable	Transient	Stable
Induction	Possible, slow	Usual, fast	No
Efficacy	Low	Moderate	High
Exposure for surrounding cells	High	Moderate	Low

^{32}P for cell destruction at short distances. This was also clearly demonstrated with calculations of tumor doses for 1-g, 1-mg, and 1-μg tumor masses (Table II). The microscropic tumors cannot be treated with ^{32}P. Instead, with ^{33}P or ^{35}S tumor doses up to 7.5-fold for 1-mg tumor mass and up to 50-fold for 1-μg tumor mass are achieved. Tumors larger than 1 g could be treated with any of these radionuclide-labeled ODNs. It would be crucial in oligoradiotherapy to achieve the highest possible uptake in the target cell and minimal radiation toxicity to surrounding normal cells. Here, ODNs carry the radioactivity source inside the cell and finally achieve close contact with target RNA macromolecules.

In the near future, RNA expression by selected target genes will be detected by cDNA microarrays in a single experiment screening thousands of genes. Thus the likelihood of the selected gene being the most important target is going to be high. This provides an excellent opportunity to modulate a tumor to be more sensitive to chemotherapeutic agents and radiation (Fig. 4). Transcriptional activation of genes by p53 may coordinately shut down cell cycle progression and induce a battery of genes involved in DNA repair. DNA damage induces p53 accumulation. Cells lacking p53 are resistant to other forms of apoptotic induction, such as that caused by chemotherapeutic agents and radiation. Tumors that have lost p53 are no longer able to respond to adverse growth conditions by initiating apoptosis. We do not know yet how efficient transfection should be to induce desired effects. Similarly, the relationship between e.g., binding activity and efficacy of antisense oligonucleotides is yet unknown *in vivo*.

When current problems in the antisense approach have been resolved, including administration, delivery, uptake, accumulation in the target, binding to the receptor molecule, and effective time, oligonucleotide radiotherapy may be integrated as a new mode of radiation therapy, partially or totally replacing conventional external radiotherapy. The possible cellular targets are presented in Table V. It seems that oligonucleotide radiotherapy is going to have many applications, e.g., in the treatment of metastases in lymph nodes, locally advanced cancer, and adjuvant therapy before and after surgical operations to reduce the size of tumors or to treat remnant tumors. In addition, true interactions between chemotherapeutic drugs and gene expression are continually being defined, which may provide an opportunity to resensitize tumor cells for already acquired resistance. Figure 5 (see color insert) shows a speculative treatment regimen consisting of nanometer-scale targeting performed by oligonucleotide radiotherapy. At that point, there will be a great practical need for optimization of the effect and for controlling radiation effects at the microcellular level. One way to optimize radiation is to select the label according to the target.

Acknowledgments

The authors are thankful to Ketil Thorstensen, Ph.D., for advisory comments and to Ms. N. Järvinen for helping us prepare the color illustrations.

[37] Antisense Oligonucleotide Therapy of Hepadnavirus Infection

By WOLF-BERNHARD OFFENSPERGER, C. THOMA, D. MORADPOUR, F. VON WEIZSÄCKER, S. OFFENSPERGER, and H. E. BLUM

Introduction

Infection with the hepatitis B virus (HBV) is endemic throughout much of the world, with an estimated 400 million persistently infected people.[1] HBV infection is associated with a wide spectrum of clinical manifestations ranging from infection without evidence of disease to acute, fulminant, or chronic hepatitis and liver cirrhosis. Further, HBV infection is associated with the development of hepatocellular carcinoma, a leading cause of death from cancer[2] worldwide.

While HBV infection can be prevented by passive and/or active vaccination, for those chronically infected there is so far no effective therapy available. The only established therapy is interferon α, with an efficacy of only 30–40% in highly selected patients.[3] While other therapeutic agents

[1] W. M. Lee, *N. Engl. J. Med.* **337,** 1733 (1997).
[2] J. R. Wands and H. E. Blum, *N. Engl. J. Med.* **325,** 729 (1991).
[3] J. H. Hoofnagle and A. M. Di Biseglie, *N. Engl. J. Med.* **336,** 347 (1997).

such as lamivudine are in clinical evaluation, the need for alternative thera-peutic strategies has provided impetus to develop novel concepts.

The structure and biology of HBV have been characterized in great detail.[1] HBV belongs to a group of hepatotropic DNA viruses (hepadnavi-ruses) that include the hepatitis virus of the woodchuck, ground squirrel, Pekin duck, and heron. HBV is a small virus, about 42 nm in diameter ("Dane particle"), and is composed of a lipid bilayer envelope containing hepatitis B surface antigen (HBsAg) and an internal nucleocapsid ("core"). The nucleocapsid consists of the hepatitis B core antigen (HBcAg) and the viral DNA genome, about 3.2 kbp in length, with an associated DNA polymerase/reverse transcriptase. The diagnosis of HBV infection is made by serologic detection of HBsAg or by the direct demonstration of viral DNA using molecular hybridization techniques. The hepadnaviral genomes are of similar size and structure, and replicate asymmetrically via reverse transcription of an RNA intermediate, a replication strategy central to the life cycle of retroviruses.

Studies of the life cycle and biology of HBV were greatly facilitated by the animal models of HBV infection mentioned above. Because Pekin ducks are readily available from commercial sources and easy to house, animals infected with the duck hepatitis B virus (DHBV) represent a highly attractive system in which to study hepadnaviruses.[4] DHBV infection seems to be endemic in most parts of the world.[5] In France 1–6% of ducklings of different breeds were found to be DHBV infected. The major route for DHBV transmission is vertical, through the eggs laid by viremic ducks. Experimentally, DHBV can be efficiently transmitted by intravenous injec-tion of DHBV-positive serum. Experimental infection during the first 5 days after hatching leads to chronic infection in almost all birds. The DHBV genome is a small circular DNA molecule about 3.0 kbp in length. The DNA is only partially double stranded. Three overlapping open reading frames are present on the DNA minus strand. These encode the viral nucleocapsid protein, the envelope proteins, and the DNA polymerase. DHBV-infected animals are useful because many aspects of the DHBV life cycle, in particular the replication pathway, are similar to those of HBV and because antiviral effects can be easily monitored. Antiviral agents can be tested *in vitro* in primary duck hepatocytes isolated from infected duck liver or *in vivo* in DHBV-infected ducks. In primary duck hepatocytes infected with DHBV many nucleoside analogs were tested. *In vivo,* adenine

[4] W. S. Mason, G. Seal, and J. Summers, *J. Virol.* **36,** 829 (1980).

[5] S. Sprengel and H. Will, *in* "Virus Diseases in Laboratory and Captive Animals" (G. Darai, ed.), p. 363. Nijhoff, Boston, 1988.

arabinoside,[6] 2',3'-dideoxycytidine,[7] 3'-azido-3'-deoxythymidine,[8] the carbocyclic analog of 2'-deoxyguanosine,[9] ganciclovir,[10] and famciclovir[11] were tested.

For diseases caused by the expression of acquired genes, such as viral genes, blocking of gene expression can be an effective therapeutic approach. Several strategies can be employed: interfering with the transcription of genes by binding of single-stranded nucleic acids to double-stranded DNA, forming a triple helix structure; by hybridization of RNA molecules possessing endoribonuclease activity (ribozymes) to target RNA molecules, resulting in sequence-specific RNA cleavage; by blocking translation through binding of antisense oligonucleotides to RNA; and by intracellular synthesis of peptides or proteins that interfere with their normal counterpart (dominant negative mutant strategy).[12]

Several groups have reported inhibition of HBV gene expression and replication by antisense oligonucleotides.[13,14] A series of oligonucleotides against HBV were evaluated using a hepatocellular carcinoma-derived cell line stably transfected with HBV DNA (Hep G 2.2.15).[15] Studies in Pekin ducks[16] and in nude mice[17] demonstrated the *in vivo* applicability of this approach. In the following sections, our own work using oligodeoxynucleotides in the Pekin duck model is presented.[16]

Duck Hepatitis B Virus-Specific Antisense Oligodeoxynucleotides in Primary Duck Hepatocytes

Phosphorothioate-modified oligodeoxynucleotides are synthesized using standard cyanoethylphosphoramidites and the sulfur transfer reagent

[6] K. Hirota, A. H. Sherker, M. Omata, O. Yokosuka, and K. Okuda, *Hepatology* **7,** 24 (1987).

[7] C. Kassianides, J. H. Hoofnagle, R. H. Miller, E. Doo, H. Ford, S. Broder, and H. Mitsuya, *Gastroenterology* **97,** 1275 (1989).

[8] H. Haritani, T. Uchida, Y. Okuda, and T. Shikata, *J. Med. Virol.* **29,** 244 (1989).

[9] W. S. Mason, J. Cullen, J. Saputelli, T.-T. Wu, C. Liu, W. London, E. Lustbader, P. Schaffer, A. O'Connell, I. Fourel, C. Aldrich, and A. Jilbert, *Hepatology* **19,** 398 (1994).

[10] Y. Wang, S. Bowden, T. Shaw, G. Civitico, Y. Chan, M. Qiao, and S. Locarnini, *Antiviral Chem. Chemother.* **2,** 107 (1991).

[11] W.-B. Offensperger, S. Offensperger, A. Keppler-Hafkemeyer, P. Hafkemeyer, and H. E. Blum, *Antiviral Ther.* **1,** 141 (1996).

[12] F. von Weizsäcker, S. Wieland, J. Köck, W.-B. Offensperger, S. Offensperger, D. Moradpour, and H. E. Blum, *Hepatology* **26,** 251 (1997).

[13] G. Goodarzi, S. C. Gross, A. Tewari, and A. Watabe, *J. Gen. Virol.* **71,** 3021 (1990).

[14] H. E. Blum, E. Galun, F. von Weizsäcker, and J. R. Wands, *Lancet* **337,** 1230 (1991).

[15] B. E. Korba and J. L. Gerin, *Antiviral Res.* **28,** 225 (1995).

[16] W.-B. Offensperger, S. Offensperger, E. Walter, K. Teubner, G. Igloi, H. E. Blum, and W. Gerok, *EMBO J.* **12,** 1257 (1993).

[17] Z. Yao, Y. Zhou, X. Feng, C. Chen, and J. Guo, *J. Viral Hepatitis* **3,** 19 (1996).

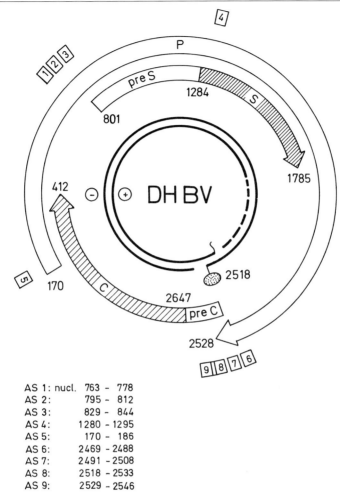

AS 1: nucl. 763 – 778
AS 2:   795 – 812
AS 3:   829 – 844
AS 4:   1280 – 1295
AS 5:   170 – 186
AS 6:   2469 – 2488
AS 7:   2491 – 2508
AS 8:   2518 – 2533
AS 9:   2529 – 2546

Fig. 1. Genetic organization of DHBV genome and location of nine antisense oligodeoxy-nucleotides.

^3H-1,2-benzodithiol-3-one 1,1-dioxide.[18] The genetic organization of the DHBV genome and the position of the nine antisense oligodeoxynucleo-tides tested are shown in Fig. 1. Four antisense (AS) oligodeoxynucleotides are located in the pre-S/S region (AS 1–4), one at the start of the polymerase

[18] R. P. Iyer, L. R. Philips, W. Egan, J. B. Regan, and S. L. J. Beaucage, *Org. Chem.* **55**, 4693 (1990).

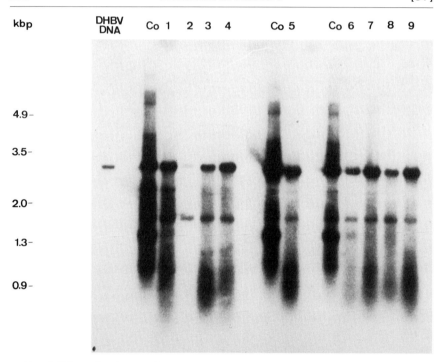

FIG. 2. Effect of antisense oligodeoxynucleotides on DHBV replication *in vitro*. Co, Untreated control cells. Numbers (*top*) indicate antisense oligodeoxynucleotides described in Fig. 1. Southern blot analysis was performed with 20 μg of total liver DNA per lane. Size markers: *Hin*dIII-cut λ DNA and cloned DHBV DNA of 3.0-kbp length.

region (AS 5), and four in the pre-C/C region (AS 6–9). To isolate primary duck hepatocytes, ~10-day-old DHBV-infected Pekin ducklings are anesthesized and the livers are perfused *in situ* with collagenase [0.5 mg/ml in Williams' medium E buffered with 20 mM HEPES (pH 7.4) and 2.5 mM CaCl$_2$].[19] The hepatocytes are seeded at a density of 2×10^5 cells/ cm^2 on Primaria tissue culture dishes (Bectom Dickinson, Franklin Lakes, NJ) in Williams' medium E supplemented with 20 mM HEPES (pH 7.4), 5 mM glutamine, 0.066 μM insulin, 10 mM dexamethasone, penicillin (100 μg/ml), streptomycin (100 μg/ml), and 1.5% dimethyl sulfoxide. The infected hepatocytes are kept in culture with daily medium changes containing the respective antisense oligodeoxynucleotide at a final concentration of 1.5 μM. After 10 days, the cells are harvested and DNA is analyzed by Southern blot hybridization (Fig. 2). AS 2 and 6 show a strong inhibitory activity in comparison with the other antisense

[19] J. S. Tuttleman, J. Pugh, and J. Summers, *J. Virol.* **58,** 17 (1986).

oligonucleotides, which all show a decrease in intracellular viral replicative intermediates. AS 2, directed against the start of the pre-S region, leads to a nearly complete inhibition of viral replication, with only residual covalently closed, circular DNA (cccDNA) molecules left. In quantifying these effects by liquid scintillation counting, the hybridization signal is found to be 3.7% after incubation with AS 2 and 9.4% after incubation with AS 6 as compared with the control cells (100%). Toxic effects can be excluded because trypan blue exclusion demonstrates nearly identical viability of control cells and cells treated with the antisense oligodeoxynucleotides. In addition, the total RNA content of hepatocytes cultured in the presence or absence of the antisense oligonucleotides is nearly identical. When the effect of the respective sense oligodeoxynucleotide is evaluated (Fig. 3), it can be demonstrated that it does not significantly alter viral replication, which may demonstrate the specificity of the inhibition by the antisense oligonucleotide.

Duck Hepatitis B Virus-Specific Antisense Oligodeoxynucleotides in Vivo

For these experiments, 1-day-old ducklings are infected with DHBV by intravenous injection of DHBV-positive duck serum. Two weeks later, treatment with AS 2 is started by daily intravenous injection for 10 days. Subsequently the ducks are sacrificed and liver DNA is analyzed for the presence of viral replicative intermediates by Southern blot. As shown in Fig. 4, treatment with AS 2 results in a dose-dependent inhibition of viral replication in vivo, with a nearly complete elimination of viral DNA forms at a daily dose of 20 μg/g body weight.

To determine the effect of AS 2 on viral gene expression in vivo, duck hepatitis B surface antigen (DHBsAg) from serum and duck hepatitis B core antigen (DHBcAg) from liver are analyzed by Western blot, using polyclonal antibodies. Figure 5 demonstrates the in vivo inhibition of viral gene expression with disappearance of viral pre-S and S antigens from serum and viral pre-C and C antigens from liver.

To detect possible side effects of this in vivo therapy several clinico-chemical parameters, including alanine aminotransferase, aspartate aminotransferase, γ-glutamyltranspeptidase, cholinesterase, total protein, and albumin, are measured in serum. No differences are observed between the control ducks and the ducks treated with AS 2.

In a further set of experiments the effect of antisense oligodeoxynucleotides on DHBV infection was tested in vivo. Two DHBV-negative ducklings are pretreated with AS 2 starting 3 days after hatching. Twelve hours later, they are infected by intravenous injection of DHBV DNA-positive serum. The ducklings are treated for 10 days with a daily dose of 20 μg of AS

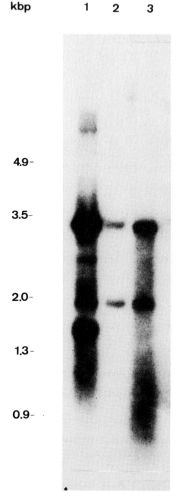

FIG. 3. Effect of AS 2 (lane 2) and sense oligodeoxynucleotide 2 (lane 3) on DHBV replication *in vitro*. For experimental details see caption to Fig. 2.

2/g body weight. At the end of the experiment, viral DNA in liver is analyzed by Southern blot hybridization and viral antigens in serum are analyzed by Western blot analyses (Fig. 6). In contrast to untreated ducks and to ducks treated with the sense or random oligodeoxynucleotides, no viral DNA can be detected in AS 2-treated ducks, demonstrating the inhibitory effect of the antisense oligonucleotide AS 2 on DHBV infection.

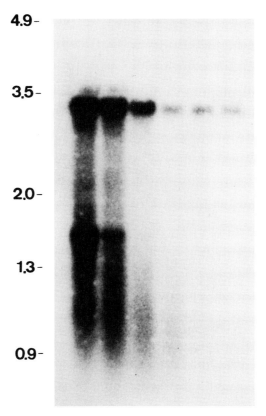

FIG. 4. Effect of the antisense oligodeoxynucleotide AS 2 on DHBV replication *in vivo* in DHBV-infected ducks. Lane 1, control duck; lanes 2–6, ducks treated for 10 days with daily intravenous injections of AS 2 at a concentration of 5 μg (lane 2), 10 μg (lane 3), and 20 μg/g body weight (lanes 4–6).

These data demonstrate that DHBV replication and gene expression in liver cells can be blocked by antisense oligodeoxynucleotides *in vitro* and *in vivo*. However, several issues remain to be resolved before this strategy may become potentially useful for the treatment of HBV infection in the clinic.

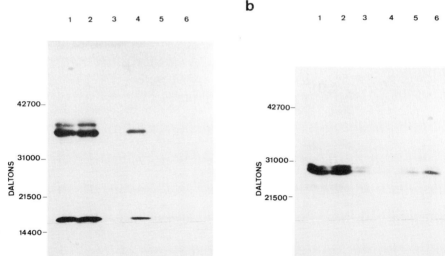

Fig. 5. (a) Western blot analysis of sera obtained from DHBV-infected ducks treated with AS 2 *in vivo* using a polyclonal antibody against native DHBsAg. Lane 1, control duck; lanes 2–6, ducks treated for 10 days with AS 2 at a concentration of 5 μg (lane 2), 10 μg (lane 3), and 20 μg (lanes 4–6) per gram body weight. (b) Western blot analysis of liver extracts obtained from DHBV-infected ducks treated with AS 2 *in vivo* using a polyclonal antibody against DHBcAg. The numbers at the top of the lanes correspond to those in (a).

Definition of Mechanisms of Action

The mechanisms of action should be understood in more detail. The hypothesis that true antisense mechanisms are involved in the activity of the antisense oligonucleotide AS 2 may be strengthened by the facts that only one of the nine different phosphorothioate-modified oligonucleotides tested was highly efficient and that the complementary sequence of sense polarity did not significantly affect viral replication and gene expression. Possible factors involved in the inhibitory activity of the oligonucleotide AS 2 in DHBV replication are the hybridization efficiency depending on secondary and tertiary structures, the significance of the targeted region for viral replication, the nonexistence of steric hindrance by proteins that may prevent oligonucleotide annealing, and the intracellular distribution of antisense oligodeoxynucleotides.[20] Because DHBV, like all hepadnaviruses, utilizes the unique pathway of reverse transcription of an RNA pregenome

[20] J. P. Leonetti, N. Mechti, G. Degols, C. Gagnor, and B. Lebleu, *Proc. Natl. Acad. Sci. U.S.A.* **88,** 2702 (1991).

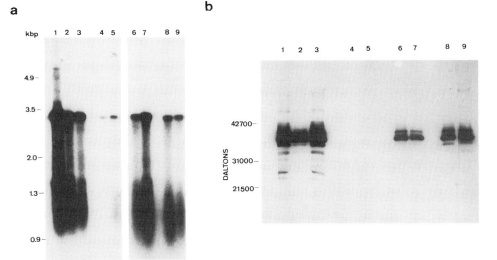

FIG. 6. *In vivo* effect of oligodeoxynucleotides on DHBV infectivity. The treatment of the ducklings with the oligodeoxynucleotides was started 12 hr before infection with DHBV-positive serum. (a) Southern blot analysis. Lanes 1–3, three control ducklings; lanes 4 and 5, two ducklings treated with AS 2; lanes 6 and 7, two ducklings treated with sense oligonucleotide 2; lanes 8 and 9, two ducklings treated with a random oligonucleotide. (b) Western blot analysis of the sera using a polyclonal antibody against DHBpre-sAg. The numbers above the lanes correspond to those in (a).

for viral DNA replication,[21] DHBV RNAs, including the viral pregenome, are intermediates in the replicative and the translational pathways. From this it follows that antisense oligonucleotides, acting via translational arrest by interfering with a number of posttranscriptional steps, e.g., RNA processing, stability, and translation, effectively block viral replication as well as viral gene expression. However, other specific and nonspecific mechanisms of oligonucleotide action[22,23] must be discussed and investigated. Our own work in progress indicates a specific destruction of the viral pregenome by RNase H (our unpublished results, 1998).

Definition of Optimal Oligodeoxynucleotide Properties

Criteria that must be met by a therapeutic oligonucleotide are easy and large-scale synthesis, *in vivo* stability, efficient uptake into the target cell,

[21] J. Summers and W. S. Mason, *Cell* **29**, 403 (1982).
[22] R. W. Wagner, *Nature (London)* **372**, 333 (1994).
[23] A. D. Branch, *Hepatology* **24**, 1517 (1996).

and sequence-specific interaction with its intracellular target. Clearly, more detailed analyses of pharmacokinetic properties of the oligonucleotides and their side effects are necessary. Great efforts are being made to optimize the chemical modification of the oligonucleotide. Phosphorothioates used in these experiments have been widely studied. They have an increased resistance to degradation as compared with unmodified oligonucleotides. A second approach to modification of the phosphodiester backbone is the replacement of the negatively charged oxygen by an uncharged group to produce a neutral species. The oligonucleotide AS 2 with the modification as benzylphosphonate leads to a 40% inhibition of viral replication at a concentration of 1.5 μM.[24]

Mixed backbone oligonucleotides are second-generation antisense therapeutics. These oligonucleotides contain appropriately placed segments of phosphorothioate-modified oligonucleotides and segments of either modified oligodeoxyribonucleotides or oligoribonucleotides.[25] The advantages of mixed backbone oligonucleotides are increased biological activity, reduced polyanionic side effects, and increased *in vivo* stability, which even makes oral administration feasible. AS 2 as a mixed backbone oligonucleotide is currently being evaluated *in vitro* and *in vivo*.

Transient Effect of Antisense Oligodeoxynucleotides

To test whether the virus was definitely eliminated after a short-term treatment *in vitro,* cells that showed complete viral inhibition after a 10-day incubation with AS 2 were kept in culture for another 3 or 6 days in the absence of AS 2. It could be seen that 6 days after stopping the antisense treatment, viral replication reemerged, suggesting that the residual covalently closed, circular (ccc) DNA acts as a template for viral reactivation. In extension of the *in vitro* studies Pekin ducks were treated with AS 2 for prolonged periods of time in an attempt to eliminate DHBV infection (our unpublished results, 1998): a total of 16 Pekin ducks chronically infected with DHBV was included in the study. Eight ducks were treated with AS 2 for 8 weeks. During the first 3 weeks AS 2 was applied daily at a dose of 10 $\mu g/g$ body weight, and during the next 5 weeks AS 2 was given thrice weekly at the same dose. Eight untreated ducks served as controls. The follow-up period after therapy was 0 (two ducks), 8 (four ducks), and 26 weeks (two ducks), respectively. The long-term treatment was well tolerated by all animals. At the end of the 8-week treatment with AS 2 viral replication was effectively inhibited in all eight ducks. No reactivation of replication

[24] W. Samstag, S. Eisenhardt, W.-B. Offensperger, and J. W. Engels, *Antisense Nucleic Acid Drug Dev.* **6,** 153 (1996).
[25] S. Agrawal and Q. Zhao, *Curr. Opin. Chem. Biol.* **2,** 519 (1998).

was observed in four of four ducks followed for 8 weeks and one of two ducks followed for 26 weeks. In contrast, the control animals had ongoing viral replication during the entire observation period. These results indicate that long-term DHBV suppression can be achieved by antisense oligodeoxynucleotides in this experimental setting.

Targeted Delivery of Antisense Oligonucleotides

To reduce the amount of oligonucleotide necessary to inhibit viral replication and to reduce undesired delivery to nontarget cells or tissues attempts were made to target the oligonucleotides to liver cells. This could be achieved by encapsidation of the antisense oligodeoxynucleotides into liposomes[26] or immunoliposomes[27] or by coupling them to an asialoglycoprotein for which liver cells have a specific receptor. Previous work from our group established a highly efficient receptor-mediated delivery system for DNA to avian liver cells.[28] The delivery system consisted of adenovirus particles, protein conjugate, and plasmid DNA. The replication-defective adenovirus mutant strain *dl*312 of the human serotype 5 was used as an endosome-disruption factor. The protein conjugate contained *N*-acetylglucosamine-modified bovine serum albumin as ligand for the avian liver cell asialoglycoprotein receptor, poly-L-lysine, and streptavidin. For preparation of protein conjugates, a simple and reproducible coupling method was developed. After protein coupling, L-lysine was added in excess to block the remaining reactive groups. The conjugates were purified by cation-exchange high-performance liquid chromatography. Fractions eluting between 1.45 and 2.1 mol/liter of NaCl contained the correct conjugate. The plasmids pCMVluc or pCMVlacZ were used to study delivery efficiency. For *in vitro* studies the chicken hepatoma cell line LMH was used. This cell line has been successfully used for the study of hepadnaviral replication and gene expression.[29] Transfection of DHBV DNA into LMH cells resulted in high levels of viral gene expression and replication with formation and export of infectious virions. Reporter assays were performed 24 hr after incubation of the LMH hepatoma cells with the complexes. The delivery complex containing adenovirus particles, the protein conjugate with *N*-acetylglucosamine-modified bovine serum albumin, and pCMVluc

[26] P. N. Soni, D. Brown, R. Saffie, K. Savage, D. Moore, G. Gregoriadis, and G. M. Dusheiko, *Hepatology* **28,** 1402 (1998).

[27] D. Moradpour, B. Compagnon, B. E. Wilson, C. Nicolau, and J. R. Wands, *Hepatology* **22,** 1527 (1995).

[28] J. Madon and H. E. Blum, *Hepatology* **24,** 474 (1996).

[29] L. D. Condreay, C. E. Aldrich, L. Coates, W. S. Mason, and T.-T. Wu, *J. Virol.* **64,** 3249 (1990).

yielded high levels of luciferase activity. In contrast, luciferase activity could not be demonstrated when one of the components of the complex was omitted or when an unspecific ligand was chosen. To determine whether the DNA delivery was mediated specifically through receptor-mediated endocytosis, the cells were incubated with the delivery complexes in the presence of increasing concentrations of free ligand N-acetylglucosamine-modified bovine serum albumin or the control protein bovine serum albumin (BSA). Delivery of pCMVluc was strongly inhibited by N-acetyl-glucosamine-modified bovine serum albumin with no luciferase activity observed at a 200-fold excess of the free ligand. No inhibition could be seen using BSA as a nonspecific competitor. To determine the delivery efficiency of antisense oligonucleotides, a fluorescein-labeled oligonucleotide Enc-Fluo was used. This oligonucleotide was phosphorothioate modified and had the following sequence: 5′ CAGTGGGACATGTACA 3′. LMH cells were incubated with the delivery complexes containing this fluorescein-labeled oligonucleotide. While incubation with the non-complexed fluorescein-labeled oligonucleotide added directly to the culture medium at a final concentration of 1.5 μmol/liter resulted in only insignificant labeling of LMH cells, complexed fluorescein-labeled oligonucleotides could be detected in more than 95% of the LMH cells, indicating a high delivery efficiency. The highest intensity of fluorescence was observed at an oligonucleotide concentration of 3.0 μmol/liter in the complex formation mixture. A further increase in the oligonucleotide concentration did not affect delivery efficiency. The amount of oligonucleotides delivered to the cells clearly depended on the time of incubation of the delivery complex with the LMH cells: longer incubation times resulted in greater oligonucleotide levels in the cells. The intracellular distribution of the delivered oligonucleotides was analyzed by confocal microscopy. The highest intensity of fluorescent oligonucleotides was found in the nuclei of the majority of cells. Ongoing studies in our laboratory arc aimed at the delivery of these molecular conjugates containing antisense oligodeoxynucleotides to liver cells in ducks *in vivo*

Acknowledgments

This study was supported by a grant from the Deutsche Forschungsgemeinschaft (Of 14/4-2). W.-B.O. is the recipient of a Heisenberg award from the Deutsche Forschungsgemeinschaft. The advice and help of F. Eckstein (Göttingen), S. Beaucage (Bethesda, MD), J. Madon (Zurich, Switzerland), and H. Will (Hamburg, Germany) as well as the technical assistance of B. Hockenjos, P. Kary, and E. Schiefermayr are gratefully acknowledged.

[38] Preclinical Antisense DNA Therapy of Cancer in Mice

By Janet B. Smith and Eric Wickstrom

Introduction

Cloning and sequencing of pathogenic genes have made possible a direct genetic approach to the treatment of disease. With knowledge of the coding or sense sequence of a pathogenic mRNA, it is possible to select a unique complementary or antisense sequence that might prevent translation of the target message into a disease-causing protein. The ability of antisense DNA to turn off an individual gene in living cells provides a powerful tool for therapeutic intervention when that gene is either overexpressed or mutated. In addition, antisense inhibition of protein expression can aid in the elucidation of the role of a particular gene. Antisense DNAs have been targeted against a wide variety of genes, in viral, bacterial, plant, and animal systems, both in whole cells[1] and in animal models.[2] Clinical trials with a variety of antisense DNAs[3] have demonstrated safety and efficacy for several diseases, allowing complementary oligonucleotides to be used as an adjuvant with or replacement for a currently used treatment regimen.

Malignant growth of tumor cells results from a multistep process of oncogene activation or suppressor gene inactivation.[4] It is therefore valuable to consider the application of antisense DNA therapy to downregulate oncogene expression in animal models. We have applied the antisense paradigm to c-*myc*,[5–7] Ha-*ras*,[8,9] Ki-*ras*,[9] and *erbB-2*.[9] In this chapter we focus on c-*myc* as a well-studied example.

The c-*myc* p65 protein Myc is a nuclear leucine zipper protein that binds with a small partner protein, Max; the resulting heterodimer binds

[1] E. Wickstrom, "Prospects for Antisense Nucleic Acid Therapy of Cancer and AIDS." Wiley-Liss, New York, 1991.

[2] S. Agrawal, "Antisense Therapeutics." Humana Press, Totowa, New Jersey, 1996.

[3] E. Wickstrom, "Clinical Trials of Genetic Therapy with Antisense DNA and DNA Vectors." Marcel Dekker, New York, 1998.

[4] J. M. Bishop, *Cell* **64,** 235 (1991).

[5] E. Wickstrom, T. A. Bacon, and E. L. Wickstrom, *Cancer Res.* **52,** 6741 (1992).

[6] Y. Huang, R. Snyder, M. Kligshteyn, and E. Wickstrom, *Mol. Med.* **1,** 647 (1995).

[7] J. B. Smith and E. Wickstrom, *J. Natl. Cancer Inst.* **90,** 1146 (1998).

[8] G. D. Gray, O. M. Hernandez, D. Hebel, M. Root, J. M. Pow-Sang, and E. Wickstrom, *Cancer Res.* **53,** 577 (1993).

[9] E. Wickstrom and F. L. Tyson, *in* "Oligonucleotides as Therapeutic Agents" (D. Chadwick and G. Cardew, eds.), Vol. 209, p. 124. Wiley, London, 1997.

TABLE I

OLIGONUCLEOTIDE SEQUENCES

Name	Complementary Sequence[a]	Target site	Ref.
MYC6	5'-dCAC GTT GAG GGG CAT	c-*myc* codons 1–5	34
SCR6	5'-dCTG CTG AGA GTC GAG	Scrambled MYC6	6
MYC61	5'-dCAC GTT CTG GGG CAT	2-nt mismatch[b]	7
MYC51	5'-dAAT TAC TAC AGC GAG	c-*myc* P2 promoter	7
MYC53	5'-dTCG AGG CTG TCT GCG	c-*myc* intron 1–exon 2	7
MYC55	5'-dCTC GTC GTT TCC TCA	c-*myc* codons 384–388	7
SCR55	5'-dAGT CCT TCC CTT GTC	Scrambled MYC55	7
MYC56	5'-dCTC GTC TCC TCC TCA	3-nt mismatch[b]	7
MYC57	5'-dCAG GTC GTT AGC TCA	2 × 2 nt mismatch[b]	7
MCG	5'-dGCA TGA CGT TGA GCT	Immunostimulation	65

[a] The nucleotides underlined represent similarities to the dPuPuCGPyPy motif.[65]
[b] Mismatched nucleotides are in boldface.

specifically to the sequence dGACCACGTGGTC, activating the transcription of a panel of proliferative genes.[10,11] Aberrant expression of the viral v-*myc* oncogene, the mouse c-*myc* protooncogene, or the human c-*myc* protooncogene has been implicated in a number of leukemias and solid tumors.[4]

Myc protein is positioned at the end of most known signal transduction pathways, inducing proliferation in response to a wide variety of normal and pathogenic growth signals. Hence, c-*myc* (like c-*myb* or c-*fos*) might be a more fundamental target for antisense ablation than genes encoding enzymes and receptors at higher levels of the pathway, such as c-Raf, c-Ras, protein kinase A and C, growth factors, and growth factor receptors. To the extent that Myc is indeed overexpressed in a broad spectrum of solid tumors,[4] c-*myc* antisense DNA may display broad-spectrum anticancer activity, without the need for genetic analysis of individual patient biopsies.

Messenger RNA Target Sites

For the examples in this chapter, several sites in c-*myc* mRNA were targeted (Table I). To date, identification of exceptionally potent RNA sites for antisense DNA inhibition has required exhaustive probing of mRNA 5' and 3' ends with dozens of antisense sequences.[12] Such message-

[10] E. M. Blackwood and R. N. Eisenman, *Science* **251,** 1211 (1991).
[11] B. Amati and H. Land, *Curr. Opin. Genet. Dev.* **4,** 102 (1994).
[12] C. F. Bennett, N. Dean, D. J. Ecker, and B. P. Monia, *in* "Antisense Therapeutics" (S. Agrawal, ed.), p. 13. Humana Press, Totowa, New Jersey, 1996.

walking experiments with antisense DNAs have not yielded any obvious correlation between efficacy of a site and its predicted secondary structure.[13,14] These studies did, however, imply that the start of transcription and the initiation codon were likely to be good starting points. For a cellular gene, as opposed to a viral gene, we and others have usually found that the 7mG-cap site at the 5' end of the mRNA, where initiation factors bind, is an effective site for antisense inhibition.[1]

The antisense DNAs utilized in this chapter (Table I) were synthesized with 15 residues. This length was selected in order to allow strong binding to single-stranded regions of mRNA, or adequate binding to regions that already have some secondary or tertiary structural limitations.[1] A probe with 15 residues is just sufficient for theoretical uniqueness in the human genome, particularly for those sequences expressed as mRNA, in which case specificity may be accomplished with as few as 12 residues.

If one chose to prepare longer probes in an effort to achieve inhibition at lower concentrations, one might encounter a loss of specificity due to hybridization to exposed mRNA regions with partial homologies to the probe. For therapeutic applications, the most rugged target should be sufficiently unique that no homologs exist with even one or two mismatches, yet is still effective with 12–16 residues. Thus, one must survey efficacy and specificity as a function of oligomer length.

A prudent selection of sequence controls should be included to test the hypothesis of an antisense mechanism.[7,15] In our studies, the sequence dependence of c-*myc* antisense therapeutic regimens was monitored by the inclusion of a number of control sequences (Table I). A variety of point mutations were created in c-*myc* antisense DNAs, and each c-*myc* antisense sequence was randomly scrambled to test for general phosphorothioate backbone effects.

Oligonucleotide Derivatives

Novel oligonucleotide analogs have been synthesized to improve the nuclease resistance, cellular uptake, intracellular trafficking, and RNase H activation of antisense agents relative to normal phosphodiesters (Fig. 1).[16,16a] The simplest oligodeoxynucleotide modification involves blocking the 3' terminus to prevent attack by 3' exonucleases, the predominant

[13] Y. Daaka and E. Wickstrom, *Oncogene Res.* **5**, 267 (1990).
[14] T. A. Bacon and E. Wickstrom, *Oncogene Res.* **6**, 13 (1991).
[15] B. P. Monia, H. Sasmor, J. F. Johnston, S. M. Freier, E. A. Lesnik, T. Geiger, K. H. Altmann, H. Moser, and D. Fabbro, *Proc. Natl. Acad. Sci. U.S.A.* **93**, 15481 (1996).
[16] E. Wickstrom, *Trends Biotechnol.* **10**, 281 (1992).
[16a] E. Wickstrom and J. B. Smith, *Cancer J. Sci. Am.* **4**(Suppl. 1), S43 (1998).

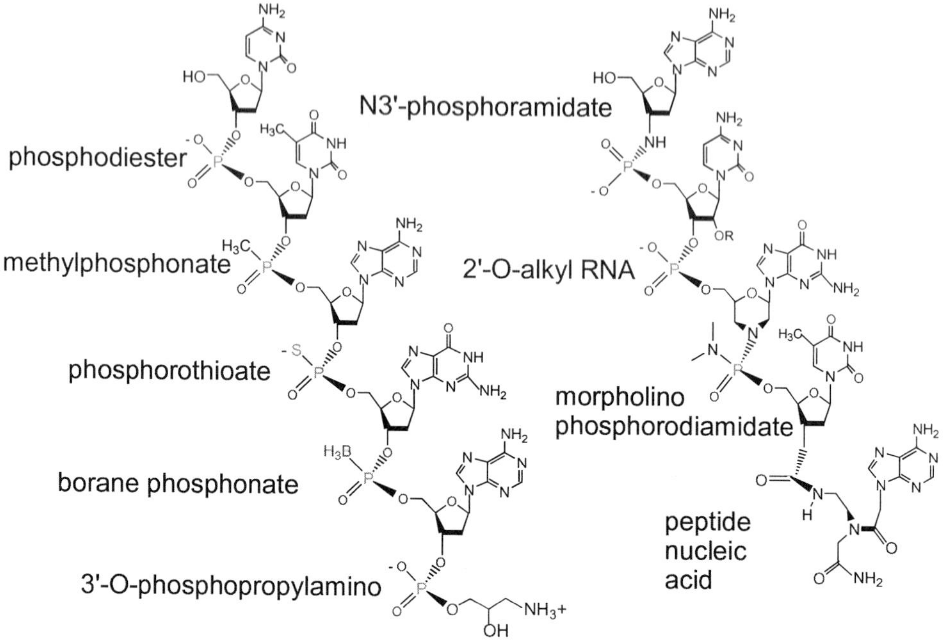

Fig. 1. Examples of oligonucleotide modifications developed for therapeutic uses.[16a]

extracellular degradative mechanism for oligodeoxynucleotides.[17] Other modifications focus on protecting the internucleoside linkage by changing the phosphodiester linkages to phosphorothioates, methyl phosphonates, or borane phosphonates. Although these modifications increase the *in vivo* half-life of oligonucleotides, they also weaken hybridization to the RNA target sites owing to the creation of chiral phosphorus diastereomers.[18]

The deoxyribose may be modified to 2'-*O*-alkyl RNAs, strengthening hybridization and resisting nuclease attack.[19] Similar improvements result from preparing 3'-aminophosphoramidates[20] or morpholinophosphorodi-amidates.[21] The most radical modifications are found in peptide nucleic

[17] J. G. Zendegui, K. M. Vasquez, J. H. Tinsley, D. J. Kessler, and M. E. Hogan, *Nucleic Acids Res.* **20**, 307 (1992).

[18] A. V. Lebedev and E. Wickstrom, *in* "Perspectives in Drug Discovery and Design" (G. Trainor, ed.), Vol. 4, p. 17. ESCOM Science, Leiden, 1996.

[19] A. M. Iribarren, B. S. Sproat, P. Neuner, I. Sulston, U. Ryder, and A. I. Lamond, *Proc. Natl. Acad. Sci. U.S.A.* **87**, 7747 (1990).

[20] S. Gryaznov, T. Skorski, C. Cucco, M. Nieborowska-Skorska, C. Y. Chiu, D. Lloyd, J. K. Chen, M. Koziolkiewicz, and B. Calabretta, *Nucleic Acids Res.* **24**, 1508 (1996).

[21] J. Summerton and D. Weller, *Antisense Nucleic Acid Drug Dev.* **7**, 187 (1997).

acids, where both the phosphodiester linkages and sugars are replaced with a peptide-like backbone of (N-2-aminoethyl)glycine units.[22]

To date, phosphorothioates are the only derivatives that have been administered to humans.[3] The known modes of polyanion toxicity by phosphorothioates, such as prolonged clotting time and complement activation,[23] imply that further modifications are desirable to reduce sulfur content in therapeutic oligonucleotides. The most potent, least toxic second-generation derivatives currently available are mixed backbone chimeras. One example includes 2'-O-methoxyethyl RNA residues at the 5' and 3' termini with phosphorothioate residues in the center,[24] while a second example includes phosphorothioate residues at the 5' and 3' termini with methyl phosphonate residues in the center.[25] Both of these chimeras exhibit sufficient stability to allow oral administration.

Oligodeoxynucleotide Synthesis

The polymerase chain reaction (PCR) primers 5'-dGCAACTTCGG-GATGAAAATG-3' (forward) and 5'-dCGAAGCGCGCATAAATTT-3' (reverse) for detecting the ΦX174 marker sequence attached to the transgene are synthesized as normal phosphodiester DNAs by the β-cyanoethyl phosphoramidite route[26] on an Applied Biosystems (Foster City, CA) 394 DNA synthesizer.

The oligodeoxynucleotide phosphorothioate sequences in Table I are synthesized by the β-cyanoethyl phosphoramidite route, using the Beaucage reagent for sulfurization.[27] Impurities in outdated reagents can lead to poor yields and multiple failure sequences, so it is important to use fresh solutions and wash out all liquid lines weekly.

MYC6 and SCR6 are synthesized at a 200-μmol scale on a Biosearch (Novato, CA) 8800 large-scale DNA synthesizer. Other sequences are prepared at a 35-μmol scale on a Biosearch 8750 DNA synthesizer. Deprotected oligonucleotides are purified by duplicate precipitation from aqueous solution with 10 volumes of n-butanol,[28] then analyzed by denaturing gel electrophoresis, and visualized with Stains-All (Bio-Rad, Hercules, CA),

[22] P. E. Nielsen, M. Egholm, R. H. Berg, and O. Buchardt, *Anticancer Drug Des.* **8,** 53 (1993).
[23] S. Agrawal and R. P. Iyer, *Curr. Opin. Biotechnol.* **6,** 12 (1995).
[24] B. P. Monia, *Anticancer Drug Des.* **12,** 327 (1997).
[25] S. Agrawal, Z. Jiang, Q. Zhao, D. Shaw, Q. Cai, A. Roskey, L. Channavajjala, C. Saxinger, and R. Zhang, *Proc. Natl. Acad. Sci. U.S.A.* **94,** 2620 (1997).
[26] N. D. Sinha, J. Biernat, J. McManus, and H. Koster, *Nucleic Acids Res.* **12,** 4539 (1984).
[27] R. P. Iyer, L. R. Phillips, W. Egan, J. B. Regan, and S. L. Beaucage, *J. Org. Chem.* **55,** 4693 (1990).
[28] M. Sawadogo and M. W. Van Dyke, *Nucleic Acids Res.* **19,** 674 (1991).

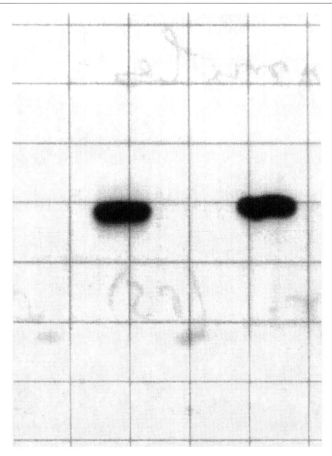

Fig. 2. Gel electrophoretic analysis of MYC6 (left) and SCR6 (right) on a 20% polyacryl-amide–7 M urea denaturing gel, visualized with Stains All, after preparative gel purification.[69]

revealing >90% full-length sequences. Preparative gel purification yields virtually homogeneous oligonucleotides (Fig. 2). Yields of purified oligonu-cleotides, measured by ultraviolet absorption on a spectrophotometer, are 40–60% of original solid support loading.

For additional quality control, random batches are tested by high-perfor-mance liquid chromatography,[29] showing >90% homogeneity (Fig. 3). Ma-trix-assisted laser desorption/ionization mass spectroscopy[30] displays full-

[29] E. L. Wickstrom, T. A. Bacon, A. Gonzalez, D. L. Freeman, G. H. Lyman, and E. Wickstrom, *Proc. Natl. Acad. Sci. U.S.A.* **85,** 1028 (1988).
[30] U. Pieles, W. Zurcher, M. Schar, and H. E. Moser, *Nucleic Acids Res.* **21,** 3191 (1993).

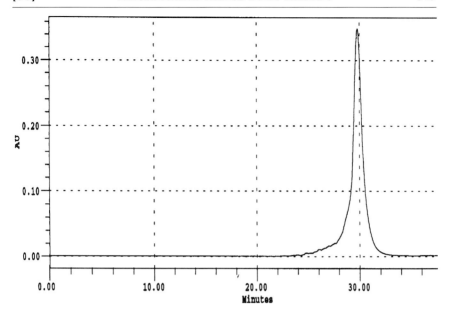

FIG. 3. HPLC analysis of MYC6 phosphorothioate on a C_{18} reversed-phase column eluted with a 0–80% gradient from 100% 0.1 M Et₃N-HOAc, pH 7, in H_2O to 80% acetonitrile–20% 0.1 M Et₃N-HOAc, pH 7, in H_2O over 60 min, at a flow rate of 1.0 ml/min.[29] Oligomers were detected by absorbance at 260 nm; the peak at 30 min represents the normal elution time of MYC6.

length oligomers (Fig. 4), and nuclear magnetic resonance[27] shows complete sulfurization (Fig. 5). Oligomers are sterilized by centrifugation through sterile 0.22-μm-pore filter units (Denville Scientific, South Plainfield, NJ).

Homogeneity of 90% is adequate for animal experiments, but not for clinical trials. To achieve >95% purity for cGMP production, high-performance ion-exchange chromatography is necessary.[31] In this situation, capillary gel electrophoresis provides the greatest resolution for quantitation of purity.[32]

Statistical Analysis

Observations of disease parameters in small groups rarely fit a normal statistical distribution, and therefore calculation of ordinary error bars is

[31] J. Gonzalez, R. G. Einig, P. Puma, T. P. Noonan, P. E. Kennedy, B. G. Sturgeon, B. H. Wang, and J.-Y. Tang, in "Clinical Trials of Genetic Therapy with Antisense DNA and DNA Vectors" (E. Wickstrom, ed.), p. 53. Marcel Dekker, New York, 1998.
[32] A. Guttman, A. S. Cohen, D. N. Heiger, and B. L. Karger, *Anal. Chem.* **62,** 137 (1990).

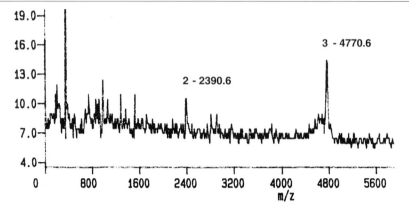

FIG. 4. MALDI-TOF mass spectroscopy of MYC6 phosphorothioate ionized from a dihy-drocinnamic acid matrix on a Hewlett-Packard LDI 1700 spectrometer.[30]

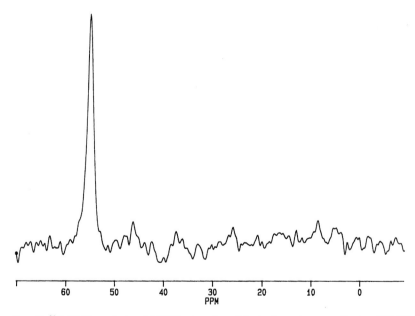

FIG. 5. ^{31}P NMR analysis of MYC6 phosphorothioate in water on a Bruker 360-MHz spectrometer, relative to phosphoric acid at 0 ppm. Note absence of phosphodiester peak in 0- to 5-ppm range.[27]

not an appropriate way to determine statistical significance in such cases. Differences in the median values of tumor onset time, tumor volumes, Western blot band densities, and Northern blot band densities among treatment groups are analyzed by Kruskal–Wallis one-way analysis of variance on ranks, using SigmaStat (Jandel Scientific, San Rafael, CA). To isolate the group or groups that differ from the others, the Student–Newman–Keuls, Dunn, or Dunnett pairwise multiple comparison procedure, and the Mann–Whitney rank sum test are used.[33]

Oligonucleotide Efficacy in Cell Culture

Antisense DNA inhibition of c-*myc* was first studied in the c-*myc*-overexpressing human promyelocytic leukemia cell line, HL-60.[29,34] A calculated secondary structure for c-*myc* mRNA placed the initiation codon in a bulge of a weakly base-paired region.[34] MYC6 phosphodiester, complementary to the initiation codon and the next four codons of c-*myc* mRNA, was synthesized along with anti-vesicular stomatitis virus (VSV) M and anti-human immunodeficiency virus (HIV) *tat* controls.

Treatment of the c-*myc*-transformed HL-60 cells with MYC6 inhibited Myc protein expression and proliferation in a sequence-specific, dose-dependent manner. Complete inhibition of proliferation was observed with 10 μM MYC6. In contrast, neither the anti-VSV M nor the anti-*tat* sequence controls inhibited proliferation or Myc levels. MYC6 had no effect, however, against the Q8 quail cell line transformed by avian MC29 virus carrying the v-*myc* oncogene, in which the first five codons include four mismatches relative to the target sequence of human MYC6.

Antisense DNA inhibition of c-*myc* expression was also studied simultaneously in normal peripheral blood lymphocytes after stimulation by the mitogen phytohemagglutinin.[35] In contrast to the HL-60 case described above,[29,34] c-*myc* inhibition in normal peripheral blood lymphocytes could not be detected below 15 μM MYC6, but was almost complete at 30 μM.[35] It is significant that normal cells required three times the dose necessary to achieve c-*myc* inhibition in HL-60 cells. We have found, in general, that transformed cells take up oligonucleotides three- to fivefold more aggressively than do normal cells.[36] Addition of 30 μM MYC6 to normal

[33] S. Siegel, "Nonparametric Statistics for the Behaviorial Sciences." McGraw-Hill, New York, 1956.

[34] E. L. Wickstrom, E. Wickstrom, G. H. Lyman, and D. L. Freeman, *Fed. Proc.* **45,** 1708 (1986).

[35] R. Heikkila, G. Schwab, E. Wickstrom, S. L. Loke, D. H. Pluznik, R. Watt, and L. M. Neckers, *Nature (London)* **328,** 445 (1987).

[36] G. D. Gray, S. Basu, and E. Wickstrom, *Biochem. Pharmacol.* **53,** 1465 (1997).

peripheral blood lymphocytes inhibited expression of Myc protein and entry into S phase. No effect was seen with the anti-VSV M oligomer, the sense version of MYC6, or a randomly scrambled version of MYC6. While MYC6 oligomer inhibited entry into S phase, it did not prevent transition from G_0 to G_1, consistent with the HL-60 results described above.

In general, oncogenic transformation, particularly by c-*myc*, blocks normal maturation.[37] In HL-60 cells, treatment with 1% dimethyl sulfoxide reduces c-*myc* expression and induces granulocytic differentiation.[38] Similarly, MYC6 was found to elicit a sequence-specific increase in HL-60 cell differentiation along the granulocytic line, as well as inhibition of colony formation in semisolid medium.[39] Daily addition of anti-c-*myc* oligomer for 5 days was as effective as 1% dimethyl sulfoxide.[40]

The dependence of antisense efficacy at other target sites in c-*myc* mRNA was then studied in HL-60 cells.[14] A variety of sequences complementary to predicted loops, bulges, and helices between the cap and initiation codon of human c-*myc* mRNA were synthesized, including the first/second exon splice junction. It was apparent that the 5' cap was three to five times more sensitive to antisense inhibition than the initiation codon bulge, while the 5' untranslated region and the coding region beyond the initiation codon presented poor targets. The physiologically critical cap and AUG sites might be those that are most exposed in the tertiary structure of c-*myc* mRNA.

One of the most frequent hematological malignancies of childhood is Burkitt's lymphoma.[41] In Burkitt's lymphoma, virtually all cases display translocation of chromosomes 8 and 14, placing the c-*myc* protooncogene from chromosome 8 under the transcriptional control of the immunoglobulin M heavy chain enhancer from chromosome 14.[41] In human ST486 Burkitt's lymphoma cells, where translocation generated unique c-*myc* transcripts lacking the first exon, but retaining part of the first intron, specific inhibition of c-*myc* expression was achieved by an antisense DNA against the first intron/second exon junction at bp 4490–4510, 5'-dGGCTGCTGGAGCGGGGCACAC-3'.[42]

The questions of sequence specificity and mechanism of antisense DNA

[37] J. Filmus and R. N. Buick, *Cancer Res.* **45,** 822 (1985).
[38] J. A. Coppola and M. D. Cole, *Nature (London)* **320,** 760 (1986).
[39] E. L. Wickstrom, T. A. Bacon, A. Gonzalez, G. H. Lyman, and E. Wickstrom, *In Vitro Cell. Dev. Biol.* **25,** 297 (1989).
[40] T. A. Bacon and E. Wickstrom, *Oncogene Res.* **6,** 21 (1991).
[41] I. Magrath, *Adv. Cancer Res.* **55,** 133 (1990).
[42] M. E. McManaway, L. M. Neckers, S. L. Loke, A. A. al-Nasser, R. L. Redner, B. T. Shiramizu, W. L. Goldschmidts, B. E. Huber, K. Bhatia, and I. T. Magrath, *Lancet* **335,** 808 (1990).

inhibition are important issues, discussed in detail later in the chapter. In the case of c-*myc* antisense phosphodiesters used to inhibit proliferation of human promyelocytic HL-60 cells, 10 different sites from the 5' end to the beginning of the coding domain in the second exon of c-*myc* mRNA were targeted.[14] All 10 antisense DNAs inhibited Myc antigen levels between 40 and 90%, under conditions where MYC6 treatment resulted in 70% inhibition. Similarly, treatment of smooth muscle cells with MYC6 and two other antisense sequences effectively inhibited proliferation in a model of restenosis, whereas scrambled, sense, reverse, and mismatch controls showed no activity.[43]

Transgenic Mouse Model for Spontaneous Tumors

Animals can provide useful therapeutic models for human diseases. Immunocompromised nude mice will accept xenografts of malignant human cell lines, allowing measurements of *in vivo* efficacy and potency of antisense DNAs.[8,9,12]

Adult mice contain about 2 ml of blood, carrying up to 10^7 lymphocytes/ml.[44] Thus, for the purpose of studying antisense DNA inhibition of c-*myc* expression, the circulating immature lymphocytes in the bloodstream of a mouse are similar to the cell number and volume of cells in a culture dish at $1-3 \times 10^6$ cells/ml. Thus we were drawn to determine whether antisense DNA therapy could downregulate c-*myc* gene expression in a mouse model.[5]

The nude mouse model, however, cannot model the normal immune responses of a healthy animal. To provide an immunocompetent animal model for childhood Burkitt's lymphoma, transgenic mice were created bearing a murine immunoglobulin M heavy chain[45] enhancer/c-*myc* fusion transgene (Eμ-*myc*) isolated from a mouse plasmacytoma.[46] Mice in these lines, including the Tg(IgH,Myc) Bri157 line used in these studies, develop aggressive multifocal lymphoma/leukemia.[45]

In these Eμ-*myc* transgenic mice an elevated constitutive expression of c-*myc* mRNA and Myc p65 antigen does not directly induce tumor onset. Overexpression of c-*myc* establishes a subpopulation of rapidly proliferating, immature B cells that may be transformed into a malignant phenotype

[43] Y. Shi, H. G. Hutchinson, D. J. Hall, and A. Zalewski, *Circulation* **88,** 1190 (1993).

[44] R. O. Jacoby and J. G. Fox, *in* "Laboratory Animal Medicine" (J. G. Fox, B. J. Cohen, and F. M. Loew, eds.), p. 31. Academic Press, New York, 1984.

[45] J. M. Adams, A. W. Harris, C. A. Pinkert, L. M. Corcoran, W. S. Alexander, S. Cory, R. D. Palmiter, and R. L. Brinster, *Nature (London)* **318,** 533 (1985).

[46] L. M. Corcoran, S. Cory, and J. M. Adams, *Cell* **40,** 71 (1985).

by a subsequent oncogene activation.[47] Thus the Eμ-myc transgenic mice provide a well-characterized animal model for studying the role of c-myc activation in B cell lymphoma.

Eμ-myc transgenic mouse embryos, line Tg(IgH,Myc) Bri157,[45] were the kind gift of R. Brinster (University of Pennsylvania). The mouse line is maintained by continually backcrossing Eμ-myc transgenic males with (C57BL/6J × SJL/J) F$_1$ hybrid females; the hemizygous offspring in each generation are screened for the presence of a ΦX174 marker sequence in the transgene by PCR amplification of DNA from toe cells at birth.[48] The reliability of this routine screening method was confirmed by Southern blot, using the c-myc probe pRyc7.4,[49] the kind gift of M. L. Veronese (Thomas Jefferson University).

Tumor onset in the Tg(IgH, Myc)Bri157 Eμ-myc transgenic line[45] was observed over several generations. A few displayed palpable tumors as early as 3 weeks of age, and half of them developed palpable tumors by the age of 8–9 weeks (Fig. 6). Fifty-five (95%) of the mice developed tumors within 16 weeks.

Because normal, unmodified DNA oligomers are rapidly hydrolyzed in the presence of serum nucleases,[50] they are a poor choice for use in animal models. To analzye the effects of a single bolus of antisense DNA, nuclease-resistant methyl phosphonate DNAs were administered to Eμ-myc transgenic mice by tail vein injection.[5] The methyl phosphonates distributed rapidly to the tissues, and showed no signs of degradation. A single treatment with MYC6, at about 50 mg/kg, significantly decreased Myc p65 expression in splenic lymphocytes when measured 3–4 hr after administration. Injection of saline vehicle or a scrambled sequence had no effect.

Having observed a decrease in Myc antigen in response to a single administration of MYC6 methyl phosphonate, it was logical to hypothesize that prolonged treatment of Eμ-myc transgenic mice with MYC6 might inhibit overproliferation of immature B cells, and interdict tumor progression. These investigations were carried out in a specific pathogen-free animal facility accredited by the Association for Assessment and Accreditation of Laboratory Animal Care (AAALAC), and are reviewed and monitored by the Institutional Animal Care and Use Committee to ensure adherence to National Institutes of Health (NIH) guidelines.

Lymphoma Cell Transplant Model for Residual Disease

The ability of MYC6 to protect Eμ-myc transgenic mice from tumor development over 6 months,[6] as described later in the chapter, suggested

[47] A. W. Harris, C. A. Pinkert, M. Crawford, W. Y. Langdon, R. L. Brinster, and J. M. Adams, J. Exp. Med. **167**, 353 (1988).
[48] M. Arita, S.-W. Li, D. C. Bertolette, and J. S. Khillan, Int. J. Anim. Sci. **10**, 1 (1995).
[49] L. K. Sun, L. C. Showe, and C. M. Croce, Nucleic Acids Res. **14**, 4037 (1986).
[50] E. Wickstrom, J. Biochem. Biophys. Methods **13**, 97 (1986).

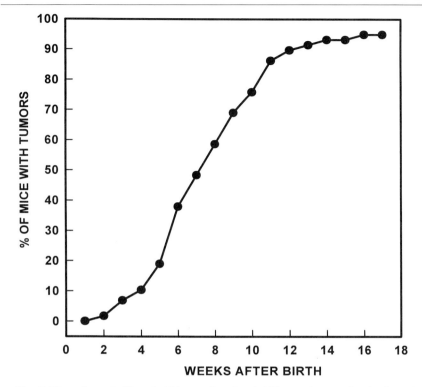

FIG. 6. Time of onset of lymphoid tumors in untreated Eμ-*myc* transgenic mice from day of birth. Presence of the transgene in these hemizygous mice was detected by PCR analysis of toe tissue at birth. Tumors were detected by palpation, and confirmed on autopsy. Over several generations, a total of 57 mice was observed.[6]

that this sequence might provide reliable consolidation therapy for lymphoma patients. The initial control of lymphoma currently relies on intensive chemotherapy using four to six agents, achieving complete remission in 60–85% of the patients, depending on the treatment strategy.[51] Analysis of the time course of failure revealed that relapse is due to the development of resistant cells, rather than the escape of sensitive cells. Lymphoma cells transplanted into an immunocompatible recipient would provide a better model for such residual disease than the spontaneous clonal tumors of the original Eμ-*myc* transgenic mice. Because the parental (C57BL/6J × SJL/J) F₁ hybrids contain all possible antigens present in the donors, they were used as the recipients of Eμ-*myc* tumor transplants.

[51] G. K. Rivera, D. Pinkel, J. V. Simone, M. L. Hancock, and W. M. Crist, *N. Engl. J. Med.* **329,** 1289 (1993).

Use of these immunocompatible hosts should eliminate tumor rejections independent of the oligonucleotide treatments.

To prepare lymphoma cells for transplantation,[52] tumors are dissected aseptically from $E\mu$-myc transgenic male mice[45] immediately after sacrifice, then placed immediately into sterile centrifugation tubes containing ice-cold sterile phosphate-buffered saline (PBS). All tissues are processed in an aseptic environment, which is maintained within a NuAire (Plymouth, MN) laminar flow biological safety hood. Tissues are minced and added to 5 ml of ice-cold PBS contained within a 7-mL Pyrex tissue homogenizer (Pyrex 7720-07, Corning, Corning, NY). Minced tissues are homogenized by careful grinding, using two or three strokes. Excessive grinding can damage the cells, thereby decreasing the viable yield. The homogenate is filtered through a sterile 70-μm nylon cell strainer (Falcon 2350; Becton Dickinson, Franklin Lakes, NJ), and the cells are pelleted by sedimentation at 1000g for 10 min. The pellets are resuspended in ice-cold sterile PBS and washed twice to remove any nontumor tissue. Excess connective tissue may result in clumping of some tumor cells. Separation of the excess nontumor tissue, clumps of cells, and debris from suspended tumor cells is accomplished by a second filtration through the cell strainer.

The final cell pellet is resuspended in RPMI 1640 plus 2 mM L-glutamine, penicillin G (100 units/ml), streptomycin (100 μg/ml), 20% fetal bovine serum, and 10% dimethyl sulfoxide to a final cell concentration of $\sim$10^8 cells/ml. Cells are slowly frozen to a final temperature of $-80°$. Stock cell lines are maintained by subsequent transplantations of the original frozen cell lines into 4- to 6-week-old (C57BL/6J $\times$ SJL/J) F$_1$ males (Jackson Laboratories, Bar Harbor, ME). Again, it is important to emphasize a critical step in this protocol: the manipulations of tumors and cells are performed aseptically within a NuAire laminar flow hood.

Proper analysis of experimental data necessitates that all experiments be carried out using an identical protocol, including the use of the identical cell line. Although all tumors originated from $E\mu$-myc transgenic male mice, the tumors in each mouse developed independently. Therefore, the possibility existed that tumors isolated from different mice developed different subsequent mutations during the tumorigenic process. Hence, tumor cell lines were prepared from a number of different $E\mu$-myc transgenic mouse tumors in order to maintain independent stock cell lines for future experiments.

For lymphoma cell implantation into immunocompatible host mice, the

[52] B. B. Mishell, S. M. Shiigi, C. Henry, E. L. Chan, J. North, R. Gallily, M. Slomich, K. Miller, J. Marbrood, D. Parks, and A. H. Good, in "Selected Methods in Cellular Immunology" (B. B. Mishell, and S. M. Shiigi, eds.), p. 3. W. H. Freeman, San Francisco, 1980.

first-passage lymphoma stock cell lines are removed from $-80°$ storage, rapidly thawed to $37°$ by gentle shaking of the frozen tube(s) in a water bath, immediately washed twice in sterile PBS to remove any residual dimethyl sulfoxide, and resuspended to 2×10^7 tumor cells/ml in sterile PBS. The transplant recipients, 4- to 6-week-old (C57BL/6J $\times$ SJL/J) F_1 males, receive subcutaneous injections of 2×10^6 tumor cells in a final volume of 0.1 ml of PBS. The mice are palpated daily to detect the development of tumors, which are first observed at the injection site. Volumes of the tumors that are detected are estimated from two perpendicular measurements ($LW^2/2 \pm$ SEM). Tumors are measured every other day after the initial observation up to the time of sacrifice.

Using this model, MYC6 and additional c-*myc* antisense oligonucleotides were tested in an adjuvant therapeutic regimen as consolidation therapy for B cell lymphoma in the transplant model.[7]

Prolonged Release Pellet Administration

Four-week prolonged release pellets (Innovative Research of America, Sarasota, FL) packed with no drug (placebo), or 4.0 μmol (20 mg) of MYC6 or SCR6, are implanted under the skin on the upper back of groups of three to five Eμ-*myc* transgenic mice within 2 weeks after tumor onset. The total amount of DNA in each pellet and the rate of release from each pellet vary widely and unpredictably. Tumor size, body mass, and lymphocyte differentiation of each mouse are analyzed weekly. Significant percentages of immature lymphocytes were not observed in tail blood smears until late in tumor development, after lymphoid tumors were prevalent throughout the body. Mice with disseminated tumors displayed primarily blasts, as well as some other immature forms, but the extent of peripheral pre-B cell populations common to these mice varied more from mouse to mouse than from treatment group to treatment group. After 4 weeks of treatment there were no significant differences in changes in tumor size increase or body mass increase among the groups of mice treated with placebo, MYC6, or SCR6. Hence, the pellet route was not pursued further.

Subcutaneous Administration

Groups of three or four Eμ-*myc* transgenic mice are selected at weaning, 3 weeks after birth, on the criterion of displaying no palpable tumor at that time. Weaning is the earliest age at which DNA could be administered to mice without danger of maternal rejection. Each mouse receives 0.1 ml of sterile saline, or saline containing 1.0 μmol (5.0 mg) of SCR6 or MYC6 subcutaneously on the upper back, twice weekly for 5 weeks. At the end of treatment, two of four mice receiving scrambled DNA displayed tumors, but none in the antisense group displayed tumors. However, the epidermal tissues surrounding the site of injection in the MYC6 group became thick, rigid, and brittle. Therefore, this mode of oligonucleotide administration was not pursued further in the transgenic mouse model. In the transplant model, however, daily subcutaneous administration of smaller doses of antisense DNA was well tolerated, without dermal irritation.

Micropump Administration

This route was substituted for subcutaneous injection in order to avoid epidermal senescence at the site of injection. Three-week-old Eμ-*myc* transgenic mice with no palpable tumors are randomly distributed into groups of 12. Siblings are assigned into all three treatment groups. Microosmotic pumps (Alzet 2002; Alza, Palo Alto CA) containing 200 μl of saline, or saline with 5.0 mM SCR6 or 5.0 mM MYC6, are implanted subcutaneously on the upper back of each mouse, anesthetized with Avertin (30 mg/kg). Each pump containing DNA therefore dispenses 1.0 μmol (5.0 mg), at 0.5 μl/hr, or 2.5 nmol/hr, over 14 days. This regimen results in a steady state serum concentration of 0.1 μM in control animals.[53]

The pumps are replaced at 5 and 7 weeks after birth, resulting in a total dose to each mouse of 3.0 μmol of oligomer, or 15 mg, administered continuously at a nearly uniform rate of about 0.36 mg/day over 6 weeks. Because the mass of 3-week-old mice is about 10 g, and the mass of 9-week-old mice is about 30 g, one estimates an initial dose of 36 mg/kg/day, a final dose of 12 mg/kg/day, an average dose of about 20 mg/kg/day, or about 1 mg/kg/hr.

Intraperitoneal Administration

In the lymphoma cell transplant model for residual disease, Eμ-*myc* transgenic tumor cells are injected subcutaneously into one rear leg of each recipient mouse. One day after the transplants were performed, administration of phosphorothioate DNAs is initiated. Initially, oligonucleotides were injected subcutaneously on the opposite rear leg. Because the efficacy of oligonucleotide treatments was found to be the same whether administered subcutaneously or intraperitoneally, intraperitoneal administration was pursued in later experiments owing to the ease of administration as well as minimization of stress.

[53] G. D. Gray and E. Wickstrom, *Antisense Nucleic Acid Drug Dev.* **7**, 133 (1997).

Oligonucleotide Pharmacokinetics

There have been a number of studies of the plasma clearance kinetics and organ distribution of antisense DNAs. Most studies to date have been done on DNA phosphorothioates.[54–56] Oligonucleotides are rapidly cleared from the bloodstream, over a time course that may be fit to a two-compartment model, distributing from the first compartment, the blood, to the second compartment, the tissues.

The initial rapid distribution after administration, the α phase, exhibits a half-life of several minutes, depending on the sequence and animal studied. The slower β phase, reflecting equilibration between tissue and blood as the drug is metabolized and excreted, exhibits a half-life of several hours. Although there is considerable heterogeneity, most studies find that high levels of oligonucleotides accumulate in the kidney and liver, and to a lesser extent in other organs. The main route of excretion is renal. For phosphorothioates, hepatic metabolism also occurs.[57,58]

In our studies, blood is collected from the retroorbital sinuses of anesthetized mice at various times after administration of phosphorothioate oligonucleotides. Plasma supernatants, obtained by sedimentation of whole blood, are frozen in 200-μl aliquots and stored for subsequent analysis. On thawing, the plasma samples are mixed with equal volumes of acetonitrile (CH_3CN). Insoluble material, mainly precipitated protein, is removed by sedimentation for 10 min at $10,000g$. The supernatant fractions are then lyophilized, redissolved in 20% CH_3CN in H_2O, passed through 0.2-μm pore size filters, and analyzed by reversed-phase liquid chromatography on a 4.6×250 mm C_{18} analytical column using a 0–80% gradient from 0.1 M triethylammonium acetate (Et_3NH-HOAc), pH 7.0, to 80% CH_3CN/20% (0.1 M Et_3NH-HOAc, pH 7.0) over 30 min, at a flow rate of 1.0 ml min.[56] Oligomers are detected by absorbance at 260 nm. The baseline runs are subtracted from the sample runs, and peak areas are quantitated by comparison with oligonucleotide standards.

A simpler method to measure plasma oligonucleotide concentrations relies on measurement of [^{14}C]oligonucleotides in plasma samples.[59] However, this method does not distinguish partially fragmented oligonucleo-

[54] P. L. Iversen, B. L. Copple, and H. K. Tewary, *Toxicol. Lett.* **82–83,** 425 (1995).

[55] R. Zhang, R. B. Diasio, Z. Lu, T. Liu, Z. Jiang, W. M. Galbraith, and S. Agrawal, *Biochem. Pharmacol.* **49,** 929 (1995).

[56] R. K. DeLong, A. Nolting, M. Fisher, Q. Chen, E. Wickstrom, M. Kligshteyn, S. Demirdji, M. Caruthers, and R. L. Juliano, *Antisense Nucleic Acid Drug Dev.* **7,** 71 (1997).

[57] S. Agrawal, J. Temsamani, and J. Y. Tang, *Proc. Natl. Acad. Sci. U.S.A.* **88,** 7595 (1991).

[58] A. Nolting, R. K. DeLong, M. H. Fisher, E. Wickstrom, G. M. Pollack, R. L. Juliano, and K. L. Brouwer, *Pharm. Res.* **14,** 516 (1997).

[59] J. A. Hughes, A. V. Avrutskaya, K. L. Brouwer, E. Wickstrom, and R. L. Juliano, *Pharm. Res.* **12,** 817 (1995).

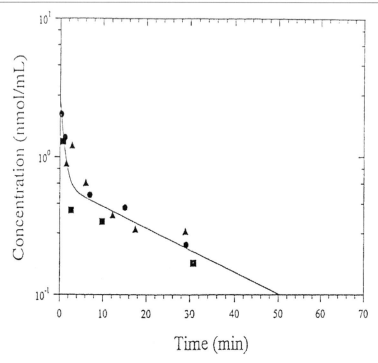

FIG. 7. Plasma concentration of oligonucleotide phosphorothioate in mice. Oligonucleotides were extracted from plasma of blood samples from three anesthetized mice (▲, ●, ■), then quantitated by reversed-phase HPLC.[56]

tides. An even simpler alternative procedure utilizes fluorescence measurements of OliGreen (Molecular Probes, Eugene, OR) added to plasma samples from mice treated with oligonucleotides.[53]

The oligonucleotide plasma concentration-versus-time profile was fitted with Scientist (Micromath Scientific Software, Salt Lake City, UT) to a two-compartment model (Fig. 7). After a short distribution phase from the blood to the tissues [$t_{1/2}(\alpha) = 0.9$ min], the oligonucleotides then cleared rapidly from the tissues [$t_{1/2}(\beta) = 19$ min]. The oligonucleotide volume of distribution (V_d) was 2.2 ml, comparable to 3.6 ml for [^{14}C]sucrose, a marker of extracellular space. At 1 and 24 hr after administration, drug levels were determined in various tissues. The rank order was kidney ≫ liver > spleen > tumor > muscle, with kidney and liver containing about 8 and 4% of the initial dose given after 1 hr, respectively. One hour after administration, the tumors accumulated about 5% of the initial dose. At 24 hr the relative tissue distribution remained similar to the 1-hr distribution, while the kidney : liver ratio was unchanged.[56]

Oligonucleotide Toxicology

At intraperitoneal bolus doses above 100 mg/kg, 3 times per week for 2 weeks, phosphorothioate DNA was found to be toxic to liver, kidney, spleen, and some other major organs of mice.[60] At this dose range, serum aspartate aminotransferase activity doubled or tripled, alanine aminotransferase activity doubled, and glucose levels decreased 30%. More than 150 mg/kg in a single bolus induced acute death in some mice.[60]

Phosphorothioate DNAs display much greater antisense efficacy *in vivo* than native phosphodiester DNA or methyl phosphonate DNA, but display non-sequence-specific antiproliferative activity.[61,62] Concerns have been raised about multiple possible modes of toxicity of phosphorothioates at high concentrations,[60] including lymphoid hyperplasia in mice[63] and leukocyte depletion, complement activation, hypotension, and prolonged clotting time in monkeys.[64] In addition, stimulation of B cell proliferation in mice by dPuPuCGPyPy motifs present in the oligonucleotide has been reported.[65] However, at therapeutic doses, no significant antigenicity, liver toxicity, kidney toxicity, or teratogenicity has been observed in animals or humans.[3,6,62]

A phase I trial of MYC6 for prophylaxis of arterial restenosis has been completed without signs of significant toxicity.[66] However, a phosphorothioate sequence targeted to the initiation codon domain of HIV *gag* reduced platelet counts excessively when administered at concentrations of 4 mg/kg, resulting in termination of its phase II trial.[67] In most other human trials with DNA phosphorothioates, efficacy was observed without significant toxicity at therapeutic doses.[3]

In our studies, phosphorothioate DNA was administered to eight Eμ-*myc* transgenic mice with established tumors by prolonged release pellet (Innovative Research of America) at approximately 30 mg/kg/day for 4

[60] U. M. Sarmiento, J. R. Perez, J. M. Becker, and R. Narayanan, *Antisense Res. Dev.* **4,** 99 (1994).

[61] P. T. Ho, K. Ishiguro, E. Wickstrom, and A. C. Sartorelli, *Antisense Res. Dev.* **1,** 329 (1991).

[62] S. Agrawal, *Trends Biotechnol.* **14,** 376 (1996).

[63] R. F. Branda, A. L. Moore, L. Mathews, J. J. McCormack, and G. Zon, *Biochem. Pharmacol.* **45,** 2037 (1993).

[64] W. M. Galbraith, W. C. Hobson, P. C. Giclas, P. J. Schechter, and S. Agrawal, *Antisense Res. Dev.* **4,** 201 (1994).

[65] A. M. Krieg, A. K. Yi, S. Matson, T. J. Waldschmidt, G. A. Bishop, R. Teasdale, G. A. Koretzky, and D. M. Klinman, *Nature (London)* **374,** 546 (1995).

[66] F. Roqué, A. Rodriguez, L. Grinfeld, D. Fischman, Y. Shi, and A. Zalewski, *J. Am. Coll. Cardiol.* **29A,** 317A (1997).

[67] S. Agrawal, *in* "Workshop on Future Needs and Developments in Antisense Technology." National Heart, Lung, and Blood Institute, Bethesda, Maryland, 1997.

TABLE 2
MAJOR PLASMA PHYSIOLOGICAL VALUES OF Eμ-myc TRANSGENIC MICE TREATED WITH
DNA PHOSPHOROTHIOATES FOR 4 TO 6 WEEKS AT DOSES UP TO 70 mg/kg/day[a]

Treatment group	AST (SGOT) (IU/liter)	ALT (SGPT) (IU/liter)	Blood urea nitrogen (mg/dl)	Blood creatinine (mg/dl)	Blood glucose (mg/dl)	Blood sodium (mEq/dl)
Normal control[b]	438 ± 241 (4)	455 ± 247 (4)	28 ± 14 (4)	0.28 ± 0.15 (4)	202 ± 15 (4)	136 ± 3[d] (4)
Saline only	518 ± 328 (8)	131 ± 120 (8)	31 ± 12 (7)	0.29 ± 0.06 (8)	144 ± 33 (3)	154 ± 14[d] (7)
Scrambled DNA	658 ± 341 (4)	154 ± 72 (4)	35 ± 14 (4)	0.35 ± 0.10 (4)	157 ± 47 (2)	148 ± 11 (2)
Antisense DNA	627 ± 306 (4)	88 ± 10 (4)	38 ± 24 (4)	0.43 ± 0.39 (4)	229 ± 30 (3)	145 ± 0 (3)
ANOVA[c]	No	No	No	No	No	Yes

[a] Numbers in parentheses indicate the number of samples for each measurement.
[b] Eμ-myc negative mice.
[c] Nonparametric comparison among groups by Kruskal–Wallis one-way analysis of variance on ranks.
[d] Significant difference between these two groups by the Dunn multiple pairwise comparison method.

weeks. In four Eμ-myc transgenic mice without palpable tumors at 3 weeks, oligonucleotides were administered by subcutaneous injection at approximately 250 mg/kg twice weekly (70 mg/kg/day) for 5 weeks. In four Eμ-myc transgenic mice without palpable tumors at 3 weeks, oligonucleotides were administered by micropump at 0.36 mg/day, or an average of 20 mg/kg/day, for up to 6 weeks, or until establishment of large tumors. At the end of each treatment, the mice were euthanized and 1 ml of blood was collected from each mouse by heart puncture. Plasma was obtained by sedimentation of whole blood, and levels of alanine aminotransferase (ALT), aspartate aminotransferase (AST), creatinine, blood urea nitrogen (BUN), creatinine, glucose, and sodium were analyzed on a Kodak (New Haven, CT) 700 chemical analyzer (AMI Town and Country Hospital, Tampa, FL). In contrast to mice given intraperitoneal boli of 100 mg/kg,[60] there were no significant differences detected among the three treatment groups in major physiological values such as AST, ALT, creatinine, BUN, glucose, and blood sodium, indicating the lack of acute toxicity of phosphorothioate DNA at the therapeutic doses (Table II). Furthermore, among the mice treated with phosphorothioate DNA at the dose rates described in this work, we observed no loss of body mass during treatment

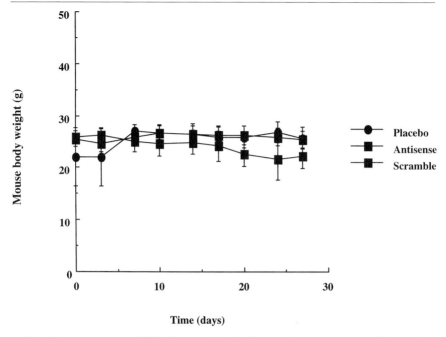

FIG. 8. Average mass ± SEM of transgenic mice from each treatment group (three to five mice per group) during 28 days of oligonucleotide treatment via cholesterol pellets.[6]

(Fig. 8). In addition, normal lymphocyte maturation was observed, providing further indication of negligible acute toxicity.

In general, the most likely mechanism for nonspecific effects at high concentrations is polyanion binding to proteins, analogous to the polyanion anticoagulant heparin binding to serum proteins.[62] As a general precaution, the most effective and least toxic oligonucleotide concentrations should be determined to minimize any potential nonspecific activity.

Aptamer Effects

As we have found, careful choice of phosphorothioate concentration minimizes nonspecific toxicity,[6,7] and careful sequence controls support the hypothesis of an antisense mechanism.[7,15] For MYC6 in particular, high *in vitro* concentrations (30–50 μM) have caused nonspecific inhibition of proliferation.[61,68] One group hypothesized that the dGGGG motif in MYC6 might have allowed tetraplex formation, which might have induced nonspecific inhibition by a nonantisense mechanism.[68] In cell culture, smooth muscle cells were treated with c-*myb* and c-*myc* antisense phosphorothioates that included dGGGG segments at 30 μM for 24 hr. Nonspecific inhibition of proliferation was observed both with the c-*myb* and c-*myc* antisense sequences and with dGGGG control sequences.[68] At this toxic dose, however, nonspecific inhibition should be expected.[61] In contrast, typical antisense DNA phosphorothioate concentrations of 0.2–2.0 μM result in sequence specific effects.[1,2] On testing of that hypothesis, however, it was found that human MYC6, and its murine counterpart, do not form detectable tetraplexes in physiological buffers at 37°.[69] In addition, MYC61, a derivative of MYC6 with two central mismatches, displays no antisense or antitumor activity, despite inclusion of the dGGGG motif.[7]

Immunostimulation

General immunostimulatory responses have been reported in murine cells exposed to particular sequences with centrally placed dPuPuCGPyPy motifs.[65,70] Because MYC6 contains a hexamer similar to the consensus dPuPuCGPyPy motif, we sought evidence of immunostimulation. In cell culture, MYC6 induced no significant spleen cell proliferation at the steady state plasma concentration of 0.1 μM, in contrast to significant spleen cell proliferation induced by the immunostimulatory sequence MCG (Table I) at the same concentration.[71] In MCG, dGACGTT occurs in the middle of the sequence. However, oligonucleotides containing a complete dPuPuCGPyPy consensus motif at either end of the sequence have been found ineffective in their immunostimulatory activity.[70]

Splenomegaly, which has been detected in mice treated with phosphorothioate DNA,[63] was also observed in the Eμ-*myc* mice treated with MYC6 or SCR6. At an age of 12–16 weeks, the mass of spleens from normal mice averaged 0.1 g, while those from Eμ-*myc* mice receiving phosphorothioates averaged 0.3 g. However, the spleens of Eμ-*myc* mice treated with the saline vehicle averaged 0.25 g, which was not significantly different (Fig. 9). Hence, the observed splenomegaly may also arise from the stress of bearing tumors and implantation of pellets or micropumps.

[68] T. L. Burgess, E. F. Fisher, S. L. Ross, J. V. Bready, Y. X. Qian, L. A. Bayewitch, A. M. Cohen, C. J. Herrera, S. S. Hu, T. B. Kramer, F. D. Loff, F. H. Martin, G. F. Pierce, L. Simonet, and C. Farrell, *Proc. Natl. Acad. Sci. U.S.A.* **92,** 4051 (1995).

[69] S. Basu and E. Wickstrom, *Nucleic Acids Res.* **25,** 1327 (1997).

[70] D. M. Klinman, A. K. Yi, S. L. Beaucage, J. Conover, and A. M. Krieg, *Proc. Natl. Acad. Sci. U.S.A.* **93,** 2879 (1996).

[71] G. D. Gray, R. Townsend, H. Hayasaka, R. Korngold, and E. Wickstrom, *Nucleosides Nucleotides* **16,** 1727 (1997).

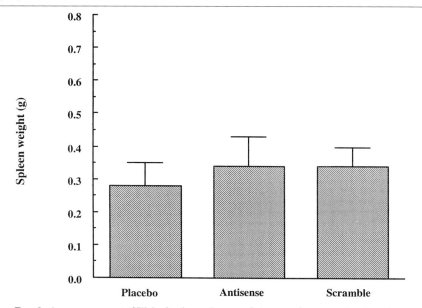

FIG. 9. Average mass ± SEM of spleens from each transgenic mouse treatment group, (three to five mice per group) after the end of 28 days of oligonucleotide treatment via cholesterol pellets.[6]

Similarly, in the residual disease model, mice treated every day with 0.76 mg of phosphorothioate DNA for 14 days after transplantation of Eμ-myc tumor cells also displayed nonspecific splenomegaly, including those treated with PBS vehicle (Fig. 10). Consistent with the known immunostim-ulatory activity of MCG, splenomegaly was greatest in mice treated with MCG. Elevation of specific cytokines indicates an immune response, yet no elevation in interferon γ (IFN-γ), tumor necrosis factor α (TNF-α), interleukin 6 (IL-6), or IL-12 could be detected in blood samples taken 1 hr after oligonucleotide administration to transplant model mice.[7]

Strong immune stimulation is toxic, as we observed in transplant model mice treated with the immunostimulatory sequence MCG. Acute ascites accumulation was observed in transplanted mice after daily administration of 0.76 mg of MCG. Among three mice treated with MCG for 14 days, one mouse died at 10 days posttransplantation, and a second at 13 days posttransplantation, prior to the end of the therapeutic schedule. These premature deaths occurred before ascites accumulation would have been detected. Ascites accumulation was detected in the single mouse that did survive to 16 days after tumor cell transplantation, 2 days after the end of MCG administration.[7] The fatal toxicity displayed by MCG suggests that

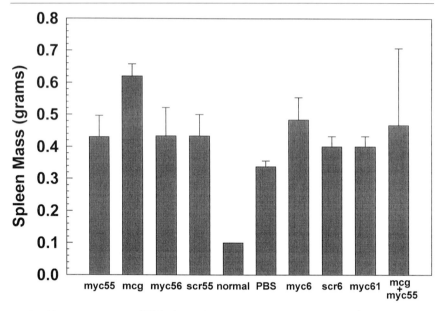

FIG. 10. Average mass ± SEM of the spleens from each treatment group (three to six mice per group) at sacrifice, 15–17 days posttransplant, after the 14-day regimen with oligonucleotide (0.76 mg/day), versus DNA sequences administered. "Normal" indicates spleens of mice that did not receive either lymphoma cell transplants or oligonucleotides.[6]

minimal amounts of this oligonucleotide should be used in future experiments.

Myc Protein Levels in Spleen and Tumor Cells

Total lymphocytes from both spleens and lymph node tumors are collected after sacrifice of Eμ-myc transgenic mice[45] treated with DNA phosphorothioates, treated transplanted mice, or untreated transgenic mice, as described in Lymphoma Cell Transplant Model for Residual Disease (above). The final cell suspension is adjusted to 5×10^7 trypan blue-excluding cells per milliliter, using sterile PBS.

For studies in the transgenic mouse model, lymphocyte nuclear protein is isolated by lysis of cells in 5 mM KCl, 5 mM MgCl$_2$, 20 mM HEPES (pH 7.4), sodium deoxycholate (1.0 g/liter), Nonidet P-40 (5 ml/liter), and aprotinin (50 μg/liter) at 4° and pelleted at 2000g for 6 min.[72] Nuclei are lysed in 150 mM NaCl, 10 mM NaH$_2$PO$_4$ (pH 7.4), 1.0% Triton X-100,

[72] K. H. Klempnauer and A. E. Sippel, *Mol. Cell. Biol.* **6**, 62 (1986).

sodium deoxycholate (5 g/liter), and sodium dodecyl sulfate (SDS, Na-DodSO$_4$, 0.1 g/liter) for 10 min at 4°, then pelleted at 10,000g for 20 min.[73]

To standardize comparison of bands from neighboring lanes, 10 μg of total nuclear protein from each sample, measured with the bicinchoninic acid (BCA) protein assay kit (Pierce, Rockford, IL), is electrophoresed on a 4–20% polyacrylamide gradient gel, then blotted onto polyvinylidene difluoride (PVDF) membrane (Bio-Rad). Coomassie blue staining of a parallel gel indicated constant intensities of prominent nuclear protein bands in each lane. Myc protein was detected by monoclonal mouse anti-Myc antibody (OM-11-904; Cambridge Research Biochemicals, Cambridge, UK) and goat anti-mouse IgG conjugated with horseradish peroxidase (A2304; Sigma, St. Louis, MO), followed by chemiluminescent detection (NEL100; NEN Life Sciences, Boston, MA).

For studies in the transplant model, 40 μg (when available) of whole-cell (not nuclear) protein lysates isolated from each spleen or tumor sample as described above is electrophoresed, blotted, and probed with monoclonal mouse anti-Myc antibody (C-33; Santa Cruz Biotechnology, Santa Cruz, CA). The blot is then probed with goat anti-mouse IgG conjugated with alkaline phosphatase, allowing Myc to be visualized by chemiluminescence (37078T; Pierce). To normalize protein loadings, membranes are either probed simultaneously with anti-β-actin antibody (clone AC-74; Sigma Chemicals), or are stripped and reprobed with the anti-β-actin antibody. Myc band intensities for each sample are normalized to the respective actin band intensities and quantitated on a densitometry system (BioImage, Ann Arbor, MI), yielding Myc:actin ratios ± SEM.

The emotional stress experienced by the mice during the course of the experiments must also be considered to affect the therapeutic response. Therefore, Myc expression levels were determined in tumors from a control group of mice that had received tumor cell transplants, but no oligonucleo-tides. PBS vehicle was administered to those tumor-bearing control mice in order to replicate similar levels of stress from handling and from injec-tions. In addition, mice that did not receive tumor cell transplants but did receive oligonucleotide injections served as an additional control for phosphorothioate DNA nonspecific effects on Myc levels in normal mouse spleen cells. To represent the Myc expression levels occurring in normal experimental spleen cells, the final control group consisted of mice that did not receive tumor transplants but did receive injections of the PBS vehicle, again to control for the stress levels experienced by the treat-ment groups.

[73] S. Biro, Y. M. Fu, Z. X. Yu, and S. E. Epstein, *Proc. Natl. Acad. Sci. U.S.A.* **90,** 654 (1993).

c-myc mRNA Levels in Spleen and Tumor Cells

Total RNA is extracted[74] from about 0.3 g of spleens and lymph node tumors from each mouse treated with DNA phosphorothioates or saline. About 20 μg of total RNA from each sample is electrophoresed on a glyoxal–dimethyl sulfoxide gel,[75] then blotted onto a nylon membrane and hybridized with two [^{33}P]dCTP-labeled probes: the c-*myc* *Pst*I restriction fragment from pRyc7.4[49] and the pTRI-GAPDH construct (7431; Ambion, Houston, TX). The c-*myc* mRNA level is calculated on the basis of the density ratio of the c-*myc* band compared with the corresponding glyceraldehyde-3-phosphate dehydrogenase (GAPDH) band, quantitated on a densitometry system (BioImage).

Pharmacodynamics in Transgenic Mouse Model for Spontaneous Tumors

Prophylactic treatment was tested by administering MYC6, SCR6, or saline vehicle to groups of 12 Eμ-*myc* transgenic mice by subcutaneous micropump continuously for 6 weeks, as described above. Animals were palpated every day to detect tumor onset. The 6 weeks of therapy represents the longest test of antisense DNA administration in animals to date.

The cohort of 12 Eμ-*myc* transgenic mice treated continuously with SCR6, and those treated with saline, displayed tumors over the usual time course (Fig. 11). In the saline group, 5 of 12 mice had palpable tumors by 8 weeks of age, and 9 of 12 had tumors at 15 weeks. In the SCR6 group, 5 of 12 mice had tumors by 10 weeks, and 9 of 12 had tumors at 15 weeks. However, in the MYC6 group only 3 of 12 mice had displayed a tumor by the time of sacrifice at 26 weeks (Fig. 11). One of the three tumors in the MYC6 group was already palpable 2 weeks after the beginning of the 6 weeks of therapy. It is possible that the 5-week tumor had already initiated prior to therapy. The second was detected at 15 weeks and the third at 17 weeks. The nine tumor-free mice treated with MYC6 were observed up until sacrifice at 48 weeks.

An overall analysis of variance of the tumor data, using the nonparametric Kruskal–Wallis procedure, indicated significant effects (H_0, $p = 0.0000851$). This was followed by multiple pairwise comparisons among the groups, using the Dunn nonparametric exact test and the Mann–Whitney rank sum test. By both tests, the MYC6-treated group differed significantly from both the saline-treated (H_0, $p < 0.05$) and SCR6-treated (H_0, $p <$

[74] P. Chomczynski and N. Sacchi, *Anal. Biochem.* **162,** 156 (1987).

[75] J. Sambrook, E. F. Fritsch, and T. Maniatis, "Molecular Cloning—A Laboratory Manual." Cold Spring Harbor Laboratory Press, Cold Spring Harbor, New York, 1989.

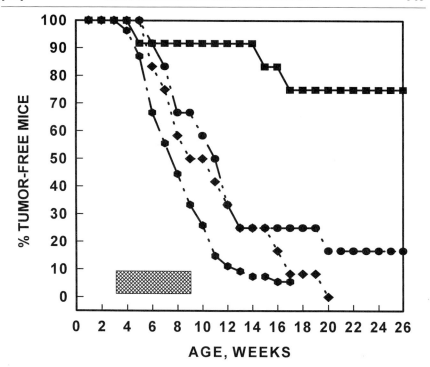

FIG. 11. Time of onset of lymphoid tumors in Eμ-*myc* transgenic mice treated with saline, or with scrambled or antisense DNA phosphorothioates. Hatched box shows period of continuous administration of DNA from microosmotic pumps. ◆, no treatment (54 mice that were tumor free at 3 weeks); ◆, saline (12 mice); ●, SCR6 (12 mice); ■, MYC6 (12 mice).[6]

0.05) groups in tumor onset, while the latter two groups did not differ significantly from each other (H_0, $p < 0.05$). Therefore, a significant degree of protection by MYC6 against tumor onset was observed in this trial, continuing out to 26 weeks of age. This is a unique demonstration of chemoprevention of tumorigenesis by antisense DNA prophylaxis.

Myc protein was virtually undetectable in splenic lymphocytes from all Eμ-*myc* transgenic mice treated with MYC6 (Fig. 12), as is the case in normal mice or Eμ-*myc* transgenic mice prior to tumor onset. In contrast, prominent Myc bands remained in splenic lymphocytes of tumor-bearing Eμ-*myc* transgenic mice treated with saline or SCR6 (Fig. 12). Analysis of lymphoid tumors of Eμ-*myc* transgenic mice revealed significant levels of Myc protein in all treatment groups (Fig. 13). The reduction of Myc antigen in the splenic lymphocytes of tumor-bearing transgenic mice treated with MYC6 correlated with antisense DNA treatment, but the tumors themselves were apparently unaffected. This reduction in Myc protein expression

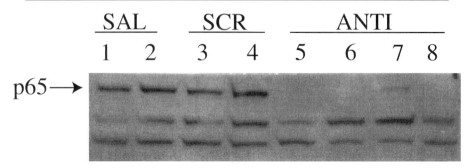

FIG. 12. Western blot analysis of Myc antigen expression in spleens of Eμ-*myc* transgenic mice with palpable tumors, treated with saline (SAL) (lanes 1, 2), SCR6 (SCR) (lanes 3, 4), or MYC6 (ANTI) (lanes 5–8) for 1 week. Ten micrograms of nuclear protein extract was applied to each well.[6]

in spleen cells, but not in tumor cells, made it plausible that prophylactic therapy with antisense DNA prior to tumor establishment might slow the proliferation of the transgenic pre-B cell population enough to lower the rate of tumor onset in Eμ-*myc* transgenic mice.

Levels of c-*myc* mRNA measured in the Eμ-*myc* transgenic mice treated with MYC6 described above were not significantly reduced, relative to control groups, in splenic lymphocytes (Fig. 14) or in lymphoid tumors (Fig. 15).

Pharmacodynamics in Transplant Model for Residual Disease

Administration of MYC6 at 0.76 mg/day for seven consecutive days after subcutaneous implantation of Eμ-*myc* tumor cells delayed tumor

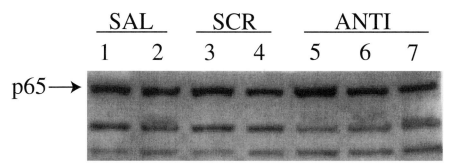

FIG. 13. Western blot analysis of Myc antigen expression in lymphoid tumors of Eμ-*myc* transgenic mice with palpable tumors, treated with saline (SAL) (lanes 1, 2), SCR6 (SCR) (lanes 3, 4), or MYC6 (ANTI) (lanes 5–7) for 1 week. Ten micrograms of nuclear protein extract was applied to each well.[6]

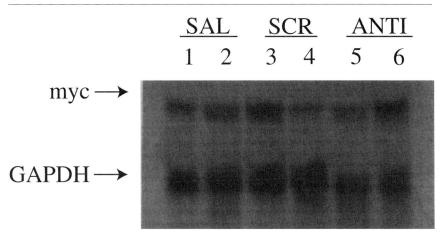

FIG. 14. Northern blot analysis of c-*myc* and GAPDH (glyceraldehyde phosphate dehydrogenase) mRNA expression in spleens of Eμ-*myc* transgenic mice with palpable tumors, treated with saline (lane 1), SCR6 (lanes 2, 3), or MYC6 (lanes 4–6) for 1 week. Twenty micrograms of total cellular RNA was applied to each well.[6]

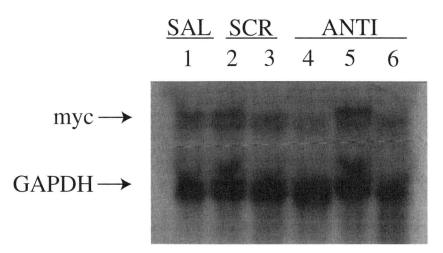

FIG. 15. Northern blot analysis of c-*myc* and GAPDH mRNA expression in lymphoid tumors of Eμ-*myc* transgenic mice with palpable tumors, treated with saline (lane 1), SCR6 (lanes 2, 3), or MYC6 (lanes 4–6) for 1 week. Twenty micrograms of total cellular RNA was applied to each well.[6]

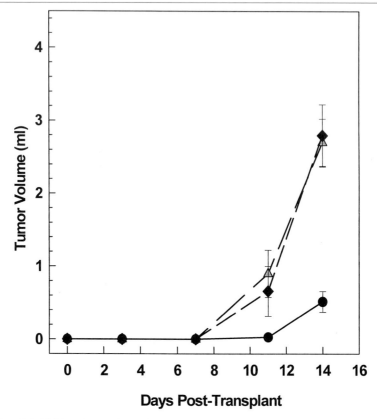

Fɪɢ. 16. Inhibition of tumorigenesis by transplanted lymphoma cells in immunocompatible mice treated with MYC6. Lymphoma cells (2×10^6) were implanted subcutaneously into six mice per treatment group, followed by daily subcutaneous administration of 0.76 mg of DNA phosphorothioate in 0.1 ml of sterile phosphate-buffered saline (PBS), or vehicle only, from day 1 to day 7 posttransplant. Tumor volumes (length $\times$ width^2/2 $\pm$ SEM) were estimated at 3, 7, 11, and 14 days posttransplant. ◆, PBS vehicle only; ▲, SCR6 scrambled control; ●, MYC6, complementary to codons 1–5.[7]

onset by 3 days, relative to control groups (Fig. 16). Differences among the MYC6-treated, SCR6-treated, and PBS-treated groups were greater than would be expected by chance ($p = 0.010$ at 11 days; $p = 0.001$ at 14 days). Multiple pairwise comparisons revealed significant differences between the MYC6-treated group and both the PBS-treated and SCR6-treated groups (all $p < 0.05$) at 11 and 14 days, whereas the PBS-treated and SCR6-treated groups did not differ significantly from each other (all $p > 0.05$). Treatment with MYC6 also decreased total tumor mass at sacri-

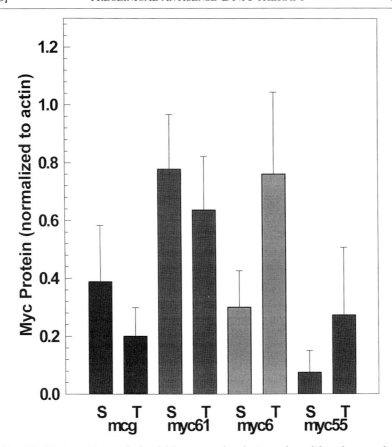

FIG. 17. Western blot analysis of Myc expression in transplanted lymphomas of mice treated with MYC6, MYC61, MYC55, or MCG for 7 days. The spleens (S) and tumors (T) of mice in each treatment group were removed at the time of sacrifice, 17 days posttransplant, and processed for Western blotting as described in Myc Protein Levels in Spleen and Tumor Cells. Ratios of Myc:β-actin band densities $\pm$ SEM are shown for each treatment group (four to six mice). DNA sequence administered is shown at the bottom of each column.[7]

fice (17 days) by 40 $\pm$ 16%, and decreased the splenic Myc:actin ratio to 0.30 $\pm$ 0.13 (Fig. 17). SCR6 had no effect.

MYC6 contains both a dGGGG motif and also a dCACGTT motif similar to the dGACGTT motif in the immunostimulatory sequence MCG, 5'-dGCATGACGTTGAGCT.[65] Therefore, MYC6 was compared with MCG to determine if tumor inhibition by MYC6 could be a result of B cell activation. Tumor inhibition by MYC6 (Fig. 16) and MCG (Fig. 18) were comparable. However, no significant inhibition of tumor growth (Fig.

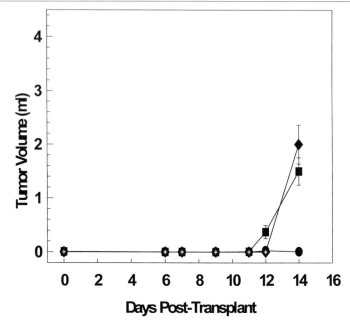

Fig. 18. Inhibition of tumorigenesis in mice treated with MCG. Lymphoma cells (2×10^6) were implanted subcutaneously into six mice per treatment group followed by daily subcutaneous administration of 0.76 mg of DNA phosphorothioate in 0.1 ml of sterile phosphate-buffered saline from day 1 to day 7 posttransplant. Tumor volumes (length $\times$ width^2/ $2 \pm$ SEM) were estimated daily for tumors. ◆, Untreated mice; ■, MYC61 mismatch control; ●, MCG, immunostimulatory sequence.[7]

18) or Myc expression (Fig. 19) was observed using MYC61, a control sequence with two mismatches in the middle, preserving both the dGGGG and the dCACGTT motifs. Differences among the MYC61-treated, MCG-treated, and untreated groups were greater than would be expected by chance ($p = 0.012$ at 14 days; $p = 0.001$ at 16 days). Multiple pairwise comparisons revealed significant differences between the MCG-treated group and both the MYC61-treated group and the untreated group (all $p < 0.05$) at 14 and 16 days. In contrast, the MYC61-treated and untreated groups failed to show any significant difference from each other (all $p > 0.05$). Thus we concluded that the entire MYC6 sequence was necessary for Myc and tumor inhibition.

MCG treatment reduced MYC:actin ratios in the tumors to 0.20 ± 0.10, but reduced the splenic MYC:actin ratio to only 0.39 ± 0.19 (Figs. 17 and 19). However, these mice developed significant ascites accumulation,

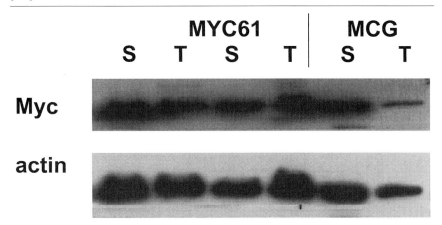

FIG. 19. Western blot analysis of mice treated with MYC61 or MCG administered subcutaneously at 0.76 mg/day for 7 days into groups of three to six mice. Spleens (S) and tumors (T) were removed at sacrifice. Cellular protein was extracted and analyzed by Western blot as described above to detect Myc relative to actin as a control. Myc levels in spleens of normal mice were below detection limits.[7]

implying blockage of the lymphatic system.[76,77] No ascites accumulation was observed during seven consecutive days of administration of 0.76 mg/day of the other sequences (Table I).

Although MYC6 significantly inhibited tumor progression (Fig. 16), complete tumor prevention was not observed. Therefore, we analyzed the inhibitory capacity of three antisense DNAs targeted against additional c-*myc* mRNA sites (MYC51, MYC53, and MYC55) that have been found susceptible to antisense DNA inhibition in HL-60 cells.[14,20,29] One day after tumor cell transplants, oligonucleotides were administered subcutaneously at 0.76 mg/day for seven consecutive days. Treatment with MYC51 and MYC53 slowed tumor growth, resulting in tumors that were about half as large as those observed in the untreated mice (Fig. 20). The differences among the four groups were greater than would be expected by chance ($p = 0.022$ at 7 days; $p = 0.020$ at 9 days; $p = 0.009$ at 11 days). Multiple pairwise comparisons revealed significant differences between the MYC55-treated group and the untreated group (all $p < 0.05$) at 7, 9, and 11 days, while the MYC51-treated, MYC53-treated, and untreated groups did not differ significantly from each other (all $p > 0.05$). Incomplete tumor inhibi-

[76] A. R. Baker and J. S. Weber, *in* "Cancer Principles and Practice of Oncology" (V. T. DeVita, S. Hellman, and S. A. Rosenberg, eds.), Vol. 2, p. 2245. Lippincott, Philadelphia, 1993.
[77] R. R. Seeley, T. D. Stephens, and P. Tate, *in* "Essentials of Anatomy and Physiology" (J. M. Smith, ed.), p. 504. Mosby-YearBook, New York, 1996.

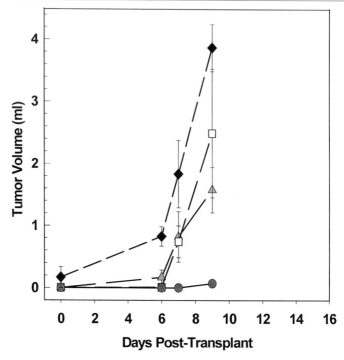

Fig. 20. Inhibition of tumorigenesis in mice treated with MYC55. Lymphoma cells (2×10^6) were implanted subcutaneously into three mice per treatment group, followed by daily subcutaneous administration of 0.76 mg of DNA phosphorothioate in 0.1 ml of sterile phosphate-buffered saline from day 1 to day 7 posttransplant. Tumor volumes (length $\times$ width^2/ $2 \pm$ SEM) were estimated at 6, 7, 9 and 11 days posttransplant. ▲, Untreated mice; □, MYC51, ▲, MYC53; ●, MYC55.[7]

tion in both the MYC51-treated and MYC53-treated groups correlated with the high levels of Myc detected in the spleens and tumors from these mice (data not shown).

Treatment with MYC55 (targeted to the carboxy-terminal end of the Myc protein) resulted in slow tumor formation during the 7-day treatment (Fig. 20). At the time of sacrifice, 17 days posttransplant, the total tumor mass was reduced by $65 \pm 6\%$, relative to control groups. In addition, these tumors developed only at the site of tumor transplant injection; no metastases to peripheral lymph nodes were detected. In all other treatment groups, however, the transplanted tumor cells metastasized from the initial injection site to peripheral lymph nodes. The Myc : actin ratio in the tumors of MYC55-treated mice was 0.27 ± 0.23, and the Myc : actin ratio in the

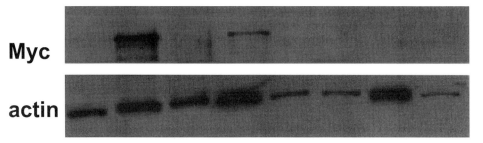

FIG. 21. Western blot analysis of mice treated with MYC55 administered subcutaneously at 0.76 mg/day for 7 days into groups of three to six mice. In the two tumor (T) lanes on the right, much less than 40 μg of total protein, was applied, because the tumors were too small to allow extraction of sufficient protein.[7] S, Spleen.

spleens was 0.08 ± 0.08 (Figs. 17 and 21), correlating with the slow rate of tumor development and the reduced tumor mass at sacrifice.

These initial antisense therapeutic regimens analyzed four c-*myc* mRNA target sites (Table I), yet the incomplete tumor inhibition resulting from each antisense sequence implies that a more analytical approach to mRNA target site accessibility is required. In addition, a more thorough determination of the appropriate therapeutic dose and treatment schedule may result in a greater inhibition of tumorigenesis, with the subsequent potential for total ablation of tumor formation.

Treatment for seven consecutive days did not ablate tumor development with any of the anti-c-*myc* sequences shown in Table I. To rule out the possibility that this partial tumor inhibition was the result of an inadequate treatment schedule, the most efficacious anti-c-*myc* DNA known at the time, MYC55, was analyzed for increased inhibitory activity by extending the therapeutic schedule from 7 consecutive days to 14 consecutive days. Surprisingly, extending the treatment schedule to 14 days with 0.76 mg (150 nmol) per day of MYC55 was no more effective than 7 days of treatment in preventing or further inhibiting tumorigenesis. The MYC55 treatment group developed tumors 11–13 days posttransplant, as observed in the 7-day regimen. Yet, extending the MYC55 therapy from 7 to 14 days fully ablated Myc protein in both the spleens and tumors from mice analyzed in this treatment group (Fig. 22). A fourfold increase in the exposure time of this Western blot revealed the presence of low levels of Myc in tumors

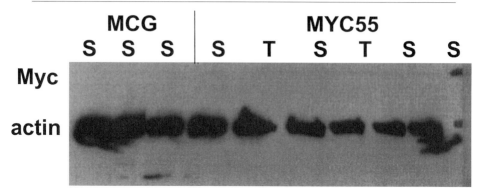

Fig. 22. Western blot analysis of mice treated with MYC55 or MCG administered intraperitoneally at 0.76 mg/day for 14 days to groups of three mice. Left three lanes: spleen (S) extracts from mice treated with MCG in the absence of tumor cell transplants; next four lanes: spleen (S) or tumor (T) extracts from mice treated with MYC55 after tumor cell transplants; right two lanes: spleen extracts (S) from mice treated with MYC55 in the absence of tumor cell transplants.[7]

from only one of the MYC55 mice (data not shown). The quantity of Myc detected in this mouse, when normalized to actin, was similar to that of the normal control group that did not receive tumor transplants, but only PBS vehicle.

Extending the therapy from 7 to 14 days resulted in 100% mortality of the MCG treatment group prior to the end of the therapeutic schedule, confirming an unacceptable level of toxicity during prolonged exposure to a known immunostimulatory sequence.[65,78] As noted previously, MCG toxicity resulted in ascites accumulation. Because the mice in this 14-day treatment regimen died before ascites accumulation would have been detected, ascites accumulation was observed only in the single mouse that survived 2 days beyond the termination of MCG injections (16 days posttransplant).

MYC56, a control sequence with the dGTCGTT motif of MYC55 disrupted (Table I), also proved to be fatally toxic: all of the mice receiving tumor cells and MYC56 died during therapy. Tumor onset in both the PBS and SCR55 control groups began 8 days posttransplant (not shown), similar to the time of onset detected in mice treated for 7 days with control oligonucleotides (Fig. 19).

Control oligonucleotides included a scrambled MYC55 sequence, SCR55, and a derivative of MYC55 containing three central mismatches, MYC56. In addition, the PBS vehicle alone was also included as a control

[78] J. P. Messina, G. S. Gilkeson, and D. S. Pisetsky, *Cell. Immunol.* **147,** 148 (1993).

for antisense specificity. Mice were also administered MYC55 or MCG in the absence of tumor cells in order to determine the effects of these sequences on immune system response.

The toxicity of MCG resulted in 100% mortality between 10 and 16 days posttransplant. These premature deaths resulted in either the absence of detectable tumors, or tumors too small to process. Therefore, the levels of Myc in this group remained undetermined. Spleen extracts from the three mice receiving MCG in the absence of tumor cells were also analyzed for Myc expression (Fig. 22). Myc was detected only in one of the mice, yielding an average Myc : actin ratio of 0.03 ± 0.03 when the Western blot was overexposed fourfold (data not shown). The MYC56 treatment group developed tumors 8 days posttransplant, similar to the PBS and SCR55 control groups. However, treatment with MYC56 was toxic, resulting in 100% mortality between 10 and 16 days posttransplant. The postmortem instability of Myc protein prevented analysis of Myc levels in spleens and tumors from these mice.

The significant, and similar, tumor inhibition resulting from treatment with MYC55, an antisense oligonucleotide, or with MCG, an immunostimulatory oligonucleotide, led us to analyze the efficacy of treatment with alternating doses of these two sequences. Because the dGTCGTT motif in MYC55 contains only one mismatch from the immunostimulatory consensus hexamer dPuPuCGPyPy, the results above cannot rule out the possibility that immunostimulation may, in part, be responsible for the tumor-inhibitory activity of this sequence. Therefore, an additional control sequence, MYC57, was synthesized. MYC57 maintains the integrity of the dGTCGTT hexamer of MYC55, but disrupts the MYC55 antisense sequence with two mismatches on the 5′ side of dGTCGTT and two mismatches on the 3′ side.

Initially, MCG at 0.76 mg/day was administered, but this proved to be fatally toxic, as described above. Therefore, 0.76 mg of MYC55, or the control sequence MYC57, was administered every other day, alternating with 0.65 mg of MCG every other day. Treatment was begun 1 day after tumor cell transplants, as previously described, and continued for 14 consecutive days, with alternating oligonucleotide injections.

Treatment with MYC57 alone, or treatment with PBS vehicle alone, resulted in tumor development between 8 and 10 days posttransplant (Fig. 23). Addition of MCG treatment to either the MYC57 or PBS groups postponed tumor development by an additional 3–4 days. Treatment with MYC55 alone, every other day, inhibited tumor development until 10–14 days posttransplant, as did MCG alone, every other day. Yet, when 0.76 mg of MYC55 was alternated every other day with 0.65 mg of MCG, no palpable tumors could be detected until 14 days posttransplant, the end of

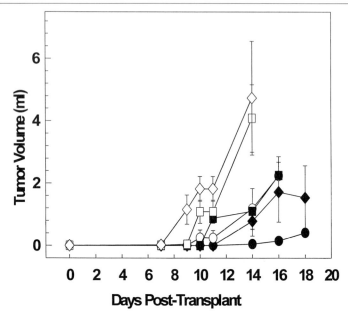

Fig. 23. Inhibition of tumorigenesis in mice treated with alternating doses of MYC55 and MCG for 14 days. Lymphoma cells (2×10^6) were implanted subcutaneously into three mice per treatment group, followed by intraperitoneal administration of 0.76 mg of MYC55 or MYC57 phosphorothioate in 0.1 ml of sterile phosphate-buffered saline (PBS) on odd-numbered days. Additional groups of mice were also treated with intraperitoneal injections of 0.65 mg of MCG in 0.1 ml of sterile PBS on even-numbered days. Tumor volumes (length × width²/2 ± SEM) were estimated at 7, 9, 10, 11, 14, 16, and 18 days posttransplant. ◇, PBS vehicle only; □, MYC57 only; ○, MYC55 only; ◆, PBS+MCG; ■, MYC57+MCG; ●, MYC55+MCG.[7]

the therapeutic regimen. The differences among the groups were greater than would be expected by chance up to 11 days ($p = 0.02$ at 9 days; $p = 0.017$ at 10 days; $p = 0.017$ at 11 days).

These results imply that treatment with an antisense oligonucleotide, MYC55, in addition to an immunostimulatory oligonucleotide, MCG, was more effective than either oligonucleotide alone. Injection of 0.65 mg of MCG was less toxic than the 0.76-mg dose administered earlier. One MYC55/MCG mouse died at the end of therapy (14 days posttransplant) and two of the MYC57/MCG mice died 11–14 days posttransplant, near the end of therapy.

Interpretation

One may hypothesize that inhibition of c-*myc* overexpression early in life, prior to secondary neoplastic transformation, might interdict tumor formation. In the Eμ-*myc* transgenic mouse model for spontaneous tumors, Myc antigen was suppressed in the spleens of tumor-bearing mice treated with antisense DNA, although no significant inhibition was observed in the corresponding lymphoid tumors from the same animals. Poor penetration by antisense DNA from the blood into the tumors[79] may be one possible explanation for the lack of Myc ablation in the tumors.

MYC6 therapy did not significantly reduce c-*myc* mRNA in either spleens or lymphoid tumors. While the antisense model presumes that RNase H would ablate c-*myc* mRNA hybridized to antisense DNA in the spleens of antisense-treated animals, a transcriptional feedback loop might replenish continuously the degraded c-*myc* mRNA. Such a feedback phenomenon has been observed in acute myelogenous leukemia cells treated with antisense DNA against p53, where the antigen level was also depressed, without concomitant reduction of the mRNA band on a Northern blot.[80]

However, 6 weeks of MYC6 therapy at 20 mg/kg/day dramatically inhibited spontaneous tumor onset when administered prophylactically, before the appearance of palpable tumors. The rather high dose that allowed chemoprevention might in principle be reduced. Dose–response experiments revealed similar protection at half the dose.[71] For prolonged or chronic therapy, intestinal uptake[59] allows oral administration, which has proved effective with chimeric oligonucleotides.[25]

Another significant question to be addressed concerns the extended tumor-free survival of 9 of the 12 Eμ-*myc* transgenic mice treated with MYC6 for only 6 weeks. The original hypothesis predicted postponement of tumorigenesis during the course of antisense therapy, i.e., until the end of the sixth week of DNA administration, 9 weeks of age, but not long-term resistance. What mechanism could one then postulate to explain the phenomenon? In the case of Eμ-*myc* mice, it may be that antisense inhibition of the expanded pre-B cell population limited the opportunities for establishment of tumor clones until the animals were sufficiently mature to eliminate transformed cells by normal immunological processes. In that case, 2 or 4 weeks of antisense therapy, up to an age of 7 weeks, might be insufficient to allow long-term resistance. Alternatively, upregulation of c-*myc* expression at the end of therapy in transformed cells may have driven

[79] R. K. Jain, *Microcirculation* **4,** 1 (1997).

[80] E. Bayever, K. M. Haines, B. L. Copple, S. S. Joshi, and P. L. Iversen, *Proc. Am. Assoc. Cancer Res.* **36,** 411 (1995).

them into apoptosis, as with promyelocytic HL-60 cells.[81] It is also possible that antisense inhibition induced sufficient differentiation of c-*myc* transformed cells[40] to expose them to immune surveillance and specific T cell activation.

In the residual disease model, tumor development was delayed and the Myc:actin ratio at sacrifice (17 days) in the spleens of mice treated with MYC6 for 7 days was reduced relative to those treated with MYC61. This phenomenon probably reflects unimpeded uptake of MYC6 by circulating lymphocytes and tumor lymphoblasts, both of which can become sequestered in the spleen, the body's largest reservoir of lymphocytes. This would result in the observed inhibition of Myc expression in the spleen, and would also explain the delay in tumor onset in MYC6-treated mice. In contrast, MYC61 treatment did not inhibit tumorigenesis, and the Myc:actin ratios in the spleens and tumors of MYC6-treated mice were not significantly different from each other ($p = 0.608$). The effect of the mismatches implies that MYC6 treatment inhibited Myc expression and subsequent tumor growth in a sequence-dependent manner, independent of either dGGGG or dCACGTT.

Inhibition of tumorigenesis by the immunostimulatory sequence MCG was accompanied by the accumulation of ascites, in part a result of blockage of the lymphatic channels after stimulation of B cell proliferation by the dGACGTT immunostimulatory motif present in this sequence. This blockage compromised drainage of the peritoneal cavity,[76] inhibiting the critical flow of lymph fluid (excess interstitial fluid) into the circulation; disruption of this flow can be fatal.[77]

In MCG, dGACGTT occurs in the middle of the sequence. However, oligonucleotides containing a complete dPuPuCGPyPy consensus motif at either end of the sequence were found ineffective in their immunostimulatory activity.[70] In MYC6 and MYC61, the dCACGTT motif is located at the 5' terminus, and has only one purine, rather then two purines, 5' to the dCG. Both the 5' terminal location and the purine-to-pyrimidine transversion should make immunostimulatory activity by MYC6 or MYC61 unlikely.[70]

The incomplete inhibition of tumors by MYC6 led us to analyze additional target sites on the mRNA in order to achieve more efficient inhibition. Tumorigenesis was inhibited similarly by treatment with either MCG or MYC55, but not by MYC57. These data, along with the decreased Myc:actin ratios observed after MYC55 treatment, imply that the efficacy of MYC55 therapy depends on the entire sequence. In mice treated with

[81] S. Kimura, T. Maekawa, K. Hirakawa, A. Murakami, and T. Abe, *Cancer Res.* **55,** 1379 (1995).

MYC55, tumors were detected only at the injection site. That is, no lymph node metastases were observed. However, we observed metastases to cervical and inguinal lymph nodes, in addition to minor lymph nodes throughout the body, in all other treatment groups, including mice receiving MCG. In addition, the accumulation of ascites in mice treated with MCG indicates a level of toxicity that requires cautious administration.

Inhibition of tumor development by MCG may, in part, be due to priming of the immune system to the tumor antigens on the tumor cells that were injected 1 day prior to administration of MCG.[65,71] This pretreatment may thus induce an immune response that might subsequently inhibit tumor development. The observation that Myc levels in the spleens of mice treated with MCG were twofold greater than those measured in their tumors (Figs. 17 and 19) further supports the hypothesis that the oligonucleotide activates B cells within the spleens of this treatment group. The lower Myc antigen levels detected in tumors of mice treated with MCG may reflect an immune response against the tumor cells. Although no Myc antigen could be detected in either circulating lymphocytes or spleens of normal mice, B cells from Eμ-myc transgenic mice display high constitutive Myc levels in circulating lymphocytes, spleens, and tumors.[5,6]

Administration of the immunostimulatory sequence MCG to normal mice in the absence of Eμ-myc lymphoma cells (Fig. 22) yielded a splenic Myc : actin ratio of 0.03 ± 0.03. This low level of Myc could be detected only when the Western blot was overexposed fourfold to allow detection. The relative increase in Myc levels due to stimulation of normal spleen cells by MCG is modest relative to the high levels of Myc expressed in Eμ-myc lymphoma cells. This modest stimulation of Myc expression by MCG would be difficult to measure above the high levels observed constitutively in Eμ-myc lymphoma cells. Scrambled and mismatched sequence controls did not affect the high levels of Myc antigen expressed in tumors or spleens.

When mice were treated with MYC55, the level of Myc expression in their tumors was similar to that in mice treated with MCG. However, the Myc levels in the spleens of mice treated with MYC55 were only one-fourth of the levels detected in their tumors. Lymphocyte proliferation occurs primarily in the lymph nodes, the spleen being the largest such lymphatic organ.[77] Expression of Myc is a prerequisite for cell cycle progression[35]; therefore the low levels of Myc detected in the spleens of the MYC55 treatment group imply an inhibition of lymphocyte proliferation. This inhibition correlates with the elimination of detectable metastases, a decreased rate of tumorigenesis, and a reduction in the total tumor mass.

The dGTCGTT motif in MYC55 differs from the dPuPuCGPyPy consensus[65,70] by having a pyrimidine (T) rather than a purine on the 5' side of the dPuPuCGPyPy, which correlates with the report that immune stimulation was significantly reduced when a purine was substituted for a pyrimidine, or vice versa.[70] The MYC57 control includes a pair of two nucleotide mismatches that disrupt complementarity to the MYC55 target mRNA, while retaining the dGTCGTT motif. Nevertheless, treatment with MYC57 failed to inhibit tumorigenesis. This result further supports the model that the effect of the MYC55 antisense oligonucleotide is dependent on its entire sequence, irrespective of the presence of the dGTCGTT motif.

Secretion of cytokines should follow immune system activation, yet no increase in TNF-α, IL-12, or IFN-γ was detected in the plasma 1 hr after the injection of MCG. The increase in cytokine levels induced by dPuPuCG-PyPy motifs is dependent on both the route of administration of dPuPuCG-PyPy-containing DNA as well as the time after administration at which the mice are bled for cytokine analysis. Intravenous injection of *Escherichia coli* DNA was reported to increase the serum IL-6 concentration by 2 hr postinjection, followed by a rapid decline to below detectable levels at 8 hr.[82] On the other hand, serum IFN-γ levels remained below detectable levels until 5 hr after the intravenous injection of *E. coli* DNA.[83] In addition, mice receiving *E. coli* DNA alone exhibited an increase in TNF-α and IL-6 levels that was only 10% of the levels achieved after administration of a combination of DNA and lipopolysaccharide. The significant increases in cytokines detected in these reports may also be due to the use of *E. coli* DNA, which contains numerous unmethylated dPuPuCGPyPy motifs. *In vitro* analysis of cytokine stimulation of dPuPuCGPyPy-containing oligonucleotides revealed that the greater the number of dPuPuCGPyPy motifs in an oligonucleotide, the greater the secretion of cytokines.[70] A study of 11 different dPuPuCGPyPy-containing sequences revealed that the presence of unmethylated dPuPuCGPyPy motifs in oligonucleotides does not necessarily result in immune stimulation.[82] In addition, both the time course of cytokine release and serum levels of cytokines depend on the route of administration of the DNA. The cytokines measured in the transplant model (TNF-α, IL-12, and IFN-γ) may have remained below detectable levels because of the route of administration, the time at which the mice were bled for samples, and/or the inability of these oligonucleotides to stimulate an immune response.

Extending MYC55 treatment to 14 days, while maintaining a dose of 0.76 mg/day, proved ineffective. Although prolonging the therapy to 14

---

[82] A. K. Yi, D. M. Klinman, T. L. Martin, S. Matson, and A. M. Krieg, *J. Immunol.* **157,** 5394 (1996).

[83] J. S. Cowdery, J. H. Chace, A. K. Yi, and A. M. Krieg, *J. Immunol.* **157,** 4570 (1996).

days did not inhibit tumorigenesis further, this result may be due to selection for a tumor cell subpopulation that does not require Myc expression for proliferation.[45,47,84] The absence of detectable Myc in the spleens and tumors of the MYC55 treatment group (Fig. 21) supports this hypothesis. The similar times of tumor onset and similar total tumor masses detected in the 7- and 14-day regimens, in addition to the existence of metastasis in the 14-day treatment group, further support the potential development of Myc-independent tumors in the 14-day treatment group.

The similar, yet incomplete, tumor inhibition resulting from treatment with MYC55 alone or with MCG alone suggested that a therapeutic regimen in which MYC55, an antisense oligonucleotide, is alternated with MCG, an immunostimulatory oligonucleotide, might prove to be more effective in preventing or limiting tumor development. Although either MYC55 or MCG alone delayed tumor onset by 3–7 days, alternating the daily therapy between 0.76 mg of MYC55 and 0.65 mg of MCG was found to be much more effective than either sequence alone. This alternating regimen postponed tumor onset by 7–9 days, which was beyond the end of the therapeutic schedule.

Applications

In view of the positive results of the micropump trial, one could postulate screening of individuals at risk for c-*myc* translocation, such as HIV-infected individuals. Although the 8; 14 chromosomal translocation is not frequent enough in children to justify screening of a large population, it does occur frequently in HIV-infected individuals, inducing lymphomas.[41] AIDS-associated non-Hodgkin's lymphomas typically display translocation of c-*myc* to an immunoglobulin locus.[85,86] If c-*myc* translocation precedes onset of lymphoma by any appreciable time, then screening might reveal positive evidence of c-*myc* activation prior to malignancy. Individuals displaying c-*myc* translocation could then be treated with antisense DNA to interdict tumor onset. It may therefore be reasonable to propose prophylactic anti-c-*myc* DNA therapy for HIV-infected individuals who display c-*myc* translocation in peripheral B cells.

Would one not, however, expect to find systemic inhibition of all proliferating cells as a result of antisense inhibition of c-*myc* expression? In mice at least, it has been observed that DNA phosphorothioates are preferen-

[84] B. Vogelstein and K. W. Kinzler, *Trends Genet.* 9, 138 (1993).
[85] G. Gaidano, N. Z. Parsa, V. Tassi, P. Della-Latta, R. S. Chaganti, D. M. Knowles, and R. Dalla-Favera, *Leukemia* 7, 1621 (1993).
[86] S. Prevot, M. Raphael, J. G. Fournier, and J. Diebold, *Histopathology* 22, 151 (1993).

tially taken up by B cells, particularly when stimulated by a B cell mitogen, such as the DNA phosphorothioate itself.[87] Furthermore, transformed cells take up oligonucleotides three- to fivefold more aggressively than do normal cells.[36] Finally, the downregulation of Myc protein observed in splenic lymphocytes suggests that c-*myc* antisense therapy might prove beneficial in those hematological malignancies in which cellular proliferation depends on elevated c-*myc* expression.

In the transplant model for residual disease, a 14-day treatment regimen consisting of the antisense oligonucleotide MYC55 alternated with the immunostimulatory oligonucleotide MCG resulted in complete inhibition of tumor formation during the therapeutic schedule. These results encourage the clinical development of MYC55 plus immunostimulatory oligonucleotide consolidation therapy for B cell lymphoma.

The broad spectrum of malignancies displaying overexpression of c-*myc* presents a compelling argument for clinical development of c-*myc* antisense DNA. In other laboratories, c-*myc* antisense DNA has been utilized to downregulate c-*myc* expression and inhibit cellular proliferation in human lymphoma cells,[42] breast cancer cells,[88] colon cancer cells,[89] lung cancer cells,[90] melanoma cells,[91] and prostate cancer cells,[92] to give just a few examples.

Acknowledgments

The authors warmly thank their associates who have worked on this investigation in earlier years: Dr. Ye Huang, Erica Wickstrom, and Thomas Bacon. We are most grateful to Dr. Rob DeLong and Dr. Rudy Juliano for their able collaboration on oligonucleotide pharmacokinetics. We thank Dr. Michael Kligstein for oligonucleotide synthesis, Stephen Cleaver for statistical assistance, and Dr. Jaspal Khillan and Dr. Tim Manser for their help in establishing the Tg(IgH,Myc)Bri157 line at Jefferson. This work was supported by NIH Grant CA42960 to E.W.

[87] P. L. Iversen, D. Crouse, G. Zon, and G. Perry, *Antisense Res. Dev.* **2,** 223 (1992).
[88] P. H. Watson, R. T. Pon, and R. P. Shiu, *Cancer Res.* **51,** 3996 (1991).
[89] J. F. Collins, P. Herman, C. Schuch, and G. C. Bagby, Jr., *J. Clin. Invest.* **89,** 1523 (1992).
[90] L. A. Robinson, L. J. Smith, M. P. Fontaine, H. D. Kay, C. P. Mountjoy, and S. J. Pirruccello, *Ann. Thorac. Surg.* **60,** 1583 (1995).
[91] C. Leonetti, I. D'Agnano, F. Lozupone, A. Valentini, T. Geiser, G. Zon, B. Calabretta, G. C. Citro, and G. Zupi, *J. Natl. Cancer Inst.* **88,** 419 (1996).
[92] K. C. Balaji, H. Koul, S. Mitra, C. Maramag, P. Reddy, M. Menon, R. K. Malhotra, and S. Laxmanan, *Urology* **50,** 1007 (1997).

[39] Retrovirally Mediated Delivery of Angiotensin II Type 1 Receptor Antisense *in Vitro* and *in Vivo*

By Hongwei Wang, Di Lu, Phyllis Y. Reaves,
Michael J. Katovich, and Mohan K. Raizada

Introduction

The renin–angiotensin system (RAS) plays an important role in the control of many cardiovascular functions including blood pressure (BP). Angiotensin II (Ang II), the active component of the RAS and one of the most potent vasoconstrictor hormones, is produced by enzymatic cleavage of Ang I by angiotensin I-converting enzyme (ACE). Ang II activates Ang II type 1 receptor (AT₁R), which is discretely distributed in various cardiovascular-relevant tissues such as adrenal, heart, kidney, blood vessels, and brain, in order to exert its physiological effects.[1,2] This traditional circulating RAS, which is responsible for "short-term" effects of Ang II, is a result of coordination of the activities of the kidney, producing renin; the liver, producing angiotensinogen; and the lungs, producing ACE.[1] In addition, tissue RAS is implicated in the "long-term" actions of Ang II involving hypertrophy, hyperplasia, and tissue remodeling. A fine cellular integration of these two systems at the level of the AT₁R must exist in order to provide a normal physiological circuitry essential for control of cardiovascular functions. The importance of the RAS is underscored by observations that alterations in its activity are associated with major pathophysiological states such as hypertension and coronary artery disease (CAD).[3] The fact that pharmacological intervention in the activity of the RAS is one of the most successful means to control these diseases further strengthens this argument. Thus, drugs that either inhibit the synthesis of Ang II (ACE inhibitors such as captopril) or the actions of Ang II by antagonizing the AT₁R (such as losartan) are effective pharmacological strategies in the control of hypertension and other CADs.

In spite of the success, pharmacological intervention for hypertension poses major limitations. They include daily administration of drugs leading

[1] M. K. Raizada, D. Lu, H. Yang, E. M. Richards, C. H. Gelband, and C. Sumners, *in* "Advances in Molecular and Cellular Endocrinology" (D. Le Roith, ed.), Vol. 3, pp. 73–79. JAI Press, Stamford, Connecticut, 1999.

[2] P. K. Whelton, *Lancet* **332,** 110 (1994).

[3] C. H. Gelband, M. J. Katovich, and M. K. Raizada, *in* "Gene Therapy and Molecular Biology," Vol. 3, pp. 249–256, 1999.

to "compliance" problems with the patients and many other side effects.[4] In addition, pharmacological therapy is useful in the control and management of hypertension but not in the cure of this disease.

We hypothesized that an interventional approach at a genetic level in the expression of the RAS may offer a significant improvement in the treatment of hypertension over the pharmacological strategy. Thus, we set out to test this concept and determine the feasibility of a virally mediated delivery system to introduce AT_1R antisense (AS) sequence for long-term control of the RAS. In this chapter we describe methods to use retroviral vector to transduce AT_1R-AS cDNA into Ang II target tissues both *in vitro* and *in vivo*. We also present evidence of cellular and physiological consequences of such antisense gene delivery.

Methods

Methods are divided into three categories: (1) preparation of retrovirus particles containing AT_1R-AS; this section describes the construction and recombination of LNSV with AT_1R-AS cDNA and the preparation of high-titer viral particles for AT_1R-AS transduction; (2) infection of Ang II target cells *in vitro* and transduction of AT_1R-AS; and (3) *in vivo* transduction of AT_1R-AS and its consequences on the physiological actions of Ang II.

Preparation of Retroviral Particles Containing Angiotensin II Type 1 Receptor Antisense

Preparation of Angiotensin II Type I Receptor cDNA

A pair of AT_1R subtype B (AT_1 B-R) specific primers (sense, 5'-CCAAGCTTGTGTCAGAGAGCAATTCACCTCACC-3'; antisense, 5'-CCAAGCTTGGTAGTGAGTGAGCTGCTTAGCCCA-3') is used to generate AT_1B-R cDNA by reverse transcriptase-polymerase chain reaction (RT-PCR). The total RNA (5 μg) from rat hypothalamus brainstem neuronal cells is subjected to RT reaction using SuperScript II (GIBCO-BRL, Gaithersburg, MD) by standard protocol.[5,6] The PCR contains 20 pmol of AT_1-R sense and antisense primers and a PCR of 35 cycles (94°

[4] T. Sambrook, E. F. Fritsch, and T. Maniatis, "Molecular Cloning: A Laboratory Manual," 2nd Ed. Cold Spring Harbor Laboratory Press, Plainview, New York, 1989.
[5] F. M. Ausubel, R. Brent, R. E. Kingston, D. D. Moore, J. G. Seidman, J. A. Smith, and K. Struhl, *in* "Short Protocols in Molecular Biology." Greene Publishing Associates and John Wiley & Sons, New York, 1992.
[6] D. Lu, K. Yu, and M. K. Raizada, *Proc. Natl. Acad. Sci. U.S.A.* **92**, 1162 (1995).

for 1 min, 58° for 1 min, and 72° for 1 min) is done by standard protocol.[7] PCR products are separated on a 1% agarose gel and the band of 1.26 kb is isolated and purified by using a QIAEX II gel extraction kit (Qiagen, Chatsworth, CA). The identity of the 1.26-kb AT$_1$-B cDNA is further confirmed by restriction enzyme analysis and by sequencing. The identity of AT$_1$R cDNA corresponding to nucleotides (nt) -132 to $+1228$ of the coding region of the AT$_1$B-R is established and confirmed. In view of the 95–98% structural homology between the AT$_1$A and AT$_1$B receptors, it is not highly relevant to be selective for a specific AT$_1$R subtype. It is particularly important in view of this fact that this level of homology would be effective in cross-influencing the transcriptional expression of either receptor subtype with the use of our proposed antisense technique. In fact, the observations presented elsewhere clearly indicate this to be the case.[6,8]

Recombination of Angiotensin II Type 1 Receptor cDNA with LNSV, Retroviral Vector

The protocol for the preparation of viral particles containing AT$_1$R-AS is outlined in Fig. 1. LNSV-AT$_1$B-R CDNA (100 ng) and LNSV (50 ng) are digested with HindIII. The fragments are purified and mixed with ligase buffer (50 mM Tris-HCl, pH 7.5, 7 mM MgCl$_2$, 1 mM DTT, 1 mM ATP) and 2 units of T$_4$ ligase (Stratagene, La Jolla, CA) and incubated overnight at 12°. Recombinant DNA is transformed into competent HB 101 bacterial cells and AT$_1$R-AS colonies are selected. The colonies that produce 1.26-kb AT$_1$R-AS are then grown in LB medium with ampicillin (100 μg/ml) and recombinant DNA is purified with a plasmid purification kit (Promega) according to the protocol provided by the company.

Various restriction enzymes are used to characterize the AT$_1$R recombinant and its antisense orientation. For example, HindIII digestion gives an anticipated 1.26-kb band corresponding to nt -132 to $+1128$ in the AT$_1$B-R sequence.[6] A ClaI digestion provides a 0.53-kb band while ClaI/SalI digestion gives two anticipated bands of 1.17 and 0.53 kb. These sizes are consistent with the location of these restriction sites.[6] FokI digestion of the 1.26-kb AT$_1$B-R band provides two bands of 0.95 and 0.30 kb, consistent with the localization of the FokI site in the AT$_1$B-R-AS. Further characterization of individual bands is carried out by the sequence analysis.

Standard protocols have been utilized for all of the molecular biological techniques used in the preparation of LNSV-AT$_1$R-AS vector.[4,5]

[7] D. Lu and M. K. Raizada, Proc. Natl. Acad. Sci. U.S.A. **92,** 2914 (1995).
[8] A. D. Miller and A. J. Rosman, BioTechniques **7,** 890 (1989).

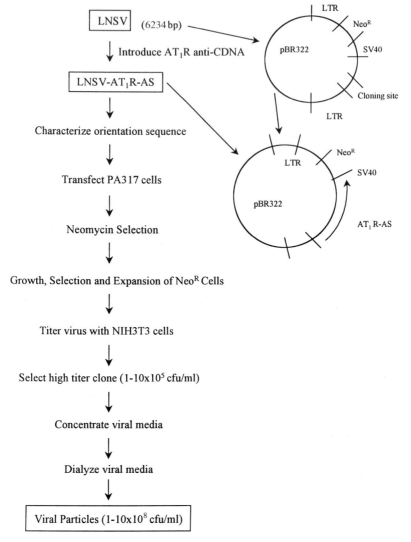

FIG. 1. Outline of protocol for the preparation of viral particles.

Preparation of Viral Medium: Transfection of PA317 Cells

Day 1. Place 2×10^5 PA317 cells (ATCC CRL 9078) in a 60-mm culture dish with 25 mM glucose containing Dulbecco's modified Eagle's medium (DMEM; GIBCO-BRL) that contains 10% fetal bovine serum (FBS; Bio-

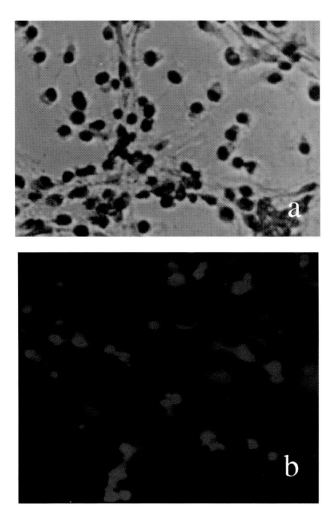

Fɪɢ. 2. RT *in situ* PCR detection of AT$_2$R-AS in neuronal and astroglial cultures in primary cultures. Neuronal (a) and astroglial cells (b) are cultured and infected with LNSV-AT$_1$R-AS essentially as outlined in the protocol. The cultures are used to carry out RT *in situ* PCR. Dig-dUTP was used in neurons while Texas Red-dUTP was used in astroglia for the detection of PCR product.

TABLE I
CONCENTRATION OF VIRAL VECTOR BY ULTRACENTRIFUGATION

Day	Step	Titer/ml
1	1. Culture supernatant (25 ml), centrifuge at 50,000g for 2 hr at 4°	1×10^5 CFU
	2. Collect pellet, dissolve in 100 μl of HBSS for 12 hr at 4°	2.5×10^7 CFU
2	Repeat step 1 and dissolve virus pellet, using virus suspension harvest in step 2 from day 1	5×10^7 CFU
3	Repeat steps 1 and 2 for further concentration as needed	1×10^8 CFU and more

cell, Rancho Dominguez, CA). Allow them to grow in 5% CO_2 at 37° for 24 hr.

Day 2. Dilute 2 μg of recombinant DNA containing LNSV-AT_1R-AS into 100 μl of serum-free DMEM. Mix 5 μl of Lipofectin reagent (GIBCO-BRL) with 100 μl of serum-free DMEM. After being allowed to stand at

TABLE II
PROTOCOL FOR INFECTION OF ANG II TARGET CELLS *in Vitro*

Neuronal cultures	Astroglia/VSMCs
Hypothalamus–brainstem of 1-day-old rat ↓	Plate 1×10^5 cells/60-mm dish in DMEM + 10% FBS ↓
Cells dissociated and cultures established in DMEM + 10% plasma-derived horse serum (PDHS) for 2 days[6] ↓	Grow for 48 hr at 37° in 5% CO_2:95% air ↓
Add 1.0 μM cytosine arabinoside for 2 days[6] ↓	Replace medium with DMEM + 10% FBS containing 1×10^5 CFU of LNSV or LNSV-AT_1RAS/ml ↓
Infect cells with 5×10^5 CFU of LNSV/ml, or LNSV-AT_1R-AS vector in 100 μl or medium ↓	Incubate for 24 hr ↓
Incubate for 48 hr at 37° ↓	Replace medium with the selection medium [DMEM + 10% FBS + G418 (800 μg/ml)] ↓
Remove medium and grow cells with G418 (800 μg/ml) in DMEM + 10% PDHS for 4–6 days ↓	Grow for 10–15 days ↓
Cultures used to measure transduction efficiency by RT *in situ* PCR	Subculture in selection medium ↓
	Grow until confluent (7–10 days) ↓
	Cultures used to measure transduction efficiency by RT *in situ* PCR

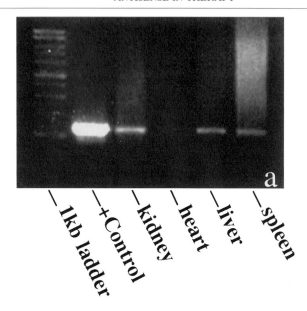

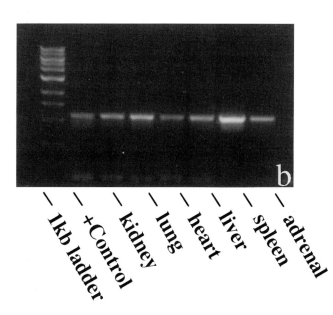

room temperature for 45 min, both solutions are mixed and incubation continued for an additional 10–15 min. Add 1.8 ml of serum-free DMEM to the Lipofectin–DNA mixture and gently overlay the mixture onto PA317 cells. Cells are incubated with this mixture for 12 hr.

Day 3. Replace the DNA-containing medium with 4 ml of DMEM containing 10% FBS.

Day 4. Subculture the cells at a 1:10 ratio in the selection medium [DMEM plus 10% FBS containing G418 (800 μg/ml)] for 10–14 days. Change the selection medium every 3 days. Further selection would depend on the rate of growth of G418-resistant colonies. G418-resistant colonies are picked and transferred into multiple six-well dishes, and cells are grown for 3–5 days until the cultures become confluent. Supernatant is harvested and filtered through a 0.45-μm pore size filter and used to determine the titer as follows.

Viral Titration. NIH 3T3 cells are used to determine the concentration of viral particles as follows.

Day 1. Plate NIH 3T3 cells at 1×10^5 cells/60-mm-diameter culture dish in DMEM plus 10% calf serum.

Day 2. Replace the medium with 2 ml of DMEM plus 10% calf serum containing Polybrene (8 μg/ml). One or 10 μl of medium containing virus particles is added to each dish and mixed carefully, and the cultures are allowed to incubate for 2.5 hr at 37°. This is followed by the addition of 2 ml of fresh DMEM plus 10% calf serum in each dish.

Day 3. Replace the virus-containing medium with fresh DMEM plus 10% calf serum and incubation is continued for an additional 24 hr at 37°.

Day 4. Subculture infected 3T3 cells at a 1:10 dilution in G418 (800 μg/ml) selection medium for 10–14 days. Count G418-resistant colonies and use a standard protocol to calculate the titer.[8] Select two to four of

[9] D. Lu, M. K. Raizada, S. Iyer, P. Reaves, H. Yang, and M. J. Katovich, *Hypertension* **30**, 363 (1997).

FIG. 3. (a) PCR detection for the presence of retroviral vector in various tissues of SH rat, 10 days postinjection of LNSV ATR-AS vector. Genomic DNA was isolated and subjected to PCR with the use of the following primers: sense (5′-GCCTCTGAGCTATTCCAGAAG-TAG-3′) and antisense (5′-AATGGCCCTTAACTCTTCTGCTGAAGATGGTATCAA-3′) (94° for 1 min, 58° for 1 min, 72° for 1 min for 40 cycles). Positive control represents DNA from PA317 cells transfected with LNSV AT$_1$R-AS. (b) Expression of AT$_1$R-AS transcript in various Ang II target tissues of SH rat, 10 days postinjection of LNSVAT$_1$R-AS vector. Total RNA is subjected to RT-PCR with the use of specific primers to detect the expression of AT$_1$R-AS.[7] Control represents RNA from CAD cells infected with AT$_1$R-AS.

the highest titer colonies (10^5–10^6 CFU/ml), subculture them once, and store them in liquid nitrogen.

Concentration of Viral Vector by Ultracentrifugation. The steps are outlined in Table I.

Protocol for Infection of Angiotensin II Target Cells: in Vitro

Vascular smooth muscle cells (VSMCs) and brain cells are well established Ang II target tissues.[1] Thus, we utilized VSMCs, neurons, and astroglial cells in culture to demonstrate the efficiency of transduction of AT_1R-AS *in vitro.* The protocol is outlined in Table II.

Figure 2a and 2b (see color insert) shows an example of AT_1R-AS transcript localization in neuron and astroglial cells by RT *in situ* PCR. The experimental protocol for RT *in situ* PCR is essentially as described elsewhere.[5,6] It is evident that almost all astroglial cells and >70% neurons in culture express AT_1R-AS transcript. The expression is confirmed by Northern analysis.[5] These experiments establish a protocol for infection of major Ang II target cells with a retroviral vector and demonstrate a stable expression of AT_1R-AS.

Protocol for in Vivo Transduction of Angiotensin II Type 1 Receptor Antisense

Five-day-old normotensive (Wistar Kyoto, WKY) and spontaneously hypertensive (SH) rats are used to determine the transduction efficiency of AT_1R-AS and physiological consequence of its expression. Rats are anesthetized and a bolus of viral particles containing 1×10^9 CFU of either LNSV (control) or LNSV AT_1R-AS in 50 μl (experimental) is injected intracardially in each rat. Rats are allowed to recover, their backs coated with peanut oil, and allowed to wean for an additional 16 days. The survival rates have been 95% once they pass the crucial period of 24–48 hr past cardiac injection.

Expression of AT_1R-AS in various Ang II target tissues is examined by RT-PCR 10 days after viral injection. The animals are allowed to grow and their blood pressure is measured indirectly by a tail cuff method.[10]

Figure 3a shows that the AT_1R-AS is integrated into the genome. Figure 3b shows the expression of AT_1R-AS transcript in various tissues 10 days postinjection. Expression of AT_1R-AS transcript is associated with an exclusive attenuation of high BP in the SH rat without a significant effect in the WKY rat. This antihypertensive effect is maintained for the duration of

[10] S. N. Iyer and M. J. Katovich, *Life Sci.* **55,** 139 (1994).

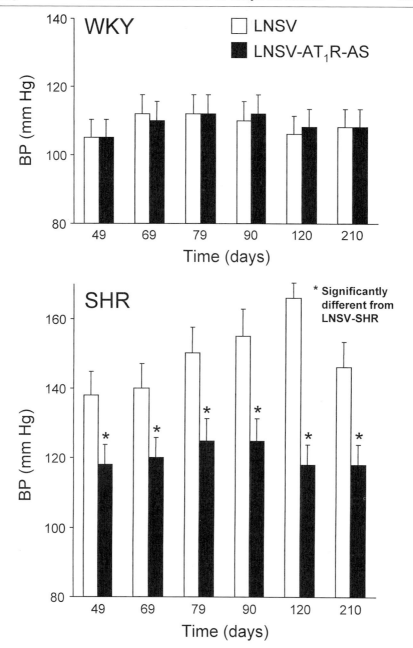

FIG. 4. Indirect blood pressure measurements in LNSV-treated (control) and LNSV-AT_1R-AS-treated (experimental) WKY and SH rats.

experiment (210 days). It indicates that Ang II target tissues demonstrate a robust expression of this transcript. The expression is maintained for at least 60 days.[6,9]

Summary and Conclusions

In spite of excellent drugs that are available for the control of hypertension, the pharmacological approach has major disadvantages including compliance, side effects, and inability to cure the disease. In the present chapter we provide evidence that a gene therapy concept based on the inhibition of the RAS at a genetic level, with the use of an antisense to the AT_1R, is an exciting and viable approach for long-term control of hypertension without the disadvantages inherent in pharmaceutical therapy. A retrovirus-based vector has been used to deliver AT_1R-AS in Ang II target tissues both *in vitro* and *in vivo*. The transduction efficiency is high and leads to the attenuation of Ang II action *in vitro* and prevention of hypertension in the SH rat, a model for primary human hypertension. These studies have unveiled a new avenue in which a similar approach could be attempted in the reversal of hypertension in adult animals.

Acknowledgment

The authors wish to thank Mrs. Marya Fancey and Mr. Kevin Fortin for the preparation of this manuscript and Ms. Liu Ling for assistance in cell culture and blood pressure measurements. The work was supported by NIH Grant HL-56921.

Author Index

Numbers in parentheses are footnote reference numbers and indicate that an author's work is referred to although the name is not cited in the text.

C

H

U

V

Subject Index

A

X